T0237018

This book provides a self-contained introduction to the cohomology theory of Lie groups and algebras and to some of its applications in physics. No previous knowledge of the mathematical theory is assumed beyond some notions of Cartan calculus and differential geometry (which are nevertheless reviewed in the book in detail). The examples, of current interest, are intended to clarify certain mathematical aspects and to show their usefulness in physical problems. The topics treated include the differential geometry of Lie groups, fibre bundles and connections, characteristic classes, index theorems, monopoles, instantons, extensions of Lie groups and algebras, some applications in supersymmetry, Chevalley–Eilenberg approach to Lie algebra cohomology, symplectic cohomology, jet-bundle approach to variational principles in mechanics, Wess–Zumino–Witten terms, infinite Lie algebras, the cohomological descent in mechanics and in gauge theories and anomalies. A list of references is given in the bibliographical notes section at the end of each chapter which allows the reader to complement and go beyond the topics covered in the book.

This book will be of interest to graduate students and researchers in theoretical physics and applied mathematics.

CAMBRIDGE MONOGRAPHS ON MATHEMATICAL PHYSICS

General Editors: P. V. Landshoff, D. R. Nelson, D. W. Sciama, S. Weinberg

LIE GROUPS, LIE ALGEBRAS, COHOMOLOGY AND SOME APPLICATIONS IN PHYSICS

Cambridge Monographs on Mathematical Physics

[†] Issued as a paperback

LIE GROUPS, LIE ALGEBRAS, COHOMOLOGY AND SOME APPLICATIONS IN PHYSICS

JOSÉ A. DE AZCÁRRAGA[*] AND JOSÉ M. IZQUIERDO[†]

[*] Departamento de Física Teórica and
IFIC, Centro Mixto CSIC-Universidad de Valencia
Facultad de Física, Valencia University

[†] Departamento de Física Teórica,
Valladolid University

CAMBRIDGE
UNIVERSITY PRESS

CAMBRIDGE UNIVERSITY PRESS
Cambridge, New York, Melbourne, Madrid, Cape Town, Singapore, São Paulo

Cambridge University Press
The Edinburgh Building, Cambridge CB2 8RU, UK

Published in the United States of America by Cambridge University Press, New York

www.cambridge.org
Information on this title: www.cambridge.org/9780521465014

© Cambridge University Press 1995

This publication is in copyright. Subject to statutory exception
and to the provisions of relevant collective licensing agreements,
no reproduction of any part may take place without the written
permission of Cambridge University Press.

First published 1995
First paperback edition 1998

A catalogue record for this publication is available from the British Library

Library of Congress Cataloguing in Publication data
Azcárraga, J. A. de, 1941–
Lie groups, Lie algebras, cohomology, and some applications in
physics / José A. de Azcárraga and José M. Izquierdo.
p. cm. – (Cambridge monographs on mathematical physics)
Includes bibliographical references.
ISBN 0 521 46501 X
1. Lie groups. 2. Lie algebras. 3. Homology theory.
4. Mathematical physics. I. Izquierdo, José M. II. Title.
III. Series
QC20.7.L54A93   1995
512′.55–dc20   94-16809 CIP

ISBN 978-0-521-46501-4 hardback
ISBN 978-0-521-59700-5 paperback

Transferred to digital printing 2008

No mirando a nuestro daño,
corremos a rienda suelta
sin parar;
des' que vemos el engaño
y queremos dar la vuelta
no hay lugar.

(Jorge Manrique, 1440–1479)

No foe, no dangerous pass, we heed.
Brook no delay – but onward speed
With loosened rein;
And, when the fatal snare is near,
We strive to check our mad career,
But strive in vain.

(Free 1833 translation by
H. Wadsworth Longfellow)

Contents

Preface

This book has its origin in a graduate course delivered at DAMTP in the Lent term of 1989 by one of the authors (J. de A.) on group cohomology. We have aimed at a self-contained presentation. Some of the background sections, however, are provided as review or reference material and occasionally they may not be suitable for first study if the reader is completely unfamiliar with their contents. There are already several good textbooks on differential geometry for physicists and the authors felt that there was no point in reproducing here topics not relevant to the main subject that could be found without much effort in a single source. Thus, a modicum of knowledge of differential manifolds, Cartan calculus and of the theory of connections is assumed. From the physics side, a knowledge of quantum field theory is also required to read certain parts of the book. Nevertheless, almost all the necessary background material is discussed in it, and often repeated in different settings of increasing complexity so that the various concepts can be assimilated more easily.

A comment should be made concerning rigour. The book uses generously the mathematical 'Theorem, Proposition, etc.' presentation style. We have found this convenient in order to stress certain key ideas or to organize and separate parts of the text that would otherwise look unreasonably long. Nevertheless, the book is primarily intended for physicists wishing to familiarize themselves with certain aspects of the theory of Lie groups, fibre bundles and a few of their physical applications, particularly to problems where cohomology and extension theory play an important rôle. Parts of the book may also be useful to mathematicians interested in the interface of physics and mathematics. Although there are good articles covering different aspects of the subjects below, we feel that a self-contained presentation in book form, including some material not readily accessible, may be useful.

The organization of the book is as follows: chapter 1, the longest one, reviews some topics in the differential geometry of Lie groups, Cartan calculus and fibre bundles. The emphasis is placed on the aspects that may be more important to physicists, and their presentation follows the order in which the need for more abstract material may likely arise rather than a more 'axiomatic' one. For instance, vector fields appear first as differential operators acting on manifolds (Lie groups in particular) and then they are also seen to be sections of a tangent bundle. We feel that this approach may be more appealing to the physicist reader than a presentation starting directly with vector bundles. After a few comments on graded Lie groups, the chapter ends by recalling some results and tables of homotopy theory and with the Poincaré polynomials for the compact simple groups.

Chapter 2 is devoted to connections, characteristic classes and the index theorem for the de Rham and the spin complexes. The selection of the topics in this chapter has been motivated by the physical applications considered (monopoles and Yang-Mills instantons in this chapter and anomalies in the last one). Thus, some important topics such as, for instance, torsion, parallel transport and holonomy have been left out; the interested reader may look for them at the references provided.

Chapter 3 introduces the cohomology of groups by discussing first some interesting physical examples, such as the relevance of projective representations in non-relativistic quantum mechanics, the Weyl-Heisenberg group, and the extended Galilei group. Group cocycles and coboundaries are introduced here by means of these examples. Using a more formal approach, a short review of the theory of dynamical groups and symplectic cohomology which also touches the problem of geometric quantization is given in this chapter, and applied to the previous examples. The appearance of two-cocycles in the projective representations naturally leads us to think about three-cocycles when the associativity is broken in a consistent manner: the problem of the appearance of the three-cocycle in quantum mechanics is discussed at the end. The chapter closes with a few remarks, one of them being the analysis of how the contraction of groups may generate non-trivial group cohomology.

After the simple examples of chapter 3, chapter 4 presents the problem of group extension theory in general terms. Exact sequences of group homomorphisms are introduced, and once the principal bundle structure of the group extension is established, its elements are given coordinates by using a trivializing section. The group law of the extended group is then expressed by means of the corresponding factor system. The simpler case of extensions by an abelian kernel (which includes the cases of chapter 3) is studied in detail in chapter 5. The cohomology of Lie algebras is given in chapter 6. This provides the treatment of cohomology theory from the 'dual' Lie algebra point of view. The Whitehead lemma, the Chevalley and Eilenberg approach to Lie algebra cohomology (which uses the invariant forms on a Lie group discussed in chapter 1), its BRS formulation and a discussion of the Van Est isomorphism between the Lie group and algebra cohomology are included in this chapter; the relation between the infinitesimal (algebra) and finite (group) cocycles and vice versa is also given. Chapter 7 is a short account of the problem of extension by non-abelian kernels, where the obstruction to the extension is given by a three-cocycle; it concludes with an example and with a few comments on how higher order cocycles may appear.

Chapter 8 is devoted to relating the mathematical problem of central extensions and two-cocycles with the property of quasi-invariance (*i.e.*, invariance but for a total derivative) of some Lagrangians in physics. This is preceded by a short review of the variational problem, including the Noether theorem, in a geometrical framework. This geometrical setting includes a pedagogical description of the mathematical notion of jet bundles (already introduced in chapter 1) and jet prolongations. In this way, cohomology and mechanics are related, and many of the basic mathematical concepts introduced in chapters 5 and 6 find their physical translation here, as might have been anticipated after the introductory discussion in chapter 3. As a result, the notion of 'Wess-Zumino term on a Lie group' (the name implies here a small terminological abuse) emerges as an important ingredient in the construction of certain physical classical actions, and the cases of the Galilean particle, the massive superparticle, the Green-Schwarz superstring and the extended supersymmetric objects are all discussed in this light. This Wess-Zumino term on a Lie (or graded Lie) group is related to a non-trivial element of the Chevalley-Eilenberg version of the Lie algebra cohomology and, as such, its presence modifies the commutators of the original symmetry algebra. This leads to the appearance of 'classical anomalies', which are discussed using the double cohomological descent approach. This provides a preparatory example for the use of this technique in the more complex setting of non-abelian Yang-Mills theories, where it leads to the geometrical description of non-abelian anomalies and Schwinger terms.

Chapter 9 treats the case of infinite-dimensional Lie groups and algebras. Although an example of an infinite-dimensional Lie group appears in chapter 8 in connection with the supersymmetry local algebra, it is here where some of the infinite-dimensional groups and algebras of interest in physics such as current algebras, Kac-Moody algebras, the two-dimensional conformal group and the Virasoro algebra are introduced, and their cohomological properties analysed in comparison with the finite-dimensional case. Chapter 10 is devoted to the geometry of the abelian and the non-abelian anomalies in Yang-Mills theories. After a discussion of the topology of their configuration space and of the gauge algebra cohomology, the origin of these anomalies is exhibited in a path integral formulation using Fujikawa's method. The geometry of the chiral and non-abelian anomalies is then discussed, including their relation with the index theorems, as well as the Wess-Zumino condition and the Schwinger terms for the non-abelian chiral anomalies. This is done by means of the cohomological descent procedure and following the pattern of the discussion for mechanics in chapter 8 since both situations present the same general structure. Nevertheless, the important differences between the two cases are stressed, one of them being the quantum origin of the chiral anomalies and the other the different nature of the Wess-Zumino term, which in this case does refer to the original one. The book closes with a few comments on the possibility of having 'consistent' anomalous theories.

References are provided at the end of each chapter in a bibliographical notes section; the reader will notice our debt to many of them. In spite of their number, no attempt has been made at completeness or at giving due credit to the earliest author(s). The references have been primarily selected by their usefulness to the text or as a possible complement to it. Whenever possible, books and review articles have been quoted; the interested reader should look at the review papers for further references. No problems are suggested in each chapter; we feel that readers are not likely to set aside their own problems to devote their attention to those suggested in a book of this nature. We ask for their understanding if they do not share this opinion. We have tried to follow a standard notation. Inevitably, a few symbols have different meanings in different contexts. A list of symbols is provided at the end of the book for the benefit of the reader; symbols appear under the chapter in which they are introduced for the first time.

The authors have benefitted from discussions on the topics of this book with many friends and colleagues; some of them also read parts of a previous draft and provided the authors with advice and criticism which have contributed to its improvement. In particular, the authors wish to thank L. Álvarez-Gaumé, M. Asorey, A.P. Balachandran, J.F. Cariñena, A. Galindo, C. Gómez, A. Guichardet, M. Henneaux, C. Isham, J. Lukierski, R. Jackiw, J. Mateos, J. Mickelsson, A. Naveira, M. del Olmo, J.C. Pérez Bueno, M. Santander, F. Scheck, P. Townsend, R. Tucker and, very specially, V. Aldaya, L.J. Boya and J. Stasheff. Needless to say, the responsibility for the text remains solely with the authors; we shall appreciate any notice about misprints or misconceptions not yet eradicated despite our efforts. Both authors wish to acknowledge the warm hospitality of the Department of Applied Mathematics and Theoretical Physics of Cambridge University, for past sabbatical (J. de A.) and present postdoctoral (J.I.) visits during the completion of this work. J. de A. also wishes to thank J.C. Taylor and Robinson College, Cambridge, for a bye-fellowship during the year 1989 and P.V. Landshoff for his interest in the earlier stages of this project. Finally, thanks are also due to the CICYT/DGES, Spain, for continuously financing the 'Geometry and Physics' research project, and to the Generalitat Valenciana (J. de A.), the Ministerio de Educación y Ciencia and the Commission of the European Community (J.I.) for partial financial support at certain stages of the work on this book.

J. A. de A. ,   J. M. I.

e-mails:
j.a.de.azcarraga@uv.es
izquierd@cpd.uva.es

1

Lie groups, fibre bundles and Cartan calculus

'It is vain to do with more what can be done with fewer'
($\approx$ entities should not be multiplied without necessity)
William of Ockham (c.1285–1349)

This chapter contains some topics of differential geometry which are needed for the rest of the book. They have been selected to serve as a reference for readers already familiar with them or as a self-contained introduction for those who encounter them for the first time.

The central concepts of Lie group, Lie algebra and their relations are analysed. This involves the introduction of left and right-invariant vector fields and forms, which in turn motivates the study of various concepts of differential geometry, such as differential forms, differential operators and de Rham cohomology. In particular, the Maurer-Cartan equations will be extensively used in the text due to their relation with Lie algebra cohomology. A review on fibre bundles is also provided due to their importance in the problem of extensions and in the study of Yang-Mills theories which will be treated in later chapters.

A few comments on how to extend some of the above Lie group concepts to graded Lie groups are made by taking the supertranslation group as an example. This requires a small knowledge of supersymmetry; those not interested in 'super' objects may skip it since it is possible to read the book omitting the few sections that contain topics related with supersymmetry.

The last two sections (Appendices) contain information and tables on the homotopy groups and the Poincaré polynomials of the simple Lie groups and on the homotopy groups of spheres.

1.1 Introduction: Lie groups and actions of a Lie group on a manifold

Lie groups are frequently introduced in physics as groups of transformations acting on a manifold* M. Let $\{x^1, ..., x^n\} = \{x^i\}$ be a local coordinate
system for M. An *r-parameter Lie group G of transformations* is a group
of transformations of M

$$x'^i = f^i(g^1, ..., g^r; x^1, ..., x^n) \quad (x' = f(g; x)) \tag{1.1.1}$$

for which the functions f^i ($i = 1, ..., n$) are smooth functions of the
r parameters g^k, which are assumed to be *essential* (*i.e.*, necessary and
sufficient to determine the transformation).

The requirement that the transformations (1.1.1) define a Lie group
implies, in particular, that $x = f(e; x)$, where e is the identity element of
the group. Also, if

$$x' = f(g; x) \quad \text{and} \quad x'' = f(g'; x') = f(g'; f(g; x)) \equiv f(g''; x), \tag{1.1.2}$$

the parameters g'' of the transformation $x'' = f(g''; x)$ are determined
from g', g by the functions

$$g''^k = g''^k(g'^1, ..., g'^r; g^1, ..., g^r) , \ k = 1, ..., r \quad (g'' = g''(g'; g)) \tag{1.1.3}$$

which express the coordinates of g'' in terms of those of g' and g and
determine the group law, $g'' = g' * g$ (or $g'' = g'g$ for short). It must also
be possible to express g' (or g) in terms of g'' and g (g'' and g'); this
requires that the Jacobian $\left| \frac{\partial g''^k}{\partial g'^r} \right|$ ($\left| \frac{\partial g''^k}{\partial g^r} \right|$) is not zero. Specifically, a group
G is a Lie group when it meets the conditions of the following

---

* We shall use the word *manifold* to indicate that M is a finite-dimensional *differentiable manifold*,
i.e. that M is a Hausdorff (: two distinct points of M admit disjoint neighbourhoods) topological
space endowed with a *differential structure*, i.e. with a class of equivalent atlases on M. Unless
otherwise stated, we shall take dim $M = n$.

A C^∞-atlas $\mathscr{A}$ on M is a set $\{(U_\alpha, \varphi_\alpha)\}$ of charts (or local coordinate systems) $\varphi_\alpha : U_\alpha \to R^n$
subordinated to the covering $\{U_\alpha\}$ of M that are compatible: if $U_\alpha \cap U_\beta \neq \emptyset$, the overlap
mappings $\varphi_\alpha \circ \varphi_\beta^{-1} : \varphi_\beta(U_\alpha \cap U_\beta) \to \varphi_\alpha(U_\alpha \cap U_\beta)$ between open sets of R^n are of class C^∞. Two
atlases $\mathscr{A}_1, \mathscr{A}_2$, are equivalent iff their union $\mathscr{A}_1 \cup \mathscr{A}_2$ is an atlas; a differentiable structure on
M is an equivalence class of atlases on M.

It is possible to define a C^k-differentiable manifold in a similar way, but here we shall use the
word differentiable (or *smooth*) to mean that k is always large enough for the statement to have
a meaning.

(1.1.1) DEFINITION (*Lie group*)

A Lie group is a (finite-dimensional*, differentiable) manifold G which is a group and where the group operations $G \times G \to G$, $(g';g) \mapsto g' * g$ and $G \to G$, $g \mapsto g^{-1}$ (inverse) (or $(g',g) \mapsto g'' = g' * g^{-1}$) are smooth.

The group law (1.1.3) can be written as

$$g'' = g'g \equiv L_{g'}g \equiv R_g g' \quad \forall g',g \in G , \tag{1.1.4}$$

where $L_g(R_g)$ indicates the left (right) translation determined by the group element g. It will be assumed that the identity element e of G has coordinates $(0,0,...,0)$. Obviously, $x = f(g^{-1};x')$ if $x' = f(g;x)$, etc. The left and right translations are diffeomorphisms of G; they clearly commute, $L_{g'}R_g = R_g L_{g'}$. Obviously,

$$L_{g'}L_g = L_{g'g} \quad , \quad R_{g'}R_g = R_{gg'} \quad . \tag{1.1.5}$$

Eqs. (1.1.1) constitute a particular example of an action of a Lie group on a manifold, since g determined f there uniquely. The properties of the action (1.1.2) are, however, general:

(1.1.2) DEFINITION (*Action of a Lie group on a manifold M*)

A left action of the group G on the manifold M is a smooth mapping f : $G \times M \to M$, $f : (g;x) \mapsto f(g;x)$, such that $f(e;x) = x$ and $f(g';f(g;x)) = f(g'g;x)$. Alternatively, we may write $f(g;x) \equiv gx$ and $f(g';f(g;x)) = g'(gx) = g'gx$. Similarly, a right action is defined by a smooth mapping $f : M \times G \to M$ such that $f(x;e) = x$ and $f(f(x;g);g') = f(x;gg')$. Thus, if $f(x;g) \equiv x^g$ this means that $f(f(x;g);g') = (x^g)^{g'} = x^{gg'}$.

In both cases we can look at f as defining a mapping $g \mapsto f_g$ by $f_g = f(g,\cdot)$ or $f_g = f(\cdot,g)$. Thus, if f_g is the mapping $f_g : M \to M$ associated with the action of g on M, it is seen that the left action satisfies the homomorphism property

$$f_{g'} \circ f_g = f_{g'g} \tag{1.1.6}$$

and that the right action satisfies the anti-homomorphism relation

$$f_{g'} \circ f_g = f_{gg'} \quad . \tag{1.1.7}$$

---

* The name 'Lie group' or 'Lie algebra' usually refers to the *finite*-dimensional case. However, infinite-dimensional Lie groups and algebras like the group of diffeomorphisms of a manifold mentioned below, the smooth mappings of a given manifold into a finite-dimensional Lie group (or algebra), the Kac-Moody algebras, etc., appear frequently in physics. Infinite-dimensional Lie groups and algebras will be introduced in chapter 9 (and in section 8.8).

Since $f_{g^{-1}} = (f_g)^{-1}$, f_g is a diffeomorphism of M. Thus, the previous definition is equivalent to saying that f is a homomorphism (or anti-homomorphism) $f : G \to Diff\,(M)$ of G into the group of diffeomorphisms of M. If f_g is a right (left) action, $f_{g^{-1}}$ is a left (right) action. Of course, for $M = G$, the above formulae reproduce (1.1.5).

Let $x \in M$. The f-orbit of x is the set $\{f_g(x)|g \in G\} \subset M$ of all the points that can be f-reached from x. Hence, the equivalence relation $x'\mathscr{R}x \Leftrightarrow \exists g$ for which $x' = gx$ characterizes $M/\mathscr{R}$ as the set of all f-orbits. Some types of action receive a specific name:

(1.1.3) DEFINITION (*Transitive, free, effective and trivial actions of a group G*)

An action is transitive if the orbit of (any) $x \in M$ is M. An action is free if the 'little' (or stability, isotropy) group $G_x \subset G$ of x, $G_x := \{g \in G | f_g(x) = x\}$, is reduced to the identity element for each $x \in M$. An action is effective (or faithful if it is a linear action) if the (invariant) subgroup $G' = \{g \in G | f_g(x) = x \;\forall x \in M\}$ (the intersection of all little groups or kernel of the action) is reduced to $e \in G$; if it is not, the action of G/G' is effective. Finally, f is trivial if $G' = G$.

If the action is free, every point of M is transformed ($f_g(x) \neq x$ if $g \neq e$) and the mapping $g \mapsto f_g(x)$ is bijective for each $x \in M$ (there is a one-to-one correspondence between G and each orbit in M). Thus, if the action is free, any given two points x', x are either unrelated or connected by a unique g, $x' = f_g(x)$. If the action is not free, it is very simple to prove that the little groups of points in the same orbit are all isomorphic. Thus, if the action of G is transitive, there is a bijective mapping between M and G/G_x and M is said to be a *homogeneous space* of G. For instance, if G is the Poincaré group $\mathscr{P}$, $M = R^4$ and we take the origin for x, G_x is the Lorentz group L and thus Minkowski space may be introduced as the homogeneous space $\mathscr{P}/L$. If the action is effective, the mapping $g \mapsto f_g$ is one-to-one since $f_g \neq f_g'$ if $g \neq g'$. For instance, the *adjoint action* of G on G ($M = G$), $g_0 \mapsto gg_0g^{-1}$, is not effective ($G' = C_G$, the centre of G), but the naturally induced action of G/C_G on G is effective. If M is a vector space and $f_g \equiv D(g)$ is a linear transformation, the mapping $g \mapsto D(g)$ defines a faithful representation if it is bijective, and a trivial one if $D(g)$ is the identity automorphism of M; the normal subgroup G' above is the kernel of the representation. A free action is necessarily effective, but the converse is not true: for instance, a linear representation $D(g)$ of G, and in particular a faithful one, cannot be free since the isotropy group of the zero vector is G itself. Finally, since the null vector is stable under the action of $D(g)$, a linear representation is never transitive.

Let us go back to eq. (1.1.1) and let now δg^k be the parameters of the Lie group which characterize an infinitesimal transformation on M. Expanding $f^i(\delta g, x)$ we obtain

$$\delta x^i = x'^i - x^i = f^i(\delta g; x) - f^i(0; x) \simeq \left.\frac{\partial f^i(g; x)}{\partial g^k}\right|_{g=e} \delta g^k , \qquad (1.1.8)$$

where $i = 1, ..., n$ and $k = 1, ..., r$. This implies that under a variation δx of x due to the infinitesimal transformation δg a function $F(x^1, ..., x^n)$ on M will change, to first order, by

$$\delta F \simeq \frac{\partial F}{\partial x^i} \delta x^i = \frac{\partial F}{\partial x^i} \left.\frac{\partial f^i(g^1, ..., g^r; x^1, ..., x^n)}{\partial g^k}\right|_{g=e} \delta g^k \qquad (1.1.9)$$

$$\equiv \delta g^k (X_{(k)}.F) \quad ; \quad \delta_{(k)} F = X_{(k)}.F .$$

Thus, the differential operator on M (or *vector field* in $\mathscr{X}(M)$, see below and section 1.3),

$$X_{(k)}(x) = \left.\frac{\partial f^i(g^1, ..., g^r; x^1, ..., x^n)}{\partial g^k}\right|_{g=e} \frac{\partial}{\partial x^i} \quad (i = 1, ..., n) , \qquad (1.1.10)$$

is the generator of the infinitesimal transformation associated with a given parameter δg^k. The fundamental theorem of S. Lie states that the r infinitesimal generators close into an algebra, the *Lie algebra* $\mathscr{G}$ of G:

$$[X_{(j)}(x), X_{(k)}(x)] = -C_{jk}^l X_{(l)}(x) \quad (j, k, l = 1, ..., r) . \qquad (1.1.11)$$

The reason for the insertion of the minus sign here is conventional; it is due to the fact that we are assuming that (1.1.1) is defining a left action, as will be explained in the next section. The real constants C_{jk}^l are the structure constants of the algebra and satisfy the Jacobi identity

$$[[X_{(i)}, X_{(j)}], X_{(k)}] + [[X_{(j)}, X_{(k)}], X_{(i)}] + [[X_{(k)}, X_{(i)}], X_{(j)}] \equiv 0 \quad , \qquad (1.1.12)$$
$$C_{ij}^l C_{lk}^m + C_{jk}^l C_{li}^m + C_{ki}^l C_{lj}^m = 0 \quad .$$

The manifold M can be, in particular, the Lie group itself and the group law (1.1.4) may now be interpreted in two ways: as the left (diffeomorphism) action $L_{g'}$ of a fixed group element g' on the manifold G parametrized by the coordinates g^k ($k = 1, ..., r$) of g, or as the result of the right translation R_g on G, its points now being determined by the coordinates of g'. In the first case, the generators X are those of the *left action*. They are given by

$$X_{(k)}^R(g) = \left.\frac{\partial g''^\ell(g', g)}{\partial g'^{(k)}}\right|_{\substack{g'=e \\ g=g}} \frac{\partial}{\partial g^\ell} \quad . \qquad (1.1.13)$$

In the second case, the generators are those of the *right action*, and their expression is

$$X^L_{(k)}(g) = \left.\frac{\partial g''^{\ell}(g',g)}{\partial g^{(k)}}\right|_{\substack{g=e \\ g'=g}} \frac{\partial}{\partial g^{\ell}} \tag{1.1.14}$$

(the reason for the superscripts R, L will be explained in the next section). Now, since $g'' = g' * e = g'$, $g'' = e * g = g$, it is clear that

$$\left.\frac{\partial g''^{\ell}(g',g)}{\partial g'^k}\right|_{\substack{g'=e \\ g=e}} = \delta^{\ell}_k \quad , \quad \left.\frac{\partial g''^{\ell}(g',g)}{\partial g^k}\right|_{\substack{g=e \\ g'=e}} = \delta^{\ell}_k \quad , \tag{1.1.15}$$

so that

$$X^R_{(k)}(e) = \delta^{\ell}_{(k)} \frac{\partial}{\partial g^{\ell}} \equiv X_{(k)}(e) \;, \; X^L_{(k)}(e) = \delta^{\ell}_{(k)} \frac{\partial}{\partial g^{\ell}} \equiv X_{(k)}(e) \;. \tag{1.1.16}$$

Thus, (1.1.13), (1.1.14) can be written in the form

$$X^R_{(k)}(g) = X^{R\ell}_{(k)}(g)\frac{\partial}{\partial g^{\ell}} \;, \; \text{and} \; X^{R\ell}_{(k)}(g) = [R^T_g(e)]^{\ell}_{\cdot k} \;, \tag{1.1.17}$$

$$X^L_{(k)}(g) = X^{L\ell}_{(k)}(g)\frac{\partial}{\partial g^{\ell}} \;, \; \text{and} \; X^{L\ell}_{(k)}(g) = [L^T_g(e)]^{\ell}_{\cdot k} \;, \tag{1.1.18}$$

where $\ell(k)$ is the row (column) index. The r vectors $\partial/\partial g^k = X_{(k)}(e)$, labelled by the index inside the brackets (k), constitute a basis of the tangent space $T_e(G)$ at the identity of G. The *linear* mappings $R^T_g(e)$, $L^T_g(e)$ take these vectors, with coordinates $\delta^j_{(k)}$, to the vectors $X^L_{(k)}(g)$, $X^R_{(k)}(g) \in T_g(G)$ tangent to G at g the coordinates of which (in terms of $\partial/\partial g^{\ell}$) are given by (1.1.17), (1.1.18). In other words, $R^T_g(e)$ and $L^T_g(e)$ are the *tangent* (or *differential*) *mappings* at e associated with the translations R_g and L_g of G and $T_e(G)$ may be identified with $\mathscr{G}$, $T_e(G) = \mathscr{G}$ (see also proposition 1.2.3 below).

The arguments in $R^T_g(e)$, $L^T_g(e)$ should be noticed. In order to see the rôle of the arguments of the tangent mappings, consider the successive action of two (left, say) translations

$$g \overset{L_{g'}}{\longmapsto} g'g \overset{L_{g''}}{\longmapsto} \tilde{g} = g''g'g$$

Diagram 1.1.1

By using the chain rule, we may write

$$[L^T_{g''g'}(g)]^{\ell}_{\cdot k} \equiv \frac{\partial \tilde{g}^{\ell}}{\partial g^k} = \frac{\partial \tilde{g}^{\ell}}{\partial(g'g)^j}\frac{\partial(g'g)^j}{\partial g^k} \equiv [L^T_{g''}(g'g)]^{\ell}_{\cdot j}\,[L^T_{g'}(g)]^j_{\cdot k} \quad . \tag{1.1.19}$$

This is a particular case of the chain rule for the tangent mapping of a composite mapping: if $\phi : M \to N$, $\psi : N \to S$ are mappings of manifolds and we write $D(\phi)$, $D(\psi)$ for ϕ^T, ψ^T, the derivative of the composite mapping $\psi \circ \phi$ reads*

$$D(\psi \circ \phi)(x) = D(\psi)(\phi(x)) \circ D(\phi)(x) \quad (x \in M) , \tag{1.1.20}$$

which, in local coordinates, expresses the multiplication of the Jacobian matrices (as in (1.1.19)) of the two transformations. Taking $g = e$, $g' = g$ and $g'' = g^{-1}$ in eq. (1.1.19) we obtain

$$L_e^T(e) = L_{g^{-1}}^T(g) \circ L_g^T(e) \quad , \quad [L_g^T(e)]^{-1} = L_{g^{-1}}^T(g) \quad , \tag{1.1.21}$$

i.e., again with $g'' = g'g$, that

$$\left. \frac{\partial g''^\ell}{\partial g^j} \right|_{\substack{g=g \\ g'=g^{-1}}} \cdot \left. \frac{\partial g''^j}{\partial g^k} \right|_{\substack{g=e \\ g'=g}} = \delta_k^\ell \tag{1.1.22}$$

and *not* that $[L_g^T(e)] [L_{g^{-1}}^T(e)]$ is 1; in fact, this last expression does not have a meaning in general (for linear mappings, however, the derivative $D(\phi)$ of a mapping ϕ clearly does not depend on the argument).

It was assumed at the beginning of this section that the Lie group G was defined through the transformations (1.1.1) or, in the terminology of Definition 1.1.3, that its action on M was effective. In this case (as obviously for the – transitive, effective and free – left and right actions of G on G above) there is an isomorphism between $\mathscr{G}$ and the Lie algebra of the infinitesimal generators of the action of G on M, and both may be identified as in (1.1.11). In general, however, we may have only a homomorphism of $\mathscr{G}$ into the algebra $\mathscr{X}(M)$ of the vector fields on M, and we may ask ourselves whether there is a Lie group action on M such that the vector fields of $\mathscr{X}(M)$ that it induces on M are the images of the elements X of the Lie algebra $\mathscr{G}$ of G. The conditions for the affirmative answer constitute the

(1.1.1) THEOREM (*Palais*)

Let G be a simply connected Lie group, M a compact manifold, and let $\tilde{f} : \mathscr{G} \to \mathscr{X}(M)$ be a Lie algebra homomorphism. Then there exists a unique

---

* Strictly speaking $D\phi$ (also written $d\phi$ and ϕ') is the *second* component of the tangent mapping ϕ^T (or $T\phi$), which is defined by $\phi^T : (x, \mathbf{v}) \mapsto (\phi(x), D\phi(x).\mathbf{v})$, where $\mathbf{v}$ and $D\phi.\mathbf{v}$ are tangent vectors to M and N and $x \in M$, $\phi(x) \in N$ are their base points. In the same way, an expression such as $L_{g'}^T(g)$, which acts on a tangent vector at g, also corresponds to the second component of the tangent mapping $L_{g'}^T$. In terms of the tangent mappings, the *composite mapping theorem* reads $(\psi \circ \phi)^T = \psi^T \circ \phi^T$ (or $T(\psi \circ \phi) = T\psi \circ T\phi$). For the second components of the tangent mappings this implies, for instance, the next formula for $D(\psi \circ \phi)(x)$ or that $(L_{g''} \circ L_{g'})^T(g) = L_{g''}^T(g'g) \circ L_{g'}^T(g)$.

action f of G on M, $f : G \to Diff\,(M)$, *the group of diffeomorphisms of* M, *such that* $\tilde{f}$ *is the tangent mapping* $f^T : \mathscr{G} \to \mathscr{X}(M)$.

Notice that $T_e(G) = \mathscr{G}$, and the tangent space to $Diff\,(M)$ may be identified with the space of vector fields on M which determine the Lie algebra of $Diff\,(M)$. The simple connectedness is required to single out a Lie group for the given Lie algebra, and on a compact manifold every vector field is complete.

1.2 Left- (X^L) and right- (X^R) invariant vector fields on a Lie Group G

A tangent vector $X(x)$ at a point x of a manifold M can be thought of as the image of x under a differentiable mapping $X : x \mapsto X(x)$. X (or X_M) is then called a *vector field** on the manifold M. Let now $\phi : M \to S$ be a differentiable mapping of manifolds (not necessarily a diffeomorphism), and let X, Y be vector fields on M and S respectively. Y is the *transformed vector field of* X *by* ϕ (i.e., X and Y are said to be ϕ-related) if

$$\phi^T \circ X = Y \circ \phi, \tag{1.2.1}$$

i.e. if the diagram

$$
\begin{array}{ccc}
T(M) & \xrightarrow{\phi^T} & T(S) \\
X \uparrow & & \uparrow Y \\
M & \xrightarrow{\phi} & S
\end{array}
$$

Diagram 1.2.1a

is commutative.

The property of being ϕ-related preserves the Lie bracket of vector fields. More precisely, the following proposition holds:

(1.2.1) PROPOSITION

Let $\phi : M \to S$ be a differentiable mapping and let X_i and Y_i be two pairs of ϕ-related vector fields on M and S respectively, $\phi^T \circ X_i = Y_i \circ \phi$, $i = 1, 2$. Then

$$\phi^T \circ ([X_1, X_2]) = [Y_1, Y_2] \circ \phi. \tag{1.2.2}$$

---

* Usually we denote vector fields on a manifold M by X. But if it is convenient to distinguish a vector field on M from the abstract algebra element $X \in \mathscr{G}$, we shall denote the former by X_M. A vector field on M may also be written $X(x)$, x not fixed (for a given x, $X(x) \in T_x(M)$ is a vector tangent to M at x).

Proof: The demonstration is simple, and makes use of the fact that for any $f \in \mathcal{F}(S)$ (smooth function on the manifold S) the equality $df(\phi^T[X_1, X_2])$ $= df([Y_1, Y_2] \circ \phi)$ holds[†]. This is in turn proved by recalling that, for any vector field Y on S, $df(Y) = Y.f$ (see section 1.4) and that, if X and Y are ϕ-related, $X.(f \circ \phi) = (Y.f)\phi$, q.e.d.

If ϕ is a *diffeomorphism* and X and Y are ϕ-related, then

$$Y = \phi^T \circ X \circ \phi^{-1} := \phi_*(X) ; \qquad (1.2.3)$$

the second equality defines the *push-forward* of X by ϕ. Similarly (see section 1.4), by replacing ϕ by ϕ^{-1} we can define the *pull-back* of a vector field Y on S as the vector field on M given by $\phi^* = \phi_*^{-1}$,

$$\phi^*(Y) = (\phi^T)^{-1} \circ Y \circ \phi \quad . \qquad (1.2.4)$$

Notice that although (1.2.1) makes sense in general, eqs. (1.2.3) and (1.2.4) require that ϕ be a *diffeomorphism*. Let now $M = S$. Then X is said to be a ϕ-*invariant vector field* if

$$X = \phi_*(X) \quad . \qquad (1.2.5)$$

Consider now the case where $M = S = G$, $\phi_* = (L_s)_*$ and let us look at Diagram 1.2.1b below,

$$X(g) \xrightarrow[L_s^T]{} (L_s^T X)(sg) \equiv Y(sg); \quad X(sg)$$

$$\uparrow \qquad\qquad\qquad\qquad\qquad \uparrow$$

$$X \qquad\qquad\qquad\qquad\qquad X$$

$$| \qquad\qquad\qquad\qquad\qquad |$$

$$g \xrightarrow[]{L_s} sg$$

Diagram 1.2.1b

If the diagram is commutative, then the image $Y(sg)$ of the tangent vector $X(g) \in T_g(G)$ by L_s^T is just the vector $X(sg)$, the image of $sg \in G$ by X, and X is an invariant vector field. Thus we can give the following

(1.2.1) Definition (*Left-invariant vector field*)

A vector field X on the Lie group G satisfying

$$X = L_s^T \circ X \circ L_s^{-1} := L_{s*}X \qquad (1.2.6)$$

is called a left-invariant vector field (LIVF) on G.

[†] The exterior derivative is introduced in section 1.4.

Left-invariant vector fields will be denoted by X^L. They determine a vector space $\mathscr{X}^L(G)$ of dimension dim G which is closed under the Lie bracket; its basis $X_{(i)}^L$ generates the algebra $\mathscr{G}$ of G. The LIVFs do not constitute an $\mathscr{F}(G)$-module, however, since $f \cdot X^L \notin \mathscr{X}^L(G)$, $f \in \mathscr{F}(G)$. The $\mathscr{F}(G)$-module of vector fields on G will be denoted $\mathscr{X}(G)$. It is a *free* $\mathscr{F}(G)$-module of dimension dim G or, in other words, if $X \in \mathscr{X}(M)$, X can be written as $X = f^i X_{(i)}^L$, $i = 1, ..., \dim G$, where $X_{(i)}^L$ is a basis of $\mathscr{X}^L(G)$ and $f^i \in \mathscr{F}(G)$.

Eq. (1.2.6) is sometimes written $X = L_s^T X$, but this is not quite correct. Eq. (1.2.6) applied to sg establishes that

$$X^L(sg) = L_s^T(g)X(g) \tag{1.2.7}$$

or, equivalently,

$$X^L(g) = L_g^T(e)X(e) \quad . \tag{1.2.8}$$

The arguments of the tangent mappings may (and often will) be left understood but, as mentioned, they cannot be ignored. Eqs. (1.2.8) or (1.2.7) constitute another form of expressing that X^L is a LIVF.

Let us go back to (1.1.18). Applying (1.2.6) it is clear that these vector fields on G are left-invariant. Similarly, a vector field is said to be *right-invariant* (RI) if

$$R_{s*}X = X \; , \quad i.e. \quad R_g^T(e)X(e) = X^R(g) \quad . \tag{1.2.9}$$

The right-invariant vector fields (RIVFs) will be denoted X^R; it is immediately seen that those given by (1.1.17) are RI (the superscripts L, R will always refer to the left and right invariance of the corresponding vector fields and *not* to the transformations that they generate). Thus, the vector fields which generate the left (right) action of G on G are RI (LI), and this justifies the notation used in (1.1.13), (1.1.14). This is now natural; for instance, for LIVFs [(1.1.14), (1.1.18)],

$$[L_{sg}^T(e)]_{\cdot k}^\ell = \left.\frac{\partial g''^\ell(g', g)}{\partial g^k}\right|_{\substack{g=e \\ g'=sg}}$$

and since $L_{sg}^T(e) = L_s^T(g)L_g^T(e)$ [(1.1.19)] it is clear that (1.2.7) is satisfied.

(1.2.2) PROPOSITION

Let the Lie algebra $\mathscr{G}$ of G be defined by

$$[X_{(i)}^L(g), \; X_{(j)}^L(g)] = C_{ij}^k X_{(k)}^L(g) \qquad (i, j, k = 1, ..., r) \; , \tag{1.2.10}$$

$X^L \in \mathscr{X}^L(G)$. *Then*

$$[X_{(i)}^R(g), \; X_{(j)}^R(g)] = -C_{ij}^k X_{(k)}^R(g) \tag{1.2.11}$$

$(X^R \in \mathscr{X}^R(G))$ and

$$[X^L_{(i)}(g),\ X^R_{(j)}(g)] = 0 \ . \tag{1.2.12}$$

Proof: Expanding (1.1.3) in series by assuming g', g small, one obtains

$$g''^k = g'^k + g^k + a^k_{ij}g'^i g^j + b^k_{ij\ell}g'^i g'^j g^\ell + d^k_{ij\ell}g'^i g^j g^\ell + O^4(g',g)^k \ . \tag{1.2.13}$$

The $O^4(g',g)$ contains the $g'^3 g$, $g'^2 g^2$ and $g' g^3$ terms; note that there are no non-linear pure g' or g terms in the expansion due to the associativity of the group law of G. Eq. (1.2.13) tells us that

a) in the first approximation, any Lie group is commutative and isomorphic to R^r;

b) to determine the structure constants of the Lie *group* in the given coordinate system,

$$C^k_{ij} \equiv a^k_{ij} - a^k_{ji} \ , \tag{1.2.14}$$

it is sufficient that G be differentiable twice;

c) the C^k_{ij} completely determine the *local* structure of G.

Writing explicitly its l.h.s., (1.2.10) reads (all arguments taken at g)

$$[X^L_{(i)},\ X^L_{(j)}]^\ell = X^{Ls}_{(i)}\frac{\partial X^{L\ell}_{(j)}}{\partial g^s} - X^{Ls}_{(j)}\frac{\partial X^{L\ell}_{(i)}}{\partial g^s} = C^k_{ij}X^{L\ell}_{(k)} \ . \tag{1.2.15}$$

Now, since $X^{Ls}_{(j)}(e) = \delta^s_j$ we obtain, using (1.1.14) and (1.2.13)

$$\left.\frac{\partial^2 g''^\ell(g',g)}{\partial g'^i \partial g^j}\right|_{\substack{g'=e\\g=e}} - \left.\frac{\partial^2 g''^\ell(g',g)}{\partial g'^j \partial g^i}\right|_{\substack{g'=e\\g=e}} = C^\ell_{ij} \ . \tag{1.2.16}$$

It is clear that repeating the above for X^R has the effect of interchanging j and i in (1.2.16), and hence (1.2.11) follows. Eq. (1.2.16) also identifies the structure constants of the algebra with those of the group, eq. (1.2.14), and is a check of Lie's second theorem: the structure constants of the algebra are, indeed, constant.

Finally, since the L and R translations commute, (1.2.12) follows (as can also be proved explicitly), *q.e.d.*

From the above discussion the following proposition is evident:

(1.2.3) PROPOSITION

Let $\{X^L_{(j)}(g)\}$, $j = 1,...,r$, be a basis of the Lie algebra of G. By using a left translation the vector space $\mathscr{X}^L(G)$ generated by $\{X^L_{(j)}(g)\}$ can be put into a one-to-one correspondence with $\mathscr{G} = T_e(G)$. This is also a Lie algebra isomorphism with the Lie bracket in $\mathcal{G}$ defined from the bracket $[X^L_{(j)}(g),\ X^L_{(k)}(g)]$ in $\mathcal{X}^L(G)$.

As is well known, a connected Lie group G has vanishing structure constants if and only if it is locally isomorphic to R^r, i.e. iff it is abelian.

Example: *Left- and right-invariant vector fields on $GL(n, R)$ and generators of its left action on R^n.*

Let $G = GL(n, R)$, and its group law given in terms of the matrix multiplication,

$$(g'')^i{}_{\cdot j} = (g')^i{}_{\cdot \ell}(g)^\ell{}_{\cdot j} \quad , \tag{1.2.17}$$

where $g^i{}_{\cdot j}$ denotes the $n \times n$ real matrix of the element g. It is convenient to keep the double index notation to label the $n \times n$ parameters. The LIVFs and RIVFs are obtained from (1.1.14), (1.1.13):

$$X^L{}_{(k)}{}^{(t)}{}_{\cdot}(g) = g^i{}_{\cdot k}\frac{\partial}{\partial g_{\cdot t}^i} \ , \quad X^R{}_{(k)}{}^{(t)}{}_{\cdot}(g) = g^t{}_{\cdot j}\frac{\partial}{\partial g_{\cdot j}^k} \tag{1.2.18}$$

and, as can be immediately checked, they generate the L and R translations respectively. Their commutation relations are given by

$$\begin{aligned} [X^L{}_{(k)}{}^{(t)}_{\cdot}, X^L{}_{(j)}{}^{(s)}_{\cdot}] &= \delta^t_j X^L{}_{(k)}{}^{(s)}_{\cdot} - \delta^s_k X^L{}_{(j)}{}^{(t)}_{\cdot} \quad , \\ [X^R{}_{(k)}{}^{(t)}_{\cdot}, X^R{}_{(j)}{}^{(s)}_{\cdot}] &= \delta^s_k X^R{}_{(j)}{}^{(t)}_{\cdot} - \delta^t_j X^R{}_{(k)}{}^{(s)}_{\cdot} \quad . \end{aligned} \tag{1.2.19}$$

The *left* action of $GL(n, R)$ on $M = R^n$ is given by $x'^i = g^i{}_{\cdot j}x^j$. The generators of this action are obtained from (1.1.10) with the result

$$X(x)_{(k)}{}^{(t)}_{\cdot} = x^t \partial_k \quad , \quad [X(x)_{(k)}{}^{(t)}_{\cdot}, X(x)_{(j)}{}^{(s)}_{\cdot}] = \delta^s_k X(x)_{(j)}{}^{(t)}_{\cdot} - \delta^t_j X(x)_{(k)}{}^{(s)}_{\cdot} \quad . \tag{1.2.20}$$

We see that these commutation relations differ in a minus sign from those of the LI generators above which are the result of the *right* action of G on G. Thus, if the algebra $\mathcal{G}$ of G is defined by (1.2.10), the algebra of the generators of the left action of G on a manifold M is anti-homomorphic to it; hence the minus sign in (1.1.11).

1.3 A summary of fibre bundles

Fibre bundles are playing an increasing rôle in the description of physical phenomena where the distinction between local and global aspects is important. In particular, fibre bundles provide the adequate mathematical framework to discuss Yang-Mills gauge theories; as we shall see Yang-Mills fields may be understood as connections on principal bundles. From the purely mathematical point of view, many familiar concepts, as *e.g.* vector fields and differential forms, naturally find their place in the fibre bundle framework. For this reason we interrupt our discussion of Lie groups and, before reviewing the essentials of Cartan calculus, we collect in this long section some basic definitions and properties of fibre bundles.

(1.3.1) DEFINITION (*Principal bundle*)

A (differentiable) principal fibre bundle $P(G, M)$ over M (the base manifold) with group G (the structure group) consists of a manifold P (the total manifold) and a (by convention) right action of G on P, $P \times G \to P$, such that

a) *the action* of G,*

$$g : p \mapsto pg \equiv R_g p \quad ,$$

  is free ($p = pg \Rightarrow g = e$, see Definition 1.1.3);

b) *M is the quotient space $P/\mathcal{R}$, where $\mathcal{R}$ is the equivalence relation induced by the action of G ($p'\mathcal{R}p \Leftrightarrow \exists g, p' = pg$) (the differentiable mapping $\pi : P \to M$ is called the projection, and $\pi(pg) = \pi(p)$);*

c) *P is locally trivial, i.e., every point $x \in M$ has an open neighbourhood U_α such that $\pi^{-1}(U_\alpha)$ is diffeomorphic to $U_\alpha \times G$. More precisely, there exists a diffeomorphism (local trivialization) Φ_α,[†]*

$$\Phi_\alpha^{-1} : U_\alpha \times G \to \pi^{-1}(U_\alpha) \quad ,$$

such that $\pi \circ \Phi_\alpha^{-1} : (x, g) \mapsto x$ and $\Phi_\alpha^{-1}(x, g)g_0 = \Phi_\alpha^{-1}(x, gg_0)$.

The inverse image of a point $x \in M$, $\pi^{-1}(x)$, is called the *fibre over x*; $\pi^{-1}(\pi(p)) = pG$ is the *fibre through p*.

(1.3.2) DEFINITION (*Local cross section*)

A local cross section over an open set $U \subset M$ is a differentiable mapping $\sigma : U \to P$ such that $(\pi \circ \sigma)(x) = x \quad \forall x \in U$.

The locally trivializing mappings Φ_α define local cross sections σ_α by

$$\sigma_\alpha(x) = \Phi_\alpha^{-1}(x, e), \quad \sigma_\alpha(x)g = \Phi_\alpha^{-1}(x, g), \quad e \text{ identity element of } G \quad .$$

Defining *id* as the mapping $id : U_\alpha \to U_\alpha \times G$, $id : x \mapsto (x, e)$, the local section defined by Φ_α is expressed by

$$\sigma_\alpha = \Phi_\alpha^{-1} \circ id \quad ; \tag{1.3.1}$$

---

* Since by convention it is a right action, $R_{g'} R_g p = R_{gg'} p = pgg'$. Thus, this action defines a group (anti)isomorphic to G, eq. (1.1.7).

[†] It is possible to define a bundle structure on a topological space; G is then a topological group and the coordinate functions Φ_α are homeomorphisms. Here we shall essentially consider differentiable bundles.

these cross sections are sometimes called *natural cross sections*. Conversely, a local section defines a local trivialization. Let ρ_α be a mapping $\rho_\alpha : \pi^{-1}(U_\alpha) \to G$ such that $\rho_\alpha(pg) = \rho_\alpha(p)g$. Then σ_α defines such a mapping ρ_α by $\sigma_\alpha(x)\rho_\alpha(p) = p$. This definition is consistent since $p[\rho_\alpha(p)]^{-1}$ does not depend on p: $p'[\rho_\alpha(p')]^{-1} \equiv pg[\rho_\alpha(pg)]^{-1} = pgg^{-1}[\rho_\alpha(p)]^{-1} = \sigma_\alpha(x)$. Then Φ_α is defined by

$$\Phi_\alpha(p) = (\pi(p), \rho_\alpha(p)) = (x, \rho_\alpha(p)) \tag{1.3.2}$$

and it follows from this expression and the defining properties of ρ_α above that $\Phi_\alpha \circ \sigma_\alpha = id$ is fulfilled. The mapping $\rho_\alpha : \pi^{-1}(x) \to G$ provides a *non-canonical* identification of the fibre over x with the structure group. Since $\Phi_\alpha(\sigma_\alpha(x)) = (x, e)$, $\rho_\alpha(\sigma_\alpha(x)) = e$.

Some of the above notions can be described pictorially by means of a (local) diagram:

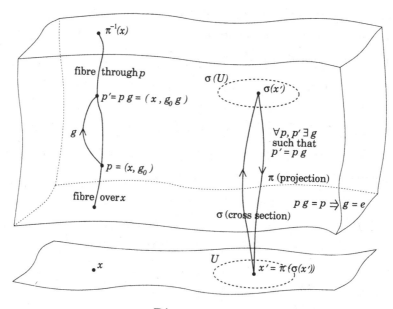

Diagram 1.3.1

From Definition 1.3.1 it follows that

– each fibre is stable under the group action, and G acts without fixed points (there is no 'little group');
– G acts transitively on each fibre and, since the action is free, the fibres are 'copies' of G.

It is important to notice that property (c) above guarantees the *local* triviality of P, but that in general the total manifold cannot be globally expressed as $P = M \times G$; in fact, the notion of fibre bundles is introduced to generalize this trivial situation. A principal bundle is *trivial* if there is an equivariant diffeomorphism $f : M \times G \to P$, $f(x, g)g_0 = f(x, gg_0)$, such that $\pi(f(x, g)) = x$ $\forall x \in M$, $\forall g \in G$.

The following result is almost intuitive and follows from Definition 1.3.1:

(1.3.1) THEOREM (*Triviality of a principal bundle*)

A principal bundle is trivial iff it admits a global cross section ($U = M$ above) or 'admits a section' for short (this condition, however, does not guarantee triviality for other bundles, see below).

Proof: Let $\sigma : M \to P$ be a section. Any $p \in \pi^{-1}(x)$ may be written in the form $\sigma(x)g$, i.e. every $p \in P$ may be 'localized' with respect to $\sigma(\pi(p))$ and characterized by the pair $(x, g) \in M \times G$. Define now the diffeomorphism $\Phi : P \to M \times G$ by $\Phi(p) = (x, g)$. Clearly, if $p' = pg_0$ we find that $\Phi(p') = \Phi(pg_0) = (x, gg_0)$, i.e. $\Phi^{-1}(x, g)g_0 = \Phi^{-1}(x, gg_0)$. Thus, Φ is a trivial chart for P and $P \sim M \times G$; the converse statement is obvious, q.e.d.

It follows from the above proof that a principal bundle is trivial iff there exists a differentiable mapping $\rho : P \to G$ such that $\rho(pg) = \rho(p)g$ $\forall p \in P$, $\forall g \in G$. Indeed, if a smooth global section exists, $\Phi(p)$ defines ρ by the identification $\Phi(p) = (x, g) := (x, \rho(p))$. Conversely, if a mapping ρ exists, the associated global cross section is defined by the mapping $\sigma : x \in M \mapsto \sigma(x) := p[\rho(p)]^{-1}$.

Let now $\{U_\alpha\}$ be an open covering of M, $\bigcup_{\alpha \in I} U_\alpha = M$. Assume now $U_\alpha \cap U_\beta \neq \emptyset$ and let Φ_α, Φ_β be the corresponding local trivializations. Define the natural cross sections by ($x \in U_\alpha \cap U_\beta$)

$$\sigma_\alpha(x) = \Phi_\alpha^{-1}(x, e) = p[\rho_\alpha(p)]^{-1} \quad , \quad \sigma_\beta(x) = \Phi_\beta^{-1}(x, e) = p[\rho_\beta(p)]^{-1} . \tag{1.3.3}$$

It follows that

$$\sigma_\alpha(x)\rho_\alpha(p) = p = \sigma_\beta(x)\rho_\beta(p) \, , \, \sigma_\beta(x) = \sigma_\alpha(x)\rho_\alpha(p)[\rho_\beta(p)]^{-1} \equiv \sigma_\alpha(x)g_{\alpha\beta}(x) \, , \tag{1.3.4}$$

since $\rho_\alpha(p)[\rho_\beta(p)]^{-1}$ depends only on $\pi(p) = x \in U_\alpha \cap U_\beta$. Thus, $g_{\alpha\beta}(x)^{-1} = g_{\beta\alpha}(x)$ and

$$\sigma_\alpha(x) = \sigma_\beta(x)g_{\beta\alpha}(x) \, , \, \rho_\alpha(p) = g_{\alpha\beta}(\pi(p))\rho_\beta(p) \, ; \, g_{\beta\alpha} : x \in U_\alpha \cap U_\beta \to G . \tag{1.3.5}$$

The g's are the *transition functions* of the principal bundle $P(G, M)$; for two local trivializations, $(\Phi_\beta \circ \Phi_\alpha^{-1})(x, \rho_\alpha(p)) = (x, g_{\beta\alpha}(x)\rho_\alpha(p))$ or, with obvious notation, $\Phi_{\beta,x} \circ \Phi_{\alpha,x}^{-1} = g_{\beta\alpha}(x)$. The '*reconstruction theorem*' guarantees that a set of mappings $g_{\beta\alpha} : U_\alpha \cap U_\beta \to G$ subordinated to the covering

$\{U_\alpha\}$, $\alpha \in I$, and satisfying the 'cocycle property'

$$g_{\gamma\beta}(x)g_{\beta\alpha}(x) = g_{\gamma\alpha}(x) \quad , \quad \forall x \in U_\alpha \cap U_\beta \cap U_\gamma \quad , \quad (1.3.6)$$

uniquely determines a principal bundle with transition functions given by the functions $g_{\beta\alpha}(x)$. The reconstruction theorem permits one to recover the abstract principal bundle from an explicit representative much in the same way as a group can be studied by using an explicit effective realization. Bundles defined by the bundle space P, the base M, the covering $\{U_\alpha\}$, the projection π, the structure group G acting on the fibre space and the coordinate functions Φ_α are called *coordinate bundles* in the sense of Steenrod. Two coordinate bundles with the same base, covering, fibre, group and with transition functions $g_{\alpha\beta}(x)$, $g'_{\alpha\beta}(x)$ related by the 'coboundary relation'

$$g'_{\alpha\beta}(x) = g_\alpha^{-1}(x)g_{\alpha\beta}(x)g_\beta(x) \quad , \quad \forall x \in U_\alpha \cap U_\beta \quad , \quad (1.3.7)$$

where g_α, g_β are mappings $g_{\alpha,\beta} : U_{\alpha,\beta} \to G$, are *equivalent* (both $g_{\alpha\beta}$ and $g'_{\alpha\beta}$ satisfy (1.3.6)) and they determine the same principal bundle: the topological information is encoded in the transition functions.

If a different cross section $\tilde{\sigma}_\alpha$ over the same chart U_α is defined, we find that on U_α

$$\tilde{\sigma}_\alpha(x) = \Phi_\alpha^{-1}(x, g) = \Phi_\alpha^{-1}(x, e)g_\alpha(x) = \sigma_\alpha(x)g_\alpha(x) \quad .$$

Similarly,

$$\tilde{\sigma}_\beta(x) = \Phi_\beta^{-1}(x, g) = \Phi_\beta^{-1}(x, e)g_\beta(x) = \sigma_\beta(x)g_\beta(x) ,$$

and then it follows that the transition functions are equivalent:

$$\tilde{\sigma}_\alpha(x) = \tilde{\sigma}_\beta(x)\tilde{g}_{\beta\alpha}(x) , \quad \sigma_\alpha(x) = \sigma_\beta(x)g_\beta(x)\tilde{g}_{\beta\alpha}(x)g_\alpha(x)^{-1}$$

$$\Rightarrow g_{\beta\alpha}(x) = g_\beta(x)\tilde{g}_{\beta\alpha}(x)g_\alpha(x)^{-1} \quad , \quad \forall x \in U_\alpha \cap U_\beta .$$

Thus, the different principal bundles correspond to the different equivalence classes induced by the previous equivalence relation. When the transition functions are equivalent to the identity mappings, $g_{\alpha\beta}(x) = e$, the bundle is *trivial*.

Examples of principal bundles

(a) *A Lie group as a principal bundle*
Let G be a Lie group and K a closed subgroup*; K acts freely on G by right translations, $k : g \in G \mapsto g * k \in G$. The factor space G/K of the

---

* A subgroup K of a topological group G is a topological subgroup if it is closed. If K is a closed subgroup of a Lie group G, then K is a submanifold of G and hence a Lie subgroup.

equivalence classes $\{g\}$, where g is the element of G which characterizes the left coset $g * K \subset G$, may be endowed with a differentiable structure. Then $G(K, G/K)$ is a principal bundle over the quotient G/K (the local triviality condition is not evident, but it can be proved in general).

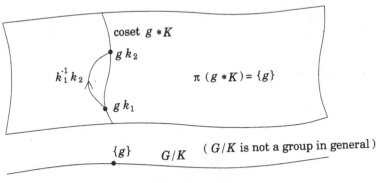

coset $g * K$

$g k_2$

$k_1^{-1} k_2$

$\pi (g * K) = \{g\}$

$g k_1$

$\{g\}$ G/K (G/K is not a group in general)

Diagram 1.3.2

For instance, $SO(n+1)$ and $SU(n+1)$ can be considered as the principal bundles $SO(n+1)(O(n), RP^n)$, $SU(n+1)(U(n), CP^n)$ respectively, over the real and complex *projective spaces* $RP^n = SO(n+1)/O(n)$ and $CP^n = SU(n+1)/U(n)$ (the case CP^1 is considered in (d) below). This construction can be generalized to the real and complex *Grassmann manifolds*, which are given by $Gr(n, k; R) = O(n)/(O(k) \otimes O(n-k))$ and $Gr(n, k; C) = U(n)/(U(k) \otimes U(n-k))$ and are compact manifolds of dimension $k(n-k)$ and $2k(n-k)$ respectively; $RP^n = Gr(n+1, 1; R)$, $CP^n = Gr(n+1, 1; C)$. Other examples of groups as principal bundles are $SO(n+1)(SO(n), S^n)$, $SU(n+1)(SU(n), S^{2n+1})$ $(n \geq 2)$, $U(n+1)(U(n), S^{2n+1})$, the sphere S^n being given by $SO(n+1)/SO(n)$ and S^{2n+1} by $SU(n+1)/SU(n) \sim U(n+1)/U(n)$. It is very easy to check in these last examples that the base manifolds are all spheres. For instance the $SO(n+1)$-orbit of the unit vector $\mathbf{x} = (1, 0, \overset{n}{\cdots}, 0) \in R^{n+1}$ is S^n, but the action is not free, the little group being $SO(n)$; thus, $SO(n+1)/SO(n) \sim S^n$.

The same construction can be extended to non-compact manifolds. For instance, the complex hyperboloid $\Omega = \left\{ \mathbf{y} \in C^{p+q} \mid |y_1|^2 + \ldots + |y_p|^2 - |y_{p+1}|^2 - \ldots - |y_{p+q}|^2 = 1 \right\}$ may be considered as the $SU(p,q)$-orbit of the vector $(1, 0, \ldots, 0) \in C^{p+q}$. The isotropy group is $SU(p-1, q)$ and thus the bundle is $SU(p,q)(SU(p-1,q), \Omega)$, where the complex hyperboloid $\Omega = SU(p,q)/SU(p-1,q)$ has real dimension $2(p+q) - 1$. Similarly, the *projective hyperboloid* $P\Omega$ is given by $P\Omega = SU(p,q)/U(p-1,q)$.

(b) *The universal covering manifold*

Let M be a connected manifold and let $\widetilde{M}$ be its universal covering; $\widetilde{M}$ is then simply connected. Let $\pi : \widetilde{M} \to M$ be the covering projection. Then the fundamental homotopy group $\pi_1(M)$ acts on $\widetilde{M}$ in such a way that $\widetilde{M}(\pi_1(M), M)$ is a principal bundle over M with structure group $\pi_1(M)$. For instance, if M is $U(1)$ it is known that its universal covering is the real line R and that the projection $\pi : R \to U(1)$ is given by $\pi : \lambda \in R \mapsto e^{i2\pi\lambda}$. Thus, the typical fibre may be identified with Z. This may be visualized by looking at the total manifold R as an infinite helix over a circle $S^1 \sim U(1)$: all the homologous points in the helix are projected on the same point of S^1.

(c) If instead of considering the universal covering R the circle $U(1)$ is covered only m times, we obtain a principal bundle over $U(1)$ with structure group Z_m (the cyclic group Z_m generated by the rotation of angle $2\pi/m$ may be considered as a zero-dimensional manifold since it is a discrete, numerable space; Z_m is not a Lie group). The bundle is trivial for the case $m = 1$ only.

(d) *A worked out example of* **(a)**: *The Hopf fibration of the three-sphere*

The Hopf bundle is the principal bundle $SU(2)(U(1), S^2)$, where S^2 is the two-dimensional sphere, $S^2 = SU(2)/U(1)$. The elements of $SU(2)$ may be described by the complex vectors $\mathbf{z}$ or the matrices g given by

$$\mathbf{z} = \begin{bmatrix} z_1 \\ z_2 \end{bmatrix} \quad , \quad g = \begin{bmatrix} z_1 & -z_2^* \\ z_2 & z_1^* \end{bmatrix} \quad , \quad z_1 z_1^* + z_2 z_2^* = 1 \quad . \tag{1.3.8}$$

If $z_1 = y_1 + iy_2$, $z_2 = y_3 + iy_4$, the condition $\det g = 1$ becomes $y_1^2 + y_2^2 + y_3^2 + y_4^2 = 1$; the group manifold of $SU(2)$ is the sphere S^3.

Let $U(1)$ be the subgroup defined by the matrices of the form

$$g_0 = \begin{bmatrix} \zeta & 0 \\ 0 & \zeta^* \end{bmatrix} \quad , \quad g_0 = \exp\frac{i}{2}\sigma_3\psi \quad . \tag{1.3.9}$$

Its (right) action on $SU(2)$ is given by

$$\mathbf{z} \mapsto \mathbf{z}g_0 = \begin{bmatrix} z_1\zeta \\ z_2\zeta \end{bmatrix} \quad \text{or} \quad g \mapsto gg_0 = \begin{bmatrix} z_1\zeta & -z_2^*\zeta^* \\ z_2\zeta & z_1^*\zeta^* \end{bmatrix} \quad . \tag{1.3.10}$$

The bundle projection π (the *Hopf mapping*) is now defined by

$$\begin{aligned} \hat{\mathbf{x}} = \pi(\mathbf{z}) &:= \mathbf{z}^\dagger \vec{\sigma} \mathbf{z} = (z_2^* z_1 + z_1^* z_2 \,,\ i(z_2^* z_1 - z_2 z_1^*)\,,\ |z_1|^2 - |z_2|^2) \\ &= (2(y_1 y_3 + y_2 y_4),\ 2(y_1 y_4 - y_2 y_3)\,,\ y_1^2 + y_2^2 - y_3^2 - y_4^2) \quad , \end{aligned} \tag{1.3.11}$$

where $\vec{\sigma} = (\sigma^1, \sigma^2, \sigma^3)$ are the Pauli matrices. The vector $\hat{\mathbf{x}} = (x_1, x_2, x_3)$ is equivalently given by[*]

$$\hat{\mathbf{x}} = \pi(g) := \frac{1}{2}\mathrm{Tr}(\vec{\sigma} X) \quad , \quad X \equiv g\sigma_3 g^\dagger = \vec{\sigma} \cdot \hat{\mathbf{x}} \quad , \qquad (1.3.12)$$

and the projection is consistent with the action (1.3.10), $\pi(gg_0) = \pi(g)$ $\forall g_0 \in U(1)$. Since $|z_1|^2 + |z_2|^2 = 1$, it follows that $\hat{\mathbf{x}}^2 = 1$; the base manifold is S^2 and this shows that the $SU(2)$ group manifold can be endowed with the principal bundle structure $SU(2)$ $(U(1), S^2)$.

Let us now find the local charts and the associated transition functions. Since the $SU(2)$ manifold is S^3, it may be parametrized by the Euler angles (ψ, θ, ϕ). Let us write an arbitrary element of $SU(2)$ in the form

$$g = \exp(\frac{i}{2}\sigma_3\phi)\exp(-\frac{i}{2}\sigma_2\theta)\exp(\frac{i}{2}\sigma_3\psi)$$

$$= \begin{bmatrix} \cos(\theta/2)\exp i[(\psi+\phi)/2] & -\sin(\theta/2)\exp i[(\phi-\psi)/2] \\ \sin(\theta/2)\exp -i[(\phi-\psi)/2] & \cos(\theta/2)\exp -i[(\phi+\psi)/2] \end{bmatrix} .$$
$$(1.3.13)$$

Identifying (1.3.13) with (1.3.8) we find

$$z_1 = \cos(\theta/2)\exp i[(\psi+\phi)/2] \quad , \quad z_2 = \sin(\theta/2)\exp i[(\psi-\phi)/2] \quad ,$$
$$(1.3.14)$$

$$y_1 = \cos(\theta/2)\cos[(\psi+\phi)/2] \quad , \quad y_2 = \cos(\theta/2)\sin[(\psi+\phi)/2] \quad ,$$
$$y_3 = \sin(\theta/2)\cos[(\psi-\phi)/2] \quad , \quad y_4 = \sin(\theta/2)\sin[(\psi-\phi)/2] \quad ,$$
$$(1.3.15)$$

and (1.3.11), (1.3.12) now characterize the points of S^2 by the polar angles (θ, ϕ),

$$x_1 = \sin\theta\cos\phi \quad , \quad x_2 = \sin\theta\sin\phi \quad , \quad x_3 = \cos\theta \quad . \qquad (1.3.16)$$

Despite the fact that $\exp i\frac{\sigma_3}{2}\psi$ appears as a factor at the right, it is clear that (1.3.13) does not trivialize $SU(2)$ as $S^2 \times U(1)$ *globally* because on S^2 the angle ϕ is undefined for $\theta = 0, \pi$. Thus, the Hopf bundle is not trivial: there is no global section. Nevertheless, the two sections determined by $\psi = -\phi$, $\psi = +\phi$,

$$\mathbf{z}_+(\theta, \phi) = \begin{bmatrix} \cos(\theta/2) \\ \sin(\theta/2)\exp -i\phi \end{bmatrix} \quad , \quad \mathbf{z}_-(\theta, \phi) = \begin{bmatrix} \cos(\theta/2)\exp i\phi \\ \sin(\theta/2) \end{bmatrix} \quad (1.3.17)$$

(or their analogous expressions in terms of g_+, g_-), are well defined when $\theta = 0$ and $\theta = \pi$ respectively, *i.e.* on $S^2_+(S^2_-)$, the two-sphere excluding

---

[*] This follows from the fact that $\vec{\sigma} \cdot \hat{\mathbf{x}} = g\sigma_3 g^\dagger$ and $(1/2)\mathrm{Tr}\sigma_i\sigma_j = \delta_{ij}$. Since the three-dimensional rotation associated with $g \in SU(2)$ is given by $R_{ij} = \frac{1}{2}\mathrm{Tr}(\sigma_i g\sigma_j g^\dagger)$, this just expresses that $\hat{\mathbf{x}}$ belongs to the orbit of $\hat{\mathbf{x}}^3 = (0, 0, 1)$.

the south (north) pole. Their associated local trivializations are defined by $S_+^2 \times U(1)$ ($S_-^2 \times U(1)$) and the corresponding local coordinates by $(\theta, \phi; \psi_+)$ $((\theta, \phi; \psi_-))$, where $\theta \in S_+(S_-)$. The transition function is given by $g_{(+-)} = \exp i \frac{\sigma_3}{2}(2\phi)$ since, obviously (see (1.3.5)),

$$\mathbf{z}_- = \mathbf{z}_+ g_{(+-)} \quad , \quad g_{(+-)} = \exp i\phi \quad , \quad \rho_\pm = \exp i(\psi \pm \phi)/2 \quad . \quad (1.3.18)$$

(e) *The group $SU(1,1)$ as a principal bundle of structure group $U(1)$*
This example may be worked out along lines similar to the previous example. The group $SU(1,1)$ (which is isomorphic to the real unimodular group $SL(2,R)$ and homomorphic to the Lorentz group of a $(1+2)$-dimensional spacetime, $SL(2,R)/Z_2 = SO_0(1,2)$) can be described by the complex matrices g of determinant one satisfying the condition $g^\dagger \sigma^3 g = \sigma^3$, i.e., by matrices of the form

$$g = \begin{pmatrix} z_1 & z_2^* \\ z_2 & z_1^* \end{pmatrix}, \quad |z_1|^2 - |z_2|^2 = 1 \; ; \quad g^{-1} = \begin{pmatrix} z_1^* & -z_2^* \\ -z_2 & z_1 \end{pmatrix} . \quad (1.3.19)$$

The action of the $U(1)$ structure subgroup (1.3.9) is defined as before, $g \mapsto g g_0$. The essential difference with respect to the previous case is that the quotient manifold $SU(1,1)/U(1)$ can be identified with the hyperboloid $\Omega = \{\mathbf{x}| \; x_3^2 - x_2^2 - x_1^2 = 1, \; x_3 \geq 1\}$ which can be covered with one single chart, $\Omega \sim R^2$. There is no need for transition functions; topologically, $SU(1,1) \sim \Omega \times U(1)$ and the bundle is trivial.

Remarks

Bundles over contractible manifolds
The above $SU(1,1)$ principal bundle provides a particular example of a general result: *any fibre bundle over a contractible base manifold (i.e., homotopic to a point) is trivial.* For instance, all principal bundles over the Minkowski space $\mathscr{M} \sim R^4$ are trivial. The possible non-triviality of a bundle rests on the non-trivial global topology of the base.

Hopf fibrings
The Hopf fibration (d) of $S^3 \sim SU(2)$ may be represented as $S^1 \to S^3 \to S^2 = CP^1$, and it is not the only Hopf fibring of a sphere. The other two *Hopf fibrations* are $S^3 \to S^7 \to S^4 = HP^1$ and $S^7 \to S^{15} \to S^8 = OP^1$, where HP^1 and OP^1 are the quaternionic and octonionic projective spaces (cf. (a) above). Indeed, the spheres S^1, S^3 and S^7 may be identified respectively with the unit complex numbers, quaternions and octonions, and S^3, S^7 and S^{15} may be identified with the complex, quaternionic and octonionic one-spheres, $|x^1|^2 + |x^2|^2 = 1$, $x \in$ C, H, O (notice that, unlike S^1 and S^3, S^7 is not a group manifold and there is no principal bundle structure; the octonions are not associative). These sphere

fibrations correspond to the only mappings $S^{2p-1} \to S^p$ ($p = 2, 4, 8$) which have a certain topological invariant, the *Hopf invariant*, equal to one. Adding $Z_2 = O(1) = S^0 \to S^1 \to RP^1$, we observe that the four fibrations of S^1, S^3, S^7 and S^{15} turn out to be associated with *the* four division algebras: R (reals), C (complex), H (Hamilton numbers or quaternions) and O (Cayley numbers or octonions) and that the fibres are the real, complex, quaternion and octonion numbers of unit modulus.

Bundles over spheres
The Hopf bundle is an example of a bundle over S^2 with structure group $U(1)$, a particular case of the more general situation of bundles over spheres S^n with a connected structure group G. In general, the inequivalent bundles over a manifold M with structure group G are determined by the Čech cohomology $\check{H}^1(M, G)$ which we shall not discuss here. In the case of spheres, the situation is particularly simple. Any sphere S^n can be covered by two charts and, as may be inferred from the simple S^2 case, the transition functions are mappings from the 'equator' S^{n-1} (the limiting intersection of the two charts of S^n) to the structure group G. From the discussion below (1.3.5) it follows that the different possible bundles are characterized by inequivalent sets of transition functions, in this case by non-equivalent mappings of $S^{n-1} \to G$, *i.e.* by $\pi_{n-1}(G)$. For bundles over S^2 with structure group $U(1)$, $\pi_1(U(1)) = Z$. The case $m = 1$ is the Hopf bundle of example (d); the other integer values ($m \in Z$ guarantees that $\exp im\phi$ is well defined when ϕ moves around a circle) correspond to other inequivalent bundles (the case $m = 0$ is the trivial one, $S_2 \times U(1)$). There are no non-trivial principal bundles over S^3 since $\pi_2(G) = 0$ for a Lie group.

The bundles over spheres have important physical applications. For the case $G = U(1)$, base S^2, they are associated with monopoles of different (quantized) charges; for $G = SU(2)$, base S^4 (compactified spacetime), they are associated with Yang-Mills instantons. In general, the instanton bundles on S^4 are determined by $\pi_3(G)$, which is Z for simple groups (see section 1.11). They will be discussed in section 2.8.

In what follows, the notion of vector bundle will appear frequently. Vector bundles are, essentially, fibre bundles which have a vector space F as the typical fibre (if the vector space is of dimension one, the bundle is called a *line bundle L*). They may be introduced, as in Def. 1.3.4 below, as bundles associated to a previously given principal bundle. Nevertheless, they may also be given the following independent definition.

(1.3.3) DEFINITION (*Vector bundles*)

A vector bundle $\eta_E(E, F, \pi_E, M)$ is a differentiable manifold E, together with a vector space F and a differentiable surjective mapping $\pi_E : E \to M$, which satisfies the local triviality condition. In other words, if $\{U_\alpha\}$ is an open covering of M, there exist diffeomorphisms $\tilde{\Phi}_\alpha^{-1} : U_\alpha \times F \to \pi_E^{-1}(U_\alpha)$ such that $\pi_E \circ \tilde{\Phi}^{-1} : (x, v) \mapsto x$.

If $U_\alpha \cap U_\beta \neq \emptyset$, then $\tilde{\Phi}_\beta \tilde{\Phi}_\alpha^{-1} : (x, v) \in (U_\alpha \cap U_\beta) \times F \mapsto (x, g_{\beta\alpha}(x)v)$, where in general $g_{\beta\alpha}(x) \in GL(k; R \text{ or } C)$ $(k = \dim(F))$ depending on whether the bundle is a real or complex vector bundle. The linear mappings $g_{\beta\alpha}(x)$ associated with the open covering $\{U_\alpha\}$ of the base manifold M are the transition functions; $\forall x \in U_\gamma \cap U_\beta \cap U_\alpha$, they satisfy the relation $g_{\gamma\beta}(x)g_{\beta\alpha}(x) = g_{\gamma\alpha}(x)$ (i.e., the 'cocycle' or compatibility condition).

As with the principal bundles, η_E in $\eta_E(E, F, \pi_E, M)$ may be used to denote the bundle itself; E indicates the total space (dim $E = n + k$, where k is the dimension of the vector space F) and, quite often, it is also used for the bundle η_E; M is the base and π_E (or simply π) is the bundle projection $E \xrightarrow{\pi} M$. The module of cross sections over $U \subset M$ will be denoted $\Gamma(U, E)$ or $\Gamma(U)$.

Typical examples of vector bundles

(a) *The tangent bundle $\tau(M)$ or $\tau(M)(T(M), \pi, M)$*
In $\tau(M)$, $T(M)$ is the total manifold, $T(M) = \bigcup_{x \in M} T_x(M)$, where $T_x(M)$ is the (vector) tangent space to M at x; π is the projection, $\pi : T(M) \to M$, $\pi : X(x) \in T_x(M) \mapsto x \in M$; the fibre over x is the vector space $T_x(M) \sim R^n$; the differentiable structure on $\tau(M)$ is induced by that on M since an atlas on M induces an atlas on $T(M)$ (if $\mathscr{A} = \{(U_\alpha, \varphi_\alpha)\}$ is an atlas on M, then $T\mathscr{A} := \{(T(U_\alpha), \varphi_\alpha^T)\}$ is a (natural) atlas on $T(M)$). If dim $M = n$, $\dim[T_x(M)] = n$ and dim $T(M) = 2n$.

In this language, a (*smooth*) *vector field* X on M is a smooth cross section of $\tau(M)$ since X associates to each point $x \in M$ a tangent vector $X(x) \in T_x(M)$. A vector field on $U \subset M$ is defined on the domain of a chart by n functions $X^i(x)$, $X = X^i(x)\partial/\partial x^i$. The set of all smooth vector fields on a manifold M will be denoted[*] by $\mathscr{X}(M)$. Although $\mathscr{X}(M) = \Gamma(M, T(M))$ is not closed under multiplication, which leads to a second order differential operator, $\mathscr{X}(M)$ is closed under the Lie bracket since, as we know, the commutator of two vector fields is a first order differential operator and again a vector field. Thus, $\mathscr{X}(M)$ has the structure

---

[*] If it is wished to be more precise and indicate explicitly the degree of smoothness (the class C^r), the notation $\mathscr{X}^r(M)$ is used; C^r sections/vector fields, etc. are defined analogously. As usual, r will be omitted; it will be assumed to be large enough or simply ∞.

of an (infinite-dimensional, see section 9.1) real Lie algebra. It can also be regarded as a module over the algebra $\mathscr{F}(M)$ of smooth functions on M, since fX is another vector field for $f \in \mathscr{F}(M)$. On the domain U of a chart, fX is given by $fX = f(x)X^i(x)\partial/\partial x^i$.

The tangent bundle always admits global sections, such as the *null section* which determines the zero vector field on M, but this does not imply the triviality of $\tau(M)$. If M admits a global coordinate system, then $\tau(M)$ is clearly trivial, but the converse is not true. We shall see below that $\tau(SU(2))$ is trivial, but as we know, the sphere $S^3 \sim SU(2)$ has no global chart.

We mention in passing that the tangent manifold $T(G)$, where G is a Lie group, is itself a Lie group with a semidirect structure.

(b) *The cotangent bundle* $\tau^*(M)$ *or* $\tau^*(M)(T^*(M), \pi, M)$
As in (a), but replacing $T_x(M)$ by $T_x^*(M)$, the dual vector space, and $T(M)$ by $T^*(M)$.

A (smooth) *one-form* $\omega(x)$ (*covector field, covariant vector field*) on M is a (smooth) differentiable cross section of $\tau^*(M)$; $\omega \in \mathscr{X}^*(M) = \Gamma(M, T^*(M))$.

(c) *General tensor bundle* $\tau_q^p(M)$
Similar to (a), (b). A (smooth) section is now a (smooth) general *tensor field* t_q^p on M of type (p, q). The above cases correspond to $(p, q) = (1, 0)$, contravariant vector field, and to $(p, q) = (0, 1)$, covariant vector field or one-form; in general, $t_q^p \in \mathscr{X}_q^p(M) = \Gamma(M, T_q^p(M))$. The subspace of $\mathscr{X}_q^0(M)$ made out of the antisymmetric covariant vector fields or *q-forms* on M will be denoted $\Lambda^q(M)$. Sections (*tensor fields*) of tensor bundles over a manifold will be used frequently in the rest of the book.

(d) *Whitney sum bundle, tensor product of vector bundles*
Let η_E and $\eta_{E'}$ be two vector bundles over the same manifold M of fibres F and F' respectively. Since F and F' are vector spaces, we can define the direct sum $F \oplus F'$ and the tensor product $F \otimes F'$. Then it is simple to obtain vector bundles over M with fibres $F \oplus F'$ and $F \otimes F'$. The *Whitney sum* of η_E and $\eta_{E'}$ is the bundle which is obtained by taking the direct sum of the fibres of η_E and $\eta_{E'}$ at each $x \in M$. If the transition functions of the original bundles are given by the $\dim F \times \dim F$ and $\dim F' \times \dim F'$ matrices g and g' respectively, the transition functions of the Whitney sum bundle are given by the direct sum $(\dim F + \dim F') \times (\dim F + \dim F')$ matrices $\begin{pmatrix} g & 0 \\ 0 & g' \end{pmatrix}$. Similarly, the *tensor product bundle* of η_E and $\eta_{E'}$ is obtained by taking the tensor product of the respective fibres at each $x \in M$; the transition functions are then the $(\dim F.\dim F') \times (\dim F.\dim F')$ matrices obtained by taking the tensor product of the transition function matrices

of η_E and $\eta_{E'}$. The tensor bundles of (c) above, which have tensor vector spaces as their fibres, may be obtained as tensor products of bundles. For instance, the skew-symmetric covariant q-tensor bundle is the tensor product $\wedge^q \tau^*(M)$; the space of sections $\Gamma(M, \wedge^q \tau^*(M))$ is the space of q-forms $\Lambda^q(M)$ on M.

(e) *Jet bundles*

Jet bundles are useful in the geometrical formulation of the variational principles of mechanics, as we shall see in section 8.1. For this reason we sketch here how they may be obtained in the simplest case. Let $\eta_E(E, F, \pi, M)$ be a vector bundle, and let Γ be the module of smooth local cross sections s. If the fibre is R^l, we can think of s as a vector-valued function on U: if E is locally parametrized by (x^i, y^α), $i = 1, ..., \dim M$ and $\alpha = 1, ..., l$, a section s is given by a vector-valued function $y^\alpha = y^\alpha(x^i)$. Let now Γ_x^r be the submodule of the cross sections s such that $y^\alpha(x)$ and its r first derivatives are zero at x, $\partial_{i_1...i_q} y^\alpha(x) = 0$, $q = 0, 1, ..., r$. Clearly this establishes an equivalence relation among cross sections that places in the same class those which have equal Taylor expansions at x up to the r-th term. Although the Taylor expansions are different in different coordinate systems (local charts), the coincidence of two expansions is a coordinate independent statement. The quotient $\Gamma / \Gamma_x^r \equiv J_x^r(E)$, the elements of which are the classes of cross sections which along with their derivatives up to order r take the same values at x, is a vector space. The equivalence class in $J_x^r(E)$ of a certain cross section s is the *r-jet of s* at x; x is the *source of the jet* and $s(x) \in E$ the *target of the jet*. By defining $\Gamma^r(E) = \bigcup_{x \in M} \Gamma_x^r(E)$, and $\pi_E^r : J^r(E) \rightarrow M$ by $\pi_E^r : J_x^r(E) \mapsto x$, $J^r(E)$ becomes a vector bundle $J^r(E) \rightarrow M$ called the *bundle of r-jets* of η_E. A section s of η_E induces a section $\bar{s}^r$ of $J^r(E)$ by means of the mapping $j^r : \Gamma(E) \rightarrow \Gamma(J^r(E))$ which assigns to s the particular section $\bar{s}^r$ of $J^r(E)$ determined by the first r derivatives of s; $\bar{s}^r$ is called the *r-jet extension* (or *prolongation*) of s, and j^r is the r-th prolongation mapping (the overbar will be used to indicate that a section $s^r \in \Gamma(J^r(E))$ is a 1-jet section $\bar{s}^r$).

Vector bundles also appear associated with principal bundles. Let $P(G, M)$ be a principal bundle, and let F be a vector space on which G acts from the left through a representation D of G,

$$g : v \in F \mapsto gv \quad (\text{short for } D(g)v) \quad,$$

i.e. F is a left G-space. Define now a right action of G on $P \times F$ by

$$g : (p, v) \mapsto (p, v)g := (pg, g^{-1}v) \quad; \tag{1.3.20}$$

the exponent minus one in $g^{-1}v$ is necessary so that the action is indeed a right one since $g'^{-1}g^{-1}v = (gg')^{-1}v$. Such an action determines an equivalence relation $\mathscr{R}_G$ on $P \times F$,

$$(p',v') \,\mathscr{R}_G\,(p,v) \Leftrightarrow (p,v)g = (p',v') \quad .$$

Let E be the set of the equivalence classes $\{p,v\}$,

$$E := (P \times F)/\mathscr{R}_G \equiv P \times_G F \quad . \tag{1.3.21}$$

The projection $\pi_E : E \to M$ given by

$$\pi_E\{p,v\} = \pi(p) \tag{1.3.22}$$

is well defined, because if another representative for the class is taken, say $\{p',v'\}$, then $\pi_E\{p',v'\} = \pi(p') = \pi(p)$. We may then give the following

(1.3.4) DEFINITION (*Associated vector bundle*)

The vector bundle $\eta_E(P)$ constructed above or, more explicitly, the bundle $\eta_E(E,F,\pi_E,M;G)$ with (standard) fibre F and (structure) group G is the vector bundle associated with $P(G,M)$.

Pictorially, an associated bundle consists of a copy of F at each point $x \in M$, all being 'glued' in a way prescribed by the topology of P and the action of G on P and on F, i.e. in the form leading to (1.3.21). The transition functions of η_E are given by the representation D of the transition functions of the principal bundle. In fact, this property provides an alternative definition: a vector bundle $\eta_E(E,F,\pi_E,M;G)$ is associated to the principal bundle $P(G,M)$ if its transition functions are the representation $D(g_{\alpha\beta})$ on F of the transition functions $g_{\alpha\beta}$ of $P(G,M)$. The following diagram may help ($\sim\sim\sim>$ indicates 'acts on'):

Diagram 1.3.3

In field theory, *matter fields* are described by sections of associated bundles; Yang-Mills fields will correspond, as we shall see, to connections on principal bundles.

Examples of associated bundles

(a) The tensor bundle $\tau_q^p(M)$ as the bundle associated with the principal bundle of linear frames on a manifold.

Let $L(M)(GL(n, R), M)$ be the *principal bundle of linear frames* on a manifold M. This is obtained as follows: at each $x \in M$, a *frame* r_x is an ordered *basis* for $T_x(M)$; given a frame r_x, any other frame is determined by a non-singular $n \times n$ matrix (linear transformation). It is clear that the structure group is $GL(n, R)$ and that π maps each frame r_x at x onto $x \in M$; also, $\dim(L(M)) = n^2 + n$. If x^i is a local coordinate system in $U \subset M$, the differentiable structure in $L(M)$ is obtained by taking (x^i, a^i_j) as a local coordinate system in $\pi^{-1}(U) \subset L(M)$, where a^i_j parametrize $GL(n, R)$

Let us obtain the bundle associated with $L(M)$ in the case where the fibre is $F = R^n$.

$$GL(n, R) \quad \rightsquigarrow \quad L(M) \xrightarrow{\ \pi\ } M$$

Diagram 1.3.4

Let $r_x = \{X_{(i)}(x)\}$ be a frame at x (a basis of $T_x(M)$). An element of $L(M)$ is given by $(x, X_{(i)}(x))$, $i = 1, ..., n$; $g: X_{(i)} \mapsto X_{(j)} a^j_i$. Then, $L(M) \times R^n$ is given by $[(x, X_{(i)}(x)), \xi^j]$, $i, j = 1, ..., n$, where $\xi^j \in R^n$ are the coordinates of $v = \xi^j e_j \in R^n$. The correspondence $[(x, X_{(i)}(x)), \xi^j] \mapsto \xi^i X_{(i)}(x) \in T_x(M)$ defines the elements of $E = L(M) \times_{GL(n, R)} R^n = T(M)$ (notice that $[(x, X_{(i)}(x))g, g^{-1}\xi^j] \mapsto \xi^i X_{(i)}(x)$ again). Thus, $\eta_E(L(M)) = \tau(M)$.

Clearly, the construction is similar if the fibre $F = R^n$ is replaced by a more general one: *in general, the tensor bundle $\tau_q^p(M)$ over M is associated with $L(M)(GL(n, R), M)$ with F of type (q, p) over R^n.*

(b) The previous example has a vector space as the fibre: it is an associated *vector* bundle as in Def. 1.3.4. But to construct an *associated bundle* it is not necessary that F be a vector space; for instance, F may be a manifold on which the structure group defines an action. The representation D is then replaced by the realization of the group on F. For instance, let $\widetilde{Ad}$ be the mapping $\widetilde{Ad} : G \to \operatorname{Int} G$ (inner automorphisms of the Lie group G) so that $\widetilde{Ad}g \equiv i_g$ is the inner conjugation mapping $i_g : g_0 \mapsto L_g R_{g^{-1}} g_0 = g g_0 g^{-1}$. Then the *associated* bundle of total space $\widetilde{Ad}(P) = P \times_{\widetilde{AdG}} G$ and fibre and group G defined by the *adjoint action* $\widetilde{Ad}$ of G on itself is called the *bundle of Lie groups*. It is not a principal bundle: in it, the group G acts on the fibre G by conjugation and this action is not free. By replacing the adjoint action of G on G by the *adjoint representation* (section 1.9) Ad of G on its Lie algebra $\mathscr{G}$, another example of an associated *vector* bundle is obtained, the *bundle of Lie algebras* $Ad(P)$, of total space $Ad(P) = P \times_{Ad G} \mathscr{G}$, fibre $\mathscr{G}$ and group G. Both $\widetilde{Ad}(P)$ and $Ad(P)$ are important in the description of gauge and infinitesimal gauge transformations. As a result, $\widetilde{Ad}(P)$ and $Ad(P)$ may also be called, respectively, the *gauge group* and *gauge algebra bundles*; their sections will correspond, respectively, to gauge and infinitesimal gauge transformations, as will be seen in section 10.1.

(c) The Möbius band may also be looked at as an associated bundle. Consider, for instance, the example (c) of principal bundles previously described for the case $m = 2$. Let F now be the line segment $I = [-1, 1]$, and let the action of Z_2 on F be the ordinary multiplication. The resulting associated bundle E over $U(1)$ is the Möbius strip; it is clearly not trivial.

Remarks

Triviality of associated bundles. Parallelizable manifolds
As we have mentioned, the existence of a global section for a *vector* bundle does not imply triviality since a vector bundle always admits the null section. But, *if the principal bundle is trivial, $P \sim M \times G$, then so is the associated bundle* η_E, $E = M \times F$. To show this, recall that a principal bundle is trivial if there is a global section or if, equivalently, there is a $\rho : P \to G$ such that $\rho(pg) = \rho(p)g \;\forall p \in P, \forall g \in G$ (Theorem 1.3.1); the natural cross section is $\sigma(x) = \Phi^{-1}(x, e)$ and $\rho(\sigma(x)) = e$. Let $\varphi : P \times F \to M \times F$ be the differentiable mapping defined by $\varphi : (p, v) \mapsto (\pi(p), \rho(p)v)$. This mapping is well defined on the equivalence classes of $P \times_G F$ since $\varphi(pg, g^{-1}v) = (\pi(pg), \rho(pg)g^{-1}v) = (\pi(p), \rho(p)v)$; all the elements in $(M \times G) \times F$ of the class $\{\sigma(x), v\}$ are mapped onto (x, v) so that $E \sim M \times F$, q.e.d.

In the example above, the triviality of a principal frame bundle $L(M)$ (existence of a global section) implies that $\tau(M)$ admits n sections (vector fields) linearly independent at each $x \in M$ (and reciprocally) or, in other words, that M is a *parallelizable manifold*: a manifold is parallelizable iff its frame bundle is trivial[*]. If M is parallelizable, $\mathscr{X}(M)$ is a free module

---

[*] We remark in passing that the isomorphism between $T_x(M)$ and a fixed vector space V_n that exists for all points x of any parallelizable manifold M is equivalent to giving a V_n-valued one-form on M.

the basis of which has dim M elements. Parallelizable manifolds are not very common. *All Lie groups are*, however, *parallelizable manifolds (G is also orientable)*: recall that at each point $g \in G$ we may define a basis for the tangent space $T_g(G)$ by $X^L_{(k)}(g) = L^T_g(e)X^L_{(k)}(e)$ (section 1.2); thus, as anticipated, $\tau(SU(2))$ is trivial. The standard example of a manifold which is not parallelizable is the sphere S^2. In fact, the only parallelizable spheres are S^1, S^3 (which are group manifolds) and S^7.

Associated principal bundles

Given a vector bundle η_E with standard fibre F and structure group G it is possible to construct an *associated principal bundle* $P(\eta_E)$ over the same base M by replacing F by G and taking the same transition functions. For instance, $P(\tau(M)) = L(M)$. Since the triviality or non-triviality of a bundle is characterized by the transition functions (which determine its twisting), we may then conclude that η_E (and $P(\eta_E)$) is trivial iff the associated principal bundle $P(\eta_E)$ admits a global section.

(1.3.1) PROPOSITION

Let $P(G, M)$ be a principal bundle, and let D be a representation of G on the vector space F. Let $\eta_E(E, F, \pi_E, M)$ be the associated bundle with standard fibre F. An F-valued function on P, $\varphi : P \to F$, is said to be an equivariant function or G-function if

$$\varphi(pg) = D(g^{-1})\varphi(p) \quad (:= g^{-1}\varphi(p)) . \tag{1.3.23}$$

Then there is a one-to-one correspondence between the F-valued G-functions on P and the cross sections of the associated bundle η_E.

Proof: Let φ be a G-function. Then $(p, \varphi(p)) \in P \times F$. But $(p, \varphi(p))$ and $(pg, \varphi(pg)) = (pg, g^{-1}\varphi(p)) \; \forall g \in G$ are equivalent under the action of G and so they determine the same element of $P \times_G F = E$. Thus, the mapping $\sigma_E : U \to E$, $\sigma_E(x) = (p, \varphi(p))$ determines an element of E which does not depend on $p \in \pi^{-1}(x)$ and accordingly defines a cross section of η_E. Conversely, σ_E determines a G-function, q.e.d.

(1.3.5) DEFINITIONS (*Bundle homomorphisms*)

(a) *Let $\eta(E, \pi, M)$ and $\eta'(E', \pi', M')$ be two fibre bundles. A smooth mapping $f_{tm} : E \to E'$ is a bundle homomorphism (or bundle mapping) if it induces* *a smooth mapping $f_b : M \to M'$ such that the diagram*

$$
\begin{array}{ccc}
E & \xrightarrow{f_{tm}} & E' \\
\pi \downarrow & & \downarrow \pi' \\
M & \xrightarrow{f_b} & M'
\end{array}
$$

Diagram 1.3.5

is commutative, i.e. if f_{tm} is fibre preserving. Alternatively we can define a fibre bundle morphism f by the pair of smooth mappings (f_{tm}, f_b) that make Diagram 1.3.5 commutative.

When the fibre bundles are *vector* bundles, f is a *vector bundle morphism* when f_{tm} defines a linear mapping on any fibre. We can use the concept of bundle mapping to characterize the triviality of the bundle. Let $\eta(E, \pi, M)$ be a fibre bundle of typical fibre F. Then η is trivial if there is a bundle morphism $f = (f_{tm}, f_b = 1)$ of η into $\eta'(M \times F, \pi, M)$ such that f_{tm} is a diffeomorphism between E and $M \times F$. However, for associated bundles there is a subtlety here, as will be shown in (c) below.

(b) *Let $P(G, M)$ and $P'(G', M')$ be two principal bundles. A principal bundle homomorphism f is a triple $f = (f_{tm}, f_{gr}, f_b)$ where $f_{gr} : G \to G'$ is a group homomorphism, such that the diagram*

$$
\begin{array}{ccc}
P \times G & \xrightarrow{\quad f_{tm} \times f_{gr} \quad} & P' \times G' \\
R \downarrow & & \downarrow R' \\
P & \xrightarrow{\quad f_{tm} \quad} & P' \\
\pi \downarrow & & \downarrow \pi' \\
M & \xrightarrow{\quad f_b \quad} & M'
\end{array}
$$

Diagram 1.3.6

where R (R') denotes the right action of G (G') on P (P'), is commutative. This means that, as expressed by the above diagram, f_{tm} 'commutes' with the right action,

$$
f_{tm}(pg) = f_{tm}(p)f_{gr}(g) \quad , \quad \forall p \in P , \quad \forall g \in G , \tag{1.3.24}
$$

and so f_{tm} is fibre preserving.

A principal bundle homomorphism f is an *embedding* when the induced mapping f_b is an embedding[*] and f_{gr} is a group monomorphism. Then $P(G, M)$ is said to be a *subbundle* of $P'(G', M')$. In particular, if $M = M'$ and $f_b = 1$, f is called a *reduction* of the structure group G' to G and the subbundle $P(G, M)$ is called a *reduced bundle*. For instance, the bundle of linear frames $L(M)(GL(n, R), M)$ can be reduced to the subbundle $O(M)(O(n), M)$ of the *orthonormal frames* on M, the structure group of which is the orthogonal group $O(n)$; this reduction is induced by the existence of a Riemannian structure on M. The group $O(n)$ can be further reduced to the special orthogonal group $SO(n)$ iff the manifold M is *orientable* (i.e., if there exists an atlas on M such that for any overlapping local charts the Jacobian determinant of the coordinate transformation is positive). The resulting reduced bundle is the bundle $SO(M)(SO(n), M)$ of the oriented frames on M.

As mentioned, an abstract bundle is defined up to equivalence: two bundles are *equivalent* when there is a bundle morphism f such that f_{tm} is a diffeomorphism and f_b is the identity map (and f_{gr} is an isomorphism when the bundle is a principal bundle). Let now $P = M \times G$, and let G act on P by $(x, g)g_0 = (x, gg_0)$. The manifold $M \times G$ with the Cartesian projection $\pi : M \times G \to M$ is obviously a principal bundle; any principal bundle isomorphic to it is a *trivial principal bundle*, $(M \times G)(G, M)$.

Let $P(G, M)$ be a principal bundle and let $Diff(P)$ be the (infinite) group of diffeomorphisms of the total manifold P. In general, the elements of $Diff(P)$ mix the fibres up. Those that are fibre preserving (eq. (1.3.24)) constitute the group $Diff_M(P)$ of bundle diffeomorphisms and define diffeomorphisms $f_b \in Diff(M)$ of the base manifold. The group $Aut_v P$,

$$Aut_v P := \{f \in Diff_M(P) | f_b = 1_{|M}\} \quad , \tag{1.3.25}$$

is the group of *vertical bundle automorphisms* of P; we shall see in section 10.1 that it is the *group of gauge transformations*. It is an invariant subgroup of $Diff_M(P)$, and

$$Diff_M(P)/Aut_v(P) = Diff(M) \quad , \tag{1.3.26}$$

the *group of coordinate transformations* $Diff(M)$, is obtained by putting in the same class all the fibre preserving diffeomorphisms which differ by one which induces the trivial action $f_b = 1$ on M.

---

[*] A smooth mapping $f : M^n \to Q^q$ is an *immersion* if it is of rank n for every point $x \in M^n$ ($f^T : T_x(M) \to T_{f(x)}(Q)$ is an injection). An immersion is not necessarily injective: for instance, the mapping of a circle S^1 onto a figure-eight-shaped curve is an immersion, but it is not injective. An injective immersion for which f is a homeomorphism $f : M \to f(M)$ is an *embedding* of M in Q. If $f : M \to Q$ is an embedding, $f(M)$ is a submanifold of Q and $f : M \to f(M)$ a diffeomorphism.

(c) Let $P(G, M)$ and $P'(G, M')$ be two principal bundles of structure group G and let η_E and $\eta_{E'}$ be the associated bundles over M and M' with standard fibre F; then $E = P \times_G F$ and $E' = P' \times_G F$ (see (1.3.21)). A principal bundle morphism f defines a morphism $\tilde{f}$ between the associated bundles η_E and $\eta_{E'}$ since

$$\tilde{f}(pg, g^{-1}v) = (f_{tm}(pg), g^{-1}v) = (f_{tm}(p)f_{gr}(g), g^{-1}v) \qquad (1.3.27)$$

induces a mapping $\tilde{f}_{tm} : E \to E'$ by

$$\tilde{f} : \{p, v\} \mapsto \{f_{tm}(p), v\} \quad . \qquad (1.3.28)$$

As was mentioned in the remarks above, if $P(G, M)$ is trivial, the associated bundle η_E is also trivial. There is a subtlety concerning the converse statement, because the bundle η_E may be trivial in the sense of Diagram 1.3.5 without being trivial as an associated bundle *i.e.*, without being associated to a trivial bundle. For instance, if $P(G, M)$ is a non-trivial principal bundle and G acts trivially on F, $gv = v$ $\forall g$, $\forall v$, then it is clear that the associated bundle is isomorphic to the trivial bundle $M \times F \xrightarrow{\pi} M$ in the sense of Diagram 1.3.5, but it is not trivial as an *associated* bundle.

1.4 Differential forms and Cartan calculus: a review

We recapitulate here some definitions and formulae concerning differential forms, vector fields, the three natural derivations d, i_X, L_X, Stokes' theorem, etc. which will be useful in the rest of the book. The reader who is already familiar with these concepts may skip this long section and come back occasionally to it for the explicit formulae quoted in the text.

Vector fields, forms and tensor fields

Let M be a manifold, and $\tau(M)$ $(\tau^*(M))$ its tangent (cotangent) bundle. Let

$$\left\{ \frac{\partial}{\partial x^i} \right\}_x \, , \, \{dx^j\}_x \; ; \; dx^j\left(\frac{\partial}{\partial x^i} \right) = \delta_i^j \; (i = 1, ..., n = \dim(M)) \quad , \qquad (1.4.1)$$

be a (dual) basis in $T_x(M)(T_x^*(M))$ induced by the local coordinates x^i at x. As we know, a *contravariant vector field* (or simply a *vector field*) on M, $X \in \mathscr{X}(M)$, has in local coordinates the general expression

$$X(x) = X^i(x) \frac{\partial}{\partial x^i} \quad . \qquad (1.4.2)$$

Similarly, a *covariant vector field* (or simply a *one-form*) on M, $\omega \in \mathscr{X}^*(M)$, is given by

$$\omega(x) = \omega_i(x) dx^i \quad . \qquad (1.4.3)$$

General tensor fields t^p_q are defined analogously; in local coordinates they are written

$$t = t^{j_1 \ldots j_p}{}_{i_1 \ldots i_q} \frac{\partial}{\partial x^{j_1}} \otimes \ldots \otimes \frac{\partial}{\partial x^{j_p}} \otimes dx^{i_1} \otimes \ldots \otimes dx^{i_q} . \qquad (1.4.4)$$

A tensor field $t^p_q \in \mathscr{X}^p_q(M)$ was defined in section 1.3 as a smooth section of $\tau^p_q(M)$. Due to the duality between $\mathscr{X}(M)$ and $\mathscr{X}^*(M)$, the module of tensor fields $\tau^p_q(M)$ may also be defined as a vector space of $\mathscr{F}(M)$-multilinear mappings taking values in $\mathscr{F}(M)$, $t^p_q : \mathscr{X}^*(M) \times \overset{p}{\ldots} \times \mathscr{X}^*(M) \times \mathscr{X}(M) \times \overset{q}{\ldots} \times \mathscr{X}(M) \to \mathscr{F}(M)$. This parallels the familiar definition of the vector space of (p,q)-tensors constructed as a vector space over a field K, only that K is replaced by a commutative ring with unit, $\mathscr{F}(M)$. An important tensor field is the metric tensor; in local coordinates, $g = g_{ij}(x) dx^i \otimes dx^j$.

The antisymmetric tensor fields of rank q, or q-forms, form the space $\Lambda^q(M)$. In local coordinates, an element of $\Lambda^q(M)$ is written

$$\alpha(x) = \alpha_{i_1 \ldots i_q}(x) \, dx^{i_1} \otimes \ldots \otimes dx^{i_q} \quad , \qquad (1.4.5)$$

where the fact that $\alpha \in \Lambda^q(M)$ is reflected in the antisymmetry of the coordinates $\alpha_{i_1 \ldots i_q}(x)$. Thus, introducing the fully antisymmetric Kronecker symbol ϵ, (1.4.5) may be rewritten as

$$\alpha(x) = \frac{1}{q!} \alpha_{i_1 \ldots i_q}(x) \epsilon^{i_1 \ldots i_q}_{j_1 \ldots j_q} \, dx^{j_1} \otimes \ldots \otimes dx^{j_q}$$

$$:= \frac{1}{q!} \alpha_{i_1 \ldots i_q}(x) dx^{i_1} \wedge \ldots \wedge dx^{i_q} \equiv \alpha_{I_1 \ldots I_q} dx^{I_1} \wedge \ldots \wedge dx^{I_q} \quad , \qquad (1.4.6)$$

where in the last equation the condition $I_1 < I_2 < \ldots < I_q$ (implied by the use of capital letters for indices, or 'minimal' basis) restricts the summation to the really independent terms (only $\binom{n}{q}$ wedge products of q different one-forms dx^i are independent; for instance, $dx \wedge dy = -dy \wedge dx$). The *exterior product* $\wedge$ is defined in (1.4.6) by antisymmetrization without weight factorials; in particular, $dx \wedge dy \equiv dx \otimes dy - dy \otimes dx$. Clearly, $\Lambda^q(M) = 0$ for $q > n = \dim(M)$; $\Lambda^0(M) \equiv \mathscr{F}(M)$. Since if $\alpha, \alpha' \in \Lambda^q(M)$ the sum $\alpha + \alpha' \in \Lambda^q(M)$ and if $f \in \mathscr{F}(M)$ $f\alpha \in \Lambda^q(M)$, $\Lambda^q(M)$ is an $\mathscr{F}(M)$-module.

The exterior product of forms again gives forms. If $\alpha \in \Lambda^q(M)$, $\beta \in \Lambda^p(M)$ then $\alpha \wedge \beta \in \Lambda^{p+q}(M)$, and it is expressed by

$$\alpha \wedge \beta = \frac{1}{p!q!} \alpha_{i_1 \ldots i_q} \beta_{j_1 \ldots j_p} dx^{i_1} \wedge \ldots \wedge dx^{i_q} \wedge dx^{j_1} \wedge \ldots \wedge dx^{j_p}$$

$$= \frac{1}{(p+q)!} (\alpha \wedge \beta)_{l_1 \ldots l_{p+q}} dx^{l_1} \wedge \ldots \wedge dx^{l_{p+q}} \qquad (1.4.7)$$

where

$$(\alpha \wedge \beta)_{l_1 \dots l_{p+q}} = \frac{1}{p!} \frac{1}{q!} \epsilon^{i_1 \dots i_q j_1 \dots j_p}_{l_1 \dots \quad \dots l_{p+q}} \alpha_{i_1 \dots i_q} \beta_{j_1 \dots j_p} \quad . \tag{1.4.8}$$

The exterior product is associative, and

$$\alpha \wedge \beta = (-)^{pq} \beta \wedge \alpha \quad . \tag{1.4.9}$$

Thus, *the real algebra* $\Lambda(M)$ *of the differential forms on M is a graded commutative* [(1.4.9)] *and associative algebra*[*].

Let $\phi : M \to S$, $\phi : x \mapsto \phi(x)$, be a differentiable mapping between two manifolds M and S. If α is a q-form on S we can pull it back to M by means of the pull-back mapping ϕ^*. Let $\{X_1, \dots, X_q\}$ be a set of vector fields on M. The *pull-back* $\phi^*\alpha$ of α by ϕ is the form on M defined by

$$(\phi^*\alpha)(x)(X_1(x), \dots, X_q(x)) = \alpha(y)(\phi^T(X_1(x)), \dots, \phi^T(X_q(x))) \quad , \tag{1.4.10}$$

where the arguments of $\alpha(y)$ in the r.h.s. are $Y_i(y)$, $y = \phi(x) \in S$ (see eq. (1.2.1)). The form $(\phi^*\alpha)$ is also called the *reciprocal image* or the form *induced* by ϕ from the form α on S. In particular, if f is a zero form, *i.e.*, a differentiable function on S, $f \in \mathscr{F}(S)$, the pull-back of f is then defined by

$$\phi^*f = f \circ \phi , \tag{1.4.11}$$

clearly, $\phi^*f \in \mathscr{F}(M)$.

Let ϕ be given in local coordinates by $y^\beta = y^\beta(x^i)$ ($i = 1, \dots, \dim M$; $\beta = 1, \dots, \dim S$) and let $\alpha \in \Lambda^q(S)$ be

$$\alpha = \frac{1}{q!} \alpha(y)_{\beta_1 \dots \beta_q} dy^{\beta_1} \wedge \dots \wedge dy^{\beta_q} , \tag{1.4.12}$$

where $y^1, \dots, y^{\dim S}$ are local coordinates on S. Then the coordinate expression for $\phi^*\alpha$ at $x \in M$ is obtained by substituting $y(x)$ for y and $(\partial y^\beta / \partial x^i) dx^i$ for dy^β ; hence the name induced form on M by ϕ from the form α on S.

On the pull-back and push-forward of arbitrary tensor fields

The pull-back ϕ^* of ϕ is the generalization to *covariant* tensor fields t^0_q of the familiar notion of the adjoint of a linear mapping between vector spaces (similarly, the tangent or differential mapping extends naturally to *contravariant* t^q_0 tensors). The pull-back goes backwards and so for a composite mapping $(\phi \circ \psi)^* = \psi^* \circ \phi^*$ in contrast with $(\phi \circ \psi)^T = \phi^T \circ \psi^T$.

---

[*] It is becoming increasingly frequent to omit the wedge product symbol $\wedge$ when there is no risk of confusion so that (1.4.9), for instance, would read $\alpha\beta = (-)^{pq}\beta\alpha$. We shall follow this practice in chapter 10.

For arbitrary ϕ, there is no natural induced mapping for mixed t_q^p tensors. The extension is possible, however, when ϕ is a *diffeomorphism*. In this case, $(\phi^{-1})^*$ exists and it goes forwards; we may then define the *push-forward* ϕ_* of ϕ by $\phi_* = (\phi^{-1})^*$; $(\phi \circ \psi)_* = \phi_* \circ \psi_*$. Clearly, if we replace ϕ by ϕ^{-1} the notions of push-forward and pull-back are interchanged, $(\phi^{-1})_* = \phi^*$; thus, the pull-back of any *tensor field* t may be defined by $\phi^* t = (\phi^{-1})_* t$.

The push-forward (pull-back) takes the appropriate objects defined on M (S) to objects defined on S (M). Let $y^l = y^l(x^i)$, $i, l = 1, ..., n$, $(x^l = x^l(y^i))$, be the local expression of the diffeomorphism $\phi : M \to S$ ($\phi^{-1} : S \to M$), and let t be the general tensor field $t_q^p \in \mathscr{X}_q^p(M)$ of eq. (1.4.4). Since ϕ is a diffeomorphism, the inverse of the Jacobian matrix exists. Then the *push-forward* $\phi_* t$ of t is the tensor field on S given by

$$(\phi_* t)^{l_1 \ldots l_p}{}_{k_1 \ldots k_q} = \frac{\partial y^{l_1}}{\partial x^{j_1}} \cdots \frac{\partial y^{l_p}}{\partial x^{j_p}} \left(\frac{\partial x^{i_1}}{\partial y^{k_1}} \circ \phi^{-1} \right) \cdots \left(\frac{\partial x^{i_q}}{\partial y^{k_q}} \circ \phi^{-1} \right) t^{j_1 \ldots j_p}{}_{i_1 \ldots i_q} \circ \phi^{-1} \ .$$

$$(1.4.13)$$

The arguments of this expression are the points of S, hence the ϕ^{-1} in the r.h.s. The matrices $\partial y^l / \partial x^j$ and $\partial x^i / \partial y^k$ are, respectively, the Jacobian and inverse Jacobian (or Jacobian of ϕ^{-1}) matrices that transform the different tensor indices. For $q = 0$ eq. (1.4.13) may be written as

$$(\phi_* t)^{l_1 \ldots l_p} = \frac{\partial y^{l_1}}{\partial x^{j_1}} \cdots \frac{\partial y^{l_p}}{\partial x^{j_p}} t^{j_1 \ldots j_p} \circ \phi^{-1} \ , \tag{1.4.14}$$

or

$$(\phi_* t) = \phi^T \circ t \circ \phi^{-1} \ , \tag{1.4.15}$$

where $\phi^T : \tau_x^p(M) \to \tau_{\phi(x)}^p(S)$. In this form, it may be considered as the expression that defines ϕ-related contravariant p-tensor fields. For $p = 1$ eq. (1.4.15) tells us that if ϕ is a diffeomorphism and X and Y are ϕ-related (eq. (1.2.1)) then $Y = \phi_* X$. The push-forward commutes with the tensor product of tensor fields: $\phi_*(t' \otimes t) = \phi_*(t') \otimes \phi_*(t)$.

Similarly, if $t \in \mathscr{X}_q^p(S)$ is a general tensor field on S and again ϕ is a diffeomorphism, its *pull-back* $(\phi^* t)$ to M is defined by

$$(\phi^* t)^{l_1 \ldots l_p}{}_{k_1 \ldots k_q} = \left(\frac{\partial x^{l_1}}{\partial y^{j_1}} \circ \phi \right) \cdots \left(\frac{\partial x^{l_p}}{\partial y^{j_p}} \circ \phi \right) \frac{\partial y^{i_1}}{\partial x^{k_1}} \cdots \frac{\partial y^{i_q}}{\partial x^{k_q}} t^{j_1 \ldots j_p}{}_{i_1 \ldots i_q} \circ \phi \ . \tag{1.4.16}$$

For instance, for $p = 0$ this gives

$$(\phi^* t)_{k_1 \ldots k_q} = \frac{\partial y^{i_1}}{\partial x^{k_1}} \cdots \frac{\partial y^{i_q}}{\partial x^{k_q}} t_{i_1 \ldots i_q} \circ \phi \ , \tag{1.4.17}$$

which on $x \in M$ is the expression that follows from the pull-back of eq. (1.4.12). If furthermore $q = 0$, (1.4.16) reproduces (1.4.11). If $p = 1$, $q = 0$, eq. (1.4.16) gives the expression for the pull-back of a vector field,[†] $\phi^* Y = (\phi^T)^{-1} \circ Y \circ \phi$. We may say that two q-covariant tensor fields $t' \in \mathcal{X}_q(M)$, $t \in \mathcal{X}_q(S)$ are ϕ-*related* if the diagram

$$
\begin{array}{ccc}
\tau_q(M) & \stackrel{(\phi^T)^*}{\longleftarrow} & \tau_q(S) \\
\uparrow t' & & \uparrow t \\
M & \stackrel{\phi}{\longrightarrow} & S
\end{array}
$$

Diagram 1.4.1

is commutative, *i.e.* if

$$ t' = (\phi^T)^* \circ t \circ \phi \equiv \phi^* t \quad . \tag{1.4.18} $$

If ϕ is omitted in eq. (1.4.18) and its rôle is left understood, we can identify the adjoint $(\phi^T)^*$ of ϕ^T and ϕ^*; its actual meaning will be clear from the expression where it appears.

The pull-back of forms is a homomorphism of the graded algebras $\Lambda(M)$ and $\Lambda(S)$ of exterior differential forms on M and S since, as is easily verified, $\phi^* : \Lambda(S) \to \Lambda(M)$ satisfies

$$ \phi^*(\alpha \wedge \alpha') = \phi^*(\alpha) \wedge \phi^*(\alpha'), \quad \alpha \in \Lambda^p(M), \ \alpha' \in \Lambda^q(M) ; \tag{1.4.19} $$

in fact, on covariant tensors $\phi^*(t \otimes t') = \phi^*(t) \otimes \phi^*(t')$.

The exterior derivative, the interior product and the Lie derivative

In what follows it will be important to be familiar with the following differential operators:

(a) *The exterior derivative d*
The operator d is a *graded derivation* (the name *antiderivation* is also common) of $\Lambda(M)$ of degree $+1$, *i.e.* for $\alpha \in \Lambda^q(M)$, $d\alpha \in \Lambda^{q+1}(M)$ and

$$ d(\alpha \wedge \beta) = (d\alpha) \wedge \beta + (-)^q \alpha \wedge (d\beta) \tag{1.4.20} $$

(Leibniz's rule). As a $(q+1)$-covariant (antisymmetric) tensor field, $d\alpha$ will be known if the result of its action on $q+1$ vector fields on M is known. This has to be given in terms of α. Explicitly, d is defined by

$$
(d\alpha)(X_1, ..., X_q, X_{q+1}) := \sum_{i=1}^{q+1} (-1)^{i+1} X_i \cdot \alpha(X_1, ..., \hat{X}_i, ..., X_{q+1}) +
$$
$$
+ \sum_{i<j} (-1)^{i+j} \alpha([X_i, X_j], X_1, ..., \hat{X}_i, ..., \hat{X}_j, ..., X_{q+1}) \quad , \tag{1.4.21}
$$

[†] Notice that eqs. (1.4.13) and (1.4.16) require that ϕ be a diffeomorphism. In particular, vector fields can only be pulled back and pushed forward by diffeomorphisms. Eq. (1.4.17) does not require $\phi : M \to S$ to be a diffeomorphism.

(*Palais formula*) where in the first term of the r.h.s. X_i acts as a derivation on $\alpha(X_1, ..., \hat{X}_i, ..., X_{q+1}) \in \mathcal{F}(M)$ and the hat $\wedge$ means that the corresponding vector field is absent.

In local coordinates, if α is given by (1.4.6),

$$d\alpha = \frac{1}{q!} \frac{\partial \alpha_{i_1 ... i_q}}{\partial x^j} dx^j \wedge dx^{i_1} \wedge ... \wedge dx^{i_q} . \tag{1.4.22}$$

Since (1.4.22) obviously implies $dd\alpha = 0$ for any α it follows that

$$dd = 0 \quad , \tag{1.4.23}$$

a fundamental result which can also be proved directly in a coordinate free way from the general definition (1.4.21). The exterior derivative is *natural* with respect to mappings, i.e. $d(\phi^* \alpha) = \phi^*(d\alpha)$. It is also natural with respect to restrictions: if U is open, $U \subset M$, and α is a q-form on M, $d(\alpha_{|U}) = (d\alpha)_{|U}$ or, equivalently, d is a *local operator*.

(b) *The interior product ('contraction') of a q-form α and a vector field X*
The interior product i_X is a *graded derivation (antiderivation)* of $\Lambda(M)$ of degree -1, i.e. for $\alpha \in \Lambda^q(M)$, $i_X \alpha \in \Lambda^{q-1}(M)$ and

$$i_X(\alpha \wedge \beta) = (i_X \alpha) \wedge \beta + (-)^q \alpha \wedge (i_X \beta) . \tag{1.4.24}$$

The interior product of X and α is also written as $X \lrcorner \alpha$ or $i(X)\alpha$. It is defined by

$$(i_X \alpha)(X_2, ..., X_q) := \alpha(X, X_2, ..., X_q) . \tag{1.4.25}$$

In local coordinates, (1.4.25) leads immediately to

$$i_Y \alpha = \frac{1}{(q-1)!} Y^j \alpha_{j i_2 ... i_q} dx^{i_2} \wedge ... \wedge dx^{i_q} ; \tag{1.4.26}$$

obviously,

$$i_X i_X = 0 . \tag{1.4.27}$$

If α is the one-form $\alpha = df$, where $f \in \mathcal{F}(M)$, then

$$i_X df = df(X) = X.f \tag{1.4.28}$$

(we shall immediately see that $X.f = L_X f$, where L_X is the Lie derivative). Also,

$$i_{X+Y} = i_X + i_Y \quad , \quad i_{fX} = f i_X . \tag{1.4.29}$$

Let $\phi : M \to N$ be a diffeomorphism, α a q-form on N and $Y \in \mathcal{X}(N)$. Then

$$\phi^*(i_Y \alpha) = i_{\phi^* Y}(\phi^* \alpha), \quad \alpha \in \Lambda^q(N), \quad Y \in \mathcal{X}(N) , \tag{1.4.30}$$

so that the interior product is natural with respect to diffeomorphisms. If ϕ is now the smooth mapping $\phi : M \to S$, $\alpha \in \Lambda^q(S)$, and $X \in \mathcal{X}(M)$ and $Y \in \mathcal{X}(S)$ are ϕ-related vector fields [(1.2.1)] eq. (1.4.30) is replaced by

$$\phi^*(i_Y \alpha) = i_X(\phi^* \alpha) \quad , \qquad (1.4.31)$$

which is a consequence of eqs. (1.4.25) and (1.4.10):

$$
\begin{aligned}
[i_X(\phi^* \alpha)(x)](X_1(x), ..., X_{q-1}(x)) &= [(\phi^* \alpha)(x)](X(x), X_1(x), ..., X_{q-1}(x)) \\
&= [\alpha(y)](\phi^T(X(x)), \phi^T(X_1(x)), ..., \phi^T(X_{q-1}(x))) \\
&= [\alpha(y)](Y(y), \phi^T(X_1(x)), ..., \phi^T(X_{q-1}(x))) \\
&= [(i_Y \alpha)(y)](\phi^T(X_1(x)), ..., \phi^T(X_{q-1}(x))) \\
&= [\phi^*(i_Y \alpha)(x)](X_1(x), ..., X_{q-1}(x)) .
\end{aligned} \qquad (1.4.32)
$$

The interior product is also natural with respect to restrictions: if U is an open subset of M, $(i_X \alpha)_{|U} = (i_{X_{|U}})(\alpha_{|U})$.

(c) *The Lie derivative*

The Lie derivative L_X is a *tensor* derivation of degree zero,

$$L_X(t \otimes t') = (L_X t) \otimes t' + t \otimes (L_X t') \quad , \qquad L_X(t + t') = L_X t + L_X t' \quad . \qquad (1.4.33)$$

In particular, on forms $L_X : \Lambda^q(M) \to \Lambda^q(M)$ and

$$L_X(\alpha \wedge \beta) = (L_X \alpha) \wedge \beta + \alpha \wedge (L_X \beta) \qquad (1.4.34)$$

(cf. (1.4.24)). Its action is defined algebraically by

$$(L_Y \alpha)(X_1, ..., X_q) := Y . \alpha(X_1, ..., X_q) - \sum_{i=1}^{q} \alpha(X_1, ..., [Y, X_i], ..., X_q) \quad , \qquad (1.4.35)$$

from which follows immediately that

$$L_{X+Y} = L_X + L_Y \quad . \qquad (1.4.36)$$

The equivalent definition (*Cartan decomposition*)

$$L_X = i_X d + d i_X \qquad (1.4.37)$$

is also useful, as well as the relations

$$[L_X, d] = 0 \ ,\ [L_X, i_Y] = i_{[X,Y]} \ \Rightarrow \ [L_X, i_X] = 0 \ ,\ [L_X, L_Y] = L_{[X,Y]} \quad . \qquad (1.4.38)$$

Also, on $\alpha \in \Lambda^q(M)$ and for $f \in \mathcal{F}(M)$,

$$L_{fX} \alpha = f L_X \alpha + df \wedge i_X \alpha \quad . \qquad (1.4.39)$$

The fact that L_X and d commute is also expressed by saying that d is natural with respect L_X. In local coordinates, (1.4.35) reads

$$(L_X \alpha)_{i_1 \ldots i_q} = X^k \frac{\partial \alpha_{i_1 \ldots i_q}}{\partial x^k} + \alpha_{k i_2 \ldots i_q} \frac{\partial X^k}{\partial x^{i_1}} + \ldots + \alpha_{i_1 \ldots i_{q-1} k} \frac{\partial X^k}{\partial x^{i_q}} \quad . \tag{1.4.40}$$

For a one-form $\alpha = \alpha_i dx^i$,

$$L_Y \alpha = (Y . \alpha_i) dx^i + \alpha_i dY^i \quad , \tag{1.4.41}$$

where $Y = Y^i \frac{\partial}{\partial x^i}$. In particular, $L_X df = d(X . f)$.

It is possible to give another definition of the Lie derivative of a q-form (in fact, *of any tensor field t*) which appeals directly to the intuitive idea of a derivative. Let X be a vector field on M, and $\phi_\tau = \exp \tau X$ the one-parameter group of diffeomorphisms of M which is generated by X (the *flow* of X). Obviously, $\phi_\tau^{-1} = \phi_{-\tau}$. Let t be a tensor field on M. Then its Lie derivative is defined by the limit

$$L_X t := \lim_{\tau \to 0} \frac{\phi_\tau^* t - t}{\tau} = \lim_{\tau \to 0} \frac{t - \phi_{-\tau}^* t}{\tau} \quad . \tag{1.4.42}$$

If $f \in \Lambda^0(M) = \mathcal{F}(M)$, $\phi^* f$ reduces to $f \circ \phi$, eq. (1.4.11), and (1.4.42) is just the definition of the ordinary (directional) derivative of f,

$$L_X f = X . f \quad , \quad f \in \mathcal{F}(M) . \tag{1.4.43}$$

For forms, the definition (1.4.42) reproduces (1.4.35). In fact, given its local character, we might start with (1.4.42) with $M \sim R^n$, set $t = f \in \Lambda^0$ and apply d to both sides to obtain that the same definition holds for the one-forms df ($\phi^* d = d \phi^*$). This is then sufficient for (1.4.42) to hold for any q-form.

For the case of a vector field Y, (1.4.42) reads, using that $\phi_\tau^* = \phi_{*(-\tau)}$ and (1.4.15),

$$L_X Y := \lim_{\tau \to 0} \frac{\phi_{-\tau}^T Y \phi_\tau - Y}{\tau} \tag{1.4.44}$$

(see also (1.2.6)). The numerator in (1.4.44) compares the vector Y at a point x with the vector at x which is obtained bringing back to x (using $\phi_{-\tau}^T$) the vector $Y(\phi_\tau(x))$. The limit gives

$$L_X Y = [X, Y] , \quad (L_X Y)^k = X^i \frac{\partial Y^k}{\partial x^i} - Y^i \frac{\partial X^k}{\partial x^i} . \tag{1.4.45}$$

This may be checked, *e.g.* by applying an arbitrary one-form ω to the r.h.s. of (1.4.44) to show that the result is given by $\omega([X, Y])$. In doing this, the relation

$$(L_X \omega)(Y) = X . \omega(Y) - \omega([X, Y]) \tag{1.4.46}$$

(see (1.4.35)) is used*. The following property follows immediately from (1.4.45):

$$L_{fX} Y = f L_X Y - (Y.f)X , \quad f \in \mathcal{F}(M) . \tag{1.4.47}$$

Let us now give explicitly the expression for the Lie derivative L_X of a general tensor field of $t_q^p \in \mathcal{X}_q^p(M)$. Since L_X is a derivation and $L_X dx^i = \frac{\partial X^i}{\partial x^k} dx^k$, $L_X \frac{\partial}{\partial x^j} = -\frac{\partial X^k}{\partial x^j}\frac{\partial}{\partial x^k}$ by (1.4.40) and (1.4.45), the application of L_X to the tensor t_q^p of eq. (1.4.4) immediately gives by (1.4.33)

$$\begin{aligned}
(L_X t)^{j_1 \ldots j_p}{}_{i_1 \ldots i_q} = {} & X^k \frac{\partial t^{j_1 \ldots j_p}{}_{i_1 \ldots i_q}}{\partial x^k} \\
& - t^{k j_2 \ldots j_p}{}_{i_1 \ldots i_q} \frac{\partial X^{j_1}}{\partial x^k} - \ldots - t^{j_1 \ldots j_{p-1} k}{}_{i_1 \ldots i_q} \frac{\partial X^{j_p}}{\partial x^k} \\
& + t^{j_1 \ldots j_p}{}_{k i_2 \ldots i_q} \frac{\partial X^k}{\partial x^{i_1}} + \ldots + t^{j_1 \ldots j_p}{}_{i_1 \ldots i_{q-1} k} \frac{\partial X^k}{\partial x^{i_q}} ,
\end{aligned} \tag{1.4.48}$$

where the terms containing a summed upper (lower) tensor index all have minus (plus) sign.

For instance, if t is a Riemannian or pseudo-Riemannian metric $g = g_{ij}dx^i \otimes dx^j$ (i.e., g is a symmetric nondegenerate covariant tensor field of rank two) on M and X is the generator of an *isometry* (i.e., a diffeomorphism of M which leaves the metric invariant), $L_X g = 0$. Then X is called a *Killing vector field*, and from (1.4.48) we read that its coordinates X^i satisfy the *Killing equations*,

$$X^k \frac{\partial g_{ij}}{\partial x^k} + g_{kj} \frac{\partial X^k}{\partial x^i} + g_{ik} \frac{\partial X^k}{\partial x^j} = 0 . \tag{1.4.49}$$

Let μ be a volume form[†] on an oriented manifold M, and let X be a vector field on M. It is easy to check explicitly that the *divergence* of X can be uniquely characterized by the expression

$$L_X \mu := (div_\mu X)\mu . \tag{1.4.50}$$

Clearly, for $\mu = dx^1 \wedge \ldots \wedge dx^n$ we find that $div_\mu X = \partial_i X^i$. The vector field is called *incompressible* if $div_\mu X = 0$.

Let $\phi : M \to M$ be a diffeomorphism of M and X a vector field on M. An important property of the Lie derivative is that L_X is *natural* with

---

* Explicitly, $\displaystyle\lim_{\tau \to 0} \omega \left(\frac{\phi^T_{-\tau} Y \phi_\tau - Y}{\tau} \right) = \lim_{\tau \to 0} \frac{[(\phi_{-\tau})^* \omega](Y) \circ \phi_\tau - \omega(Y)}{\tau} \equiv \lim_{\tau \to 0} \frac{[(\phi_{-\tau})^* \omega - \omega](Y)}{\tau} \circ \phi_\tau$

$+ \displaystyle\lim_{\tau \to 0} \frac{\omega(Y) \phi_\tau - \omega(Y)}{\tau} = -(L_X \omega)(Y) + X.\omega(Y) = \omega([X, Y])$.

† On an n-dimensional *orientable manifold* a volume element is an n-form $\mu(x)$ such that $\mu(x) \neq 0$ $\forall x \in M$. Two volume forms μ and μ' are equivalent if there is a strictly positive function $f \in \mathcal{F}(M)$ such that $\mu' = f\mu$. An equivalence class is an *orientation* of M; the classes represented by μ and $-\mu$ are the only possible orientations iff M is connected.

respect to the push-forward by ϕ, that is, if t_q^p is a tensor field on M, $t \in \mathcal{X}_q^p(M)$,

$$L_{\phi_* X}(\phi_* t) = \phi_*(L_X t) \tag{1.4.51}$$

or, equivalently, the following diagram is commutative:

$$
\begin{array}{ccc}
\mathcal{X}_q^p(M) & \xrightarrow{\phi_*} & \mathcal{X}_q^p(M) \\
L_X \downarrow & & \downarrow L_{\phi_* X} \\
\mathcal{X}_q^p(M) & \xrightarrow{\phi_*} & \mathcal{X}_q^p(M)
\end{array}
$$

Diagram 1.4.2

Let now $\phi : M \to S$ be a differentiable mapping and $\alpha \in \Lambda^q(S)$. In this case, the expression (1.4.51) above is replaced by

$$\phi^*(L_Y \alpha) = L_X(\phi^* \alpha) \,, \tag{1.4.52}$$

where X and Y are ϕ-related vector fields on M and S respectively, *i.e.,* the diagram

$$
\begin{array}{ccc}
\Lambda(M) & \xleftarrow{\phi^*} & \Lambda(S) \\
L_X \downarrow & & \downarrow L_Y \\
\Lambda(M) & \xleftarrow{\phi^*} & \Lambda(S)
\end{array}
$$

Diagram 1.4.3

is commutative. The Lie derivative is also natural with respect to restrictions: if U is an open subset of M, $(L_X \omega)_{|U} = (L_{X_{|U}}) \omega_{|U}$.

To conclude this description of the Lie derivative we mention that the properties that characterize its behaviour on vector fields may be summarized by saying that[*]

$$L \in \mathrm{Hom}_K(\mathcal{X}(M), \mathrm{Hom}_K(\mathcal{X}(M), \mathcal{X}(M))) \,, \tag{1.4.53}$$

where L is the mapping $L : X \mapsto L_X$. The first Hom_K tells us that $L : \lambda X + \mu Y \mapsto L_{\lambda X + \mu Y} = \lambda L_X + \mu L_Y$ [(1.4.36)] $\forall \lambda, \mu \in K, X, Y \in \mathcal{X}(M)$,

---

[*] We remark here in passing that, in contrast, the *covariant differentiation* ∇ satisfies $\nabla_{fX} = f \nabla_X$, $f \in \mathcal{F}(M)$, or, in other words,

$$\nabla \in \mathrm{Hom}_{\mathcal{F}(M)}(\mathcal{X}(M), \mathrm{Hom}_K(\mathcal{X}(M), \mathcal{X}(M))) \,.$$

This allows us to define covariant derivatives *along a curve* $x(\tau)$. Notice that, although (1.4.43) does define X as a directional derivative acting on f, $L_X t_q^p$ is not a directional derivative. For instance, $L_X Y = [X, Y]$ depends *also* on the derivatives of the components of X. The covariant derivative is not natural with respect to diffeomorphisms.

although $L_{fX} \neq f L_X$, eq. (1.4.47); the second one implies that L_X :
$\lambda Y + \mu Z \mapsto L_X(\lambda Y + \mu Z) = \lambda L_X Y + \mu L_X Z$.

Integration of forms and Stokes' theorem

Let α be the n-form on a compact subset of the open set $V \subset R^n$ given by

$$\alpha = \frac{1}{n!} \alpha_{i_1 \dots i_n}(y) dy^{i_1} \wedge \dots \wedge dy^{i_n} = \alpha_{1 \dots n}(y) dy^1 \wedge \dots \wedge dy^n \quad . \tag{1.4.54}$$

Then the integral $\int_V \alpha$ of the differential form α is defined by the integral

$$\int_V \alpha = \int d^n y \, \alpha_{1 \dots n}(y) \tag{1.4.55}$$

over any domain of R^n containing the support of $\alpha_{1 \dots n}(x)$. Let now ϕ : $U \to V$ be an orientation preserving diffeomorphism between two open subsets U, V of R^n. Then

$$\int_V \alpha = \int_U \phi^* \alpha \quad . \tag{1.4.56}$$

Eq. (1.4.56) immediately leads to the familiar expression for the change of variables under integration since, if the mapping ϕ corresponds to the change of variables expressed by $y^i = y^i(x^j)$, eq. (1.4.17) gives

$$(\phi^* \alpha)(x) = \frac{1}{n!} \left(\frac{\partial y^{i_1}}{\partial x^{k_1}} \dots \frac{\partial y^{i_n}}{\partial x^{k_n}} \alpha_{i_1 \dots i_n} \circ \phi \right) dx^{k_1} \wedge \dots \wedge dx^{k_n} \tag{1.4.57}$$

$$= \det \left(\frac{\partial y^i}{\partial x^j} \right) \alpha_{1 \dots n}(y(x)) dx^1 \wedge \dots \wedge dx^n \quad .$$

It is now simple to move from R^n to M in general. Let M be an orientable n-dimensional manifold with a given orientation, and let (V, φ) be an orientation preserving local chart. Then the integral of a form α with compact support $C \subset V$ is defined as above by pushing forward α to $\varphi(V)$, i.e. by

$$\int_{(\varphi)} \alpha = \int \varphi_*(\alpha_{|V}) \tag{1.4.58}$$

where $\alpha_{|V}$ is the restriction of α to V (if φ reverses the orientation a minus sign is added). This definition does not depend on the chart: if the support of α is contained in the intersection $V \cap V'$ of two local charts $(V, \varphi), (V', \varphi)$ then $\int_{(\varphi)} \alpha = \int_{(\varphi')} \alpha$. The extension of this definition to any

n-form on M is made by using a partition $\{(V_\sigma, \eta_\sigma)\}$ of unity* subordinate to the atlas $\mathscr{A}$ of orientation preserving charts on M. The form α is then an *integrable form* on M if $\sum_\sigma \int_{V_\sigma} \alpha_\sigma$, where $\alpha_\sigma = \eta_\sigma \alpha$, exists; this defines $\int_M \alpha$ (the definition does not depend on the atlas and subordinate partition of unity chosen). This is the case when α has compact support since then the sum reduces to a finite number of terms. Finally, the integral of α on a subset $N \subset M$ is defined by

$$\int_N \alpha = \int_M \chi\alpha \qquad (1.4.59)$$

where χ is the *characteristic function* of N, $\chi(x) = 1$ if $x \in N$ and zero otherwise; this is tantamount to appropriately restricting the integration limits.

The well-known integral theorems are included in the elegant *Stokes theorem* for the integrals of forms. Let α be an $(n-1)$-form with compact support on an oriented n-dimensional manifold M with a non-empty boundary† ∂M; $d\alpha \in \Lambda^n(M)$. Let $i : \partial M \to M$ be the inclusion map and i^* its pull-back. Then Stokes' theorem states that

$$\int_{\partial M} i^*\alpha = \int_M d\alpha \quad, \qquad (1.4.60)$$

where, for short, one simply writes α for the $(n-1)$-form $i^*\alpha$ of $\Lambda^{n-1}(\partial M)$. If $\partial M = \emptyset$, then the integral is set equal to zero.

Let μ now be a volume form of M. It is now easy to see using (1.4.37) that, if X is a vector field on M, $L_X \mu = d i_X \mu = (div_\mu X)\mu$. Then (1.4.60)

---

* A *partition of unity* on a manifold M is a collection $\{(V_\sigma, \eta_\sigma)\}$, where $\{V_\sigma\}$ is a locally finite open covering of M and the η_σ are (smooth) positive real functions ($\eta_\sigma(x) \geq 0 \ \forall x \in M$) with compact support $supp \ \eta_\sigma \subset V_\sigma$ such that, for each $x \in M$, $\sum_\sigma \eta_\sigma(x) = 1$ (the sum is necessarily finite). A partition of unity is *subordinate* to an atlas $\mathscr{A}\{(U_\alpha, \varphi_\alpha)\}$ on M if each open set V_σ is a subset of a chart domain $U_{\alpha(\sigma)}$. Since any open covering of a paracompact space admits a locally finite refinement, on a paracompact M it is always possible to define a partition of unity subordinate to a locally finite covering. M will always be assumed paracompact.

A manifold is *paracompact* if it is Hausdorff and every open covering $\{U_\alpha\}$ admits a refinement which is locally finite, *i.e.* $\forall x \in M$ there exists a neighbourhood which has non-empty intersection with only a finite number of U_α. This is not a strong condition; all metric spaces, for instance, are paracompact.

† The definition of an n-dimensional *manifold with boundary* M is similar to that of a *manifold*; the essential difference is that, if (U, φ) is a chart, now $\varphi(U) \subset R_+^n$, where $R_+^n = \{x \in R^n | x^n \geq 0\}$. In the upper half space R_+^n the interior (boundary) is the set of all points such that $x^n > 0$ ($x^n = 0$); this naturally defines the interior and the boundary of U and $\varphi(U)$. Int U (∂U) is open (closed) in U, and Int $U \cap \partial U = \emptyset$. Then, given an atlas $\mathscr{A}(U_\alpha, \varphi_\alpha)$ on M, the *interior* of M is defined by Int $M = \cup_\alpha \varphi_\alpha^{-1}(\text{Int}\varphi(U))$, and its *boundary* by $\partial M = \cup_\alpha \varphi^{-1}(\partial(\varphi(U)))$. Int M is an n-dimensional manifold open in M and $\partial(\text{Int } M) = \emptyset$; ∂M is an $(n-1)$-dimensional manifold if it is not empty, and $\partial(\partial M) = \emptyset$, $\partial M \cap \text{Int } M = \emptyset$. If $M = N \times S$, $\partial M = (\partial M \times S) \cup (N \times \partial S)$; a diffeomorphism $f : M \to N$ between two manifolds with boundary induces two diffeomorphisms Int $M \to$ Int N and $\partial M \to \partial N$.

for $\alpha = i_X \mu$ is the expression of Gauss' theorem

$$\int_M (div_\mu X)\mu = \int_{\partial M} i_X \mu \ . \tag{1.4.61}$$

The formula of integration by parts also finds its generalization in the context of Stokes' theorem: if α and β are p- and q-forms respectively, $d(\alpha \wedge \beta) = d\alpha \wedge \beta + (-1)^p \alpha \wedge d\beta$. Hence,

$$\int_N (d\alpha)\wedge\beta = \int_N d(\alpha\wedge\beta) - \int_N (-1)^p\alpha\wedge d\beta = \int_{\partial N} \alpha\wedge\beta - \int_N (-1)^p\alpha\wedge d\beta \ . \tag{1.4.62}$$

1.5 De Rham cohomology and Hodge-de Rham theory

Let us describe briefly in this section the Hodge-de Rham theory. It will reappear again in section 2.10, as the basic ingredient for the index theorem for the de Rham complex.

(a) *De Rham cohomology on a manifold M and homology*
Let us start by stating the essential definitions of the de Rham cohomology for the real forms on an n-dimensional manifold M. A q-form α on M is called *closed* if $d\alpha = 0$; a q-form α is *exact* if there is a $(q-1)$-form b such that $\alpha = db$ (we may say then that b is a '*potential form*' of α). Clearly, all exact forms are closed since $dd = 0$ [(1.4.23)], but the converse property depends on the *global* properties of the manifold M. Although *Poincaré's lemma* guarantees that on R^n all closed forms are exact (R^n is topologically trivial) this is not the case in general. The simplest example is provided when $M = S^1$ and α is the one-form $d\varphi$; despite the way it is written, $d\varphi$ is not exact since φ is not a periodic function and thus is not globally defined on S^1 (otherwise, by Stokes' theorem, the length of the unit circle would be zero instead of 2π since $\partial S^1 = \emptyset$). The *closed* q-forms on M are called *de Rham q-cocycles*; those of them that are *exact* are called *de Rham q-coboundaries*. The equivalence relation that places in the same class q-cocycles that differ by a q-coboundary defines the de Rham cohomology:
The *de Rham cohomology groups*, $H_{DR}^q(M,R)$, are the quotient vector spaces

$$H_{DR}^q(M,R) := Z_{DR}^q(M,R)/B_{DR}^q(M,R) \tag{1.5.1}$$

where Z_{DR}^q is the vector space of the q-cocycles and B_{DR}^q the vector subspace of the q-coboundaries (the argument R will be omitted frequently). If we imagine d as the operator $d_q : \Lambda^q(M) \to \Lambda^{q+1}(M)$, the above is equivalent to saying that $H_{DR}^q(M) = \ker d_q / \text{range } d_{q-1}$. Clearly, $0 \le q \le \dim M$. The dimension of $H_{DR}^q(M)$ is called the q-th *Betti number*,

$$b^q(M) := \dim H_{DR}^q(M) \ , \tag{1.5.2}$$

and it is a topological invariant. The *Euler-Poincaré characteristic* of the manifold M is the integer

$$\chi(M) = \sum_{q=0}^{n}(-1)^q b^q(M) \quad . \tag{1.5.3}$$

The Euler-Poincaré characteristic is also a topological property of the manifold; as long as M and N are homeomorphic, $\chi(M) = \chi(N)$ (see below and section 2.9). On a manifold M there exists a continuous non-zero vector field iff $\chi(M) = 0$. Thus, parallelizable manifolds have zero Euler-Poincaré characteristic, but $\chi(M) = 0$ does not imply parallelizability in general.

Since there are no exact zero-forms, $B^0(M) = 0$; since a closed zero-form is just a constant, $Z_{DR}^0(M) = R$ for M connected. Thus, $H_{DR}^0(M) = R$ if M is connected and, in general, $H_{DR}^0(M) = R \oplus R \oplus \overset{l}{...} \oplus R$ where l is the number of connected components or 'pieces' of the manifold M. For R^n, $H_{DR}^q(R^n) = 0$ for $q \neq 0$; this is the content of *Poincaré's lemma* which establishes that on a contractible manifold any closed form is exact. This is clearly a consequence of the fact that R^n can be covered by a single coordinate patch, and also tells us that a manifold M will have a non-trivial de Rham cohomology when the local coordinate neighbourhoods cannot be replaced by a single, global chart. Some other results for the real de Rham cohomology of an n-dimensional manifold M are:

– If M is a compact, connected, orientable manifold, $H_{DR}^n(M) = R$;
– If M is a compact, connected, non-orientable manifold, $H_{DR}^n(M) = 0$;
– If M is a non-compact, connected manifold, $H_{DR}^n(M) = 0$.

For the de Rham cohomology on spheres, we have

$$H_{DR}^0(S^n) = R \quad , \quad H_{DR}^q(S^n) = 0, \ 1 \leq q < n \quad , \quad H_{DR}^n(S^n) = R \quad .$$

Let M be a compact, oriented manifold and let $\alpha \in H_{DR}^p(M, R)$ and $\omega \in H_{DR}^{n-p}(M, R)$. Since (see below) $\Lambda^p(M)$ and $\Lambda^{n-p}(M)$ have the same dimension and $\alpha \wedge \omega \in \Lambda^n(M)$ is a volume form on M, the product $(\alpha, \omega) = \int_M \alpha \wedge \omega$ establishes a bilinear mapping $H_{DR}^p(M, R) \times H_{DR}^{n-p}(M, R) \to R$ which is non-singular since, if $\alpha \neq 0$ (similarly ω), $(\alpha, \cdot)$ cannot be zero. The product $(,)$ defines a duality, the *Poincaré duality*, which states that H_{DR}^p is dual to H_{DR}^{n-p} with respect to $(,)$. Consequently they are isomorphic as vector spaces and

$$b^p = b^{n-p} \quad , \quad n = \dim M \quad . \tag{1.5.4}$$

As a result, the Euler-Poincaré characteristic (1.5.3) of an odd-dimensional manifold is zero, $\chi(M^{odd}) = 0$ (this will be shown again in the context of the Euler class in section 2.9). For the spheres, $\chi(S^n) = 2$ (n even) or 0 (n odd). All compact connected orientable two-dimensional manifolds are homeomorphic to a two-dimensional sphere S_g^2 with a number g of handles, the *genus* of the manifold[*]. For them, $b^0 = 1 = b^2$, $b^1 = 2g$ and $\chi(S_g^2) = 2 - 2g$. The fact that $\chi(S^2) = 2 \neq 0$ explains why there cannot be a *universal* storm on Earth as stated by the 'hairy ball theorem': on S^2 there are no globally non-zero vector fields.

Since d is natural with respect to mappings, the pull-back ϕ^* of a smooth mapping $\phi : M \to S$ takes the de Rham-equivalent q-cocycles on S to de Rham-equivalent q-cocycles on M. Thus, ϕ induces a linear mapping $\phi^* : H_{DR}^q(S) \to H_{DR}^q(M)$. Indeed, if β is a closed form on S, $\phi^*\beta$ is a closed form on M since $d(\phi^*\beta) = \phi^*(d\beta) = 0$. Also, if β_1 and β_2 belong to the same cohomology class on N, $\beta_1 - \beta_2 = d\gamma$. Then $\phi^*(\beta_1) - \phi^*(\beta_2) = \phi^*(d\gamma) = d(\phi^*\gamma)$, and their pull-backs to M are also cohomologous. Thus, the cohomology groups of M and N are homomorphically related by the pull-back of the mapping between M and N. Since these cohomology groups are a consequence of the topological properties of the respective manifolds, it is accordingly not surprising that if ϕ and ϕ' are two smooth mappings ϕ, $\phi' : M \to N$ that can be continuously deformed into each other (i.e., *homotopic*) the induced homomorphisms ϕ^*, $\phi'^* : H_{DR}^q(N) \to H_{DR}^q(M)$ turn out to be the same. This is the *homotopy axiom for the de Rham cohomology*: homotopic maps induce the same map in cohomology.

Let $[\alpha]$ ($[\beta]$) be an element of $H_{DR}^q(M)$ ($H_{DR}^r(M)$) (a *class* of closed $q(r)$-forms). Due to (1.4.20), the wedge product of two closed q and r-forms, respectively, is always a closed $(q + r)$-form. Thus, the product of *classes* is well defined, $[\alpha] \wedge [\beta] = [\alpha \wedge \beta]$ (notice that if $\alpha = d\gamma$ and ρ is closed $\alpha \wedge \rho = d(\gamma \wedge \rho)$ and thus it is exact) and $\wedge$ defines a mapping $\wedge : H_{DR}^q(M) \times H_{DR}^r(M) \to H_{DR}^{q+r}(M)$. The *de Rham cohomology ring* $H_{DR}^*(M)$ is then defined by the direct sum $H_{DR}^*(M) = \oplus_r H_{DR}^r(M)$: the addition is provided by the formal sum of elements of $H_{DR}^r(M)$, and the product is the exterior product of forms, $\wedge : H_{DR}^*(M) \times H_{DR}^*(M) \to H_{DR}^*(M)$.

Let M be the product of two manifolds, $M = M_1 \times M_2$. Then it is not difficult to prove the *Künneth formula* for the de Rham cohomology,

$$H_{DR}^r(M, R) \approx \bigoplus_{p+q=r} H_{DR}^p(M_1, R) \otimes H_{DR}^q(M_2, R) \quad i.e.$$

$$H_{DR}^*(M, R) \approx H_{DR}^*(M_1, R) \otimes H_{DR}^*(M_2, R) ,$$

$$(1.5.5)$$

which for the respective Betti numbers and the Euler-Poincaré characteristic gives

$$b^r(M) = \sum_{p+q=r} b^p(M_1)b^q(M_2) , \quad \chi(M) = \chi(M_1)\chi(M_2) . \quad (1.5.6)$$

---

[*] The *genus* of a surface is the maximum number of disjoint closed curves which can be drawn on it without separating the surface.

The proof of (1.5.5) is based on the observation that if $\alpha \in H^p_{DR}(M_1, R)$ and $\beta \in H^q_{DR}(M_2, R)$ then $\alpha \wedge \beta \in H^{p+q}_{DR}(M, R)$. For instance, for the torus $T^2 = S^1 \times S^1$ we find $b^0(T^2) = b^0(S^1)b^0(S^1) = 1^2 = 1$, $b^1(T^2) = b^1(S^1)b^0(S^1) + b^0(S^1)b^1(S^1) = 1^2 + 1^2 = 2$ (i.e. $2g$ for $g = 1$), $b^2(T^2) = b^1(S^1)b^1(S^1) = 1$. Thus, $\chi(T^2) = 1 - 2 + 1 = 0$ and $\chi(T^2) = \chi(S^1)\chi(S^1) = 0$. For the n-dimensional torus $T^n = S^1 \times ... \times S^1$ it is found that $b^r(T^n) = \binom{n}{r}$ ($H^r_{DR}(T^n, R)$ is generated by the $\binom{n}{r}$ possible different r-forms $d\varphi^{i_1} \wedge ... \wedge d\varphi^{i_r}$). Again, the direct evaluation $\chi(T^n) = \sum_{r=0}^{n}(-1)^r \binom{n}{r} = (1-1)^n = 0$ confirms the obvious result from (1.5.6).

The additive property of the definite integral, $\int_a^c f(x)dx = \int_a^b f(x)dx + \int_b^c f(x)dx$, suggests that the class of submanifolds of M be extended in general by defining linear combinations of submanifolds with coefficients in $A = Z_2, Z, R, C$. In this way, a *q-chain* c is a formal sum $\lambda^i c_i$ of q-dimensional oriented submanifolds of M. If ∂ denotes the operation of taking the oriented boundary (and $\partial \circ \partial \equiv 0$, see footnote after (1.4.59)), a *q-cycle* is a q-chain c_q with no boundary, $\partial c_q = 0$, and a *q-boundary* is a q-chain which is the boundary of a $(q+1)$-chain, $c_q = \partial c_{q+1}$. Thus (and in contrast with d which moves forward by increasing the order of a form) the *boundary operator* ∂ goes backwards. The *simplicial homology* (with coefficients in A) of M is given by

$$H_q(M, A) = Z_q(M, A)/B_q(M, A) \quad . \tag{1.5.7}$$

If A is a field, $A = (Z_2, R, C)$, $H_q(M, A)$ is a vector space over A; its dual will be denoted $H^q(M, A)$. Let $A = R$. The (bilinear) product of a q-cycle $c_q \in Z_q(M)$ and a closed q-form $\alpha \in Z^q_{DR}(M)$ is defined by

$$<c, \alpha> = \int_{c_q} \alpha \quad ; \tag{1.5.8}$$

$<c, \alpha>$ is called a *period*. For instance, in this language Stokes' theorem reads $<c, d\alpha> = <\partial c, \alpha>$: the exterior derivative is the adjoint of the boundary operator and vice versa. It is easy to check that the period of a q-form over a q-cycle is class independent. Thus, there is a mapping

$$<,> : H_q(M) \times H^q_{DR}(M) \to R \quad . \tag{1.5.9}$$

More precisely, if M is a compact manifold without boundary, H_q and H^q_{DR} are finite-dimensional, the period matrix $<c_i, \omega^j>$ is invertible and the *simplicial cohomology* $H^q(M)$ and the *de Rham cohomology* $H^q_{DR}(M)$ are naturally isomorphic. A closed form is exact if all its periods vanish.

(b) *The Hodge or star operator and the codifferential* δ

Due to the familiar combinatorial identity $\binom{n}{p} = \binom{n}{n-p} = \frac{n!}{(n-p)!p!}$, the $\Lambda^p(U)$

and $\Lambda^{n-p}(U)$ $\mathscr{F}(U)$-modules of differential forms on the n-dimensional manifold M ($U \subset M$) have the same dimension over $\mathscr{F}(U)$ and are therefore isomorphic. When M is an oriented Riemannian or pseudo-Riemannian manifold there is a linear mapping, the *Hodge dual* or *star mapping*, which establishes the canonical isomorphism

$$* : \Lambda^p(M) \to \Lambda^{n-p}(M) \quad . \tag{1.5.10}$$

Notice that, unlike the differential operators d, i_X, L_X, the existence of $*$ requires a metric g on M. Given an oriented, orthonormal basis $X_1, ..., X_n \in T_x(M)$, the 'dual' $*\alpha$ of $\alpha \in \Lambda^p(M)$ is pointwise defined $((*\alpha)(x) = *(\alpha(x)))$ by

$$* : \alpha \mapsto *\alpha \quad , \quad (*\alpha)(X_{p+1}, ..., X_n) = \alpha(X_1, ..., X_p) \quad . \tag{1.5.11}$$

This definition does not depend on the chosen basis. We see clearly that the mapping is injective ($\alpha \neq 0 \Rightarrow *\alpha \neq 0$) and, since both spaces have the same (finite) dimension, is also surjective. Let $\{x^i\}$ be a local coordinate system, g_{ij} the coordinates of the metric, $g = g_{ij}dx^i \otimes dx^j$, $g^{ij}g_{jl} = \delta^i_l$, and let g also denote (there will be no confusion with the metric itself) its (non-zero) determinant. Then the action of $*$ on an element of a local basis of $\Lambda^p(M)$ is given by

$$*(dx^{i_1} \wedge ... \wedge dx^{i_p}) = \frac{\sqrt{|g|}}{(n-p)!} g^{i_1 j_1}...g^{i_p j_p}\epsilon_{j_1...j_n}dx^{j_{p+1}} \wedge ... \wedge dx^{j_n} \quad , \tag{1.5.12}$$

where the totally antisymmetric Levi-Civita tensor (density) $\epsilon_{j_1...j_n}$ is defined from

$$\epsilon_{12...n} = +1 \quad ; \quad \epsilon^{i_1...i_n} = g^{i_1 j_1}...g^{i_n j_n}\epsilon_{j_1...j_n} = \frac{1}{g}\epsilon_{i_1...i_n} \tag{1.5.13}$$

(notice that in many physics books the convention $\epsilon^{1...n} = 1$ is followed). In particular, the Hodge dual of 1 ($p = 0$) is the *invariant volume element* $\mu(x)$ on M,

$$\mu(x) = *1 = \frac{\sqrt{|g|}}{n!}\epsilon_{j_1...j_n}dx^{j_1} \wedge ... \wedge dx^{j_n} = \sqrt{|g|}dx^1 \wedge ... \wedge dx^n \quad , \tag{1.5.14}$$

which is easily seen to be invariant under changes of coordinates that are orientation preserving. The coefficients of the dual $(n-p)$-form $*\alpha$ are given in terms of those of α by

$$(*\alpha)_{j_{p+1}...j_n} = g^{i_1 j_1}...g^{i_p j_p}\epsilon_{j_1...j_n}\frac{\sqrt{|g|}}{p!}\alpha_{i_1...i_p} \tag{1.5.15}$$

or, equivalently,

$$*\alpha = \frac{1}{(n-p)!\,p!}\sqrt{g}\, g^{i_1 j_1}...g^{i_p j_p}\epsilon_{j_1...j_p j_{p+1}...j_n}\alpha_{i_1...i_p}dx^{j_{p+1}} \wedge ... \wedge dx^{j_n} \quad . \tag{1.5.16}$$

For instance, the dual of the Minkowski ($n = 4$) electromagnetic field strength tensor $F = \frac{1}{2}F_{\mu\nu}dx^\mu \wedge dx^\nu$ is given by $(*F) = \frac{1}{2}\mathscr{F}_{\mu\nu}dx^\mu \wedge dx^\nu$ where $\mathscr{F}_{\mu\nu} = \frac{\sqrt{|g|}}{2}\epsilon_{\mu\nu\rho\sigma}F^{\rho\sigma}$. Notice that, although in this case the physical dimensions of F and $*F$ are the same, $[F] = [*F]$, in general *the Hodge operator 'carries' physical dimensions.*

It is not difficult to check that, *on* $\Lambda^p(M)$,

$$** = (-1)^{p(n-p)+s}\mathbf{1} \quad , \quad (-1)^s = \frac{g}{|g|} \quad , \quad *^{-1} = (-1)^{p(n-p)+s}* \quad , \quad (1.5.17)$$

so that, e.g., $*\mu = (-1)^s\mathbf{1}$ where s is zero for a Riemannian metric; notice that the factor $p(n-p)$ is the same for a p-form and an $(n-p)$-form. To find the proportionality factor in $**$ it suffices to apply the star operator again to (1.5.12) to obtain[†]

$$* * (dx^{i_1} \wedge ... \wedge dx^{i_p}) = \left(\frac{\sqrt{|g|}}{(n-p)!}g^{i_1j_1}...g^{i_pj_p}\epsilon_{j_1...j_n} \right) \cdot$$
$$\left(\frac{\sqrt{|g|}}{p!}g^{j_{p+1}k_{p+1}}...g^{j_nk_n}\epsilon_{k_{p+1}...k_nk_1...k_p}dx^{k_1} \wedge ... \wedge dx^{k_p} \right)$$

$$= \frac{|g|}{(n-p)!p!}g^{i_1j_1}...g^{i_pj_p}\epsilon_{j_1...j_n}g^{j_{p+1}k_{p+1}}...g^{j_nk_n}(-1)^{(n-p)p}\epsilon_{k_1...k_n}dx^{k_1} \wedge ... \wedge dx^{k_p}$$

$$= \frac{|g|(-1)^{(n-p)p}}{(n-p)!p!}g^{j_1i_1}...g^{j_pi_p}g^{j_{p+1}k_{p+1}}...g^{j_nk_n}\epsilon_{j_1...j_n}\epsilon_{k_1...k_n}dx^{k_1} \wedge ... \wedge dx^{k_p}$$

$$= \frac{|g|(-1)^{(n-p)p}}{(n-p)!p!}\frac{1}{g}\sum_{k_{p+1},...,k_n}\epsilon_{i_1...i_pk_{p+1}...k_n}\epsilon_{k_1...k_n}dx^{k_1} \wedge ... \wedge dx^{k_p}$$

$$= \frac{(-1)^{(n-p)p+s}}{p!}\sum_P \delta_{i_1k_{P_1}}...\delta_{i_pk_{P_p}}dx^{k_1} \wedge ... \wedge dx^{k_p}$$

$$= (-1)^{(n-p)p+s}dx^{i_1} \wedge ... \wedge dx^{i_p} .$$

$$(1.5.18)$$

Thus, for Euclidean manifolds $*^2 = \mathbf{1}$ $(-\mathbf{1})$ for p even (odd) if n is even, and $*^2 = \mathbf{1}$ (irrespective of p) if they are odd-dimensional. If $s = 1$, the sign of $*^2$ changes. For instance, let $F = \frac{1}{2}F_{\mu\nu}dx^\mu \wedge dx^\nu$ be the two-form associated with the electromagnetic field strength tensor (gauge fields will be discussed in next chapter). Since $p = 2$, $* * F = -F$ in Minkowski space ($s = 1$) and $* * F = F$ in Euclidean space ($s = 0$). Thus, the condition $*F = \lambda F$ may be satisfied for $\lambda = \pm 1$ in Euclidean space ($\lambda = \pm i$ in Minkowski space); in fact, since only the signature s is relevant, these

---

[†] Recall that $\epsilon_{j_1...j_pi_{p+1}...i_n}dx^{i_1} \wedge ... \wedge dx^{i_n} \equiv \sum_{\{i_{p+1},...,i_n\}} \epsilon_{j_1...j_pi_{p+1}...i_n}\epsilon_{i_1...i_pi_{p+1}...i_n}dx^1 \wedge ... \wedge dx^n$ and that $\sum_{\{i_{p+1},...,i_n\}} \epsilon_{j_1...j_pi_{p+1}...i_n}\epsilon_{i_1...i_pi_{p+1}...i_n} = (n-p)!\sum_P\{(-1)^P \delta_{i_1j_{P_1}}...\delta_{i_pj_{P_p}}\}$ where $(-1)^P$ is the signature of the permutation P.

conditions hold for general elliptic and hyperbolic metrics respectively. In a four-dimensional Euclidean manifold there exist *self-dual* gauge fields, $F = *F$ and *anti-self-dual* gauge fields, $F = -(*F)$ (it is easy to see that on a Riemannian manifold there exist self-dual and anti-self-dual forms only if $n = 4k$, k integer).

Using the $*$ operator, it is possible to define an inner product of forms. Let $\alpha, \beta \in \Lambda^p(M)$ be given by

$$\alpha = \frac{1}{p!}\alpha_{i_1...i_p}dx^{i_1} \wedge ... \wedge dx^{i_p} \quad , \quad \beta = \frac{1}{p!}\beta_{j_1...j_p}dx^{j_1} \wedge ... \wedge dx^{j_p} \quad . \quad (1.5.19)$$

First, we notice that $\alpha \wedge *\beta$ is an n-form and that this expression is symmetric under the interchange of α and β, $\alpha \wedge *\beta = \beta \wedge *\alpha$, since

$$\alpha \wedge *\beta = \frac{1}{(p!)^2}\frac{1}{(n-p)!}\sqrt{|g|}\,\alpha_{i_1...i_p}\beta^{j_1...j_p}\epsilon_{j_1...j_p i_{p+1}...i_n}dx^{i_1} \wedge ... \wedge dx^{i_n}$$

$$= \frac{1}{p!}\alpha_{i_1...i_p}\beta^{i_1...i_p}\sqrt{|g|}dx^1 \wedge ... \wedge dx^n = \beta \wedge (*\alpha) \quad . \tag{1.5.20}$$

If we denote

$$< \alpha, \beta >_x := \frac{1}{p!}\alpha_{i_1...i_p}(x)\beta^{i_1...i_p}(x) \quad , \tag{1.5.21}$$

the above can be written as

$$\alpha \wedge *\beta = \beta \wedge *\alpha = <\alpha, \beta > \mu \quad , \quad *(\alpha \wedge *\beta) = (-1)^s <\alpha, \beta > \quad , \tag{1.5.22}$$

where $\mu(x)$ is the volume on M (the first equation in (1.5.22) can be taken, using (1.5.21), as an equivalent definition of the $*$ operator). Using this, it is possible to endow $\Lambda^p(M)$ with an $L^2(M)$ inner product. Given $\alpha, \beta \in \Lambda^p(M)$, $\alpha \wedge *\beta$ is an n-form on M and thus the *scalar product of two forms of* $L^2(\Lambda^p(M))$ may be defined by

$$(\alpha, \beta) = (\beta, \alpha) = \int_M \alpha \wedge *\beta \quad . \tag{1.5.23}$$

Since the integral is not necessarily convergent, (α, β) will be well defined if M is compact or if one of the forms has compact support; it will be positive definite $((\alpha, \alpha) \geq 0$ and $(\alpha, \alpha) = 0 \Rightarrow \alpha = 0)$ if M is a Riemannian manifold.

Using the star operator, we can now define the *metric adjoint* of the exterior differential d. It is called the *codifferential* δ, and it is the linear operator $\delta : \Lambda^p(M) \to \Lambda^{p-1}(M)$ defined by[†]

$$\delta = (-1)^{n(p+1)+1+s}*d* = (-1)^p *^{-1} d* \quad , \tag{1.5.24}$$

where the second equality follows from the fact that, on an $(n-p+1)$-form, $*^{-1} = (-1)^{(n+1)(p-1)+s}*$. Thus, for n even, $\delta = (-1)^{1+s} * d*$ on any form and, for n odd, $\delta = (-1)^{p+s} * d*$. In terms of δ, the differential d of a p-form is expressed by

$$d = (-1)^{n(p+1)+s}*\delta* \quad . \tag{1.5.25}$$

The above expressions are summarized by saying that the diagram

$$
\begin{array}{ccc}
\Lambda^p(M) & \xrightarrow{\quad\delta\quad} & \Lambda^{p-1}(M) \\
\downarrow * & & \uparrow * \\
\Lambda^{n-p}(M) & \xrightarrow{(-1)^{n(p+1)+1+s}d} & \Lambda^{n-p+1}(M)
\end{array}
$$

Diagram 1.5.1

is commutative. Since d is nilpotent and $*^2 \propto 1$, the operator δ is also nilpotent,

$$\delta^2 = 0 \quad . \tag{1.5.26}$$

The codifferential may be considered as a generalization of the divergence. To check this on an arbitrary p-form is slightly cumbersome and we shall restrict ourselves to the simple case $\alpha = \alpha_i dx^i$. On a one-form, $\delta = (-1)^{1+s} * d*$ irrespective of n. Thus, using (1.5.12) we find

$$\delta\alpha = (-1)^{1+s} * \partial_l \left\{ \frac{\sqrt{|g|}}{(n-1)!} \alpha_{i_1} g^{i_1 j_1} \epsilon_{j_1 \dots j_n} \right\} dx^l \wedge dx^{j_2} \wedge \dots \wedge dx^{j_n}$$

$$= (-1)^{1+s} \frac{\sqrt{|g|}}{(n-1)!} \partial_l \{ \sqrt{|g|} \alpha_{i_1} g^{i_1 j_1} \} \epsilon_{j_1 \dots j_n} g^{l s_1} g^{j_2 s_2} \dots g^{j_n s_n} \epsilon_{s_1 \dots s_n} \tag{1.5.27}$$

$$= -\frac{1}{\sqrt{|g|}} \partial_l \{ \sqrt{|g|} \alpha^l \}$$

since $\det g^{ij} = 1/g$. Thus, ignoring the sign, $\delta\alpha$ is nothing but the familiar *generally covariant* divergence $\nabla_i \alpha^i$ used in general relativity, since for a (pseudo-)Riemannian connection $\nabla_i \alpha^i = \partial_i \alpha^i + \Gamma^i_{ij}\alpha^j$ and the Christoffel symbols give $\Gamma^i_{ij} = \frac{1}{\sqrt{|g|}}\partial_j \sqrt{|g|}$. Using (1.5.27), (1.5.14) and (1.5.17) it also follows that $d * \alpha = (\nabla_i \alpha^i)\mu$. We note in passing that this example shows

---

[†] We shall use the same symbol δ later on to denote the coboundary operator for the group cohomology without risk of confusion. We note that, in contrast with d, L and i, which are derivations of degrees +1, 0 and −1 respectively, δ is *not* a derivation.

that δ 'carries' physical dimensions (if the coordinates have dimensions of length, $[\delta] = L^{-2}$) in contrast with d, which is dimensionless (L_X and i_X are also dimensionless differential operators if the vector field X is itself dimensionless).

Let us now check that δ is indeed the adjoint of the exterior derivative d with respect to the inner product (1.5.23). If $\alpha \in \Lambda^{p-1}(M)$, $\beta \in \Lambda^p(M)$ and M is a manifold *without* boundary we find

$$(d\alpha, \beta) = \int_M d\alpha \wedge *\beta = \int_M d(\alpha \wedge *\beta) + (-1)^p \int_M \alpha \wedge d(*\beta)$$

$$= (-1)^p \int_M \alpha \wedge **^{-1} d*\beta = \int_M \alpha \wedge *\delta\beta = (\alpha, \delta\beta) \qquad (1.5.28)$$

where we have used Stokes' theorem in the third equality and the definition (1.5.24) in the last. If the manifold M has a boundary, (1.5.28) is then replaced by

$$(d\alpha, \beta) = (\alpha, \delta\beta) + \int_{\partial M} \alpha \wedge *\beta \quad . \qquad (1.5.29)$$

The sum of the differential and codifferential operators is sometimes called the *Hodge-de Rham operator* $\dslash$,

$$\dslash := d + \delta \quad . \qquad (1.5.30)$$

It is not a homogeneous operator, since it is the sum of an operator of degree 1 (d) and another of degree -1 (δ). Its square is the Laplace-de Rham operator or simply the *Laplacian* Δ,

$$\dslash^2 := \Delta = d\delta + \delta d \quad ; \qquad (1.5.31)$$

clearly, Δ maps p-forms into p-forms. The following properties relating d, δ and Δ on p-forms hold irrespective of whether the metric is Riemannian or pseudo-Riemannian:

$$\begin{aligned} \delta* &= (-1)^{p+1}*d \quad , &\quad *\delta &= (-1)^p d* \quad , \\ d\delta* &= *\delta d \quad , &\quad *d\delta &= \delta d* \quad , \\ d\Delta &= \Delta d \quad , &\quad \delta\Delta &= \Delta\delta \quad , &\quad *\Delta &= \Delta* \quad ; \end{aligned} \qquad (1.5.32)$$

for instance, and always on p-forms, $\delta* = (-1)^{n(n-p+1)+1+s} * d ** = (-1)^{p+1} * d$ by using (1.5.24) and (1.5.17); $*d\delta = (-1)^p \delta * \delta = \delta d*$.

To see that Δ is related to (minus) the Laplacian we easily check, using (1.5.27), that

$$\Delta f = \delta df = \delta(\partial_k f dx^k) = -\frac{1}{\sqrt{|g|}} \partial_l (\sqrt{|g|} g^{lk} \partial_k f) \qquad (1.5.33)$$

for $f \in \mathscr{F}(M)$. Thus, for $g_{lk} = \eta_{lk}$ we find $\Delta f = -\partial^l \partial_l f$ (notice the minus sign). The Laplace-de Rham operator Δ is self-adjoint; for any two forms $\alpha_1, \alpha_2 \in \Lambda^p(M)$

$$(\Delta \alpha_1, \alpha_2) = (\alpha_1, \Delta \alpha_2) \tag{1.5.34}$$

since, using the fact that d and δ are adjoints of each other, both sides are equal to $(\delta \alpha_1, \delta \alpha_2) + (d \alpha_1, d \alpha_2)$. This also shows that on a Riemannian manifold Δ is positive, $(\Delta \alpha, \alpha) \geq 0$.

Let M, M' be two n-dimensional manifolds endowed with respective metric tensors g and g'. A diffeomorphism $f : M \to M'$ is called an *isometry* if $f^* g' = g$. A mapping $f : M \to M'$ is a *local isometry* if there exist neighbourhoods $U \ni x$ and $V \ni f(x)$ for each x in M such that $f : U \to V$ is an isometry. On M and M', the operators $*$ and δ have different definitions, since they depend on the metric. What are their transformation properties under f? The answer is provided by

(1.5.1) THEOREM

Let $f : M \to M'$ be an orientation preserving local isometry and α' a p-form on M'. Then

– the pull-back f^* commutes with the Hodge mapping in the sense that

$$*(f^* \alpha') = f^*(*'\alpha') \tag{1.5.35}$$

– f^* commutes with δ and Δ (which follows from (1.5.35) and the fact that $f^* \circ d = d \circ f^*$)
– if f is an isometry it preserves the scalar product between p-forms,

$$(\alpha'_1, \alpha'_2) = (f^* \alpha'_1, f^* \alpha'_2) \quad . \tag{1.5.36}$$

(c) Hodge-de Rham theory

A differential form $\gamma \in \Lambda^p(M)$ is called *harmonic* if

$$\Delta \gamma = 0 \quad . \tag{1.5.37}$$

Let M be a compact boundaryless oriented Riemannian manifold. If γ is harmonic $(\Delta \gamma, \gamma) = 0$. This means that $(\delta \gamma, \delta \gamma) + (d \gamma, d \gamma) = 0$ and, since the scalar product is positive definite, that $d \gamma = 0 = \delta \gamma$. Thus, a form γ is harmonic iff it is closed and coclosed. Moreover, if γ is a harmonic *function*, then γ is obviously constant (this is also implied by the *Liouville theorem*: on a compact oriented Riemannian manifold, any harmonic function $\Delta f = 0$ is constant). The restriction to M compact is important: for instance, on $M = R^3 \setminus \{0\}$ (which is non-compact) the Coulomb or gravitational potential function $\gamma = k\frac{1}{r}$ has $d\gamma \neq 0$, but it is harmonic. The terminology of de Rham cohomology is also used for δ: a p-form α is *coclosed* if $\delta\alpha = 0$ and it is called *coexact* if there exists a $(p+1)$-form β such that $\alpha = \delta\beta$. A p-form is called *primitively harmonic* if it is closed and coclosed; thus, harmonic and primitively harmonic are equivalent concepts on a compact Riemannian manifold. The (unique) decomposition of a form α in its exact, coexact and harmonic parts is the content of

(1.5.2) THEOREM (*Hodge decomposition theorem*)

Let M be a compact, oriented, Riemannian manifold without boundary. Let $d\Lambda^{p-1}(M)$ be the space of the exact p-forms, $\delta\Lambda^{p+1}(M)$ the space of the coexact p-forms and $\mathrm{Harm}^p(M)$ the space of the harmonic p-forms. Then any p-form α on a compact Riemannian manifold admits a unique, global, decomposition as

$$\alpha = d\omega + \delta\beta + \gamma \tag{1.5.38}$$

where $\omega \in \Lambda^{p-1}(M)$, $\beta \in \Lambda^{p+1}(M)$ and $\gamma \in \mathrm{Harm}^p(M)$, i.e. α is the sum of an exact, a coexact and a harmonic form. Moreover,

$$(d\omega, \delta\beta) = 0 \quad , \quad (d\omega, \gamma) = 0 \quad , \quad (\delta\beta, \gamma) = 0 \quad . \tag{1.5.39}$$

Proof: The existence of ω, β and γ is difficult to prove and will not be discussed here, but the uniqueness of a given decomposition follows easily from considering the implications of $d\tilde{\omega} + \delta\tilde{\beta} + \tilde{\gamma} = 0$, which would be the form of the equation giving the difference between two possible decompositions of the same α. Acting with d we would find $d\delta\tilde{\beta} = 0$ and hence $(d\delta\tilde{\beta}, \tilde{\beta}) = 0$, $\delta\tilde{\beta} = 0$; then $d\tilde{\omega} + \tilde{\gamma} = 0$ would similarly imply, acting with δ, that $d\tilde{\omega} = 0$ and hence that $\tilde{\gamma} = 0$ also. In fact, using $(d\alpha, \beta) = (\alpha, \delta\beta)$ for a boundaryless M it is easy to check that the spaces $d\Lambda^{p-1}(M)$, $\delta\Lambda^{p+1}(M)$ and $\mathrm{Harm}^p(M)$ of the Hodge decomposition are mutually orthogonal, q.e.d.

Clearly, $(\Delta\alpha, \gamma) = 0$ for any harmonic γ. The converse statement is also true: a p-form λ can be written as $\lambda = \Delta\alpha$ iff $(\lambda, \gamma) = 0 \; \forall \gamma \in \mathrm{Harm}^p(M)$. Finally, let us assume that α is closed. Then it follows that the term $\delta\beta$ in the decomposition (1.5.38) is absent, and since $d\omega$ is exact, α and γ belong to the same de Rham cohomology class. Moreover, $H^p_{DR}(M)$ and $\mathrm{Harm}^p(M)$ are isomorphic. Indeed, if α in (1.5.38) is closed,

$$0 = (d\alpha, \beta) = (d\delta\beta, \beta) = (\delta\beta, \delta\beta) \quad , \tag{1.5.40}$$

and thus $\alpha = d\omega + \gamma$. Thus, each de Rham cohomology class contains an harmonic representative. Besides, if α is harmonic, $\delta\alpha = 0 = \delta d\omega \Rightarrow (\delta d\omega, \omega) = 0 \Rightarrow d\omega = 0$, $\alpha = \gamma$. This is the content of the Hodge

(1.5.3) THEOREM

On a compact Riemannian orientable manifold M, $H_{DR}^p(M)$ is isomorphic to the space $\mathrm{Harm}^p(M)$ of harmonic forms: $\dim H_{DR}^p(M) = \dim \mathrm{Harm}^p(M) = b^p(M)$, the p-th Betti number of M.

This is a striking result: the Laplacian depends on the Riemannian metric, whereas the Betti number is a topological quantity associated with M. This is an anticipation of the index theorem for the de Rham complex to be considered in next chapter. As an example, on S^n only $H_{DR}^0(S^n) = R = H_{DR}^n(S^n)$ are non-zero, as has been already mentioned. The elements of H^n are the constant multiples of the volume (hypersurface) element on S^n, and these are the harmonic forms; those of H^0 are the constant functions.

(d) *Equivariant cohomology*

Let M be a manifold on which a Lie group G acts on the left, and let X_M be the vector field on M generated by $X \in \mathcal{G}$. The (inhomogeneous) operator

$$d_X := d - i_{X_M} \tag{1.5.41}$$

defines an antiderivation of $\Lambda(M)$ which preserves its grading into the spaces of even and odd forms. Its square is given by

$$(d_X)^2 = -(i_{X_M}d + di_{X_M}) = -L_{X_M} \quad . \tag{1.5.42}$$

Thus, d_X is nilpotent on the subalgebra $\Lambda_X(M)$ of invariant forms α i.e. such that $L_{X_M}\alpha = 0$. If we now define d_X-closed and d_X-exact elements in $\Lambda_X(M)$, the equivariant cohomology ring is defined, with obvious notation, as $H_X(M) = Z_X(M)/B_X(M)$. Clearly, if α is d_X-exact its top-form component is exact in the ordinary sense (notice that the i_X part in (1.5.41) lowers the degree of the form by one). In the trivial case, $X_M = 0$, $H_X(M)$ becomes the usual de Rham cohomology ring $H_{DR}^*(M)$.

1.6 The dual aspect of Lie groups: invariant differential forms. Invariant integration measure on G

Let us come back to the differential geometry of Lie groups and let X^L be a left-invariant vector field on G. It is natural to say that the one-form ω^L on G is left-invariant (hence the superscript L also on ω) if $\omega^L(g)$ and $\omega^L(sg)$ are such that

$$\omega^L(sg)(X^L(sg)) = \omega^L(g)(X^L(g)) \quad , \tag{1.6.1}$$

i.e. when $\omega^L(g)(X^L(g))$ is a *constant* function on G (see Diagram 1.6.1 below in which $(L_s)^*(sg)$, L_s^* inside brackets, is the adjoint of the mapping $L_s^T(g)$).

$$X^L(g) \xrightarrow{\quad L_s^T \quad} X^L(sg)$$

$$\omega^L(g) \underset{(L_{s^{-1}})^*}{\overset{(L_s)^*}{\rightleftharpoons}} \omega^L(sg)$$

Diagram 1.6.1

Thus, we have

(1.6.1) DEFINITION (*Left-invariant form on a Lie group*)

A one-form ω on G is a left-invariant form (LIF) if

$$(L_s)^* \circ \omega \circ L_s := \omega \tag{1.6.2}$$

(cf. (1.2.6)) or, equivalently,

$$(L_{s^{-1}})^*(g)\omega(g) = \omega(sg) \quad , \quad (L_{g^{-1}})^*(e)\omega(e) = \omega(g) \tag{1.6.3}$$

(cf. (1.2.7), (1.2.8)). Eq. (1.6.2) is often written simply as

$$L_s^*\omega = \omega \quad . \tag{1.6.4}$$

The definition extends trivially to q-forms on G. When it is convenient to stress the left-invariant (LI) character of a form, we shall write ω^L by adding a superscript L as in eq. (1.6.1).

Looking at Diagram 1.6.1 above, and using the convention by which the coordinates of vector fields (one-forms) are represented by columns (rows), it is clear that, as matrices,

$$[(L_{g^{-1}})^*(e)]^i{}_j [L_g^T(e)]^j{}_k = \delta^i_k \quad (i,j,k = 1,...,r) \tag{1.6.5}$$

in order to have (1.6.1) fulfilled. We then have

$$[(L_{g^{-1}})^*(e)]^i{}_j = [L_g^T(e)^{-1}]^i{}_j = [L_{g^{-1}}^T(g)]^i{}_j \tag{1.6.6}$$

or, equivalently,

$$[(L_{g^{-1}})^*(e)]^i{}_j = \left[\left(\left. \frac{\partial g''(g',g)}{\partial g} \right|_{\substack{g=e \\ g'=g}} \right)^{-1} \right]^i{}_{\cdot j} = \left. \frac{\partial g'''^i(g',g)}{\partial g^j} \right|_{\substack{g=g \\ g'=g^{-1}}} \quad . \tag{1.6.7}$$

Much in the same way as (1.1.13), (1.1.14) provide a basis for vector fields on G in terms of right and left-invariant vector fields, RIVFs and LIVFs respectively, the right and left-invariant one-forms (RIFs and LIFs)

$$\omega^{R,L(k)}(g) = \omega_\ell^{R,L(k)}(g)dg^\ell \tag{1.6.8}$$

where

$$\omega_\ell^{R(k)}(g) = \left.\frac{\partial g''^k(g',g)}{\partial g'^\ell}\right|_{\substack{g'=g \\ g=g^{-1}}} \quad , \tag{1.6.9}$$

$$\omega_\ell^{L(k)}(g) = \left.\frac{\partial g''^k(g',g)}{\partial g^\ell}\right|_{\substack{g=g \\ g'=g^{-1}}} \quad , \tag{1.6.10}$$

constitute a basis for the one-forms on G. In $\omega_\ell^{R(k)}$, say, the superscript k in brackets indicates a specific one-form of the basis $\{\omega^{R(k)}\}$ of RIFs, and the subscript ℓ, without brackets, labels the functions of $\mathscr{F}(G)$ which express $\omega^{R(k)}$ in terms of dg^ℓ. Again, at the identity $e \in G$,

$$\omega^{R,L(k)}(e) = dg^{(k)} \quad . \tag{1.6.11}$$

Let ω be an LI (RI) form. Since left and right translations commute, $L_s^* \circ R_g^* = R_g^* \circ L_s^*$. Hence, the form transformed by R^* (L^*) is also LI (RI). The following Proposition relates the left and right invariance of q-forms:

(1.6.1) PROPOSITION

Let $a : G \to G$ be the antipodal mapping, $a : g \mapsto g^{-1}$. Then a form ω is LI (RI) iff $a^*(\omega)$ is RI (LI).

Proof: Let ω be an LI form. Clearly $a(R_g s) = a(sg) = g^{-1}s^{-1} = L_{g^{-1}}a(s)$. Thus, $a \circ R_g = L_{g^{-1}} \circ a$ and $R_g^* \circ a^* = a^* \circ L_{g^{-1}}^*$. Then $R_g^*(a^*\omega) = a^*(L_{g^{-1}}^*\omega) = a^*\omega$ and thus $a^*\omega$ is right-invariant. Reciprocally, if $a^*\omega$ is RI then ω is LI: since $a \circ a = e$, $a^*(a^*\omega) = \omega$, we find $a^* R_g^*(a^*\omega) = L_{g^{-1}}^*\omega = \omega$, q.e.d.

Now let ω be an LI q-form on G, $q \leq r$. Then

$$((L_{g^{-1}})^*(e)\omega(e))(X_1^L(g),...,X_q^L(g)) = \omega(e)(L_{g^{-1}}^T(g)X_1^L(g),...,L_{g^{-1}}^T(g)X_q^L(g))$$

$$= \omega(e)(X_1^L(e),...,X_1^L(e)) \tag{1.6.12}$$

as a consequence of the definition (1.4.10) of the pull-back. Thus, the following proposition follows:

(1.6.2) PROPOSITION

There is a one-to-one correspondence between multilinear antisymmetric mappings of $T_e(G) \times ... \times T_e(G) \equiv \mathscr{G} \times ... \times \mathscr{G}$ to R and LI q-forms on G.

Also, the following theorem holds true:

(1.6.1) THEOREM

A left-invariant q-form $\alpha(g)$ on G has the expression

$$\alpha(g) = \alpha_{i_1...i_q}\omega^{L(i_1)}(g) \wedge ... \wedge \omega^{L(i_q)}(g)$$

where $\omega^{L(i)}$ is a basis of LI one-forms on G and the coefficients $\alpha_{i_1...i_q}$ are constants. Exchanging left and right, the same result holds for RI q-forms on G.

Let now $\omega(e)$ be the *r*-linear antisymmetric mapping defined by $dg^{(1)} \wedge ... \wedge dg^{(r)}$ where dg^i, $i = 1, ..., r$, is a basis of $T_e^*(G)$. Then

$$((L_{g^{-1}})^*(e))(dg^{(1)} \wedge ... \wedge dg^{(r)}) = \omega^{L(1)}(g) \wedge ... \wedge \omega^{L(r)}(g) \qquad (1.6.13)$$

is an LI *r*-form on G. Since all invariant *r*-forms (volume forms) on G are unique up to a constant factor, we may give

(1.6.2) DEFINITION (*LI Haar measure on a Lie group*)

The form

$$\mu^L(g) = \omega^{L(1)}(g) \wedge ... \wedge \omega^{L(r)}(g) \qquad (1.6.14)$$

is called the LI Haar measure on G.

From the definition it follows that

$$\mu^L(g) = \det\left[\frac{\partial g''^i(g',g)}{\partial g^j}\bigg|_{\substack{g=g \\ g'=g^{-1}}}\right] dg^{(1)} \wedge ... \wedge dg^{(r)}$$

$$= \det^{-1}\left[\frac{\partial g''^i(g',g)}{\partial g^j}\bigg|_{\substack{g=e \\ g'=g}}\right] dg^{(1)} \wedge ... \wedge dg^{(r)} \quad . \qquad (1.6.15)$$

A parallel definition can be given for the RI Haar measure on G, from which it follows that

$$\mu^R(g) = \det\left[\frac{\partial g''^i(g',g)}{\partial g'^j}\bigg|_{\substack{g'=g \\ g=g^{-1}}}\right] dg^{(1)} \wedge ... \wedge dg^{(r)}$$

$$= \det^{-1}\left[\frac{\partial g''^i(g',g)}{\partial g'^j}\bigg|_{\substack{g'=e \\ g=g}}\right] dg^{(1)} \wedge ... \wedge dg^{(r)} \quad . \qquad (1.6.16)$$

For many groups of physical interest (compact – see section 1.8 – semisimple, real connected algebraic groups* with determinant +1, groups

---

* The *algebraic groups* are the subgroups S of $GL(n, C)$ defined by conditions expressed by polynomials in their n^2 matrix elements that are satisfied for all the elements of S. The classical matrix groups are algebraic groups.

with semidirect structure such as the Poincaré group, etc.) the LI measure is also RI. Of course, if the group is non-compact, the domain of integration has to be suitably defined so that the integral has a meaning. In the theory of integration on topological groups, the groups admitting a *bi-invariant* integration measure are called *unimodular* (see Theorem 1.7.1).

1.7 The Maurer-Cartan equations and the canonical form on a Lie group G. Bi-invariant measure

Let $\omega^{L(k)}(g)$ and $X^L_{(k)}(g)$ be dual bases of LIFs and LIVFs on G,

$$\omega^{L(k)}(g)\, X^L_{(k')}(g) = \delta^k_{k'} \quad (k, k' = 1, ..., r) . \tag{1.7.1}$$

For any two-form $d\omega$,

$$d\omega(X, Y) = X.\omega(Y) - Y.\omega(X) - \omega([X, Y]) \tag{1.7.2}$$

(see (1.4.21)). Then

$$d\omega^{L(\ell)}(X^L_{(j)}, X^L_{(k)}) = -\omega^{L(\ell)}([X^L_{(j)}, X^L_{(k)}]) \tag{1.7.3}$$

since the first two terms in the r.h.s. of (1.7.2) are zero [(1.7.1)]. Thus, we may write

$$d\omega^{L(\ell)} = -\frac{1}{2} C^\ell_{jk} \omega^{L(j)} \wedge \omega^{L(k)} \tag{1.7.4}$$

which are the *Maurer-Cartan structure equations*, where C^l_{jk} are the structure constants relative to the basis $\omega^{L(i)}$ (or $X^L_{(i)}$). These equations are equivalent to the Lie bracket relations for the basis $X^L_{(i)}$ of the algebra. They also imply the Jacobi identity, which follows by taking the exterior derivative of (1.7.4) and again using the Maurer-Cartan equations on the r.h.s. In terms of the structure constants, the Jacobi identity (1.1.12) appears by noticing that the coefficient of $\omega^{L(i)} \wedge \omega^{L(j)} \wedge \omega^{L(k)}$ has to be zero since these exterior products constitute a basis for the vector space of the LI three-forms on G by Theorem 1.6.1. Nevertheless, the fact that $d^2 \equiv 0$ implies the Jacobi identity for the elements of the Lie algebra is a basis independent statement which follows immediately from (1.7.3) and (1.4.21).

Clearly, the equivalent relation for the RIFs is

$$d\omega^{R(l)} = \frac{1}{2} C^l_{jk} \omega^{R(j)} \wedge \omega^{R(k)} \tag{1.7.5}$$

where the difference in sign is due to the change $C^l_{jk} \to -C^l_{jk}$ when moving from $[X^L_{(j)}, X^L_{(k)}]$ to $[X^R_{(j)}, X^R_{(k)}]$.

Let us now introduce the LI canonical form θ on a Lie group.

(1.7.1) DEFINITION (*Canonical form on a Lie group*)

The (LI) canonical form θ on a Lie group G is the $\mathcal{G}$-valued one-form on
G $\theta(g) : T_g(G) \to T_e(G) \sim \mathcal{G}$, defined by

$$\theta(g)(X(g)) = X(e) \in \mathcal{G} \tag{1.7.6}$$

or, equivalently, by $\theta(g)(X(g)) = L^T_{g^{-1}}(g)X(g)$.

(1.7.1) PROPOSITION

a) The canonical form θ is left-invariant and
b) under right translations,

$$R^*_s \theta = Ad(s^{-1}) \circ \theta \quad . \tag{1.7.7}$$

Proof: Property a) is easy to check using the definition of θ:

$$((L_s)^* \theta(g))(X(s^{-1}g)) = \theta(g)(L^T_s X(s^{-1}g)) = L^T_{g^{-1}} L^T_s X(s^{-1}g)$$

$$= L^T_{g^{-1}s} X(s^{-1}g) = L^T_{(s^{-1}g)^{-1}} X(s^{-1}g) = \theta(s^{-1}g)X(s^{-1}g) \; . \tag{1.7.8}$$

Thus, $(L_s)^* \theta(g) = \theta(s^{-1}g)$ and θ is LI, eq. (1.6.2). Similarly, the proof of
b) is simple:

$$(R^*_s \theta(g))(X(gs^{-1})) = \theta(g)(R^T_s X(gs^{-1})) = L^T_{g^{-1}} R^T_s X(gs^{-1}) =$$

$$L^T_{s^{-1}} L^T_{sg^{-1}} R^T_s X(gs^{-1}) = L^T_{s^{-1}} R^T_s L^T_{(gs^{-1})^{-1}} X(gs^{-1}) =$$

$$L^T_{s^{-1}} R^T_s \theta(gs^{-1})(X(gs^{-1})) \equiv L^T_{s^{-1}}[\theta(gs^{-1})X(gs^{-1})]R^T_s \quad , \tag{1.7.9}$$

which defines *Ad* in (1.7.7), q.e.d. In fact, in section 1.9 we shall identify
Ad with the adjoint representation of the group on its algebra, hence the
notation in (1.7.7).

Given a basis of LI one-forms on G, the canonical form θ may be
expressed as $\theta(g) = \omega^{L(k)}(g) \otimes X_{(k)}$. Indeed, if $X(g) = \lambda^i X^{L(k)}_{(i)}(g)$ we find
that $\theta(X) = \lambda^i X_{(i)} = X$ by virtue of the duality relations (1.7.1).

(1.7.2) DEFINITION (*Brackets of $\mathcal{G}$-valued differential forms*)

Let α (β) be a $\mathcal{G}$-valued p- (q-)form, $\alpha = \alpha^{(i)} \otimes X_{(i)}$ ($\beta = \beta^{(i)} \otimes X_{(i)}$). Then
the commutator $[\alpha, \beta]$ defines a $\mathcal{G}$-valued $(p+q)$-form by

$$[\alpha, \beta] := \alpha \wedge \beta - (-1)^{pq} \beta \wedge \alpha = \alpha^{(i)} \wedge \beta^{(j)} \otimes [X_{(i)}, X_{(j)}] \tag{1.7.10}$$

so that $[\alpha, \beta]^{(k)} = C^k_{ij} \alpha^{(i)} \wedge \beta^{(j)}$.

The following properties follow from Definition 1.7.2:

$$[\alpha, \beta] = (-1)^{pq+1}[\beta, \alpha] \,, \quad d[\alpha, \beta] = [d\alpha, \beta] + (-1)^p[\alpha, d\beta] \,. \tag{1.7.11}$$

Thus, if α and β are one-forms (as ω above), $[\alpha, \beta] = [\beta, \alpha]$; if α is an even form, $[\alpha, \alpha] = 0$. Let γ be a $\mathscr{G}$-valued r-form. Then

$$(-1)^{pr}[\alpha, [\beta, \gamma]] + (-1)^{qp}[\beta, [\gamma, \alpha]] + (-1)^{rq}[\gamma, [\alpha, \beta]] = 0 \,. \tag{1.7.12}$$

Let α be odd; then $[\alpha, [\alpha, \alpha]] = 0$.

Similarly, the anticommutator of $\mathscr{G}$-valued forms is defined by

$$\{\alpha, \beta\} = \alpha \wedge \beta + (-)^{pq}\beta \wedge \alpha \tag{1.7.13}$$

and then

$$\{\alpha, \beta\} = (\alpha^i \wedge \beta^j + (-)^{pq}\beta^i \wedge \alpha^j)X_{(i)}X_{(j)} = \alpha^i \wedge \beta^j \otimes \{X_{(i)}, X_{(j)}\} \,. \tag{1.7.14}$$

Using (1.7.10) the Maurer–Cartan equations (1.7.4) can be expressed in terms of the (LI) canonical form $\theta(g) \equiv \theta^L(g)$ as

$$d\theta = -\frac{1}{2}[\theta, \theta] = -\theta \wedge \theta \,, \quad \theta(g) = \omega^{L(i)} \otimes X_{(i)} \,. \tag{1.7.15}$$

Exterior differentiation of (1.7.15) gives the Jacobi identity, which now reads

$$[\theta, [\theta, \theta]] = 0 \,. \tag{1.7.16}$$

Had we defined an RI canonical form θ^R, the structure equations would have read

$$d\theta^R = \frac{1}{2}[\theta^R, \theta^R] = \theta^R \wedge \theta^R \,, \quad \theta^R = \omega^{R(i)}(g) \otimes X_{(i)} \,. \tag{1.7.17}$$

Recalling the definition of a parallelizable manifold (section 1.3), the previous discussion of invariant forms shows that all Lie groups G are endowed with a *canonical parallelism* which is defined by the canonical form on G. We shall see that Proposition 1.7.1 also characterizes the properties of the connection form on a principal bundle (Definition 2.1.2b).

Using the Maurer–Cartan equations and the invariance properties of the LI and RI forms, it is not difficult to establish when a Lie group G admits a bi-invariant measure (*i.e.*, when G is *unimodular*) in terms of the structure constants of its Lie algebra. This is the content of

(1.7.1) THEOREM (*Bi-invariant measure, unimodular group*)

A connected Lie group G admits a left-invariant measure which is also right-invariant, i.e. G is unimodular, iff the structure constants of its Lie algebra satisfy

$$C^i_{ji} = 0 \tag{1.7.18}$$

or, equivalently, if the adjoint representation adX $((adX)Y := [X, Y]$, sections 1.8 and 1.9) has trace zero for all $X \in \mathscr{G}$ (notice that, although the structure constants depend on the basis chosen for $\mathscr{G}$, the trace property $C^k_{jk} = 0$ is basis independent and thus is an intrinsic property of $\mathscr{G}$).

Proof: Let $\{\omega^{L(i)}\}$ and $\{\omega^{R(j)}\}$ be two bases for the LI and RI vector spaces $\mathscr{X}^{*L}(G)$, $\mathscr{X}^{*R}(G)$ respectively. Since they also constitute a basis for the module of one-forms $\mathscr{X}^*(G)$ on G, the LI forms, say, can be written as

$$\omega^{L(i)}(g) = a^i_{\ j}(g)\omega^{R(j)}(g) \tag{1.7.19}$$

where the $a^i_{\ j} \in \mathscr{F}(G)$ (eq. (1.7.19) just expresses that $\mathscr{X}^*(G)$, as an $\mathscr{F}(G)$-module, is a free module of dimension $r=\dim G$). If we now use (1.7.19) to relate the LI and RI measures on G, we find that

$$\mu^L(g) = \Delta(g)\mu^R(g) \tag{1.7.20}$$

where $\Delta(g) = \det(a(g))$. Thus, the measure will be bi-invariant if $\Delta(g) = 1$. Since the RIVFs $X^R_{(m)}$ generate left translations and the $\omega^{L(i)}$'s are left-invariant, it is clear that $L_{X^R_{(m)}} \omega^{L(i)} = 0$. Then, taking the Lie derivative of (1.7.19), we obtain

$$(da^i_{\ k})(X^R_{(m)}) + a^i_{\ s}C^s_{mk} = 0 \quad , \tag{1.7.21}$$

an expression which is easily found by using (1.4.37) and (1.7.5) or from $L_{X^R_{(m)}} \omega^{R(j)} = C^j_{ms}\omega^{R(s)}$ (a relation that will be proved in the next section). Since the matrix $\alpha \equiv \Delta a^{-1}$ satisfies $a^i_{\ s}\alpha^s_{\ j} = \Delta\delta^i_{\ j}$, $d\Delta = \alpha^s_{\ j}da^j_{\ s}$ (recall that $d(\ln\Delta) = \mathrm{Tr}(a^{-1}da)$). Thus, multiplying eq. (1.7.21) by $\alpha^k_{\ i}$ we obtain

$$(d\Delta)(X^R_{(m)}) + C^k_{mk}\Delta = 0 \quad \forall m = 1, ..., r \quad , \tag{1.7.22}$$

an equation that, due to the independence of the $X^R_{(m)}$'s, implies that

$$\frac{d\Delta}{\Delta} = -C^k_{jk}\omega^{R(j)} \quad . \tag{1.7.23}$$

Now, if $\Delta = 1$, eq. (1.7.23) implies $C^k_{jk}\omega^{R(j)}=0$ and then $C^k_{jk}=0$ (sum over k !) due to the independence of the $\omega^{R(j)}$. Conversely, if $C^k_{jk}=0$, the determinant Δ is constant and a suitable normalization renders $\mu^L(g)$ equal to $\mu^R(g)$, q.e.d.

Taking into account that $\det(e^A) = e^{\mathrm{Tr}\,A}$, we find for $A = adX$ that another necessary and sufficient condition for the existence of a bi-invariant measure is that $\det(Adg) = 1 \ \forall g \in G$ where Ad is the adjoint representation of G (section 1.9). Finally, we mention that a simple modification of eq. (1.7.18) gives a necessary and sufficient condition for the existence of a G-invariant m-form (density) on the m-dimensional homogeneous space

G/H, where H is a closed subgroup of G, the parameters of which run from $m+1$ to r (since G acts transitively on G/H, such a density, if it exists, is unique up to a constant factor). It is given by Chern's criterion:

$$C^k_{jk} = 0 \quad , \quad j = m+1, ..., r \,, \, k = 1, ..., m \quad . \tag{1.7.24}$$

1.8 Left-invariance and bi-invariance. Bi-invariant metric tensor field on the group manifold

As we have discussed, the LI (RI) vector fields on G generate the R (L) translations. Thus,

$$L_{X^{L,R}} \, \omega^{R,L} = 0 \quad . \tag{1.8.1}$$

Let us prove this explicitly. Let ω^L be an LI form, and $L_{s(t)} := \exp t X^R$ the one-parameter subgroup generated by X^R. From (1.4.42) and (1.6.2) we obtain

$$L_{X^R} \, \omega^L := \lim_{t \to 0} \frac{(L_{s(t)})^* \omega^L L_{s(t)} - \omega^L}{t} = 0 \quad . \tag{1.8.2}$$

Of course, this also follows from the equivalent definition (1.4.35),

$$(L_{X^R}\omega^L)(X^L_1, ..., X^L_q) = X^R.(\omega^L(X^L_1, ..., X^L_q))$$

$$- \sum_{i=1}^{q} \omega^L (X^L_1, ..., [X^R, X^L_i], ..., X^L_q) = 0 \quad \forall X^L_1, ..., X^L_p \in \mathscr{X}^L(G) \quad , \tag{1.8.3}$$

since the first term in the r.h.s. is zero because $\omega^L(X^L_1, ..., X^L_p)$ is constant on G (eq. (1.6.1) for a q-form) and the second one because of (1.2.12). (It is sufficient to use in (1.8.3) above an arbitrary set of q independent LIVFs because, as we know, $\mathscr{X}(G)$ is (as $\mathscr{X}^*(G)$) a free $\mathscr{F}(G)$-module of dimension $r = \dim G$.) By interchanging L and R we similarly find that $L_{X^L}\omega^R = 0$.

The action of X^L, X^R on ω^L, ω^R respectively involves the structure constants of $\mathscr{G}$, because in $(L_{X^L}\omega^L)(X^L_1...X^L_q)$, say, again the first contribution is zero, but the second is a sum of terms involving commutators of LIVFs. When the ω's are the LI one-forms $\{\omega^{(i)}\}$ of the basis dual to $\{X^L_{(j)}\}$ ($i, j = 1, ..., \dim \mathscr{G} = r$), the expression for $L_{X^L_j} \omega^{L(k)}$ is particularly simple, since

$$(L_{X^L_{(j)}} \, \omega^{L(k)}) \, (X^L_{(\ell)}) = -\omega^{L(k)}([X^L_{(j)}, X^L_{(\ell)}]) = -C^k_{j\ell} \quad .$$

Thus, for the canonical dual basis of LI and RI one-forms and vector fields we find

$$L_{X^L_{(j)}} \, \omega^{R(k)} = 0 \,, \quad L_{X^R_{(j)}} \, \omega^{L(k)} = 0 \quad ,$$

$$L_{X^L_{(j)}} \, \omega^{L(k)} = -C^k_{j\ell} \omega^{L(\ell)} \quad , \quad L_{X^R_{(j)}} \, \omega^{R(k)} = C^k_{j\ell} \, \omega^{R(\ell)} \quad . \tag{1.8.4}$$

The following theorem is now proved:

(1.8.1) THEOREM

Let ω be an LI q-form on G. Then a) the LIF ω is bi-invariant (is both LI and RI) if $L_{X^L}\,\omega = 0$, i.e. iff

$$\sum_{i=1}^{q}\omega(X_1^L,...,[Z^L,X_i^L],...,X_q^L) = 0 \quad \forall\, Z^L, X_1^L,...,X_q^L \in \mathscr{X}^L(G) \ . \quad (1.8.5)$$

Moreover, b) every bi-invariant form is closed.

Proof: The first part follows immediately from the definition of right-invariance of the (already LI) q-form. For the second, it is sufficient to recall the action (1.4.21) of d:

$$(d\omega)(X_1^L,...,X_{q+1}^L) = \sum_{i=1}^{q+1}(-)^{i+1}X_i^L.\omega(X_1^L,...,\hat{X}_i^L,...,X_{q+1}^L)$$

$$+ \sum_{\substack{i,j=1\\i<j}}^{q+1}(-)^{i+j}\omega([X_i^L,X_j^L],\,X_1^L,...,\hat{X}_i^L,...,\hat{X}_j^L,...,X_{q+1}^L) \qquad (1.8.6)$$

$$\forall X_1^L,...,X_{q+1}^L \in \mathscr{X}^L(G) \ .$$

Again the first term is zero because X_i^L acts on a constant. The second may be written as

$$\frac{1}{2}\sum_{i,j=1}^{q+1}(-)^{i+j}\epsilon^{ij}\omega([X_i^L,X_j^L],\,X_1^L,...,\hat{X}_i^L,...,\hat{X}_j^L,...,X_{q+1}^L) \qquad (1.8.7)$$

where $\epsilon^{ij} = 1\ (-1)$ if $i < j\ (i > j)$. For each fixed $j = 1,...,q+1$ (1.8.7) is of the form (1.8.5) and hence zero because ω is also RI, *q.e.d.*

This theorem will be important when introducing the Chevalley-Eilenberg cohomology (section 6.7). The following statement is also true: let G be an r-dimensional *compact* connected Lie group. Then, any LI r-form on G is bi-invariant. This follows from the fact that, if G is compact, $\det(Adg) = 1$. In fact, right-invariance follows from left-invariance and invariance under $i_g = L_g R_g^{-1}$: $(i_g^*\omega)(X_1,...,X_r) = \omega((Adg)X_1,...,(Adg)X_r) = \det(Adg)\omega(X_1,...,X_r) = \omega(X_1,...,X_r)$ for $\det(Adg) = 1$.

To conclude this section, let us introduce a bi-invariant metric tensor field on G. A fundamental theorem of the theory of Lie algebras (Cartan's criterion) states that a Lie algebra $\mathscr{G}$ is *semisimple* iff the Cartan-Killing tensor k,

$$k : (X,Y) \mapsto k(X,Y) = \mathrm{Tr}\,(adX\,adY) \ , \qquad (1.8.8)$$

where *ad* is the adjoint representation (section 1.9) of $\mathscr{G}$ on itself, is nonsingular[*]. If moreover G is compact, k is then (negative) definite. It is simple to find the form of the matrices $adX_{(i)}$ since $[ad\,X_{(i)}]\,Y = C_{i\cdot j}^{\ k}\lambda^j X_{(k)}$ if $Y = \lambda^j X_{(j)}$, which means that the matrix elements of $ad\,X_{(i)}$ are given by

$$(ad\,X_{(i)})_{\cdot j}^{k} = C_{(i)\cdot j}^{\ \ k} \equiv C_{i\ j}^{\ k} \quad . \tag{1.8.9}$$

To check that $ad\,X_{(i)}$ satisfies the commutation relations of $\mathscr{G}$ it is sufficient to rewrite the Jacobi identity for the indices i, j, k in the form $C_{(i)\cdot s}^{\ \ l}C_{(k)\cdot j}^{\ \ s} - C_{(k)\cdot s}^{\ \ l}C_{(i)\cdot j}^{\ \ s} = C_{i\cdot k}^{\ s}C_{(s)\cdot j}^{\ \ l}$ where the brackets and the dots in the structure constants have been added to read the expression easily in matrix notation. Thus, the $r \times r$ matrix of the Cartan-Killing metric is given by

$$k_{ij} := C_{i\ s}^{\ l}C_{j\ l}^{\ s} = k_{ji} \quad . \tag{1.8.10}$$

Let G now be semisimple for the remainder of this section. It is easy to prove, by using the Jacobi identity, that

$$k_{\ell s}C_{i\ j}^{\ s} = -C_{i\ \ell}^{\ r}k_{rj} \tag{1.8.11}$$

(in fact, this corresponds to saying that the adjoint and the coadjoint representations are equivalent, and that k_{ij} is the nonsingular matrix which realizes the equivalence). If the metric k is used to raise and lower indices, $k^{si}k_{ij} = \delta_j^s$ and (1.8.10) imply, respectively,

$$C_{\cdot lk}^{s}C_{j}^{\cdot kl} = \delta_j^s \quad , \quad C_{i\ell j} = -C_{ij\ell} \quad , \tag{1.8.12}$$

i.e., total antisymmetry for the C_{ijk}. It is simple to find an expression for C_{ijk} in terms of the matrices of the adjoint representation:

$$\begin{aligned} C_{ilj} &= k_{lk}C_{i\cdot j}^{\ k} = \mathrm{Tr}(adX_{(l)}adX_{(k)}C_{i\cdot j}^{\ k}) \\ &= \mathrm{Tr}(adX_{(l)}[adX_{(i)}adX_{(j)} - adX_{(j)}adX_{(i)}]) \\ &= \mathrm{Tr}(adX_{(i)}adX_{(j)}adX_{(l)}) - \mathrm{Tr}(adX_{(j)}adX_{(i)}adX_{(l)}) \ , \end{aligned} \tag{1.8.13}$$

which exhibits the full antisymmetry of C_{ilj}.

Using this fact and the LI one-forms on G, we may now define a bi-invariant metric on the group manifold G:

---

[*] Let G be simple. It is possible to generalize (1.8.8) for an arbitrary representation ρ: $(X,Y)_\rho$ $:= \mathrm{Tr}(\rho(X), \rho(Y))$ is the inner product of two elements within the representation ρ, and (1.8.8) corresponds to $\rho = ad$. Then it is found that $(X,Y)_\rho = f(\rho)k(X,Y)$ where $f(\rho)$, related to the *Dynkin index*, depends only on the representation ρ.

(1.8.2) THEOREM (*Bi-invariant metric tensor field on G*)

Let *G* be a semisimple Lie group and $\{\omega^{L(i)}\}_{i=1,\dots,r}$ a basis for the LI one-forms on *G*. The symmetric $\binom{0}{2}$-type tensor field *k* on *G*,

$$k = k_{ij}\omega^{L(i)} \otimes \omega^{L(j)} \quad , \tag{1.8.14}$$

where k_{ij} is the Killing matrix (1.8.10), is bi-invariant.

Proof: It is sufficient to recall that L_X is a tensor derivation. Thus, by construction, (1.8.14) is already LI; the right invariance follows from

$$
\begin{aligned}
L_{X_{(\ell)}^L} k &= k_{ij}(L_{X_{(\ell)}^L}\omega^{L(i)}) \otimes \omega^{L(j)} + k_{ij}\omega^{L(i)} \otimes (L_{X_{(\ell)}^L}\omega^{L(j)}) \\
&= -k_{ij}C_\ell{}^i{}_s\omega^{L(s)} \otimes \omega^{L(j)} - k_{ij}C_\ell{}^j{}_s\omega^{L(i)} \otimes \omega^{L(s)} = 0 ,
\end{aligned}
\tag{1.8.15}
$$

which is a consequence of (1.8.4) and (1.8.11), q.e.d.

In fact, this is a particular case of the following: Let ρ be an LI metric on *G*. Then ρ is bi-invariant iff

$$\rho([X,Y],Z) = \rho(X,[Y,Z]) \quad \forall X,Y,Z \quad , \tag{1.8.16}$$

a relation that includes (1.8.15) and which follows by imposing the right-invariance, $L_{X^L}\rho = 0$, on ρ.

1.9 Applications and examples for Lie groups

(a) *The adjoint (coadjoint) representations of a Lie group G and its Lie algebra $\mathscr{G}$ (coalgebra $\mathscr{G}^*$)*

The *adjoint representation Ad* of *G* is given by

$$[Adg]^i{}_j = [R_{g^{-1}}^T(g)L_g^T(0)]^i{}_j \quad . \tag{1.9.1}$$

Let *G* be a Lie group and $\mathscr{G}$ its Lie algebra, and let $i_g : G \to G$ be the inner automorphism $i_g : g_0 \mapsto gg_0g^{-1}$; thus, $i_g = L_g R_g^{-1}$. Then the adjoint representation *Ad* of *G* on $\mathscr{G}$ is the mapping $Ad : \mathscr{G} \to \text{Aut}\mathscr{G}$ (automorphisms of $\mathscr{G}$) defined by

$$Adg := (i_g^T)(e) \quad . \tag{1.9.2}$$

Thus, $Adg : \mathscr{G} \to \mathscr{G}$ is given by

$$(Adg)X(e) = (R_{g^{-1}}L_g)^T(e)X(e) \quad , \quad X(e) \in \mathscr{G} \quad , \tag{1.9.3}$$

and using the chain rule, eq. (1.1.20), we find that $(R_{g^{-1}}L_g)^T(e) = R_{g^{-1}}^T(g) \circ L_g^T(e)$. (Note that, since *R* and *L* translations commute, we could have equally well applied the chain rule to $(L_g R_{g^{-1}})^T(e) = L_g^T(g^{-1})R_{g^{-1}}^T(e) = [L_{g^{-1}}^T(e)]^{-1}[R_g^T(g^{-1})]^{-1} = [R_g^T(g^{-1})L_{g^{-1}}^T(e)]^{-1} \equiv M^{-1}(g)$, which produces the same result since $M^{-1}(g) = M(g^{-1})$ and $M(g^{-1}) = Adg$.) Symbolically we may write $Adg\, X = gXg^{-1}$, $X \in \mathscr{G}$, an equation that is literally correct if the Lie group G is $GL(n, R)$ or is given by any of its matrix subgroups since then $R_{g^{-1}}$ and L_g are linear mappings and, as we have seen, the tangent of a linear mapping can be identified with the mapping itself.

The *adjoint representation of* $\mathscr{G}$ on $\mathscr{G}$, already encountered, may be defined from Ad. Since we may view Ad as the group homomorphism $G \to GL(r, R)$, $Ad^T(e)$ is a mapping $\mathscr{G} \mapsto \mathscr{G}(GL(r, R)) = L(\mathscr{G}, \mathscr{G})$ ($r \times r$ matrices). Then, ad is the algebra homomorphism $ad : \mathscr{G} \to L(\mathscr{G}, \mathscr{G})$ (endomorphisms of $\mathscr{G}$) defined by $ad : X \mapsto adX = (Ad^T)(e)X$. Since $exp(Adg\, Y) = g(exp\, Y)g^{-1}$ it follows that $(adX)(Y) = [X, Y]$.

The *coadjoint representation Ad^* of* G on the dual algebra $\mathscr{G}^*$ of $\mathscr{G}$ is defined, e.g., by noticing that the pull-back $(Adg)^* : \mathscr{G}^* \to \mathscr{G}^*$ of Adg satisfies $\mu[(Adg)(X)] = [(Adg)^*\mu](X)$, $\forall X \in \mathscr{G}$, $\forall \mu \in \mathscr{G}^*$. Then the *coadjoint representation Ad^* of* G on $\mathscr{G}^*$ is defined by

$$Ad^*g \equiv (Adg^{-1})^* \quad \forall g \in G \quad . \tag{1.9.4}$$

Notice the different rôle of the stars in (1.9.4). The star in Ad^* just indicates *coadjoint representation*, and that outside the bracket in $(Adg^{-1})^*$ the adjoint mapping; thus, $Ad^*g'g = (Ad^*g')(Ad^*g) \equiv (Adg'^{-1})^*(Adg^{-1})^* = (Adg^{-1}Adg'^{-1})^* = (Ad(g'g)^{-1})^*$. As a matrix, $(Adg^{-1})^* = (Adg^{-1})^t$ so that $Ad^*g'g = Ad^*g'Ad^*g$, but the transposition may be omitted if $(Adg) : X \mapsto X'$ and $(Ad^*g) : \mu \mapsto \mu'$, respectively, correspond to the transformations $\lambda'^i = (Adg)^i{}_j\lambda^j$ and $\mu'_i = \mu_j(Adg^{-1})^j{}_i$ of the coordinates of $X \in \mathscr{G}$ and $\mu \in \mathscr{G}^*$ when they are written as a column and as a row respectively.

The *coadjoint representation ad^* of* $\mathscr{G}$ on the coalgebra $\mathscr{G}^*$ is defined similarly from the adjoint representation ad of $\mathscr{G}$ on itself by the condition $\mu((adX)X') \equiv \mu([X, X']) = ((adX)^*\mu)(X')$. Thus $(adX)^*\mu = \mu \circ adX$ on any $X' \in \mathscr{G}$. The matrix of the coadjoint representation ad^* of X is then given by

$$ad^*X = -(adX)^* \quad , \tag{1.9.5}$$

so that $ad^*X = -(adX)^t$ if the elements of $\mathscr{G}$ and $\mathscr{G}^*$ are both written as column vectors. Again, the minus sign *and* the transposition guarantee that ad^* is a Lie algebra representation. They translate to $\mathscr{G}$ the analogous situation for G: given a representation $D(g)$ of a group, the inverse D^{-1} and the transpose D^t are not representations, but $(D^{-1})^t$, the contragredient representation of D, defines a representation of G.

(b) *Invariant forms on the group $GL(n, R)$ and bi-invariant measure*
The LI and RI vector fields were already given as an example at the end of section 1.2. Let us now compute the n^2 LIFs $\omega^{L(i)}{}_{(k)}$ on G from the

$GL(n, R)$ group law (1.2.17) and (1.6.10) (see also (1.6.5)):

$$[L^*_{g-1}(e)]^i_{.k\ n.}{}^\ell = \left.\frac{\partial g'''^i_{.k}}{\partial g^n_{.\ell}}\right|_{\substack{g=g \\ g'=g^{-1}}} = (g^{-1})^i_{.n}\delta^\ell_k \ ,$$

$$(\omega^L(g))^{(i)}_{(k)} = (g^{-1})^i_{.n}\delta^\ell_k dg^n_{.\ell} = (g^{-1})^i_{.n}dg^n_{.k} \ , \tag{1.9.6}$$

which is written simply (and similarly for the RIFs) as

$$\theta(g) \equiv \theta^L(g) = g^{-1}dg \qquad (\theta^R(g) = dgg^{-1}) \ . \tag{1.9.7}$$

These expressions show explicitly that the antipodal mapping interchanges θ^L and θ^R (Proposition 1.6.1); since $d(gg^{-1}) = 0$ gives $dg^{-1} = -g^{-1}dgg^{-1}$ we find, specifically,

$$\theta(g) = -\theta^R(g^{-1}) \ . \tag{1.9.8}$$

They also show that $d\theta = dg^{-1} \wedge dg = -\theta \wedge \theta$ and that $d\theta^R = \theta^R \wedge \theta^R$, as required by eqs. (1.7.15) and (1.7.17).

It is worth mentioning that eqs. (1.9.7), here derived for the group $GL(n, R)$, are valid for all matrix groups. This means that, for a group of matrices G, $\dim G = r$, there are exactly r linearly independent one-forms among the n^2 one-forms that are obtained by following the procedure that led to (1.9.6). This has to be so, because at the identity $e \in G$ eq. (1.9.6) determines a basis for the cotangent space $T^*_e(G)$ (see eq. (1.6.11)) which can be used to generate a basis of r linearly independent LI one-forms on G. As a result, eqs. (1.9.7) can be used to find the r LI and RI forms of any r-dimensional matrix Lie group G, providing an alternative (but equivalent) procedure to formulae (1.6.10), (1.6.9).

Let us now compute the invariant measure on $GL(n, R)$. We only need to compute $\det^{-1}[\partial(g'g)/\partial g]_{\substack{g=e \\ g'=g}}$ (see (1.6.15)). The expression $\left(\partial g'''^i_{.j}/\partial g^m_{.n}\right)\big|_{\substack{g=e \\ g'=g}}$

$$= \left[\frac{\partial g''}{\partial g}\right]^{i\ n}_{.m\cdot j} = g^i_{.m}\delta^n_{.j}$$ can be looked at as the matrix which results from taking the Kronecker product $g \otimes \mathbf{1}$ of the $n \times n$ matrix $g^i_{.j}$ and the n-dimensional unit matrix $\mathbf{1}$. Then, since $\det(A \otimes B) = (\det A)^{\dim B}$ $(\det B)^{\dim A}$, we find that $\det(g \otimes \mathbf{1}) = (\det g)^n$ and the invariant (in fact, bi-invariant) measure is given by

$$\mu(g) = \frac{dg_{11} \wedge ... \wedge dg_{1n} \wedge dg_{21} \wedge ... \wedge dg_{nn}}{(\det g)^n} \ . \tag{1.9.9}$$

(c) *Invariant measures on triangular matrix groups*
Although all groups of finite volume admit a bi-invariant measure, this is not a necessary condition as the previous example of $GL(n, R)$ shows.

Let us consider now an example of a group with different left and right measures. Let G be the group of the real triangular matrices

$$g = \begin{pmatrix} g_{11} & g_{12} \\ 0 & g_{22} \end{pmatrix} \quad , \quad g^{-1} = \begin{pmatrix} g_{11}^{-1} & -g_{12}/g_{11}g_{22} \\ 0 & g_{22}^{-1} \end{pmatrix} \quad . \tag{1.9.10}$$

From (1.9.7) one easily finds

$$
\begin{aligned}
\omega^{L(g_{11})} &= g_{11}^{-1} dg_{11}, & \omega^{R(g_{11})} &= g_{11}^{-1} dg_{11}, \\
\omega^{L(g_{12})} &= g_{11}^{-1} dg_{12} - \frac{g_{12}}{g_{11}g_{22}} dg_{22}, & \omega^{R(g_{12})} &= \frac{-g_{12}}{g_{11}g_{22}} dg_{11} + g_{22}^{-1} dg_{12}, \\
\omega^{L(g_{22})} &= g_{22}^{-1} dg_{22}, & \omega^{R(g_{22})} &= g_{22}^{-1} dg_{22} \quad .
\end{aligned}
$$
$$\tag{1.9.11}$$

Then, using these results or directly from (1.6.15) and (1.6.16), we find

$$\mu^L(g) = \frac{dg_{11} \wedge dg_{12} \wedge dg_{22}}{|g_{11}^2 g_{22}|} \quad , \quad \mu^R(g) = \frac{dg_{11} \wedge dg_{12} \wedge dg_{22}}{|g_{11} g_{22}^2|} \quad ; \tag{1.9.12}$$

as expected from Proposition 1.6.1, the antipodal mapping converts $\mu^{L,R}$ into $\mu^{R,L}$. The generalization to the group of arbitrary triangular matrices follows the same pattern:

$$
\begin{aligned}
\mu^L(g) &= \frac{dg_{11} \wedge dg_{12} \wedge ... \wedge dg_{n-1,n} \wedge dg_{nn}}{|g_{11}^n g_{22}^{n-1} ... g_{n-1,n-1}^2 g_{nn}|} \quad , \\
\mu^R(g) &= \frac{dg_{11} \wedge dg_{12} \wedge ... \wedge dg_{n-1,n} \wedge dg_{nn}}{|g_{11} g_{22}^2 ... g_{n-1,n-1}^{n-1} g_{nn}^n|} \quad ,
\end{aligned}
$$
$$\tag{1.9.13}$$

which are also obtained directly by noticing that $g^{-1}dg$ and dgg^{-1} are triangular.

A particular case of (1.9.10) is the non-compact affine group of transformations of the real line, $x' = c^{-1}x + a$ $(c > 0)$; all non-abelian two-parameter Lie groups are equivalent to this group. It corresponds to taking in (1.9.10) $g_{11} = c^{-1}$, $g_{12} = a$ and $g_{22} = 1$ so that in this case

$$\mu^L(g) = da \wedge dc \quad , \quad \mu^R(g) = \frac{1}{c} da \wedge dc \quad . \tag{1.9.14}$$

(d) *The rotation group SO(3)*

Because in example (b) above the group law is linear in either the primed or the unprimed parameters, there is actually no need to worry about the argument of L^T, L^*, etc. But, as mentioned, this is not always so. This may be checked with the rotation group which we shall now discuss in some detail. The group law of the rotation group can be given in a neighbourhood of the identity element in the form

$$\epsilon''^i = \sqrt{1 - \frac{\vec{\epsilon}^2}{4}} \, \epsilon'^i + \sqrt{1 - \frac{\vec{\epsilon}'^2}{4}} \, \epsilon^i - \frac{1}{2} \eta^i_{.jk} \, \epsilon'^j \epsilon^k \quad , \tag{1.9.15}$$

where $\vec{\epsilon}$ is the vector which determines the rotation axis and the rotation angle ψ by $|\vec{\epsilon}| = 2\sin(\psi/2)$; η denotes here the totally antisymmetric tensor. Let us compute first the invariant vector fields. From (1.9.15) and using (1.1.14) and (1.1.13) we obtain

$$
X^L_{(i)}(\epsilon) = \left(\sqrt{1 - \frac{\vec{\epsilon}^2}{4}} \delta^j_{\cdot i} + \frac{1}{2} \eta^j_{\cdot ik} \epsilon^k \right) \frac{\partial}{\partial \epsilon^j} ,
$$

$$
X^R_{(i)}(\epsilon) = \left(\sqrt{1 - \frac{\vec{\epsilon}^2}{4}} \delta^j_{\cdot i} - \frac{1}{2} \eta^j_{\cdot ik} \epsilon^k \right) \frac{\partial}{\partial \epsilon^j} ,
$$
$$
[X^L_{(i)}, X^L_{(j)}] = -\eta_{ij\cdot}{}^k X^L_{(k)} , \quad [X^R_{(i)}, X^R_{(j)}] = \eta_{ij\cdot}{}^k X^R_{(k)} .
$$

(1.9.16)

The LI and RI forms can be computed in various equivalent ways (see section 1.6). The result is

$$
\omega^{L(i)}(\epsilon) = \sqrt{1 - \frac{\vec{\epsilon}^2}{4}} d\epsilon^i + \frac{1}{4} \frac{\epsilon^i(\vec{\epsilon} \cdot d\vec{\epsilon})}{\sqrt{1 - \frac{\vec{\epsilon}^2}{4}}} + \frac{1}{2}(\vec{\epsilon} \wedge d\vec{\epsilon})^i ,
$$

$$
\omega^{R(i)}(\epsilon) = \sqrt{1 - \frac{\vec{\epsilon}^2}{4}} d\epsilon^i + \frac{1}{4} \frac{\epsilon^i(\vec{\epsilon} \cdot d\vec{\epsilon})}{\sqrt{1 - \frac{\vec{\epsilon}^2}{4}}} - \frac{1}{2}(\vec{\epsilon} \wedge d\vec{\epsilon})^i .
$$

(1.9.17)

(It is also possible to extract the LIFs and RIFs from the algebra-valued forms (1.9.7).) The expression (1.9.16) also tells us that the bracket in the r.h.s. of $X^L_{(i)}(\epsilon)$ is the matrix $[L^T_\epsilon(0)]^j_{\cdot i}$ by (1.1.18) and that similarly (cf. (1.6.6) and (1.6.7)), $[R^T_{-\epsilon}(\epsilon)]^i_{\cdot j} = (\partial g''^i(g',g)/\partial g'^j)|_{\substack{g'=\epsilon \\ g=-\epsilon}}$ is also determined by the r.h.s. of $\omega^{R(i)}(\epsilon)$ $([(R_{-\epsilon})^*(0)]^i_{\cdot j})$:

$$
[L^T_\epsilon(0)]^j_{\cdot l} = \left[\sqrt{1 - \frac{\vec{\epsilon}^2}{4}} \delta^j_{\cdot l} + \frac{1}{2} \eta^j_{\cdot lk} \epsilon^k \right] ,
$$

$$
[R^T_{-\epsilon}(\epsilon)]^i_{\cdot j} = \sqrt{1 - \frac{\vec{\epsilon}^2}{4}} \delta^i_{\cdot j} + \frac{1}{4} \frac{\epsilon^i \epsilon_j}{\sqrt{1 - \frac{\vec{\epsilon}^2}{4}}} + \frac{1}{2} \eta^i_{\cdot js} \epsilon^s .
$$

(1.9.18)

Once the expressions (1.9.18) are known, it is easy to compute the adjoint representation of the group on its algebra, which is given by $[R^T_{-\epsilon}(\epsilon) L^T_\epsilon(0)]^i_{\cdot l}, i, l = 1, 2, 3$, and it is simply the ordinary three-dimensional representation. Multiplying the two matrices (1.9.18) we obtain that the matrix for the element $\vec{\epsilon}$ in AdG is given by[*]

$$R^i_{.l}(\epsilon) = (1 - \frac{\vec{\epsilon}^2}{2})\delta^i_{.l} + \sqrt{1 - \frac{\vec{\epsilon}^2}{4}}\eta^i_{.ls}\epsilon^s + \frac{1}{2}\epsilon^i\epsilon_l \qquad (1.9.19)$$

or, equivalently ($\vec{\epsilon} = 2\sin\frac{\psi}{2}\hat{n}^i$),

$$R(\psi)^i_{.l} = \cos\psi\delta^i_{.l} + \sin\psi\eta^i_{.ls}\hat{n}^s + (1 - \cos\psi)\hat{n}^i\hat{n}_l . \qquad (1.9.20)$$

The Killing form [(1.8.10)] is given by $k_{ij} = \eta_{ik}{}^l\eta_{jl}{}^k = -2\delta_{ij}$; it is negative definite as it corresponds to a compact group. Finally, it is simple to compute the invariant measure on the group, which is given by $d^3\epsilon/\sqrt{1 - \frac{\vec{\epsilon}^2}{4}}$. From this the volume of $SO(3, R)$ is found to be $8\pi^2$.

1.10 The case of super Lie groups: the supertranslation group as an example

We conclude this chapter with some comments on supergroups[†]. Many of the concepts previously discussed for Lie groups extend easily, at least at a formal level, when 'manifold' and 'Lie group' are replaced by 'supermanifold' and 'super Lie group' respectively. Without entering into the subtleties of supermanifold theory, it is sufficient for practical purposes here to bear in mind the anticommuting properties of the (odd) Grassmann variables (often denoted by θ) and to keep track of the resulting additional minus signs arising in computations due to their interchange. In general, for any two variables ζ and η, $\zeta\eta = (-1)^{ab}\eta\zeta$, where a and b denote the *grading* (0, even; 1, odd) of ζ and η respectively. As a result, a two-form such as $d\theta' \wedge d\theta$ is *symmetric* rather than antisymmetric if θ', θ are ('Fermi' or 'odd') anticommuting variables and the finite group transformation associated with an odd generator is found by exponentiation in an easier way (the series necessarily terminates since its terms will eventually include the square of an odd variable). In particular, it is possible to define matrix generators that multiplied by the corresponding parameters can be exponentiated into a matrix supergroup, etc. Some of the concepts built on ordinary manifolds cannot, however, be extended in a trivial way to supermanifolds, as is the case with the integration and the definition of the (Berezin) integral measure. We shall not discuss these questions here; the reader who feels uneasy about this rather simple approach to graded Lie groups and encounters them here for the first time may find further

---

[*] In connection with the relevance of the arguments the reader may check here, for instance, that $[R^T_{-\epsilon}(\epsilon)]^i_{.j}[R^T_\epsilon(0)]^j_{.l} = \delta^i_{.l}$, where $[R^T_\epsilon(0)]^j_{.l} = \sqrt{1 - \frac{\vec{\epsilon}^2}{4}}\delta^j_{.l} - \frac{1}{2}\eta^j_{.lk}\epsilon^k$, as can be read from the second expression in (1.9.16).

[†] 'Super'-objects will appear only in this section and in sections 5.3, 8.7 and 8.8.

guidance in the references quoted in the bibliographical notes at the end of the chapter.

Let us consider the important example of *superspace* Σ which is the basic notion in supersymmetric theories. As an illustration of how to extend some of the previous formulae to the graded Lie group case, we shall compute the invariant vector fields and forms on Σ considered as the *supertranslation group* manifold. Its elements are parametrized by $g \equiv z^A = (x^\mu, \theta^\alpha)$, where x^μ is even and θ^α odd, and the supergroup law is given by

$$x''^\mu = x'^\mu + x^\mu + i\bar\theta'^\alpha(\gamma^\mu)_{\alpha\beta}\theta^\beta \quad , \tag{1.10.1}$$

$$\theta''^\alpha = \theta'^\alpha + \theta^\alpha \quad (\mu = 0, 1, 2, 3; \quad \alpha, \beta = 1, 2, 3, 4) \; .$$

We are considering for definiteness the ordinary four-dimensional space-time ($D = 4$, Minkowski spacetime) and thus the anticommuting variables (spinors) θ^α are also four-dimensional. The γ^μ are the 4×4 Dirac matrices and $\bar\theta = \theta^\dagger\gamma^0$ is the Dirac adjoint. In the Weyl realization, which is specially convenient to exhibit the opposite chirality components of the spinors, the Dirac matrices and the γ_5 matrix are given by

$$\gamma^\mu = \begin{pmatrix} 0 & \sigma^\mu \\ \tilde\sigma^\mu & 0 \end{pmatrix} \quad , \quad \gamma_5 = i\gamma_0\gamma_1\gamma_2\gamma_3 = \begin{pmatrix} 1 & 0 \\ 0 & -1 \end{pmatrix} \quad , \quad \{\gamma^\mu, \gamma^\nu\} = 2\eta^{\mu\nu} \quad , \tag{1.10.2}$$

where $\eta^{\mu\nu} = (1, -1, -1, -1)$, $\sigma^\mu = (\sigma^0 \equiv 1_2, \sigma^i)$, $\tilde\sigma^\mu = (\sigma^0, -\sigma^i)$ and σ^i are the Pauli matrices,

$$\sigma^1 = \begin{pmatrix} 0 & 1 \\ 1 & 0 \end{pmatrix} \quad , \quad \sigma^2 = \begin{pmatrix} 0 & -i \\ i & 0 \end{pmatrix} \quad , \quad \sigma^3 = \begin{pmatrix} 1 & 0 \\ 0 & -1 \end{pmatrix} \; . \tag{1.10.3}$$

The projectors $\frac{1}{2}(1 \pm \gamma_5)$ project out the positive and negative chirality components respectively.

The θ's in (1.10.1) are taken to be Majorana spinors, which corresponds to the so-called '$N = 1$ supersymmetry'. Majorana spinors are defined by the condition that their Dirac adjoint $\bar\theta \equiv \theta^\dagger\gamma^0$ equals their Majorana adjoint $\theta^T C$, i.e., by the property that $\bar\theta = \theta^T C$. The matrix C is the charge (particle–antiparticle) conjugation matrix; it satisfies $C\gamma^\mu C^{-1} = -\gamma^{\mu T}$, $C = -C^{-1}$, $C = -C^T$. In Minkowski space, Majorana spinors have four real independent components. They are accordingly 'real' spinors, although in the Weyl realization above a Majorana spinor is given in terms of a two-dimensional complex (undotted) Pauli spinor ξ_δ and its complex conjugate (dotted) one, $\bar\xi^{\dot\delta}$, which constitute its chiral components: $\theta^\alpha = (\xi_\delta, \bar\xi^{\dot\delta})$, $\alpha = 1, ..., 4$, $\delta, \dot\delta = 1, 2$. They are explicitly real in the Majorana realization of the γ matrices, in which $C = \gamma^0$. The vector bilinear made out of two Grassmann Majorana spinors is antisymmetric, $\bar\theta'\gamma^\mu\theta = -\bar\theta\gamma^\mu\theta'$; this property will be important for the discussion in section 5.3. The structure of the group Σ shows that the odd translations θ define an abelian subgroup, and that they generate (through the term $\bar{\theta}'\gamma^\mu\theta$) spacetime translations; these constitute an invariant subgroup of Σ (the group extension character of Σ will be discussed in section 5.3).

The generators $Q_{(A)}$ of the left action of the group on Σ are given by (1.1.13),

$$Q_{(\alpha)} = \frac{\partial}{\partial\theta^\alpha} + i(\bar{\theta}\gamma^\mu)_\alpha\partial_\mu \ \ (= C\bar{Q}) \ , \ \ \ X_{(\mu)} = \partial_\mu \ , \tag{1.10.4}$$

and satisfy the commutation ([,]) or anticommutation ({ , }) (*i.e.*, Z_2-*graded algebra commutation*) rules[*]

$$\{Q_{(\alpha)}, Q_{(\beta)}\} = 2i(C\gamma^\mu)_{\alpha\beta}\partial_\mu \ \ , \ \ \ [X_{(\mu)}, Q] = 0 \ . \tag{1.10.5}$$

It is simple to check that the supersymmetry algebra generators $Q_{(A)}$ are RI and that $\bar{\epsilon}Q : (x, \theta) \to (i\bar{\epsilon}\gamma^\mu\theta, \ \epsilon) \equiv \delta_\epsilon(x, \theta)$; thus, Q generates the left odd translation characterized by the constant Majorana spinor ϵ.

Similarly, the LIVFs $D_{(A)}$ are given by

$$D_{(\alpha)} = \frac{\partial}{\partial\theta^\alpha} - i(\bar{\theta}\gamma^\mu)_\alpha\partial_\mu \ \ , \ \ \ D_{(\mu)} = \partial_\mu = X_{(\mu)} \ . \tag{1.10.6}$$

When computing differentials, it is natural to adopt the convention that the differentials are placed to the left of the derivatives, *i.e.*,

$$d = dz^A\frac{\partial}{\partial z^A} \ \ , \ \ \ A = (\mu, \alpha) \ . \tag{1.10.7}$$

This is now especially convenient since some of the coordinates $z^A = (x^\mu, \theta^\alpha)$ are odd and hence anticommute. With this convention, (1.10.6) may be expressed as in (1.1.18) by exchanging the rôle of rows and columns as

$$\begin{bmatrix} D_{(\mu)} \\ D_{(\alpha)} \end{bmatrix} = \begin{bmatrix} \delta_\mu^\nu & 0 \\ -i(\bar{\theta}\gamma^\nu)_\alpha & \delta_\alpha^\beta \end{bmatrix} \begin{bmatrix} \partial_\nu \\ \partial_\beta \end{bmatrix} \ . \tag{1.10.8}$$

Similarly, the LIFs can be obtained, *e.g.* from (1.6.10):

$$[\Pi^{(\rho)}, \ \Pi^{(\gamma)}] = [dx^\nu, \ d\theta^\beta] \begin{bmatrix} \delta_\nu^\rho & 0 \\ i(\bar{\theta}\gamma^\rho)_\beta & \delta_\beta^\gamma \end{bmatrix} \ , \tag{1.10.9}$$

---

[*] In a *graded Lie group*, the generators associated with the odd parameters are consequently odd, and as a result its algebra includes anticommutators when the generators of two odd transformations are involved: it is a *graded Lie algebra*. It is easy to see how the graded structure ([even,even], [even,odd], {odd,odd}) comes about. If $X_{(i)}$ are the generators and α^i the parameters, the first order transformation is produced by the action of $\alpha^i X_{(i)}$ which is even irrespective of the grading of the parameter i since α^i and $X_{(i)}$ have the same grading. Including α^i, we are as in the ordinary Lie group case. If we now compute the *commutator* $[\alpha^i X_{(i)}, \beta^j X_{(j)}]$ we find $[\alpha^i X_{(i)}, \beta^j X_{(j)}] = \alpha^i X_{(i)}\beta^j X_{(j)} - \beta^j X_{(j)}\alpha^i X_{(i)} = \alpha^i\beta^j[(-1)^{ij}X_{(i)}X_{(j)} - X_{(j)}X_{(i)}]$. This gives $\alpha^i\beta^j[X_{(i)}, X_{(j)}]$ if i or j is even, and $-\alpha^i\beta^j\{X_{(i)}, X_{(j)}\}$ if both are odd.

$$\Pi^{(\rho)} = dx^\rho - i\bar{\theta}\gamma^\rho d\theta \quad , \quad \Pi^{(\gamma)} = d\theta^\gamma \; . \tag{1.10.10}$$

The duality condition $\Pi^{(A)}(D_{(B)}) = \delta^A_{\cdot B}$ $(A = \mu, \alpha)$ follows from the fact that the matrices in (1.10.8) and (1.10.9) are the inverse of each other (they may be understood as the (rigid superspace) 'supervielbein' $E_{(A)}^{\;\;B}$, which defines LI reference frames for $T(\Sigma)$, and its inverse $E_B^{(C)}$). The odd generators $D_{(\alpha)}$ in (1.10.8) satisfy the obvious relation

$$d = dz^A \frac{\partial}{\partial z^A} = \Pi^{(A)}(g) \circ D_{(A)} \tag{1.10.11}$$

since $E_{A\cdot}^{(B)} E_{(B)\cdot}^{\;\;C} = \delta_A^{\;C} .$*

As we saw, LIVFs and RIVFs commute for a Lie group. This property, when applied to supergroups, means that LIVFs and RIVFs commute or anticommute, the latter case appearing when the two elements in the bracket are odd. Thus $\{Q_{(\alpha)}, D_{(\beta)}\} = 0$ and the (chiral components of the) LI vector fields $D_{(\alpha)}$ can be used to impose constraints on the functions on superspace or *superfields* $\phi(x, \theta)$ that constitute the generalization to Σ of the relativistic fields $\phi(x)$ on Minkowski spacetime. These constraints are compatible with the group transformations since they commute/anticommute with the generators of the group action, which is taken to be generated by the RIVF $Q_{(\alpha)}$'s. In supersymmetric theories the chiral components of the $D_{(\alpha)}$'s are called *covariant derivatives* and the constraints they impose on the superfields are the *chirality conditions*.

The *supersymmetry algebra* (1.10.5) is the simplest superalgebra used in the description of $D = 4$ supersymmetry. Other extensions or generalizations of this algebra are also interesting in physics. The N-extended $(D = 4)$ super-Poincaré algebra contains $4N$ odd generators $Q^i_{(\alpha)}$ $(i = 1, ..., N)$ and may include $\binom{N}{2}$ central charges (cf. section 6.4; the case $N = 2$, which allows for one central charge, will be discussed in section 5.3). Other interesting examples are the $D = 4$ superconformal and the $D = 4$ superDe-Sitter algebras, but they will not be considered here.

1.11 Appendix A: some homotopy groups

We give below some examples of homotopy groups, a few of which have already appeared or will appear in the text. The k-th *homotopy group* $\pi_k(G)$

---

* When there are Grassmann (odd) variables, the order of the indices has to be taken into account. Here it will suffice to note that for the duality relation we take $dz^A\left(\frac{\partial}{\partial z^B}\right) = \Pi^{(A)}(D_{(B)}) = \delta^A_{\;B}$ and that the Kronecker delta satisfies $\delta^A_{\;B} = (-1)^{AB} \delta_B^{\;A}$. This means that $E_{A\cdot}^{(B)} E_{(B)\cdot}^{\;\;C} = \delta_A^{\;C} = E_{(A)\cdot}^{\;\;B} E_B^{(C)}$.

Table 1.11.1. Bott's periodicity table

k	$\pi_k(U(n))$, $n \geq (k+1)/2$	$\pi_k(O(n))$, $n \geq k+2$	$\pi_k(Sp(n))$, $n \geq (k-1)/4$
0	0	Z_2	0
1	Z	Z_2	0
2	0	0	0
3	Z	Z	Z
4	0	0	Z_2
5	Z	0	Z_2
6	0	0	0
7	Z	Z	Z
8	0	Z_2	0
period:	2	8	8

of G is the group of (classes of) continuous mappings of a k-sphere S^k into G under a certain composition law which generalizes the operation of the fundamental group $\pi_1(G)$. The first (*fundamental* or *Poincaré*) group $\pi_1(G)$ is always abelian and is finite for compact semisimple groups. The second homotopy group $\pi_2(G)$ of any Lie group is zero (and hence every continuous mapping of a two-sphere into G may be continuously deformed into a point). The homotopy groups $\pi_k(G)$ of the classical compact groups $U(n)$, $SO(n)$ and $Sp(n)$ ($Sp(n) \equiv SU(2n) \cap Sp(2n, C) \equiv USp(2n)$) become *stable* if n is sufficiently large; specifically (*and for $k \geq 2$*), $\pi_k(Sp) = \pi_k(Sp(n))$ for $n \geq (k-1)/4$; $\pi_k(U) = \pi_k(U(n)) \approx \pi_k(SU(n))$ for $n \geq (k+1)/2$ and $\pi_k(O) = \pi_k(SO(n))$ for $n \geq (k+2)$. These *stable homotopy groups* are given by Bott's periodicity theorem, which leads to Table 1.11.1. The stable range in which the periodicity appears is indicated in the top line. As read from the table, Bott's periodicity theorem states that, provided that the group is large enough, $\pi_k(U(n)) = (0, Z)$ for $k = 0 \pmod 2$, and $k = 1 \pmod 2$ respectively; $\pi_k(O(n)) = (Z_2, 0, Z)$ for $k = 0, 1 \pmod 8$, $k = 2, 4, 5, 6 \pmod 8$ and $k = 3, 7 \pmod 8$ respectively; and, finally, that $\pi_k(Sp(n)) = (0, Z, Z_2)$ for $k = 0, 1, 2, 6 \pmod 8$, $k = 3, 7 \pmod 8$, and $k = 4, 5 \pmod 8$ respectively. Notice that (removing the last line, which reproduces the first) the homotopy groups of $Sp(n)$ are the same as those of $O(n)$ shifted by four. Some physically interesting groups may fall outside the stable homotopy range: for instance, $\pi_3(SO(4)) = Z \oplus Z$ ($SO(4) \approx [SU(2) \otimes SU(2)]/Z_2$), $\pi_7(SO(8)) = Z \oplus Z$.

Let G *now* be $SO(n)$ $(n > 1)$, $\mathrm{Spin}(n)$ $(n > 2)$,[*] $U(n)$, $SU(n)$ $(n > 1)$, $Sp(n)$, or any of the exceptional groups G_2, F_4, E_6, E_7, E_8. The first few homotopy groups are the following: $\pi_1(G)$, the fundamental homotopy group, is given by

$$\pi_1(G) = Z \quad (G = U(n), SO(2)) \quad , \quad \pi_1(G) = Z_2 \quad (SO(n), n > 2)$$
$$\pi_1(G) = 0 \quad \text{(all others)} \quad .$$
$$(1.11.1)$$

The second is always zero, $\pi_2(G) = 0$, and $\pi_3(G) = Z$ provided G is different from $SO(2)$, $U(1)$, $SO(4)$ and $\mathrm{Spin}(4)$ (*i.e.*, provided G is simple). The fourth homotopy group, $\pi_4(G)$, is given by

$$\pi_4(G) = Z_2 \oplus Z_2 \quad (SO(4), \mathrm{Spin}(4)) \quad ;$$
$$\pi_4(G) = Z_2 \quad (SU(2), SO(3), SO(5), \mathrm{Spin}(3), \mathrm{Spin}(5), Sp(n)) \quad ;$$
$$\pi_4(G) = 0 \quad (SU(n), n > 2, SO(n), n > 5, \text{ and the exceptional groups}) \quad .$$
$$(1.11.2)$$

As a final example, we give $\pi_5(G)$:

$$\pi_5(G) = Z_2 \oplus Z_2 \quad (SO(4), \mathrm{Spin}(4)) \quad ;$$
$$\pi_5(G) = Z_2 \quad (Sp(n), SU(2), SO(3), SO(5), \mathrm{Spin}(3), \mathrm{Spin}(5)) \quad ; \quad (1.11.3)$$
$$\pi_5(G) = Z \quad (SU(n)\,(n \geq 3), SO(6), \mathrm{Spin}(6)) \quad ;$$
$$\pi_5(G) = 0 \quad (SO(n), \mathrm{Spin}(n)\,(n \geq 7); G_2, F_4, E_6, E_7, E_8) \quad .$$

Some isomorphisms between homotopy groups are worth noticing $(k \geq 2)$:

$$\pi_k(U(n)) \approx \pi_k(SU(n)) \,(n \geq 2) \quad ;$$
$$\pi_k(U(1)) \approx \pi_k(SO(2)) = 0 \quad ; \quad \pi_k(\mathrm{Spin}(n)) \approx \pi_k(SO(n)) \,(n \geq 3) \quad .$$
$$(1.11.4)$$

The above results will be useful in chapter 10 to discuss the topological aspects of anomalies.

Let us now list some *stable homotopy groups for the n-dimensional spheres* S^n. The homotopy groups $\pi_{n+k}(S^n)$ do not depend on n for $n \geq k + 2$ and become stable. For $k \leq 15$ they are given in Table 1.11.2. It follows from the table that $\pi_n(S^n) = Z$ (recall that $\pi_1(U(1)) = Z$). The table shows that for $l > n$, $\pi_l(S^n)$ is not zero in general; in contrast, $\pi_l(S^n) = 0$ for $l < n$ (*i.e.*, $\pi_{n+k}(S^n) = 0$ for $k < 0$). Other homotopy groups outside the stable range are $\pi_{k+1}(S^1) = 0$ for $k \geq 1$ (in agreement with $\pi_l(U(1)) = 0$ for

---

[*] $\mathrm{Spin}(3) = SU(2)$; $\mathrm{Spin}(4) = SU(2) \otimes SU(2)$; $\mathrm{Spin}(5) = USp(4)$ (unitary symplectic group in four dimensions or $Sp(2)$); $\mathrm{Spin}(6) = SU(4)$. For $n > 6$, $\mathrm{Spin}(n)$ is not isomorphic with a classical group. As groups, $U(n) \approx [SU(n) \otimes U(1)]/Z_n$ where Z_n is 'diagonal' (in the centre of both factors).

Table 1.11.2. Some stable sphere homotopy groups

k	$\pi_{n+k}(S^n)$	k	$\pi_{n+k}(S^n)$
0	Z	8	$Z_2 \oplus Z_2$
1	Z_2	9	$Z_2 \oplus Z_2 \oplus Z_2$
2	Z_2	10	Z_6
3	Z_{24}	11	Z_{504}
4	0	12	0
5	0	13	Z_3
6	Z_2	14	$Z_2 \oplus Z_2$
7	Z_{240}	15	$Z_{480} \oplus Z_2$

$l \geq 2$, eq. (1.11.4)); $\pi_3(S^2) = Z$; $\pi_4(S^2) = Z_2 = \pi_5(S^2)$; $\pi_5(S^3) = Z_2$, etc. The following isomorphism is the *Freudenthal suspension theorem*:

$$\pi_l(S^n) = \pi_{l+1}(S^{n+1}) \quad \text{for} \quad l \leq 2n - 2 \quad . \tag{1.11.5}$$

The computation of the homotopy groups is a difficult task in general. However, the exact homotopy sequence provides a useful tool in computing higher homotopy groups when the lower ones are known. Consider a principal bundle $P(G, M)$; then we may write

$$G \xrightarrow{i} P \xrightarrow{\pi} M \tag{1.11.6}$$

where i and π indicate inclusion and projection respectively. It is an important result of homotopy theory that this induces the following *(long) exact sequence of homotopy groups*:

$$\dots \to \pi_n(G) \to \pi_n(P) \to \pi_n(M) \to \pi_{n-1}(G) \to \pi_{n-1}(P) \to \pi_{n-1}(M) \to \dots \quad , \tag{1.11.7}$$

where the word exact indicates that the kernel of each mapping is the range of the previous one (exact sequences of groups will be used often in this book; they are introduced in section 4.1). The sequence (1.11.7) is very useful since it relates different homotopy groups of the total space P, fibre G and base M. For instance, if the bundle is trivial $P = G \times M$, and this implies that $\pi_n(P) = \pi_n(G) + \pi_n(M)$ (the homotopy group $\pi_n(M \times N)$ of the product space is isomorphic to the direct sum $\pi_n(M) + \pi_n(N)$; moreover, the Cartesian projections onto M, N induce projections of $\pi_n(M \times N)$ onto $\pi_n(M)$, $\pi_n(N)$). As an example of (1.11.7) consider the Hopf bundle $SU(2)(U(1), S^2)$ of section 1.3. A part of the homotopy sequence it gives rise to is

$$\pi_3(U(1)) \to \pi_3(SU(2)) \to \pi_3(S^2) \to \pi_2(U(1)) \to \pi_2(SU(2)) \to \pi_2(S^2)$$
$$\to \pi_1(U(1)) \to \pi_1(SU(2)) \to \pi_1(S^2) \quad , \tag{1.11.8}$$

i.e.,

$$0 \to Z \to Z \to 0 \to 0 \to Z \to Z \to 0 \to 0 \quad ,$$

which is indeed exact. The homotopy properties of $SU(2)$ also show that the Hopf bundle cannot be trivial; otherwise $\pi_1(SU(2))$ would be equal to $\pi_1(U(1)) + \pi_1(S^2)$ which is not the case. Clearly, the homotopy sequence is specially useful when it contains a short exact sub-sequence beginning and ending with zero since $0 \to A \to B \to 0$ implies that A and B are in one-to-one correspondence. For instance, consider $Z \to R \to U(1) \simeq S^1$. This implies that

$$0 = \pi_n(R) \to \pi_n(S^1) \to \pi_{n-1}(Z) \to \pi_{n-1}(R) = 0 \quad , \tag{1.11.9}$$

i.e. that $\pi_n(S^1) = \pi_{n-1}(Z)$. Since Z is discrete, $\pi_0(Z) = Z$ and $\pi_{j>0}(Z) = 0$; thus, (1.11.9) recovers the known previous results $\pi_1(S^1) = Z$, $\pi_n(S^1) = 0$ $(n > 1)$. A less trivial example follows from considering the bundle $SU(n+1)(SU(n), S^{2n+1})$ which gives

$$0 = \pi_2(S^{2n+1}) \to \pi_1(SU(n)) \to \pi_1(SU(n+1)) \to \pi_1(S^{2n+1}) = 0 \quad (n \geq 1) \quad . \tag{1.11.10}$$

This implies $\pi_1(SU(n)) = \pi_1(SU(n+1))$ and then $\pi_1(SU(n))$=const= 0 as stated in (1.11.1). Similarly, $U(n)(SU(n), U(1))$ gives

$$0 = \pi_1(SU(n)) \to \pi_1(U(n)) \to \pi_1(U(1)) \to \pi_0(SU(n)) = 0$$

since $SU(n)$ is connected; then $\pi_1(U(n)) = \pi_1(U(1)) = Z$. As a last simple example, consider $SO(n+1)(SO(n), S^n)$. It gives rise to

$$0 = \pi_3(S^n) \to \pi_2(SO(n)) \to \pi_2(SO(n+1)) \to \pi_2(S^n) = 0 \quad (n > 3) \tag{1.11.11}$$

so that again the homotopy group, here $\pi_2(SO(n))$, does not depend on n $(n > 3)$. Now, from $SU(2)/Z_2 \approx SO(3)$, we immediately obtain $\pi_2(SO(3)) = 0$, and from the fact that $SO(4)$ is the product of the rotation groups we get $\pi_2(SO(4)) = 0$. Since $\pi_2(SO(2)) = 0$ also, we get the result $\pi_2(SO(n)) = 0$, a particular case of the general one $\pi_2(G) = 0$ quoted above. A judicious use of parts of the homotopy sequence induced by a given bundle permits us to obtain many of the homotopy groups given above.

1.12 Appendix B: the Poincaré
polynomials of the compact simple groups

In this appendix we discuss *intuitively* the Poincaré polynomials for the simple groups. The *Poincaré polynomial* of an n-dimensional manifold M is defined as the polynomial in t

$$P(M, t) = \sum_{q=0}^{n} b_q(M) t^q \quad , \tag{1.12.1}$$

the coefficients of which are the Betti numbers of the manifold; thus, they contain the information about its cohomological (or homological) properties. Because of (1.5.6), the Poincaré polynomial of the Cartesian product $M_1 \times M_2$ is the product of the polynomials of M_1, M_2, $P(M_1 \times M_2) = P(M_1)P(M_2)$. Notice that, for $t = -1$, $P(M, -1) = \sum_{q=0}^{n}(-1)^q b_q(M)$, i.e. the Poincaré polynomial gives the Euler-Poincaré characteristic (1.5.3).

Consider first the case of the unitary groups $SU(n)$ (algebra A_{n-1}). Since $SU(2) \sim S^3$, and only $b_0(S^n) = 1 = b_n(S^n)$ are non-zero (section 1.5(a)), it follows that $P(SU(2)) = (1 + t^3)$. To find the Poincaré polynomial for higher dimensions we may proceed by recalling (Examples of principal bundles (a) in section 1.3) that $SU(n)/SU(n-1) \sim S^{2n-1}$. Thus, since $SU(3)/SU(2) \sim S^5$, $P(SU(3)) = (1 + t^3)(1 + t^5)$ and, continuing by induction ($SU(l+1) \sim [SU(l+1)/SU(l)] \times [SU(l)/SU(l-1)] \times ... \times [SU(2)/1]$), it is found that

$$P(SU(l+1)) = (1 + t^3)(1 + t^5)(1 + t^7)...(1 + t^{2l+1}) = \prod_{p=1}^{l}(1 + t^{2p+1})$$

$$(P(A_l) \; l \geq 1 \; ; \dim(A_l) = (l+1)^2 - 1) \quad . \tag{1.12.2}$$

The above argument does not imply *e.g.*, that $SU(3)$ is $S^3 \times S^5$; the bundles are non-trivial. But, as far as the cohomology is concerned, $SU(n)$ behaves as the product of spheres $S^3 \times S^5 \times S^7 \times ... \times S^{2n-1}$. Indeed, for $SU(3)$ the Poincaré polynomial is $(1 + t^3 + t^5 + t^8)$; this means that the nine Betti numbers $(b_0, ..., b_8)$ are $(1, 0, 0, 1, 0, 1, 0, 0, 1)$, and this is what is found by applying the Künneth formula (1.5.6) to $S^3 \times S^5$.

The case of the compact form $Sp(l)$ of the symplectic group in dimension $n = 2l$ (also written $USp(2l)$ and even $Sp(2l)$) follows similarly by induction based on the fact that $Sp(l)/Sp(l-1) = S^{4l-1}$ and $Sp(1) \approx SU(2) \sim S^3$. In this way one may obtain

$$P(Sp(l)) = (1 + t^3)(1 + t^7)(1 + t^{11})...(1 + t^{4l-1}) = \prod_{p=1}^{l}(1 + t^{4p-1})$$

$$(P(C_l) \; l \geq 3 \; ; \dim(C_l) = l(2l+1)) \quad . \tag{1.12.3}$$

This means that, again as far as the cycles are concerned, $Sp(l)$ behaves as the product of spheres $S^3 \times S^7 \times S^{11} \times ... \times S^{4l-1}$.

The derivation of the Poincaré polynomial for the orthogonal groups is more involved than the naïve consideration of $SO(n)/SO(n-1) \sim S^{n-1}$ might suggest[*]. We shall just quote here the results

$$P(SO(2l+1)) = (1+t^3)(1+t^7)(1+t^{11})...(1+t^{4l-1}) = \prod_{p=1}^{l}(1+t^{4p-1})$$

$$(P(B_l)\ l \geq 2\ ; \dim(B_l) = l(2l+1)) \quad , \tag{1.12.4}$$

$$P(SO(2l)) = (1+t^{2l-1})(1+t^3)(1+t^7)(1+t^{11})...(1+t^{4l-5})$$

$$= (1+t^{2l-1})\prod_{p=1}^{l-1}(1+t^{4p-1}) \tag{1.12.5}$$

$$(P(D_l)\ l \geq 4\ ; \dim(D_l) = l(2l-1)) \quad ,$$

$(A_1 \approx B_1 \approx C_1,\ B_2 \approx C_2,\ A_3 \approx D_3,\ D_2 \approx A_1 \times A_1)$. Thus, and without being precise about the meaning of the equivalence, the odd orthogonal groups $SO(2l+1)$ 'behave' as $S^3 \times S^7 \times S^{11} \times ... \times S^{4l-1}$, and the even ones $SO(2l)$ as $S^{2l-1} \times S^3 \times S^7 \times S^{11} \times ... \times S^{4l-5}$. For the covering groups of the orthogonal groups, the $\mathrm{Spin}(2l+1)$ and $\mathrm{Spin}(2l)$ groups, equal results hold since they 'differ' from the orthogonal ones by the finite fundamental group (the Betti numbers of a compact group and its covering group are the same).

For the exceptional groups, the dimensions of the 'primitive cycles' are

$$
\begin{aligned}
&G_2\ (14)\ :\ 3,\ 11 \\
&F_4\ (52)\ :\ 3,\ 11,\ 15,\ 23 \\
&E_6\ (78)\ :\ 3,\ 9,\ 11,\ 15,\ 17,\ 23 \\
&E_7\ (133)\ :\ 3,\ 11,\ 15,\ 19,\ 23,\ 27,\ 35 \\
&E_8\ (248)\ :\ 3,\ 15,\ 23,\ 27,\ 35,\ 39,\ 47,\ 59 \quad ,
\end{aligned}
\tag{1.12.6}
$$

where the dimension is indicated in brackets and is of course equal to the sum of the numbers at the right. Again intuitively we may find the result for G_2 by noticing that $\mathrm{Spin}(7)/G_2 = S^7$, but there is no simple procedure in general. The Poincaré polynomials follow from (1.12.6); for instance, $P(F_4) = (1+t^3)(1+t^{11})(1+t^{15})(1+t^{23})$.

The underlying structure in the above discussion is that the compact forms of simple groups behave like products of odd-dimensional spheres

---

[*] We note in passing that the above considered cosets may be jointly written in the form $G(n,r)/G(n-1,r) = S^{rn-1} \subset R^{rn}$ where $r = 1,2,4$ (reals, complex, quaternions) characterize the cases $SO(n)$, $SU(n)$ and $Sp(n)$.

from the point of view of real homology, *i.e.* ignoring finite groups or 'torsion' which will not be discussed here[†]. This is no problem since de Rham theory, being a cohomology theory with coefficients in R, necessarily ignores torsion effects (torsion does enter, however, in homotopy theory). We might have guessed that only odd-dimensional spheres should appear here since spheres of even dimension do not admit a single non-zero vector field, and Lie groups are parallelizable: no even-dimensional sphere can be a Lie group and, in fact, only S^1 and S^3 are groups (the proof is based on the fact that $G = S^n$ must be compact and semisimple for $n > 1$; then $b_3(S^n)$ non-zero requires $n = 3$). It follows as well from the above analysis that the Betti numbers $b_1(G)$, $b_2(G)$ of a simple compact group are zero and, moreover, that the Euler-Poincaré characteristic $\chi(G)$ of a Lie group[‡] is zero (for the cases above, $\chi(G) = 0$ is trivially found by setting $t = -1$). Also, looking at eqs. (1.12.2)–(1.12.6) we may find quickly the (infinite part of) the homotopy groups for the simple groups by recalling that $\pi_k(S^k) = Z$, as may be checked with the information contained in the previous appendix. This allows us to go outside the stable homotopy range: for instance, from (1.12.5) we obtain (here $n \not\geq k + 2$)

$$\pi_7(SO(8)) = \pi_{11}(SO(12)) = \pi_{15}(SO(16)) = ... = Z \oplus Z \quad . \tag{1.12.7}$$

Thus, the cohomology ring $H^*_{DR}(G)$ over R of a compact connected Lie group is isomorphic to the direct sum of the cohomology rings of l (= rank G) odd spheres, as shown by Hopf in 1941. A way of determining the l generators of this exterior algebra is, obviously, by looking at the differential forms on the group manifold G. For compact groups it turns out that the relevant generators of $H^*_{DR}(G)$ are the bi-invariant differential ones. This, in turn, is related to constructing invariant symmetric multilineal forms on $\mathscr{G}$, a problem which is closely related to the (generalized) Casimir invariants of Racah. We shall not discuss the Chevalley-Weil approach to this problem here, although we shall come back to it in section 6.7.

[†] Homology (H_q) and cohomology (H^q) groups depend on their coefficients; for instance, we may consider $H_q(M, R)$, $H_q(M, C)$ or $H_q(M, Z)$ (section 1.5a). When A is a field, the cohomology and homology groups are vector spaces dual to each other, $H^q(M, A) \approx H_q(M, A)$ and their dimension is the q-th Betti number (we take $A = R$ and omit R in $H^q(M)$). The finite part of the integer homology is the *torsion*, and it may be seen that torsion subgroups disappear when moving from the integer to the real homology. The $U(n)$, $SU(n)$ and $Sp(n)$ groups have no torsion. The torsion elements of the orthogonal groups are of order two; their covering groups Spin(n) have torsion for $n \geq 7$. The exceptional groups also have torsion.

[‡] Recall that any connected group is homeomorphic to the Cartesian product of a maximal compact subgroup and a subset homeomorphic with an Euclidean space (and therefore the non-compact part is of a trivial character). For instance, $GL(n, R)$ is homeomorphic to $O(n) \times R^{n(n+1)/2}$ and the connected Lorentz group to $SO(3) \times R^3$.

Bibliographical notes for chapter 1

There is a large amount of differential geometry books with special emphasis on physical applications. The book of Godbillon (1969) provides a brief introduction to the basic aspects of differential geometry and its applications to mechanics. A short and excellent review of many topics of differential geometry of relevance to modern physics can be found in the review paper by Eguchi *et al.* (1980). The paper by Ol'shanetskiĭ (1982) gives a list of definitions of mathematical concepts now used in mathematical physics. The books of Flanders (1963), Felsager (1981), Nash and Sen (1983), Von Westenholz (1986), Trautman (1984), Göckeler and Schücker (1987), Benn and Tucker (1987), Isham (1989), Nakahara (1990) and especially Choquet-Bruhat *et al.* (1982) may be used to complement some of the topics covered in this chapter. The treatise of Abraham and Marsden (1978) covers in detail the applications to mechanics and includes a self-contained mathematical introduction. At a more general level, the reader may look at the mathematical physics course by Thirring (1978, 1979), which also covers the case of classical field theory, and at Dubrovin *et al.* (1992). All these references are written with the applications to physics in mind. Among the purely mathematical literature see, *e.g.*, the books of Warner (1971) and Sternberg (1983). The book of Abraham *et al.* (1988) contains an extensive treatment of manifolds, differential calculus and integration which is an enlargement of the first part of Abraham and Marsden (1978); it has influenced some parts of our presentation. The classical text by Hodge (1941) is still useful reading. For details and applications of the equivariant cohomology just sketched at the end of section 1.5 the reader may look at Berline and Vergne (1982, 1983) and Atiyah and Bott (1984); see also Mathai and Quillen (1986) for the connection between the two approaches. Physical applications to field theory may be found in Duistermaat and Heckman (1982), Morozov *et al.* (1992) and Witten (1992a); see also Guillemin *et al.* (1996).

A classic reference on differential geometry and fibre bundles is the book of Kobayashi and Nomizu (1963). More specific (and advanced) references on fibre bundles are the books of Steenrod (1951), Husemoller (1966) and Hirzebruch (1966). The Hopf fibrings are discussed from a physical point of view in Trautman (1977), Boya *et al.* (1978), Boya and Mateos (1980), Minami (1980) and Grossman *et al.* (1989).

Standard references on the theory of Lie groups and Lie algebras are the textbooks of Hamermesh (1962), Hochschild (1965), Bacry (1967), Jacobson (1962), Pontryagin (1966), Humphreys (1972), Wybourne (1974), Gilmore (1974) and Barut and Raczka (1980); see also Kahan (1960). For an introduction to the differential geometry of Lie groups see the classic texts of Weyl (1946) and Chevalley (1946). The books of Gilmore (1974),

Greub *et al.* (1976) and Helgason (1978) (see also Barut and Raczka (1980)) contain a detailed discussion of homogeneous spaces. The book of Kirillov (1976) may be useful in connection with the theory of geometric quantization to be briefly discussed in chapter 3. For a further study of the invariant measures mentioned in section 1.7 the reader may look at the paper by Chern (1942) and at Santaló (1976). The invariant metrics on Lie groups are discussed in Milnor (1976); see also Lichnerowicz and Medina (1988). The Lie group structure of $T(G)$ is described in Milnor (1984). Modern texts covering different aspects of the differential geometry of Lie groups are the books of Helgason (1978), Warner (1971), and Gray (1992). A class of non-associative algebras which generalizes the Lie algebras, the Mal'čev algebras (not treated in the text), has sometimes appeared in physics in the context of anomalies (chapter 10); for an exposition see Sagle (1961).

There is already an extensive mathematical literature on supermanifolds, although it seems that a consensus has not been reached yet on the basic definitions. From the point of view of the physical applications, all modern quantum theory books include a treatment of Grassmann variables; see *e.g.* Ramond (1989), Zinn-Justin (1989), Brown (1992) or Bailin and Love (1993); the pioneering book by Berezin (1966) contains the first textbook discussion. The interested reader may look there to compensate for the all too brief remarks at the beginning of section 1.10. The mathematical theory of supermanifolds is discussed in the books of De Witt (1992) and Berezin (1987) and in the review of Leĭtes (1980), which contain further references; a brief account is given in Choquet-Bruhat (1989) and in Choquet-Bruhat and De Witt-Morette (1989). For the integration theory (only briefly used in chapter 10) the reader may also wish to read the classic paper of Berezin (1979) and the review by Voronov (1992). Lie superalgebras and groups are treated in the review of Corwin *et al.* (1975), in the books of De Witt (1992) and Berezin (1987) already mentioned, in Kac (1977) and in Cornwell (1989); the interested reader may find in them further references. For the Haar measure on Lie supergroups see Williams and Cornwell (1984). Further generalizations of Lie algebras are discussed by Rittenberg and Wyler (1978) and Scheunert (1979).

For an introduction to supersymmetry see, *e.g.*, Salam and Strathdee (1978), Sohnius (1985), West (1990) and Wess and Bagger (1991). For the geometry of supersymmetric theories the reader may look at Wess and Bagger (1991) and Gieres (1988).

Introductions to homotopy theory for physicists are given in, *e.g.*, Boya *et al.* (1978), Nash and Sen (1983), Nakahara (1990) and Morandi (1992). Basic purely mathematical texts are those of Hilton (1953) and Steenrod (1951), where the proof of the homotopy sequence can be found; see also Bott and Tu (1986) and Dubrovin *et al.* (1985). For the topology of classical groups see Hodge (1941), Weyl (1946), Pontryagin (1966), Samelson (1952), Serre (1953), Borel (1955) and Bott (1959); see also Chevalley and Eilenberg (1948), Milnor (1969), Greub *et al.* (1976) and Boya (1991). More information on the cohomology and homotopy properties of groups and spheres can be found in the text by Dubrovin *et al.* (1990) and in the mathematical tables edited by Itô (1987), which contain a wealth of data. A recent mathematical overview of homotopy theory is given in Whitehead (1983).

Morse theory (not treated in the text), which relates the number of critical points of a (Morse) function on a manifold with the Betti numbers of a manifold, has become important in theoretical physics. The interested reader may find an introduction in Bott and Tu (1986) and in Dubrovin *et al.* (1990); see Milnor (1969) for a detailed account. In connection with physics see Witten (1982a) (see also Henniart (1985)).

2

Connections and characteristic classes

This chapter is devoted to describing the elements of the theory of connections on principal bundles and a few of their physical applications, including the Yang-Mills theories, the Wu-Yang description of the monopole, and $SU(2)$-instantons. These, and further applications to gauge anomalies that will be made in chapter 10, motivate the sections dealing with characteristic classes. The chapter also includes an elementary discussion of some index theorems, of importance in the theory of gauge anomalies.

2.1 Connections on a principal bundle: an outline

A geometrical object such as a vector is moved by *parallel transport* on a manifold M if its components with respect to a parallelly transferred frame are kept constant. But how is the parallel transport of frames defined? We saw in section 1.3 that it is convenient to look jointly at the manifold M and at the set of all frames r_x (bases of $T_x(M)$) at the different points $x \in M$ as *the* bundle of linear frames $L(M)(GL(n,R),M)$. However, given the frame bundle $L(M)$, there is no canonical way of relating a certain frame r_x at a point $x \in M$ to another frame $r_{x'}$ at another point $x' \in M$ or, in other words, given a curve on the base manifold M, starting at $x \in M$, there is no intrinsic way of lifting it to a curve in $L(M)$ starting at a certain $p \in L(M)$ over x, $\pi(p) = x$. Since each point p of such a lifted curve, $p = (x, r_x)$ in local coordinates, would determine a frame at each $x = \pi(p)$, it is clear that the determination of the above set of frames (parallelly transported, by definition, from the initial one r_x) would allow us to define parallel transport of vectors, tensors, etc. But since the frame bundle $L(M)$ does not, by itself, provide us with a set of parallelly transported frames, it follows that an *additional* geometrical ingredient

84

needs to be given. This rôle is played by the connection which in this case is called a linear connection; there are as many parallel transport rules as connections can be given (of course, if the manifold is a Riemannian manifold, the Levi-Civita connection is singled out). But the problem we have sketched above, *i.e.* the problem of relating different fibres, is not restricted to the frame bundle $L(M)$. In its general form, it leads to the idea of connection on a principal bundle, the essentials of which will be discussed below.

Let $P(G, M)$ be a principal bundle. In it, each fibre is diffeomorphic to the typical fibre G, but the diffeomorphism is not canonical: it depends on the choice of the locally trivializing mappings Φ_α. Thus, to define parallel transport we have to compare neighbouring fibres in a way that does not depend on the local trivialization chosen. A connection provides such a correspondence between two fibres along a curve c on M: a point p of the fibre $\pi^{-1}(x)$ over x is then said to be transported by parallelism by means of this correspondence and the curve $\tilde{c}$ described by the parallel transport of c is called the *horizontal lift* of c. To introduce the horizontal lift of curves, it is sufficient to define its infinitesimal counterpart; this is achieved by associating to each tangent vector to M at x and to each point $p \in \pi^{-1}(x)$ a tangent vector to P at p, its *horizontal lift* at p. The horizontal lift $\tilde{c}$ through the point $p \in \pi^{-1}(x)$ of the curve c on M through x is then the integral curve of these horizontal vectors; the vectors tangent to $\tilde{c}$ are projected by $\pi^T : T_p(P) \to T_{\pi(p)}(M)$ onto vectors tangent to c. The above is, essentially, the original definition by Ehresmann of a connection on $P(G, M)$: clearly, the lifting and parallel transport have to be compatible with the differential structure defined on $P(G, M)$ and, in particular, with the action of G on P. This requires the lifting to be smooth and that, if $p' = pg \equiv R_g p$, the lift of a vector of $T_x(M)$ to $T_{p'}(P)$ be obtained from the lift to $T_p(P)$ by the tangent mapping R_g^T.

The horizontal lift of a basis of $T_x(M)$ to $T_p(P)$ determines a subspace $H_p(P) \subset T_p(P)$ which is called the *horizontal subspace* at p; clearly, $\pi^T(H_p(P)) = T_x(M)\ \forall p \in \pi^{-1}(x)$. The vectors of $H_p(P)$ 'point' from one fibre to another, and this justifies the name of horizontal vectors that is given to them. As has been stressed, $H_p(P)$ is not given by the bundle $P(G, M)$; it is obtained by the horizontal lift of $T_x(M)$, which is determined by the connection. There is a subspace in $T_p(P)$, however, which *is* determined by the bundle itself; it is the *vertical subspace* $V_p(P)$, the elements of which are the vectors $Y(p) \in T_p(P)$ which satisfy the condition $\pi^T(Y(p)) = 0$. Thus, $V_p(P) = ker\ \pi^T$, and its elements are tangent to the fibre $\pi^{-1}(x)$ at p. The notion of vertical vector is easily extended to

(2.1.1) DEFINITION (*Vertical vector field*)

A vector field $Y \in \mathscr{X}(P)$ *is called* vertical *or* tangential to the fibres *iff* $\pi^T(Y(p)) = 0 \; \forall p \in P$.

Clearly, if $f \in \mathscr{F}(M)$, $\pi^* f$ [(1.4.11)] is constant along the fibres. Thus, an alternative definition establishes that Y is vertical iff $Y.(\pi^* f) = 0$ $\forall f \in \mathscr{F}(M)$ since Y moves along the fibres. It follows immediately from this second definition that the space $V(P)$ of vertical vector fields is an $\mathscr{F}(P)$-submodule of the module $\mathscr{X}(P)$ of vector fields on P. Since, given $H_p(P)$, any vector $Y(p) \in T_p(P)$ admits a unique decomposition as the sum of its vertical and horizontal components, $Y(p) = Y^v(p) + Y^h(p)$, we can now give the following precise

(2.1.2a) DEFINITION (*Connection on a principal bundle*)

A connection on $P(G,M)$ *is a smooth assignment of a subspace* $H_p(P)$ *of* $T_p(P)$, *the* horizontal subspace *of* $T_p(P)$, *such that*

a) $T_p(P) = H_p(P) \oplus V_p(P)$ *(i.e., the horizontal subspace is complementary to the vertical subspace* $V_p(P)$ *at each* $p \in P$ *and the connection defines a unique splitting of* $T_p(P)$*), and*
b) $H_{pg}(P) = R_g^T H_p(P) \; \forall p \in P, \; \forall g \in G$.

Condition (b) guarantees that the family of horizontal subspaces $H_p(P)$ is defined in a way compatible with the group action, *i.e.* that it is possible to move from $H_p(P)$ to $H_{p'}(P)$ with p and $p' = pg$ on the same fibre by means of the linear mapping R_g^T.

Given a vector field X on M there is a *unique* vector field Y on P, its *horizontal lift*, such that X and Y are π-related [(1.2.1)] and $Y^v = 0$ (Y is horizontal); the vector field Y is invariant under G. Conversely, if a vector field Y on P satisfies $Y^v = 0$ *and* $R_g^T Y(p) = Y(pg) \; \forall p \in P$ and $\forall g \in G$ (right invariance of Y), then it is the horizontal lift of the vector field X on M determined by its π^T-projection. Thus, if Y is the horizontal lift of X, (a) $Y^v = 0$ and (b) Y is a *projectable* vector field, *i.e.* a vector field for which all the vectors at points in the same fibre have the same projection, $\pi^T(Y(pg)) = \pi^T(Y(p)) \; \forall g \in G$.[†] Clearly, although by definition the horizontal lift of a vector field is a projectable vector field, the converse is not true since a vector field which is tangential to the fibres (vertical) is projected to zero (see the Remark below).

It follows from this discussion that given a vector field X on M its horizontal lift is invariant under $R_g \; \forall g \in G$ and that, conversely, every

---

[†] This is tantamount to saying that a vector field is *projectable* iff $R_g^T X - X$ is vertical.

R_g-invariant horizontal vector field Y on P is the horizontal lift of a vector field X on M.

(2.1.1) PROPOSITION

The horizontal lift vector fields on P constitute a submodule of $\mathcal{X}(P)$ which is an $\mathcal{F}(M)$-module, where $\mathcal{F}(M)$ is identified with $\pi^*(\mathcal{F}(M)) \subset \mathcal{F}(P)$ (since π is a projection, π^* is an injective mapping). Moreover, if Y_1, Y_2 are the horizontal lifts of X_1, X_2, the horizontal part $[Y_1, Y_2]^h$ of their commutator is the horizontal lift of $[X_1, X_2]$.

Proof: Clearly, the horizontal lift of a sum is the sum of the horizontal lifts. Also, if Y is the horizontal lift of X and $f \in \mathcal{F}(M)$, $(\pi^* f)Y = (f \circ \pi)Y$ is the horizontal lift of fX. Finally, since $\pi^T([Y_1, Y_2]^h) = \pi^T([Y_1, Y_2])$ it follows from proposition 1.2.1 that $\pi^T([Y_1, Y_2]^h) = [\pi^T Y_1, \pi^T Y_2] = [X_1, X_2]$, q.e.d.

Remark

Let $\mathcal{P}(P)$ be the $\mathcal{F}(M)$-module of projectable (not necessarily horizontal) vector fields on P. Then π^T is a homomorphism of $\mathcal{P}(P)$ onto the Lie algebra $\mathcal{X}(M)$ of the vector fields on M. Thus, if Y_1, Y_2 are horizontal vector lifts, $[Y_1, Y_2]$ is projectable but not necessarily horizontal. We may summarize the properties of the different vector fields introduced by stating that

$$\{projectable\ vector\ fields\} = \{vertical\ vector\ fields,\ V(P)\} \oplus$$
$$\{horizontal\ lifts = (projectable\ vector\ fields)$$
$$\cap (horizontal\ vector\ fields)\} \ ;$$

the mapping $\mathcal{P}(P) \to \mathcal{X}(M)$ induced by the projection π is an $\mathcal{F}(M)$-homomorphism the kernel of which is $V(P)$; its restriction to the horizontal lift vector fields is a bijection.

From the previous discussion it follows that the rôle of a connection is to determine an equivariant splitting $T_p(P) = H_p(P) \oplus V_p(P)$ which depends differentiably on p; this permits selecting, among the vector fields projectable on a certain vector field on M, the horizontal vector field defining its horizontal lift. Thus, another possible way of defining a connection consists in extracting (again, in a way consistent with the right action of G) the vertical component of a vector $Y(P)$ since then $Y^h(p) = Y(p) - Y^v(p)$; $Y(p)$ will be characterized as horizontal when $Y^v(p) = 0$. This second definition is best given in terms of the connection form and requires the following

(2.1.2) PROPOSITION (*Fundamental vector fields*)

The right action of G on P defines a natural homomorphism of $\mathscr{G}$ into $\mathscr{X}(P)$; the vector field $Y^A \in \mathscr{X}(P)$ which is the image of the element $A \in \mathscr{G}$ is called the fundamental vector field on P associated with A, and $Y^A(p)$ is tangent to the fibre at each $p \in P$.

Proof: The action of G on P allows us to construct the associated vector fields which reproduce the algebra of $\mathscr{G}$. Since the action is effective (a free action always is), the homomorphism $\mathscr{G} \to \mathscr{X}(P)$ is injective; since it is free, the fundamental vector field Y^A corresponding to $A \in \mathscr{G}$, $A \neq 0$, does not vanish at any point, $Y^A(p) \neq 0$. Finally, since the action of G moves along the fibres, a fundamental vector field Y^A is vertical. Thus, there is a canonical isomorphism between $\mathscr{G}$ and $V_p(P)$: if $A_{(i)} \in \mathscr{G}$ is a basis for $\mathscr{G}$, $Y^A_{(i)}(p)$ is a basis for $V_p(P)$, *q.e.d.*

(2.1.3) PROPOSITION

The Lie bracket of a horizontal vector field Y and a fundamental vector field Y^A is horizontal.

Proof: By definition, $L_{Y^A}Y = [Y^A, Y]$ is given by the limit (1.4.44) in which ϕ represents the right action $R_{g(t)}$ of $g(t) \in G$. The fact that Y is horizontal implies that $R_g^T Y$ is also horizontal and hence (1.4.44) shows that $[Y^A, Y]$ is horizontal, *q.e.d.*

We are now in a position to give an alternative definition of a connection in terms of the connection form.

(2.1.2b) DEFINITION (*Connection form*)

A connection on P(G, M) is a linear mapping $\omega(p) : T_p(P) \to \mathscr{G}$, i.e. is a $\mathscr{G}$-valued one-form ω on P such that

a) ω is a smooth function of p,
b) $\omega(p)(Y^v(p)) = Y^A(p) = A$, where $Y^v(p) \in V_p(P)$ and A are related by the canonical isomorphism above,
c) $(R_g^* \omega(pg))Y(p) = \omega(pg)(R_g^T Y(p)) = Ad(g^{-1})\omega(p)(Y(p)) = Ad(g^{-1})A$.

The horizontal subspace $H_p(P)$ is then the kernel of the mapping $\omega(p) : T_p(P) \to \mathscr{G}$, and the connection form $\omega(p)$ is a projection of $T_p(P)$ onto $V_p(P) \sim \mathscr{G}$.

This definition is equivalent to Definition 2.1.2a; in fact, it is its 'dual' version. Property (c) guarantees that the decomposition of $T_p(P)$ into vertical and horizontal subspaces is consistent with the action of G, and the appearance of g^{-1} in $Ad(g^{-1})$ is due to the fact that it is a right action. In fact, the restriction of ω to a fibre $\pi^{-1}(x)$ defines a form which can be identified with the canonical form on G on account of the fact that $\pi^{-1}(x) \approx G$, and thus a reasoning similar to that of Proposition 1.7.1 applies here.

Let ω be a connection form on $P(G, M)$. Can we obtain a $\mathscr{G}$-valued form on the base manifold M from it? We can pull back ω to M by means of a cross section, but if the bundle is not trivial all we can obtain are the local representatives $\bar{\omega}_\alpha = \sigma_\alpha^*(\omega)$ which are $\mathscr{G}$-valued forms on the open sets of the covering $\{U_\alpha\}$ of M. The fact that all the $\bar{\omega}_\alpha$ come from the connection ω, and that conversely they should determine ω, imposes a compatibility condition on them which clearly has to involve the transition functions $g_{\alpha\beta}$ which relate the cross sections σ_α, σ_β on the overlapping domain of definition $U_\alpha \cap U_\beta \neq \emptyset$. The precise wording of this fact constitutes the content of the following fundamental result:

(2.1.1) THEOREM

Let $P(G, M)$ be a principal bundle endowed with a connection ω. The local $\mathscr{G}$-valued one-forms $\bar{\omega}_\alpha = \sigma_\alpha^*(\omega)$ or local representatives on $U_\alpha \subset M$ are subjected on $U_\alpha \cap U_\beta \neq \emptyset$ to the following compatibility condition:

$$\bar{\omega}_\beta = Ad(g_{\alpha\beta}^{-1})\bar{\omega}_\alpha + \theta_{\alpha\beta} , \tag{2.1.1}$$

where Ad is the adjoint representation of G and $\theta_{\alpha\beta}$ is the form on $U_\alpha \cap U_\beta$ induced from the canonical form θ on G by the transition function mappings $g_{\alpha\beta} : U_\alpha \cap U_\beta \to G$; $\theta_{\alpha\beta}$ can be written as $g_{\alpha\beta}^{-1} dg_{\alpha\beta}$.

Conversely, for each family $\{\bar{\omega}_\alpha\}$ of $\mathscr{G}$-valued forms subordinated to the covering $\{U_\alpha\}$ of M satisfying (2.1.1), there is a unique connection form ω on $P(G, M)$ such that $\sigma_\alpha^*(\omega) = \bar{\omega}_\alpha$.

Proof: Let $X \in T_x(U_\alpha \cap U_\beta)$. Then

$$\bar{\omega}_\beta(X) \equiv (\sigma_\beta^*\omega)(X) = \omega(\sigma_\beta^T X) = \omega((\sigma_\alpha g_{\alpha\beta})^T X) .$$

To evaluate the derivative of $\sigma_\beta(x) = \sigma_\alpha(x)g_{\alpha\beta}(x)$ [(1.3.5)] one uses the Leibniz formula for the product, which states that

$$(\sigma_\alpha g_{\alpha\beta})^T(X) = {}'\sigma_\alpha^T(X) \circ g_{\alpha\beta}(x) + \sigma_\alpha(x)^T g_{\alpha\beta}^T(X) , \tag{2.1.2}$$

where in the r.h.s. $g_{\alpha\beta}$ should be understood as the right action $R_{g_{\alpha\beta}(x)}^T$ of $g_{\alpha\beta}$ on $\sigma_\alpha^T(X)$, and $\sigma_\alpha(x)^T$ is the tangent of the mapping $\sigma_\alpha(x) :$ $G \to P$ $(\sigma_\alpha(x) : g_{\alpha\beta}(x) \mapsto \sigma_\alpha(x)g_{\alpha\beta}(x))^*$. Then, and adding subscripts and

---

* We can check the consistency of the expression (2.1.2) by looking at the rôle of the different mappings: $\sigma_\alpha^T : T_x(M) \to T_p(P)$ and thus $\sigma_\alpha^T(X) \circ g_{\alpha\beta} \in T_{pg}(P)$; $g_{\alpha\beta}^T : T_x(M) \to T_{g_{\alpha\beta}(x)}(G)$ and $\sigma_\alpha(x)^T : T_{g_{\alpha\beta}(x)}(G) \to T_{p'}(P)$.

arguments occasionally to keep track of them, we find

$$\bar{\omega}_\beta(X) = \omega_{p'}(\sigma_\beta^T X) = \omega_{p'}((\sigma_\alpha^T(X))R^T_{g_{\alpha\beta}(x)}) + \omega_{p'}((\sigma_\alpha(x))^T g^T_{\alpha\beta}(X)) \quad (2.1.3)$$

where $\sigma_\alpha(x) = p$, $\sigma_\beta(x) = pg_{\alpha\beta}(x) = p'$. The second term determines an element of $\mathscr{G}$ associated with the vector at p' determined by the fundamental vector field and so it is equal to $\theta_g(g^T_{\alpha\beta}(X))$. Thus, eq. (2.1.3) leads to

$$\begin{aligned}
\bar{\omega}_\beta(X) &= (R^*_g \omega_{p'})_p(\sigma_\alpha^T(X)) + \theta_g(g^T_{\alpha\beta}(X)) \\
&= Ad\, g^{-1}_{\alpha\beta}(x)(\omega_p(\sigma_\alpha^T(X))) + (g^*_{\alpha\beta}\theta_g)(X) \quad (2.1.4) \\
&= (Ad\, g^{-1}_{\alpha\beta}(x)\bar{\omega}_\alpha + \theta_{\alpha\beta})(X)
\end{aligned}$$

and, since X is arbitrary, (2.1.1) follows. It is worth repeating that the connection form is defined on P and that the local representatives $\bar{\omega}_\alpha$ are defined on $U_\alpha \subset M$.

The converse theorem can be proved by reversing the argument, *q.e.d.*

Expression (2.1.1) relates the local forms $\bar{\omega}_\alpha$, $\bar{\omega}_\beta$ on $U_\alpha \cap U_\beta$ and shows that in general it does not make sense to talk about the 'zero connection'. We can also follow the same reasoning for two local sections σ' and σ over the same U. Clearly, $\sigma'(x) = \sigma(x)g(x)$, and repeating the argument we would conclude that, on $U \subset M$,

$$\bar{\omega}'(x) = g^{-1}(x)\bar{\omega}(x)g(x) + g(x)^{-1}dg(x) = g^{-1}(x)(\bar{\omega}(x) + d)g(x). \quad (2.1.5)$$

In contrast with (2.1.1), which relates local representatives $\bar{\omega}_\alpha$, $\bar{\omega}_\beta$ on *different* U_α, U_β of M, this equation relates representatives on the *same* $U \subset M$ (if the bundle is trivial, (2.1.5) will be valid on M). Expression (2.1.5) shows that the transformation law consists of two parts; the first one is tensorial, but the second is not. This has, in fact, the same structure as the transformation law of the Christoffel symbols and the Yang-Mills potentials, the geometrical meaning of which will be discussed in the examples.

When is there a connection? The following theorem, which will be stated without proof, guarantees the existence of connections:

(2.1.2) THEOREM

On a principal bundle $P(G, M)$ with a paracompact base there exist infinitely many connections.

It is worth noticing that the space of connections is an *affine* space and not a vector space. Let ω_1 and ω_2 be two connection forms; thus, they satisfy condition (b) of Definition 2.1.2b. As a result, it can be immediately seen that $\omega_1 + \omega_2$ cannot be a connection; in fact, $\lambda_1\omega_1 + \lambda_2\omega_2$ can be a connection only if $\lambda_1 = 1 - \lambda_2$ because, if Y^A is a fundamental vector field, $\omega_1(Y^A) = A = \omega_2(Y^A)$ and only in this case does $\lambda_1 A + \lambda_2 A = A$ (see below). Thus, if ω_1 and ω_2 are connections, $\omega_1 + \lambda(\omega_2 - \omega_1)$ is a connection.

The nontensorial character of the transformation (2.1.1) is encompassed in the following general

(2.1.3) DEFINITION (*Pseudotensorial and tensorial forms*)

Let $P(G, M)$ be a principal bundle and D a representation of G on a vector space V. A pseudotensorial or equivariant form of degree q on P of type (D, V) is a V-valued q-form α on P such that

$$R_g^* \alpha = D(g^{-1})\alpha . \tag{2.1.6}$$

The form α is moreover called tensorial if it is horizontal, i.e. if $\alpha(Y^1, ..., Y^q) = 0$ if one or more of the tangent vectors to P is vertical. The inhomogeneous term in the transformation law (2.1.1) or (2.1.5) is a consequence of the fact that the connection form, which is pseudotensorial of type $(Ad, \mathscr{G})$, is not horizontal.

Let β be a tensorial one-form of adjoint type, i.e., $R_g\beta = Ad(g^{-1})\beta$ and $\beta(Y) = 0$ if Y is vertical. Then, if ω is a connection, $\omega' = \omega + \beta$ is a connection. In fact, given a connection form ω, any other connection can be obtained by adding to ω a tensorial one-form of type $(Ad, \mathscr{G})$.

(2.1.4) DEFINITION (*Exterior covariant derivative*)

Let α be a pseudotensorial q-form on P. The exterior covariant derivative $\mathscr{D}\alpha$ of α is given by

$$\mathscr{D}\alpha = d\alpha \circ h \quad , \tag{2.1.7}$$

where h has the effect of taking the horizontal part of the vectors that constitute the arguments of α (for instance, $(\beta \circ h)(X_1, X_2, ...) := \beta(X_1^h, X_2^h, ...))$. The operator $\mathscr{D}$ is called exterior covariant differentiation.

Clearly, since h takes the horizontal part of the arguments, $\mathscr{D}\alpha$ is a tensorial $(q + 1)$-form. In particular, we have the following

(2.1.5) DEFINITIONS (*Curvature, flat connection*)

Let ω be a connection form on $P(G, M)$. The $\mathscr{G}$-valued two-form $\Omega = \mathscr{D}\omega$ is called the curvature of ω; a connection is flat if $\Omega = 0$.

The curvature Ω is an $(AdG, \mathcal{G})$ tensorial form. Thus,

$$R_g^* \Omega = Ad(g^{-1})\Omega \ , \tag{2.1.8}$$

and $\Omega(Y_1, Y_2) = 0$ if one of the vectors is vertical. The definition of Ω implies that the curvature two-form satisfies the *Bianchi identity*

$$\mathscr{D}\Omega = 0 \tag{2.1.9}$$

since $\mathscr{D}\mathscr{D}\omega = 0$ (this follows from the structural equation below which gives $d\Omega(X, Y, Z) = 0$ whenever X, Y, Z are horizontal). Notice, however, that in contrast with $dd \equiv 0$, $\mathscr{D}\mathscr{D}$ is not zero in general.

(2.1.3) THEOREM (*Cartan structural equation*)

Let ω be a connection and Ω its curvature. Then, for an arbitrary pair of vector fields of $\mathscr{X}(P)$,

$$\Omega(Y_1, Y_2) = d\omega(Y_1, Y_2) + [\omega(Y_1), \omega(Y_2)] \ , \tag{2.1.10}$$

or, equivalently, since $[\omega(Y_1), \omega(Y_2)] = \frac{1}{2}[\omega, \omega](Y_1, Y_2)$ (see Definition 1.7.2),

$$\Omega = d\omega + \frac{1}{2}[\omega, \omega] = d\omega + \omega \wedge \omega \ . \tag{2.1.11}$$

Clearly, if the curvature is zero, (2.1.11) has the form of (1.7.15).

Proof: It will be sufficient to check (2.1.10) for the three possibilities for the arguments which can present themselves since both sides of eq. (2.1.10) are bilinear in them:

– both vector fields are horizontal
  Then $\omega(Y_1) = 0 = \omega(Y_2)$ and, since h acts trivially, $D\omega = d\omega$.

– both vector fields are vertical

Let us take them as the fundamental vector fields Y^A, Y^B. Clearly, the l.h.s. is zero. Also,

$$d\omega(Y^A, Y^B) = Y^A.\omega(Y^B) - Y^B.\omega(Y^A) - \omega([Y^A, Y^B]) \ . \tag{2.1.12}$$

The first two terms are zero because $\omega(Y^A)$ and $\omega(Y^B)$ are the constant Lie algebra elements A, B. Since $[Y^A, Y^B] = Y^{[A,B]}$ (Proposition 2.1.2) we now have $-\omega([Y^A, Y^B]) = -\omega(Y^{[A,B]}) = -[A, B]$ which cancels the last term in (2.1.10).

– Y_1 is horizontal and Y_2 is vertical

The l.h.s. is again zero, and the r.h.s reduces to $d\omega(Y_1, Y_2)$ since $[\omega(Y_1), \omega(Y_2)] = 0$ because $\omega(Y_1) = 0$. Now, proceeding as for (2.1.12), we find that now the only term in $d\omega$ not obviously zero is $\omega([Y_1, Y^4])$. But the commutator of a horizontal vector field and the fundamental vector field is horizontal (Proposition 2.1.3) and hence this term is also zero, *q.e.d.*

Expression (2.1.11) also shows that, on a connection ω,

$$\mathscr{D}\omega = d\omega + \omega \wedge \omega \quad . \tag{2.1.13}$$

It was already mentioned that the commutator of two horizontal vector fields is not necessarily horizontal. The following theorem relates the failure of this commutator to remain horizontal to the curvature:

(2.1.4) THEOREM

Let $P(G, M)$ be a principal bundle endowed with a connection ω. Then the following three statements are equivalent:

a) the connection is flat, $\Omega = 0$;
b) the commutator of two horizontal vector fields is a horizontal vector field;
c) for each $x \in M$ there exists an open neighbourhood U and a local section σ such that $\sigma^*\omega \equiv \bar{\omega} = 0$.

Proof:

a) $\Rightarrow$ b): Using eqs. (2.1.10), (2.1.12) we get $\Omega(Y_1, Y_2) = -\omega([Y_1, Y_2]) = 0$ since $\Omega = 0$ and the vector fields are horizontal. Thus, $[Y_1, Y_2]$ is also horizontal.

b) $\Rightarrow$ c): The statement b) means that the differential system given by the connection is stable under the Lie bracket and hence integrable. Thus, each point $p \in P$ is contained in an integral manifold of dimension dim $N = $ dim $H_p(P) = $ dim M. For this integral manifold, the canonical projection π is a one-to-one mapping of N into an open neighbourhood U of x that allows us to construct the local section $\sigma : U \to P$, $\sigma^T(X_x) \in H_p(P)$. Then $0 = \omega(\sigma^T X) = (\sigma^*\omega)(X)$ from which it follows that $\bar{\omega} = 0$.

c) $\Rightarrow$ a): The pull-back of (2.1.11) to U by σ gives $\bar{\Omega} = d\bar{\omega} + \frac{1}{2}[\bar{\omega}, \bar{\omega}] = 0$. Thus, $\Omega(Y_{1p}, Y_{2p}) = 0$ if the two vectors Y_{1p}, Y_{2p} at $p \in \sigma(x)$ are horizontal. But, since Ω is tensorial, this is also true if any of the vectors is vertical. Thus, $\Omega(Y_p, Y'_p) = 0$ for any two vectors at p and hence Ω is zero at p. But this is also true at any pg, $g \in G$; since it is also true for any $p = \sigma(x)$, $x \in U$, it follows that $\Omega = 0$, *q.e.d.*

The *exterior covariant derivative of a tensorial form* α of type (D, V) is given by

$$\mathscr{D}\alpha = d\alpha + \rho(\omega) \wedge \alpha \quad , \tag{2.1.14}$$

where ρ is the representation of $\mathscr{G}$ associated with D and $\rho(\omega)$ means $\omega^i \rho(X_{(i)})$ where $\rho(X_{(i)})$ represents the action of $X_{(i)}$ on the vector space V. If $\alpha = \alpha^s e_s$, $s = 1, ..., \dim V$, eq. (2.1.14) reads

$$\mathscr{D}\alpha^s = d\alpha^s + \rho_{(i)}{}^s{}_t \omega^i \wedge \alpha^t \tag{2.1.15}$$

and, under $g \in G$,

$$(\mathscr{D}\alpha)' = D(g^{-1})(\mathscr{D}\alpha) \quad . \tag{2.1.16}$$

For instance, let α be a tensorial p-form of type $(Ad, \mathscr{G})$. Then $(adX_{(i)})(X_{(j)}) = [X_{(i)}, X_{(j)}]$, and equation (2.1.14) above reads, using (1.7.10),

$$\mathscr{D}\alpha = d\alpha + [\omega, \alpha] = d\alpha + \omega \wedge \alpha + (-1)^{p+1}\alpha \wedge \omega \quad , \tag{2.1.17}$$

and $(\mathscr{D}\alpha)' = g^{-1}(\mathscr{D}\alpha)g$. Eq. (2.1.17) gives the *action of $\mathscr{D}$ on an Ad-tensorial p-form* (cf. (2.1.13)). If α is a one-form, this expression may also be written in a way similar to the structural equation (2.1.10),

$$(\mathscr{D}\alpha)(Y_1, Y_2) = d\alpha(Y_1, Y_2) + [\omega(Y_1), \alpha(Y_2)] + [\alpha(Y_1), \omega(Y_2)] \quad , \tag{2.1.18}$$

and can be checked by looking at the three cases for (Y_1, Y_2) also considered there. If, for instance, $\alpha = \Omega$, eq. (2.1.17) reproduces the Bianchi identity since

$$\mathscr{D}\Omega = d\Omega + [\omega, \Omega] = d\Omega + \omega \wedge \Omega - \Omega \wedge \omega = 0 \tag{2.1.19}$$

on account of (2.1.11). In contrast, eq. (2.1.14) shows that $(\rho(\omega) \wedge \rho(\omega) = \frac{1}{2}\rho([\omega, \omega]))$

$$\mathscr{D}\mathscr{D}\alpha = \rho(d\omega) \wedge \alpha + \rho(\tfrac{1}{2}[\omega, \omega]) \wedge \alpha = \rho(\Omega) \wedge \alpha \tag{2.1.20}$$

which is non-zero unless the curvature is zero. In particular, if α is of type $(Ad, \mathscr{G})$,

$$\mathscr{D}\mathscr{D}\alpha = [d\omega + \tfrac{1}{2}[\omega, \omega], \alpha] = [\Omega, \alpha] \tag{2.1.21}$$

(*i.e.*, on a tensor quantity, $[\mathscr{D}_\mu, \mathscr{D}_\nu]\alpha = [\Omega_{\mu\nu}, \alpha]$).

If α and β are tensorial p- and q-forms respectively, $\alpha \wedge \beta$ is a tensorial $(p + q)$-form, and

$$\mathscr{D}(\alpha \wedge \beta) = (\mathscr{D}\alpha) \wedge \beta + (-1)^p \alpha \wedge (\mathscr{D}\beta) \tag{2.1.22}$$

(Leibniz's rule) which shows that $\mathscr{D}$ is a skew-derivation on the algebra of tensorial forms.

The above formulae allow us to compute the exterior covariant derivative of the forms satisfying (2.1.6). This does not yet solve the problem of taking the covariant derivatives of sections of associated vector bundles of fibre V, a subject that we shall not treat here (in physics, the matter fields are sections of associated vector bundles). But, due to Proposition 1.3.1, these sections are in one-to-one correspondence with the V-valued G-functions on $P(G, M)$, which in turn may be considered as zero-forms. Then, if $\alpha = \psi$ is a zero-form of type (D, V), $\rho(X_{(i)}) = (T_{(i)})^t._s$, the covariant derivative can be defined by

$$\mathscr{D}\psi^s = d\psi^s + \omega^{(i)}(T_{(i)})^t._s\psi^s \quad (\mathscr{D}_\mu = \partial_\mu + \omega^{(i)}_\mu T_{(i)}) \quad . \tag{2.1.23}$$

If the representation is the trivial one, $T = 0$ and $\mathscr{D} = d$.

2.2 Examples of connections

Let us consider some explicit examples of the theory sketched in the previous section.

(a) *Linear connections*
A linear connection on a manifold M is a connection on $L(M)(GL(n, R), M)$, the frame bundle of an n-dimensional manifold M. Let $g^i._j$ $(i, j = 1, ..., n)$ be a local coordinate system on $GL(n, R)$. The canonical form on $GL(n, R)$ is then given by

$$\theta = (g^{-1})^i._k\, dg^k._j \otimes e^j._i \tag{2.2.1}$$

(see (1.9.6)), where $e^i._j$ is the natural basis of $\mathscr{G}l(n, R) \sim T_e(GL(n, R))$. Since $\mathscr{G}l(n, R)$ may be identified with $L(R^n, R^n)$ (and with real $n \times n$ matrices in an obvious way) the basis $e^i._j$ may be given, e.g., by the matrices $(e^i._j)^k._l = \delta^{ik}\delta_{jl}$; $[e^i._j, e^s._t] = \delta^s_j e^i._t - \delta^i_t e^s._j$. Let $\{x^i_\alpha\}$ and $\{x^j_\beta\}$ $(i, j = 1, ..., n)$ be two coordinate systems on U_α, $U_\beta \subset M$, $U_\alpha \cap U_\beta \neq \emptyset$. On the overlapping domain of the two charts, we have $x^i_\alpha = x^i_\alpha(x^j_\beta)$. From the relation between the two bases at $T_x(U_\alpha)$, $T_x(U_\beta)$,

$$\frac{\partial}{\partial x^j_\beta} = \frac{\partial x^i_\alpha}{\partial x^j_\beta}\frac{\partial}{\partial x^i_\alpha} \equiv (g_{\alpha\beta})^i._j\frac{\partial}{\partial x^i_\alpha} \quad , \tag{2.2.2}$$

we read the expression for the transition functions, $(g_{\alpha\beta})^i._j = \partial x^i_\alpha/\partial x^j_\beta$. Hence,

$$(\theta_{\alpha\beta})^i._j = (g^*_{\alpha\beta}\theta)^i._j = (g^{-1}_{\alpha\beta})^i._k(dg_{\alpha\beta})^k._j = \frac{\partial x^i_\beta}{\partial x^k_\alpha}\frac{\partial^2 x^k_\alpha}{\partial x^l_\beta \partial x^j_\beta}dx^l_\beta \quad . \tag{2.2.3}$$

The local representatives $\bar{\omega}_\alpha$ are given directly by the Christoffel symbols,

$$\bar{\omega}_\alpha = \Gamma^i_{kj}\, dx^k_\alpha\, e^j._i \quad , \tag{2.2.4}$$

and thus the relation (2.1.5) leads, putting primes for the β and no primes for the α, to

$$\Gamma'^i_{lj} = \frac{\partial x'^i}{\partial x^m} \Gamma^m_{kn} \frac{\partial x^k}{\partial x'^l} \frac{\partial x^n}{\partial x'^j} + \frac{\partial x'^i}{\partial x^k} \frac{\partial^2 x^k}{\partial x'^l \partial x'^j} , \qquad (2.2.5)$$

which is the familiar transformation law for the Christoffel symbols which exhibits an inhomogeneous term besides the (first) tensorial one. It is worth mentioning, as clearly seen in (2.2.4), that although all indices in this example take values from 1 to n (the dimension of the manifold), the indices i, j in Γ^i_{kj} are group or algebra (fibre) indices, while the index k is the base manifold coordinates index; we could in general write $(\Gamma_k)^i_j$ to stress this point.

Let M be a Riemannian manifold with metric g. We can use the metric to introduce the bundle of oriented *orthonormal frames* on M, $O(M)(SO(n, R), M)$. Clearly, there is a bundle homomorphism (Definitions 1.3.5) $f : O(M)(SO(n, R), M) \to L(M)(GL(n, R), M)$ for which f_{gr} is injective and f_b the identity; thus, the bundle of orthonormal frames is a *reduction* or *restriction* of the bundle of linear frames. It is now intuitively clear that the morphism f maps* every connection in $O(M)$ into a connection in $L(M)$: a *metric connection* is determined by a connection on $O(M)$. Every Riemannian manifold admits a unique metric connection such that the torsion is zero, *i.e.* a *Riemannian connection*.

(b) Yang-Mills connections

b_1) $P(G, M)$ trivial

Let M now be spacetime and G a compact group; let us assume initially that M is topologically trivial (contractible) and hence that $P(G, M)$ is trivial (see the first remark in section 1.3). Yang-Mills fields are introduced in physics as solutions of the corresponding equations, which have to be given beforehand (see sections 2.7 and 2.8). Looking at the familiar gauge transformation properties of a $\mathcal{G}$-valued Yang-Mills field $A_\mu(x)$ on

---

* What follows is a particular example of the following situation. Let $f : P(G, M) \to P'(G', M')$ be a bundle homomorphism such that f_{gr} is injective and $f_b = \mathbf{1}$ $(M = M')$, and let the connection ω be mapped into ω'. Then it is said that the connection form ω' on P' is *reducible* to the connection ω on P. The action of the mapping f on ω is understood in the following sense: given ω, there is a unique connection ω' on P' such that $f^*\omega' = f \circ \omega$, where $(f \circ \omega)(Y) = f^T_{gr}(\omega(Y))$ and f^T_{gr} is the algebra homomorphism $f^T_{gr} : \mathcal{G} \to \mathcal{G}'$ induced by $f_{gr} : G \to G'$ (see Def. 1.3.5b)); clearly, f maps horizontal subspaces into horizontal subspaces. For the curvature $f^*\Omega' = f \circ \Omega$, and $f \circ \Omega$ is defined similarly. Note, however, that it is not necessary that f be a reduction to define a mapping of connections.

spacetime M,

$$A'_\mu(x) \equiv A^g_\mu(x) = g^{-1}(x)A_\mu(x)g(x) + g^{-1}(x)\partial_\mu g(x) = g^{-1}(A_\mu + \partial_\mu)g \quad,$$
$$A_\mu(x) = A^i_\mu(x)T_i \quad, \quad T_i \in \mathscr{G} \quad (i = 1, ..., r = \dim \mathscr{G}) \quad,$$
$$\text{(2.2.6)}$$

where for definiteness we take the representation matrices of $\mathscr{G}$ to satisfy

$$T_i = -T_i^\dagger \,, \ [T_i, T_j] = C^k_{ij}T_k \quad, \ \text{Tr}(T_iT_j) = -\frac{1}{2}\delta_{ij} \,, \ A^i_\mu = -2\text{Tr}(T^iA_\mu)\,,$$
$$\text{(2.2.7)}$$

it is seen that the *Yang-Mills or potential fields* can be interpreted as a $\mathscr{G}$-valued connection on the principal bundle $P(G, M)$ and that eqs. (2.2.6) correspond to eqs. (2.1.5) with $\bar{\omega}(x) = A(x) \equiv A^i_\mu(x)dx^\mu T_i$. Notice that, although $A^i_\mu = A^{i\,\dagger}_\mu$, $A^\dagger_\mu = -A_\mu$ because $T_i = -T_i^\dagger$ (different representation matrices may be used; if the normalization changes, the last two equations in (2.2.7) have to be modified accordingly). The group of elements $g(x)$ associated with the structure group G corresponds to the *gauge group*. The cross sections (which are global since we have assumed $P(G, M)$ trivial) are used to define the Yang-Mills potentials $A(x)$ on spacetime, and accordingly may be interpreted as the specific *gauges* in which the potentials are given. The functions $g(x)$, or *gauge functions*, are elements of the (infinite-dimensional) group $G(M)$ of mappings $g : M \to G$ (section 9.2), and define the transition functions between two different cross sections or gauges. Specifically, eq. (2.2.6) (eq. (2.1.5)) gives for a transformation $g(x) = exp \, \zeta^i(x)T_i \simeq 1 + \zeta^i(x)T_i$, to first order in the *infinitesimal gauge transformation** functions $\zeta^i(x)$,

$$A'^i_\mu - A^i_\mu \simeq \delta_\zeta A^i_\mu = \partial_\mu \zeta^i(x) + C^i_{jk}A^j_\mu(x)\zeta^k(x) = (\mathscr{D}_\mu)^i_j\zeta^j(x)\,, \quad \text{(2.2.8)}$$

since here

$$(\mathscr{D}_\mu)^i_j = \delta^i_j\partial_\mu + C^i_{kj}A^k_\mu(x) \quad\quad\quad\quad \text{(2.2.9)}$$

---

* Usually, the gauge transformation includes the coupling constant λ explicitly (if the group is simple, there can only be one coupling constant; Yang-Mills theories starting with more than one coupling constant are based on non-simple groups as is the case for the $SU(2) \otimes U(1)$ standard model). If the matter fields (sections of the associated vector bundle) transform as $\psi'(x) = g^{-1}(x)\psi(x)$, and the covariant derivative of ψ is given by $\mathscr{D}_\mu\psi = (\partial_\mu + \lambda A_\mu)\psi$ (eq. (2.1.23) but for the factor λ), then $\delta_\zeta A^i_\mu(x) = \frac{1}{\lambda}\partial_\mu\zeta^i(x) + C^i_{jk}A^j(x)\zeta^k(x)$ leads to $(\mathscr{D}_\mu\psi)' = g^{-1}(x)(\mathscr{D}_\mu\psi)$. Explicitly, $(\mathscr{D}\psi)' = (\partial_\mu + \lambda A'_\mu)\psi' = (\partial_\mu + \lambda(g^{-1}A_\mu g + \frac{1}{\lambda}g^{-1}\partial_\mu g))g^{-1}\psi = (-g^{-1}\partial_\mu g g^{-1} + g^{-1}\partial_\mu + \lambda g^{-1}A_\mu + g^{-1}\partial_\mu g g^{-1})\psi = g^{-1}(\mathscr{D}_\mu\psi)$. We shall set $\lambda = 1$ throughout or, what amounts to the same thing, absorb λ in the definition of $A_\mu(x)$.

Due to the transformation $\psi'(x) = g^{-1}(x)\psi(x)$, the gauge group is sometimes referred to as the *local gauge group*; the word local is *here* used in a physical sense and just expresses the freedom of choosing the phase $\zeta(x)$ at each spacetime point x (see the last footnote in section 9.1). The precise definition of gauge group is given in Def. 10.1.1.

and C^i_{kj} may be understood as the matrix $(adX_{(k)})^i{}_j$ (under constant transformations, gauge fields transform according to the adjoint representation). Eq. (2.2.8) may be rewritten as

$$\delta_\zeta A_\mu(x) = \partial_\mu \zeta(x) + [A_\mu(x), \zeta(x)] \quad , \tag{2.2.10}$$

where we have introduced $\zeta(x) \equiv \zeta^i(x) T_i$. The $\mathscr{G}$-valued functions $\zeta(x)$ are given by mappings $\zeta : M \to \mathscr{G}$; they define the infinite-dimensional Lie algebra of the infinitesimal gauge transformations (see section 9.2). In terms of the one-form A, the previous formulae read

$$A' \equiv A^g = g^{-1}(d + A)g \quad , \quad \delta_\zeta A = d\zeta + [A, \zeta] \equiv \mathscr{D}\zeta \quad . \tag{2.2.11}$$

The curvature of the potentials $A(x)$ gives the *field strength* F on the spacetime. Writing $F = \frac{1}{2} F_{\mu\nu} dx^\mu \wedge dx^\nu$, $F = F^i T_i$, the field components of

$$F = \mathscr{D}A = dA + A \wedge A \tag{2.2.12}$$

(eq. (2.1.13)) are given by

$$F^i_{\mu\nu} = -2\mathrm{Tr}(T^i F_{\mu\nu}) \quad , \quad F^i_{\mu\nu} = \partial_\mu A^i_\nu - \partial_\nu A^i_\mu + C^i_{jk} A^j_\mu A^k_\nu \quad . \tag{2.2.13}$$

The field strength behaves tensorially (eq. (2.1.8)) under a gauge transformation,

$$F^g = dA^g + A^g \wedge A^g = g^{-1}Fg \quad , \quad F'_{\mu\nu}(x) = g^{-1}(x)F_{\mu\nu}(x)g(x) \quad . \tag{2.2.14}$$

Then the variation $\delta_\zeta F$ of F under an infinitesimal gauge transformation is given by

$$\delta_\zeta F(x) = [F(x), \zeta(x)] \quad , \quad \delta_\zeta F^i_{\mu\nu} = C^k_{ij} F^i_{\mu\nu} \zeta^j \quad . \tag{2.2.15}$$

The two-form F satisfies the Bianchi identities (2.1.19),

$$\mathscr{D}F \equiv dF + A \wedge F - F \wedge A = dF + [A, F] = 0 \quad (dF = -[A, F]) \quad . \tag{2.2.16}$$

For $G = U(1)$ (ordinary electromagnetism) this reduces to $dF = 0$. In components eq. (2.2.16) gives $(\partial_\mu F^i_{\nu\rho} + C^i_{jk} A^j_\mu F^k_{\nu\rho}) dx^\mu \wedge dx^\nu \wedge dx^\rho = 0$, i.e.

$$\partial_{(\mu} F^i_{\nu\rho)} + C^i_{jk} A^j_{(\mu} F^k_{\nu\rho)} = (\delta^i_k \partial_{(\mu} + C^i_{jk} A^j_{(\mu)}) F^k_{\nu\rho)} = (\mathscr{D}_{(\mu} F_{\nu\rho)})^i = 0 \, , \tag{2.2.17}$$

where the bracket in $(\mu\nu\rho)$ indicates sum over cyclic permutations. Since this expression is antisymmetric in μ, ν, ρ, it is empty in two dimensions. Using the dual tensor $*F = \frac{1}{2} \mathscr{F}_{\rho\sigma} dx^\rho \wedge dx^\sigma$, $\mathscr{F}_{\rho\sigma} := \frac{1}{2} \epsilon_{\rho\sigma\mu\nu} F^{\mu\nu}$, the Bianchi identity reads

$$\mathscr{D}_\mu \mathscr{F}^{\mu\nu} = 0 \quad ; \tag{2.2.18}$$

it also follows as a consequence of the Jacobi identity for $\mathscr{D}_\mu$,

$$[\mathscr{D}_{(\mu}, [\mathscr{D}_\nu, \mathscr{D}_{\rho)}]] = 0 \quad . \tag{2.2.19}$$

In both Maxwell ($G = U(1)$) and general Yang-Mills theories, the Bianchi identities do not involve the source current j. For instance, in ordinary electromagnetism (no magnetic monopoles) the Bianchi identity ($dF = 0$ or $\partial_{[\mu} F_{\rho\sigma]} = 0$) includes Faraday's law and the statement that the magnetic field is divergence free. The electric source current appears in the other two equations which, as in Yang-Mills theories, involve the Hodge dual operator defined with the help of the (pseudo-)Riemannian metric on spacetime. It is interesting to note that, much in the same way as the energy-momentum tensor has no immediate geometric interpretation in the Einstein equations where it appears as the matter source for the gravitational field (although it may have a rôle from the point of view of the symplectic structure of the manifold of solutions in lower dimensional theories), the source current of Yang-Mills theories also has no obvious geometric meaning*.

b₂) $P(G, M)$ *not trivial*

In general, the principal bundle $P(G, M)$ will not be trivial and it will not be possible to describe a connection by a single representative ($\mathscr{G}$-valued Yang-Mills field A) on M. In this case eq. (2.1.1) applies with $\bar{\omega}_\alpha = A_\alpha$, the Yang-Mills potential on U_α, etc.,

$$A_\beta = g_{\alpha\beta}^{-1}(d + A_\alpha)g_{\alpha\beta} \quad , \tag{2.2.20}$$

and the gauge transformations are locally represented by functions $g_{\alpha\beta}(x)$ that satisfy the cocycle or gluing conditions $g_{\alpha\beta}(x)g_{\beta\gamma}(x) = g_{\alpha\gamma}(x)$ on $U_\alpha \cap U_\beta \cap U_\gamma$.

A particularly interesting case of non-triviality may arise when the base manifold is a sphere. Physically, this may appear in field theory when the fields have suitable boundary conditions at infinity, which effectively produces the one-point compactification of (Euclidean) spacetime into a sphere. Since any sphere S^n can be covered by two charts $U_\pm$ excluding the south and north poles, a gauge field configuration can be characterized by a single transition function g_{-+} relating the local Yang-Mills potentials

---

* The theory of gravitation can also be considered as a gauge theory where the gauge group is the Lorentz group, although the correspondence between the physical and geometrical entities does not completely parallel the Yang-Mills case. Both theories are the result of applying the dynamical principle of gauge invariance (called of *general covariance* in the case of gravitation). This requires replacing the ordinary derivatives by the corresponding covariant ones to achieve covariance under general coordinate or gauge transformations. The Yang-Mills gauge group, the Yang-Mills potentials and the field strengths have their analogues in the group of general coordinate transformations, the Christoffel symbols and the spacetime curvature. However, in general relativity the connection coefficients are given in terms of the Riemannian metric. Thus, it is the metric (and not the connection itself) that primarily describes the gravitational field. In contrast, the Yang-Mills fields are themselves the connection.

on the two different charts,

$$A_+ = g_{-+}^{-1}(d + A_-)g_{-+} \quad ; \tag{2.2.21}$$

g_{-+} defines a mapping of the equator of S^{n-1} into the gauge group G. These mappings (see the third remark after (1.3.19)) are classified by $\pi_{n-1}(G)$; thus, *a connection on S^n can be topologically non-trivial if and only if $\pi_{n-1}(G) \neq 0$*. A simple example of a topologically non-trivial connection is the potential for the magnetic monopole to be discussed in section 2.7, which corresponds to $G = U(1)$ (the electromagnetic field, no algebra index i) and $S^n = S^2$. The non-triviality appears when the world line of the monopole is removed from spacetime $M \sim R^4$. Topologically, $R^4 \backslash \{$world line$\}$ is homeomorphic to $S^2 \times R^2$ (equivalently, if the monopole is at the origin, $R^3 \backslash \{0\} = S^2 \times R$). Since R^2 is contractible, the different non-trivial $U(1)$-bundles are given by $\pi_1(U(1)) = Z$. As a result, it is not possible to give a (singularity free) monopole potential $A(x)$ on spacetime, and two of them (which are given on two local charts in such a way that the north and south poles are excluded) are needed. The bundle of a monopole of half unit strength is the Hopf bundle. Another example of a non-trivial bundle is the instanton bundle (which corresponds to the quaternionic Hopf bundle), to be discussed in section 2.8.

(c) *Connections on a Lie group G*
Consider now the case of a linear connection on G, with parallel transport defined by the left translations L_g. The curvature is then zero; in fact the zero-curvature condition is equivalent to the Jacobi identity for $\mathscr{G}$. The torsion is given by the structure constants and is not zero unless the group is abelian. Given an invariant metric on a Lie group G it is also simple to define a Riemannian connection on G. We shall omit the discussion since these connections will not be used in this book.

(d) *Invariant connections*
We shall restrict ourselves to the special case which is of interest in our applications. Let G be a connected Lie group with a closed subgroup V with Lie algebra $\mathscr{V}$. Then, as we mentioned in section 1.3, $G(V, G/V)$ is a principal bundle. Let us assume that there is in $\mathscr{G}$ a subspace $\mathscr{M}$ such that it is a representation space for the adjoint representation of V, $Ad(V)\mathscr{M} = \mathscr{M}$, and that $\mathscr{G} = \mathscr{M} \oplus \mathscr{V}$. We can now consider the $\mathscr{V}$-valued form Θ which is obtained from the $\mathscr{V}$-components of the left-invariant canonical form θ on G. The one-form Θ is then, by construction, a *left-invariant* (under the left translations of G) *connection* on $G(V, G/V)$.

Conversely, if such a left-invariant connection on $G(V, G/V)$ exists, it determines a subspace $\mathscr{M}$ supplementary to $\mathscr{V}$ so that $\mathscr{G} = \mathscr{M} \oplus \mathscr{V}$. $\mathscr{M}$ is the horizontal subspace; thus, a LI vector field is horizontal iff it belongs to $\mathcal{M}$.

The curvature Ω of the connection Θ is given now by

$$-\Omega(X_1, X_2) = [X_1, X_2]|_{\mathscr{V}} = \Theta([X_1, X_2]) \qquad (2.2.22)$$

for any two LIVFs X_1, X_2 of $\mathscr{G}$ belonging to the subspace $\mathcal{M}$. Eq. (2.2.22) follows from the structural equations (2.1.10) since $\Theta(X) = 0 \ \forall X \in \mathcal{M}$ because Θ is the part of the canonical form θ 'dual' to $\mathscr{V}$; the connection is flat if $\Omega = 0$, which will be the case iff $\mathcal{M}$ is an ideal in $\mathscr{G}$.

Consider in particular a Lie group $\tilde{G}$ with a central one-dimensional subgroup, the quotient by which is another group G. The previous splitting now becomes $\tilde{\mathscr{G}} = \mathcal{M} \oplus \mathscr{R}$. Let Θ be the LI one-form dual to the central generator Ξ of $\tilde{\mathscr{G}}$ or 'vertical' one-form, $\Theta(\Xi) = 1$. Then, if Y_1, Y_2 are two LIVFs of the algebra $\mathscr{G}$ of G, we may define their horizontal lifts $\tilde{Y}_i$ ($i = 1, 2$) by $\tilde{Y}_i = Y_i - \Theta(Y_i)\Xi$ since, obviously, $\Theta(\tilde{Y}) = 0$ (in the previous expression, Y_i is considered as a vector field on $\tilde{G}$ without component in Ξ; we might look at it as the result of transforming Y_i by the injective mapping $s : G \to \tilde{G}$). The vector fields $\tilde{Y}_i$ belong to the subspace $\mathcal{M}$, which is not a subalgebra since their commutator is not necessarily horizontal (see Proposition 2.1.1). The vertical part of the commutator in (2.2.22), which now is written $[\tilde{Y}_1, \tilde{Y}_2]_{|\mathscr{R}}$, is given by

$$([\tilde{Y}_1, \tilde{Y}_2] - [\widetilde{Y_1, Y_2}]) = \{\Theta([Y_1, Y_2]) - Y_1.\Theta(Y_2) - Y_2.\Theta(Y_1)\}\Xi$$
$$= -d\Theta(Y_1, Y_2)\Xi \ , \qquad (2.2.23)$$

which, ignoring Ξ, is of course equal to $\Theta([\tilde{Y}_1, \tilde{Y}_2]) = -d\Theta(\tilde{Y}_1, \tilde{Y}_2)$. Since, as we shall see, $\tilde{G}$ may be considered as a central extension of G by the one-dimensional group, this example illustrates how such an extension may be characterized by the curvature two-form $d\Theta$ associated with the LI connection Θ.

2.3 Equivariant forms on a Lie group

Let us now generalize the idea of invariant forms on a Lie group (Definition 1.6.1) to the case where the q-form Ω on G takes values in a vector space V of dimension $p > 1$ which is the carrier space of a representation $D(g)$ of G. This leads naturally to the notion of equivariant forms on G. As we shall see, it is convenient to introduce them here once the ideas of connection and of covariant derivative have been considered.

(2.3.1) DEFINITION (*Equivariant form on a Lie group G*)

Let Ω be a V-valued q-form on G. Ω is called an equivariant form *on* G *with respect to the representation* D *of* G *if*

$$\Omega L_g = D(g)(L_{g^{-1}})^*\Omega \qquad (2.3.1)$$

or, equivalently, if $(L_g)^*\Omega L_g = L_g^*\Omega = D(g)\Omega$ *(cf. eq. (2.1.6) where* $D(g^{-1})$ *appears due to the fact that* R_g *enters there). If* $V=R$, $D(g)=1$, *eq. (2.3.1) reproduces (1.6.2).*

It is simple to write Ω in terms of LI ordinary (*i.e.*, R-valued) q-forms. A V-valued q-form may be written as

$$\Omega = e_s\tilde{\omega}^s \qquad (s = 1,...,p) \qquad (2.3.2)$$

where $\{e_s\}$ is a basis of V. Let ω^s be R-valued LI q-forms[*]. The fact that Ω is equivariant implies that the following equality in V holds:

$$\tilde{\omega}^s(g) = D^s_{.t}(g)\,(L_{g^{-1}})^*(e)\omega^t(e) \qquad (2.3.3)$$

(recall that $(L_{g^{-1}})^*(e)\omega^s(e)$ is just $\omega^s(g)$ since ω^s is LI, eq. (1.6.3)). Because D is a representation,

$$
\begin{aligned}
\tilde{\omega}^s(g'g) &= D^s_{.\ell}(g')D^\ell_{.t}(g)(L_{g^{-1}g'^{-1}})^*(e)\omega^t(e) \\
&= D^s_{.\ell}(g')D^\ell_{.t}(g)(L_{g'^{-1}})^*(g)(L_{g^{-1}})^*(e)\omega^t(e) \\
&= D^s_{.t}(g')(L_{g'^{-1}})^*(g)\tilde{\omega}^t(g) \quad ,
\end{aligned}
\qquad (2.3.4)
$$

as implied by (2.3.1). If D is trivial, (2.3.3) tells us that $\tilde{\omega}^s = \omega^s$ and that the $\tilde{\omega}^s$ are LI, as also follows from (2.3.4) with $D^t_{.s} = \delta^t_s$. In particular, this is also the case when $V = R$ and the index s is absent.

For equivariant forms, Proposition 1.6.2 is replaced by

(2.3.1) THEOREM

The correspondence

$$\Omega(g) = D(g)(L_{g^{-1}})^*(e)\Omega(e) \qquad (2.3.5)$$

is an isomorphism between the space of equivariant q-forms on the group G with values in V and the space of antisymmetric q-linear mappings on its Lie algebra $\mathcal{G}$ with values in V.

---

[*] There should be no confusion with this new index s, which refers to V and *not* to the local coordinates on the group manifold G (indices $i, j, ...$).

(2.3.2) THEOREM

Let Ω be an equivariant q-form and let ρ be the representation of $\mathscr{G}$ corresponding to the representation D of G. Then

$$(d\Omega)(X_1, X_2, ..., X_{q+1}) = \sum_{i=1}^{q+1}(-1)^{i+1}D(g)\rho(X_i)D(g^{-1})\Omega(X_1, ..., \hat{X}_i, ..., X_{q+1})$$

$$+ \sum_{\substack{i,j=1 \\ i<j}}^{q+1}(-1)^{i+j}\Omega([X_i, X_j], X_1, ..., \hat{X}_i, ..., \hat{X}_j, ..., X_{q+1}) \qquad \cdot ^{\cdot}$$

$$(2.3.6)$$

Proof: We may write (2.3.5) in the form

$$\Omega(g) = D(g)\omega(g) = e_s D^s_{.t}(g)\omega^t(g) \tag{2.3.7}$$

where $\omega(g)$ is the V-valued LI q-form obtained from $\Omega(e)$. Then

$$d\Omega(g) = dD(g) \wedge \omega(g) + D(g)d\omega(g) \cdot \tag{2.3.8}$$

Acting on a set of LI vector fields $(X_1, X_2, ..., X_{q+1})$, the second term of the r.h.s. of (2.3.8) is obtained from the second term in the r.h.s. of (1.8.6) since the ω^t are LI:

$$D(g)d\omega(g)(X_1, ..., X_{q+1}) =$$
$$D(g)\sum_{i<j}(-1)^{i+j}\omega(g)([X_i, X_j], ..., \hat{X}_i, ..., \hat{X}_j, ..., X_{q+1}) ,$$

and it gives the second term in (2.3.6), again using (2.3.7). The first term in (2.3.8) gives the corresponding one in (2.3.6) recalling that $D(g)$ is obtained by exponentiation of the Lie algebra representation $\rho(X)$, q.e.d.

The expression (2.3.8) for $d\Omega(g)$ can also be given in terms of a covariant derivative of the V-valued LI form $\omega(g)$. Multiplying from the left by $D^{-1}(g)$ we get

$$D^{-1}(g)d\Omega(g) = D^{-1}(g)dD(g) \wedge \omega(g) + d\omega(g)$$
$$= \omega^{L(i)} \wedge \rho(X_{(i)})\omega(g) + d\omega(g) \tag{2.3.9}$$

since $D^{-1}(g)dD(g)$ is the $\rho(\mathscr{G})$-valued canonical form on G (cf. (1.9.7)). Thus, the above equation gives

$$d\Omega(g) = D(g)\nabla\omega(g), \quad \nabla := d + \omega^{L(i)}\rho(X_{(i)}), \tag{2.3.10}$$

where the connection is given by $\omega^{L(i)}\rho(X_{(i)})$ (and, since the $\omega^{L(i)}$ are LI, is of zero curvature). Notice, finally, that (2.3.10) implies that

$$\nabla\nabla = 0 \tag{2.3.11}$$

since the $\omega^{L(i)}$ satisfy the Maurer-Cartan equations.

2.4 Characteristic classes

Certain quantities constructed from the curvature Ω on a differentiable principal bundle $P(G, M)$ turn out to be inherent to the bundle and not dependent on the specific Ω (and hence ω) used for their definition. They are, therefore, *characteristic* of the bundle; they are preserved by bundle diffeomorphisms and define *topological invariants* associated with them. They measure the obstruction to extending a local product structure $(U \times G, U \subset M)$ to a global product (*i.e.*, $M \times G$) structure. We shall see that these characteristic classes are given by various closed differential forms on M which are invariant polynomials in the curvature of a connection or, more precisely, by the de Rham cohomology generators defined by these forms on the base manifold M. Being closed, they are locally exact on each chart $U_i \subset M$ so that by Stokes' theorem their integrals over M depend only on the transition functions which contain the topological information on the bundle. Properly used, the term *characteristic class* refers to the cohomology class of the de Rham equivalent characteristic forms, although in the literature it is common to refer to the characteristic forms as 'classes'. The integration of a class over the base manifold gives rise to the corresponding *characteristic number*. The *Chern, Euler* and *Pontrjagin characteristic classes* are all examples of this. The *Stiefel-Whitney classes* w_i are not de Rham cohomology classes and are not given in terms of the curvature; they will not be discussed in this book. They have nevertheless important physical applications: the vanishing of the first $w_1(\tau(M))$ and second $w_2(\tau(M))$ classes are, respectively, necessary and sufficient conditions for a manifold M being orientable or for admitting a spin structure when M is orientable. (M is a *spin manifold* if *both* classes vanish, $w_i(\tau(M)) = 0, i = 1, 2$).

The following (Chern-Weil) procedure used to obtain them goes through three steps. First, the set $I^l(G)$ of symmetric AdG-invariant polynomials of degree l is introduced, and $\sum_{l=0}^{\infty} I^l(G) = I(G)$ is shown to have a graded ring structure. Secondly, it is found that these invariant polynomials can be used to define closed forms $P(\bar{\Omega})$ on M, the cohomology classes of which do not depend on the connection. Finally, the algebra Weil homomorphism $I(G) \to H_{DR}^{even}(M)$ which associates de Rham cohomology classes with the closed forms on M previously obtained is constructed.

(2.4.1) DEFINITION (*AdG-invariant symmetric multilinear mapping*)

Let AdG be the adjoint representation of G on $\mathscr{G}$ (section 1.9). A symmetric l-linear mapping (a polynomial on $\mathscr{G}$)

$$P : \mathscr{G}^l = \mathscr{G} \times ... \times \mathscr{G} \to R \qquad (2.4.1)$$

is AdG-invariant if

$$P(Adg\, X_1, ..., Adg\, X_l) = P(X_1, ..., X_l) \quad , \quad X_1, ..., X_l \in \mathcal{G} \quad ; \qquad (2.4.2)$$

the polynomial is then called a characteristic or invariant polynomial.

Let $I^l(G)$ be the vector space of all these *AdG*-invariant polynomials[*] on $\mathcal{G}$ and let

$$I(G) = \sum_{l=0}^{\infty} I^l(G) \quad . \qquad (2.4.3)$$

$I(G)$ can be endowed with a graded ring structure by defining the product of two polynomials $P \in I^l(G)$, $P' \in I^k(G)$, $PP' \in I^{l+k}(G)$, by

$$(PP')(X_1, ..., X_{l+k}) = \frac{1}{(k+l)!} \sum_{\text{perm } \sigma} P(X_{\sigma(1)}, ..., X_{\sigma(l)}) P'(X_{\sigma(l+1)}, ..., X_{\sigma(l+k)}) \cdot$$
$$(2.4.4)$$

The next step is to relate $I(G)$ with the de Rham cohomology ring $H_{DR}^{even}(M)$ where M is the base manifold of $P(G, M)$. This requires the proof of a simple lemma.

(2.4.1) LEMMA (*Forms β on P projectable to forms $\bar{\beta}$ on M*)

Let β be a q-form on the total space P of a principal bundle. Then it projects to (it is the pull-back $\pi^*(\bar{\beta})$ of) a unique form $\bar{\beta}$ on the base manifold M (we shall use the overbar to indicate that the form is defined on the base manifold) if

a) $R_g^* \beta = \beta$ (β is invariant under the right action of G),
b) $\beta(Y_1, ..., Y_q) = 0$ if any of the arguments is vertical.

Proof: Let $\pi^T(Y_i(p)) = X_i(x)$ be the projections on $T_x(M)$ of the vectors $Y_i(p) \in T_p(P)$, $i = 1, ..., q$. Then, if β is projectable, its projection $\bar{\beta}$ is defined by

$$\bar{\beta}(x)(X_1(x), ..., X_q(x)) := \beta(p)(Y_1(p), ..., Y_q(p)) , \qquad (2.4.5)$$
$$X_i(x) = \pi^T Y_i(p) , \quad \pi(p) = x .$$

This will indeed define $\bar{\beta}$ provided that β gives the same value on any set of vectors $Z_i(p') \in T_{p'}(P)$, $p' \in \pi^{-1}(x)$, that project onto a given set $X_i(x) \in T_x(M)$, and this is what (a) and (b) above guarantee. Let $Z_i(p')$ be

---

[*] *Polynomial functions* on $\mathcal{G}$ are defined by covariant *l*-tensors of the form $P = P_{i_1...i_l} \alpha^{i_1} \otimes ... \otimes \alpha^{i_l}$, where the coefficients $P_{i_1...i_l}$ are symmetric and $\{\alpha^{(i)}\}$ is a basis of the coalgebra vector space $\mathcal{G}^*$ dual to $\mathcal{G}$, $\alpha^{(i)}(X_{(j)}) = \delta^i_j$, *i.e.* by *l*-linear symmetric mappings. Thus, there is an isomorphism between the symmetric, *AdG*-invariant *l*-linear mappings and the *AdG*-invariant polynomials of order *l*.

such a set, and $p' = pg$. Then, if we denote by $Z'_i(p)$ the tangent vectors to which $Z_i(pg)$ are R_g^T-related,

$$\beta(pg)(Z_1(pg), ..., Z_q(pg)) = \beta(pg)(R_g^T Z'_1(p), ..., R_g^T Z'_q(p))$$
$$= (R_g)^* \beta(pg)(Z'_1(p), ..., Z'_q(p)) = \beta(p)(Z'_1(p), ..., Z'_q(p)) \quad ,$$

$$(2.4.6)$$

where (a) has been used in the last equality. But this is the same as $\beta(p)(Y_1(p), ..., Y_q(p))$ because the differences $Z'_i(p) - Y_i(p)$ are vertical and (b) holds. Thus, the definition (2.4.5) of $\bar\beta$ does not depend on the specific $p \in \pi^{-1}(x)$, $Y_i(p)$ considered, q.e.d.

(2.4.1) THEOREM *(Chern-Weil)*

Let $P(G, M)$ be a principal bundle endowed with a connection ω of curvature Ω. Given an invariant polynomial $P \in I^l(G)$, construct the $2l$-differential form $P(\Omega)$ on P by

$$P(\Omega)(Y_1, ..., Y_{2l}) = \frac{1}{2l!} \sum_{perm\ \sigma} sign(\sigma)\, P\left[\Omega(Y_{\sigma(1)}, Y_{\sigma(2)}), ..., \Omega(Y_{\sigma(2l-1)}, Y_{\sigma(2l)})\right] ,$$

$$\forall Y_1, ..., Y_{2l} \in T_p(P) \quad ,$$

$$(2.4.7)$$

where $sign(\sigma)$ is the signature of the permutation σ. Then

a) *for each $P \in I^l(G)$, the $2l$-form $P(\Omega)$ on P projects to a unique closed $2l$-form on M denoted $P(\bar\Omega)$, i.e. $\pi^*(P(\bar\Omega)) = P(\Omega)$;*
b) *the element of the de Rham cohomology group characterized by the co-homology class of the form $P(\bar\Omega)$ on M is independent of the choice of the connection (and thus $P(\bar\Omega)$ has topologically invariant integrals);*
c) *the mapping $I(G) \to H_{DR}^{even}(M)$ defined by $P \mapsto$ (cohomology class of $P(\bar\Omega)$) is an algebra homomorphism, the Weil homomorphism.*

Proof :

(a) The form defined in (2.4.7) satisfies the two conditions of lemma 2.4.1: since Ω is tensorial, $R_g^*(\Omega) = Ad(g^{-1})\Omega$ [(2.1.8)] implies $R_g^* P(\Omega) = P(\Omega)$ (P is AdG-invariant, eq. (2.4.2)) and also that $P(\Omega)(Y_1, ..., Y_{2l}) = 0$ whenever a vector Y is vertical. Thus, there is a unique $2l$-form $\bar P(\Omega) = P(\bar\Omega)$ which is the projection of $P(\Omega)$ on P.

To see that $P(\bar\Omega)$ is closed, it suffices to prove that $P(\Omega)$ is closed. This follows from the fact that $dP(\Omega) = \mathcal{D}P(\Omega)$, a property which is true for any form β on P which projects to a form $\bar{\beta}$ on M since $\beta = \pi^*\bar{\beta}$ implies

$$
\begin{aligned}
(d\beta)(Y_1, ..., Y_q) &= (d\pi^*\bar{\beta})(Y_1, ..., Y_q) = (\pi^* d\bar{\beta})(Y_1, ..., Y_q) \\
&= (d\bar{\beta})(\pi^T Y_1, ..., \pi^T Y_q) = (d\bar{\beta})(\pi^T Y_1^h, ..., \pi^T Y_q^h) \\
&= (\pi^* d\bar{\beta})(Y_1^h, ..., Y_q^h) \\
&= (d\pi^*\bar{\beta})(Y_1^h, ..., Y_q^h) = (\mathscr{D}\beta)(Y_1, ..., Y_q) ,
\end{aligned}
\tag{2.4.8}
$$

where $\pi^T(Y_i) = \pi^T(Y_i^h)$ and (2.1.7) have been used. Thus, using the Bianchi identity (2.1.9) and (2.1.22) it follows that $\mathscr{D}P(\Omega) = 0$ and thus that $P(\Omega)$ and $P(\bar{\Omega})$ are closed.

(b) Let ω_0, ω be two connections on $P(G, M)$. Their difference $\eta = \omega - \omega_0$ is a tensorial one-form of type $(AdG, \mathscr{G})$ and

$$
\omega_t = \omega_0 + t\eta , \quad 0 \le t \le 1 , \tag{2.4.9}
$$

is the interpolating one-parameter family of connections. Let $\mathscr{D}_t$ and Ω_t be the exterior covariant differentiation and curvature associated with ω_t. Then the Cartan structural equation (2.1.11) for the connection ω_t reads

$$
\Omega_t = d\omega_t + \frac{1}{2}[\omega_t, \omega_t] = \Omega_0 + t\mathscr{D}_0\eta + \frac{t^2}{2}[\eta, \eta] , \tag{2.4.10}
$$

where the last term may also be written $t^2\eta \wedge \eta$. Then

$$
\frac{d\Omega_t}{dt} = \mathscr{D}_0\eta + t[\eta, \eta] = d\eta + [\omega_t, \eta] = \mathscr{D}_t\eta , \tag{2.4.11}
$$

i.e. $d\Omega_t/dt$ is the covariant derivative $\mathscr{D}_t$ of the tensorial form η (see (2.1.17)).

To prove now that $P(\bar{\Omega})$ defines a de Rham cohomology class on M independent of the connection ω it suffices to show that the closed forms $P(\bar{\Omega})$ and $P(\bar{\Omega}_0)$ differ by an *exact* form: they are cohomologous forms on M. This is indeed the case since

$$
P(\Omega, \overset{l}{\cdots}, \Omega) - P(\Omega_0, \overset{l}{\cdots}, \Omega_0) = \int_0^1 dt\, \frac{d}{dt} P(\Omega_t, \overset{l}{\cdots}, \Omega_t)
$$

$$
= l \int_0^1 dt\, P\left(\frac{d\Omega_t}{dt}, \Omega_t, \overset{l-1}{\cdots}, \Omega_t\right) = l \int_0^1 dt\, P(\mathscr{D}_t \eta, \Omega_t, \overset{l-1}{\cdots}, \Omega_t)
$$

$$
= l \int_0^1 dt\, \mathscr{D}_t P(\eta, \Omega_t, \overset{l-1}{\cdots}, \Omega_t) = l \int_0^1 dt\, dP(\eta, \Omega_t, \overset{l-1}{\cdots}, \Omega_t)
$$

$$
= d[l \int_0^1 dt\, P(\eta, \Omega_t, \overset{l-1}{\cdots}, \Omega_t)] \equiv dQ^{(2l-1)}(\omega, \omega_0) \quad,
$$

(2.4.12)

where in the third (fourth) equality eq. (2.4.11) ($\mathscr{D}_t \Omega_t = 0$) has been used, and in the fifth the fact that a covariant derivative is equivalent to d on a projectable form (the notation $Q^{(2l-1)}$ in the r.h.s. of (2.4.12) has been introduced for convenience, see below). Equation (2.4.12) on P projects to M with the result

$$
P(\bar{\Omega}, \overset{l}{\cdots}, \bar{\Omega}) - P(\bar{\Omega}_0, \overset{l}{\cdots}, \bar{\Omega}_0) = d[l \int_0^1 dt\, P(\bar{\eta}, \bar{\Omega}_t, \overset{l-1}{\cdots}, \bar{\Omega}_t)] \equiv d\bar{Q}^{(2l-1)} \quad \text{(2.4.13)}
$$

which proves (b) above. The class in $H_{DR}^{2l}(M, R)$ defined by $P(\bar{\Omega})$ is a characteristic class; the integrals of $P(\bar{\Omega})$ and $P(\bar{\Omega}_0)$ over a compact boundaryless M are the same.

(c) Finally, the mapping $w : I(G) \mapsto H_{DR}^{even}(M, R)$ that associates to each invariant polynomial $P \in I(G)$ the de Rham cohomology class of the form $P(\bar{\Omega})$ is a ring homomorphism, the *Weil homomorphism*. Writing $\Omega = \Omega^i X_{(i)}$ where $\{X_{(i)}\}$ is a basis of $\mathscr{G}$, eq. (2.4.4) implies that, if P, $P' \in I^{l,l'}(G)$,

$$
(PP')(\bar{\Omega})
$$

$$
= \frac{1}{(l+l')!} P(X_{i_1}, ..., X_{i_l}) P'(X_{j_1}, ..., X_{j_{l'}}) \bar{\Omega}^{i_1} \wedge ... \wedge \bar{\Omega}^{i_l} \wedge \bar{\Omega}^{j_1} \wedge ... \wedge \bar{\Omega}^{j_{l'}}
$$

$$
= P(\bar{\Omega}) \wedge P'(\bar{\Omega})
$$

(2.4.14)

since the exterior product of even forms is commutative, q.e.d.

On the total space P of $P(G, M)$, $P(\Omega)$ is an exact form. To show this, it is sufficient to note that the derivation that led to (2.4.12) could be repeated again starting from a form ω_t given by an expression *like* (2.4.9) but where *now* ω and ω_0 are arbitrary $\mathscr{G}$-valued one-forms on P and Ω_t is the 'curvature' defined by $\Omega_t := d\omega_t + \frac{1}{2}[\omega_t, \omega_t]$. Indeed, the 'curvature' Ω_t satisfies the 'Bianchi' identity $\mathscr{D}_t \Omega_t = d\Omega_t + [\omega_t, \Omega_t] = 0$, which is derived in the same way as (2.1.19). If this is done, we can set $\omega_0 = 0$ in (2.4.12) (something that was not allowed when ω_0 was a connection, since a zero-form on P cannot satisfy property (b) of Definition 2.1.2b) and obtain, taking ω to be an arbitrary connection on P,

$$P(\Omega) = d[TP(\omega, \Omega)] = d\left[l \int_0^1 dt P(\omega, \overset{l-1}{\Omega_t, \cdots, \Omega_t}) \right] \quad . \qquad (2.4.15)$$

This is (in Chern's sense) the *transgression formula*, and T is called the *transgression operator*; it enables P to be written as an exact form in a canonical way. It follows from eq. (2.4.15) that when $P(\Omega)$ is identically zero $TP(\omega, \Omega)$ is a closed $(2l-1)$-form on P which defines a cohomology class in $H_{DR}^{2l-1}(P, R)$, the *Chern-Simons secondary class*; it depends on the connection ω.

The above $(2l-1)$-form $TP(\omega, \Omega)$ on P does not satisfy the condition (b) of lemma 2.4.1 and accordingly cannot be projected to the base manifold M. In contrast, the previous form $Q^{(2l-1)}(\omega, \omega_0)$ on P (which will also be written $TP(\omega, \omega_0)$, eq. (2.4.12)) is a projectable form:

(2.4.2) DEFINITION (*Chern-Simons form of $P(\Omega)$*)

The Chern-Simons $(2l-1)$-form associated with the symmetric polynomial $P(\Omega)$ is the projection on M of the form $Q^{(2l-1)}(\omega, \omega_0)$ given by (2.4.12).

Its expression is given by

$$Q^{(2l-1)}(\bar{\omega}, \bar{\omega}_0) = TP(\bar{\omega}, \bar{\omega}_0) = l \int_0^1 dt P(\bar{\omega} - \bar{\omega}_0, \overset{l-1}{\bar{\Omega}_t, \cdots, \bar{\Omega}_t}) \quad ; \qquad (2.4.16)$$

by extension, we shall also refer to (2.4.16) as the *transgression formula*. We might naïvely think of setting $\bar{\omega}_0 = 0$ on M itself (and consequently $P(\bar{\Omega}_0) = 0$) to derive a potential form for the $2l$-form $P(\bar{\Omega})$ on M from (2.4.13). But this would be wrong in general, because only if $P(G, M)$ is trivial can we find a representative of a flat connection that is zero on M as the transformation law (2.1.1) shows. On a local chart $U \subset M$, however, there always exists a flat connection for which $\bar{\omega}_0 = 0$ and then it is possible to write

$$P(\bar{\Omega}) = dQ^{(2l-1)}(\bar{\omega}, \bar{\Omega}) \quad , \quad Q^{(2l-1)}(\bar{\omega}, \bar{\Omega}) = l \int_0^1 dt\, P(\bar{\omega}, \overset{l-1}{\bar{\Omega}_t, \; \cdots \;, \bar{\Omega}_t}) \quad .$$

(2.4.17)

The potential form $Q^{(2l-1)}(\bar{\omega}, \bar{\Omega})$ on $U \subset M$ is called a *Chern-Simons form* of $P(\bar{\Omega})$.

It is worth mentioning that eq. (2.4.12) can be generalized by replacing the 'segment' (2.4.9) in the affine line $\omega_t = \omega_0 + t\eta$ of connections by e.g. a 'triangle' of interpolating connections

$$\omega_{t_1 t_2} = \omega_0 + t_1(\omega_1 - \omega_0) + t_2(\omega_2 - \omega_0) \quad , \quad t_1 + t_2 \leq 1 \quad , \quad (2.4.18)$$

determined by three connections ω_0, ω_1, ω_2 on $P(G, M)$. Then, if $Q^{(2l-1),1}(\omega, \omega_0)$ is defined as in (2.4.12), the following *triangle equality* holds:

$$Q^{(2l-1),1}(\omega_0, \omega_1) + Q^{(2l-1),1}(\omega_1, \omega_2) + Q^{(2l-1),1}(\omega_2, \omega_0) = dQ^{(2l-2),2}(\omega_0, \omega_1, \omega_2)$$

(2.4.19)

where

$$Q^{(2l-2),2}(\omega_0, \omega_1, \omega_2) =$$
$$l(l-1) \int_0^1 dt_1 \int_0^{1-t_1} dt_2\, P(\omega_1 - \omega_0, \omega_2 - \omega_0, \overset{l-2}{\Omega_{t_1 t_2}, \; \cdots \;, \Omega_{t_1 t_2}}) \quad (2.4.20)$$

and $\Omega_{t_1 t_2}$ is the curvature of the connection $\omega_{t_1 t_2}$. By extending the procedure, it is possible to find an expression for $dQ^{(2l-r),r}$ in terms of the different $Q^{(2l-r+1),(r-1)}$'s by introducing $r+1$ connections $\omega_0, \omega_1, ..., \omega_r$ and defining $\omega_{t_1...t_r} = \omega_0 + t_1(\omega_1 - \omega_0) + ... + t_r(\omega_r - \omega_0), \sum_{i=1}^r t_i \leq 1$. The resulting polynomials $P(\omega_1 - \omega_0, ..., \omega_r - \omega_0, \Omega_{t_1...t_r}^{l-r})$ may be used to define the Chern-Simons-like $(2l - r)$-forms $Q^{(2l-r),r}$ of r-th type on M by extending Definition 2.4.2 to them (cf. eq. (2.4.20), $r = 2$).

The Cartan homotopy operator and the Cartan homotopy formula

To conclude this section, let us rewrite the transgression equation (2.4.13) by introducing the Cartan homotopy operator. Let the operator k_t be defined by being a graded derivation on forms, $k_t(\alpha \wedge \beta) = k_t\alpha \wedge \beta + (-1)^p \alpha \wedge k_t\beta$ (cf. (1.4.20)) which satisfies

$$k_t\omega_t = 0 \quad , \quad k_t\Omega_t = \eta \quad . \tag{2.4.21}$$

Then, on ω_t, Ω_t the operator $(k_t d + dk_t)$ gives (d does not act on the parameter t)

$$(k_t d + dk_t)\omega_t = k_t d\omega_t = k_t(\Omega_t - \omega_t \wedge \omega_t) = \eta = \frac{d}{dt}\omega_t \quad,$$

$$(k_t d + dk_t)\Omega_t = k_t d(\omega_t \wedge \omega_t) + dk_t d\omega_t \qquad (2.4.22)$$

$$= \eta \wedge \omega_t + \omega_t \wedge \eta + d\eta = \mathscr{D}_t \eta = \frac{d}{dt}\Omega_t ,$$

where (2.4.11) has been used. Thus, $(k_t d + dk_t)$ acts as a derivation with respect to t. Then, on an arbitrary polynomial $S(\omega_t, \Omega_t)$

$$\delta t(k_t d + dk_t)S(\omega_t, \Omega_t) = \delta t \frac{d}{dt}S(\omega_t, \Omega_t) \quad . \qquad (2.4.23)$$

The integration of (2.4.23) from zero to one gives the *Cartan homotopy formula*,

$$S(\omega, \Omega) - S(\omega_0, \Omega_0) = (k_{01}d + dk_{01})S(\omega_t, \Omega_t) \quad , \qquad (2.4.24)$$

where in the right hand side the action of the *Cartan homotopy operator* is defined by

$$k_{01}S(\omega_t, \Omega_t) := \int_0^1 \delta t \, k_t S(\omega_t, \Omega_t) \quad . \qquad (2.4.25)$$

Let now S be the polynomial $P(\Omega)$. Then, since $P(\Omega)$ is closed, (2.4.24) immediately reproduces (2.4.12) in the form

$$Q^{(2l-1)}(\omega, \omega_0) = \int_0^1 \delta t \, k_t P(\Omega_t) \quad . \qquad (2.4.26)$$

2.5 Chern classes and Chern characters

The Chern classes $c_1, ..., c_n$ are defined for a complex vector bundle η_E over M of fibre C^n and structure group $GL(n, C)$ or, equivalently, for a $P(GL(n, C), M)$ principal bundle. Thus, Chern classes are associated with bundles with $GL(n, C)$-valued transition functions (see, however, the remark below). The Lie algebra $\mathscr{Gl}(n, C)$ can be identified with the vector space of the $n \times n$ complex matrices X. Clearly, the polynomial in λ,

$$P(X, \lambda) := \det(\lambda 1_n + X) = \sum_{l=0}^n \lambda^{n-l} P_l(X) \quad , \qquad (2.5.1)$$

is *AdG*-invariant since $\det(\ ) = \det[g^{-1}(\ )g]$; $P_0(X) = 1$, $P_1(X) = \text{Tr}(X)$, $P_2(X) = \frac{1}{2}\{(\text{Tr}X)^2 - \text{Tr}X^2\}$, etc. In fact, the coefficients of λ^{n-l} determine the characteristic equation for the matrix $-X$, the solutions of which are its eigenvalues. Let us now replace X in (2.5.1) by $(i/2\pi)\Omega$ rather than by Ω. The sum $P(\Omega,1)$,

$$c(\Omega) \equiv \det\left(1 + \frac{i}{2\pi}\Omega\right) = 1 + P_1(\Omega) + P_2(\Omega) + \dots \quad , \tag{2.5.2}$$

is called the *total Chern form*; the factor $(i/2\pi)$ has been introduced for normalization purposes so that the Chern numbers to be defined below are integers. The finite series (2.5.2) can be projected on M. The resulting expansion leads to

(2.5.1) DEFINITION (*Chern forms, Chern classes*)

The Chern classes $c_l(E)$ of a complex vector bundle E are the elements of $H^{2l}_{DR}(M,R)$ determined by the cohomology classes of the closed Chern $2l$-forms $c_l(\bar\Omega)$ on M that are the projections of the forms $P_l(\Omega)$ on P, $\pi^* c_l(\bar\Omega) = P_l(\Omega)$. The sum of all Chern classes is the total Chern class $c(E)$. *When E is the tangent bundle $\tau(M)$ of a complex manifold, the Chern classes $c_l(\tau(M))$ are called* Chern classes of the manifold M *(and frequently written $c_l(M)$).*

Let $\bar\Omega^i_{\ j}$ be the components of the curvature two-form on M. Then the Chern forms $c_l(\bar\Omega)$ on M are the closed $2l$-forms given by

$$c_l(\bar\Omega) = \frac{1}{l!}\left(\frac{i}{2\pi}\right)^l \epsilon^{i_1\dots i_l}_{j_1\dots j_l} \bar\Omega^{j_1}_{\ i_1} \wedge \dots \wedge \bar\Omega^{j_l}_{\ i_l} \quad . \tag{2.5.3}$$

The first four examples of (2.5.3) are

$$c_0(\bar\Omega) = 1 \quad ,$$

$$c_1(\bar\Omega) = \frac{i}{2\pi}\mathrm{Tr}\bar\Omega \quad ,$$

$$c_2(\bar\Omega) = \frac{-1}{8\pi^2}\{\mathrm{Tr}\bar\Omega \wedge \mathrm{Tr}\bar\Omega - \mathrm{Tr}(\bar\Omega \wedge \bar\Omega)\} \quad ,$$

$$c_3(\bar\Omega) = \frac{-i}{48\pi^3}\{\mathrm{Tr}\bar\Omega \wedge \mathrm{Tr}\bar\Omega \wedge \mathrm{Tr}\bar\Omega - 3\mathrm{Tr}(\bar\Omega \wedge \bar\Omega) \wedge \mathrm{Tr}\bar\Omega + 2\mathrm{Tr}(\bar\Omega \wedge \bar\Omega \wedge \bar\Omega)\} \quad .$$

$$\tag{2.5.4}$$

Clearly, $c_l(\bar\Omega) \equiv 0$ if $2l > \dim M$ but, if $2n \leq \dim M$, the last term of the series is

$$c_n(\bar\Omega) = \left(\frac{i}{2\pi}\right)^n \frac{1}{n!}\epsilon^{i_1\dots i_n}_{j_1\dots j_n} \bar\Omega^{j_1}_{\ i_1} \wedge \dots \wedge \bar\Omega^{j_n}_{\ i_n} = \det\left(\frac{i}{2\pi}\bar\Omega\right) \quad . \tag{2.5.5}$$

We see in these expressions how the AdG invariance is tied to the fact that the characteristic forms are closed: the invariance requires the presence of the trace, and this makes the form closed. If the connection is flat, $\bar\Omega = 0$, the Chern (and also the Pontrjagin and Euler) classes vanish. If, conversely, a Chern class is non-zero for a certain connection, then the bundle does not admit a flat connection since otherwise the Chern form would not define a characteristic class.

The above expressions for $c_l(\bar{\Omega})$ are, strictly speaking, just representatives of a cohomology class in $H^{2l}_{DR}(M, R)$. With the normalization introduced by the factor $i/2\pi$ in (2.5.2), *the Chern forms $c_l(\bar{\Omega})$ are actually of integer class*: their integral over any $2l$-cycle on M with integer coefficients is an integer that does not depend on $\bar{\Omega}$. When elements of the algebra generated by the Chern forms of degree dim M are integrated over the manifold M itself, the resulting topological invariants receive the name of Chern numbers:

(2.5.2) DEFINITION (*Chern numbers*)

The Chern numbers are those obtained by integrating characteristic polynomials of degree dim M *over the entire manifold M.*

For instance, if dim $M = 4$, there are only two independent Chern numbers given by

$$C_1 = \int_M c_1(\bar{\Omega}) \wedge c_1(\bar{\Omega}) \quad \text{and} \quad C_2 = \int_M c_2(\bar{\Omega}) \ . \tag{2.5.6}$$

Remark

The structure group of a complex vector bundle may be restricted to $U(n)$ by introducing a hermitian metric on η_E. It turns out that the $GL(n, C)$ and $U(n)$ characteristic polynomials can be identified and hence their characteristic classes are the same. Since the polynomials $P_l(X)$ are real when $X \in u(n)$, it follows that the Chern classes are elements of $H^{2l}_{DR}(M, R)$, $l = 0, 1, ..., (\text{dim } M/2)$. The same can be said of $SU(n)$; the characteristic classes are simply given by $(c_1 = 0, c_2, c_3, ..., c_n)$ where (cf. (2.5.4))

$$c_2 = \frac{1}{8\pi^2} \text{Tr}(\bar{\Omega} \wedge \bar{\Omega}) \quad , \quad c_3 = \frac{-i}{24\pi^3} \text{Tr}(\bar{\Omega} \wedge \bar{\Omega} \wedge \bar{\Omega}) \tag{2.5.7}$$

etc., since det $g = 1$ implies Tr $X = 0$ (in particular, this also means that if an arbitrary bundle with $G = GL(n, C)$ reduces to an $SU(n)$ bundle, $c_1(\bar{\Omega})$ must be zero).

Of special interest in the theory of anomalies to be discussed in chapter 10 and in the Atiyah-Singer theorem are the *Chern characters* $ch(\bar{\Omega})$. They are obtained from the obviously *AdG*-invariant polynomial Tr(*exp X*) by replacing X by $\frac{i}{2\pi}\Omega$. The form on P

$$\text{Tr}\left\{\exp\left(\frac{i}{2\pi}\Omega\right)\right\} \tag{2.5.8}$$

has the expansion

$$ch(\Omega) = \mathrm{Tr}\left\{ \sum_{l=0}^{\infty} \frac{1}{l!}\left(\frac{i}{2\pi}\right)^{l} \Omega \wedge \overset{l}{\cdots} \wedge \Omega \right\} \equiv \sum_{l=0}^{\infty} ch_l(\Omega) \tag{2.5.9}$$

and projects onto a non-homogeneous, closed form on M, which is the total Chern character $ch(E)$ of the complex vector bundle E. It is given by

$$ch(\bar{\Omega}) = n + \frac{i}{2\pi}\mathrm{Tr}\bar{\Omega} + \frac{1}{2}\left(\frac{i}{2\pi}\right)^{2}\mathrm{Tr}(\bar{\Omega}\wedge\bar{\Omega}) + \dots \tag{2.5.10}$$

since $\mathrm{Tr}\mathbf{1} = n$. For instance, for a complex line bundle L

$$ch(L) = exp\frac{i}{2\pi}\bar{\Omega} \quad . \tag{2.5.11}$$

(2.5.3) DEFINITION (*Chern characters*)

The Chern character forms are the closed $2l$-forms on M given by

$$ch_l(\bar{\Omega}) = \frac{1}{l!}\left(\frac{i}{2\pi}\right)^{l}\mathrm{Tr}(\bar{\Omega}\wedge\overset{l}{\cdots}\wedge\bar{\Omega}) \ ; \tag{2.5.12}$$

again, $ch_l(\bar{\Omega}) = 0$ *if* $2l > \dim M$.

It is not difficult to express the Chern character forms in terms of the Chern forms. For this, let us assume that X has been written in diagonal form, with eigenvalues $\lambda_1, \dots, \lambda_n$. Then the expansion (2.5.1) that leads to the Chern forms is

$$\det(\mathbf{1} + X) = \prod_{i=1}^{n}(1 + \lambda_i) = 1 + \sum_{i}\lambda_i + \sum_{i<j}\lambda_i\lambda_j + \dots + \lambda_1\dots\lambda_n$$

$$\equiv c_0(\lambda_i) + c_1(\lambda_i) + \dots + c_n(\lambda_i) \tag{2.5.13}$$

$$\equiv 1 + \mathrm{Tr}\,X + \frac{1}{2}\{(\mathrm{Tr}\,X)^2 - \mathrm{Tr}\,X^2\} + \dots + \det X \quad ;$$

obviously, the last equality does not require X to be in diagonal form. Similarly, the invariant polynomial $\mathrm{Tr}(exp\,X)$ gives, using (2.5.13),

$$\mathrm{Tr}(exp\,X) = \sum_{i=1}^{n}exp\,\lambda_i = \sum_{i=1}^{n}\left(\sum_{l}\frac{1}{l!}\lambda_i^l\right)$$

$$= n + c_1(\lambda_i) + \frac{1}{2}[c_1(\lambda_i)^2 - 2c_2(\lambda_i)] + \dots \quad . \tag{2.5.14}$$

Since the polynomials are invariant, we can use for them the expression in terms of X (or, rather, in terms of $\frac{i}{2\pi}\Omega$) to identify (2.5.14) with (2.5.9).

In this way, we get

$$ch_0(\bar{\Omega}) = n \quad,$$
$$ch_1(\bar{\Omega}) = c_1(\bar{\Omega}) \quad, \tag{2.5.15}$$
$$ch_2(\bar{\Omega}) = \frac{1}{2}[c_1(\bar{\Omega})^2 - 2c_2(\bar{\Omega})] \quad, \quad \text{etc.}$$

This provides a particular example of a more general result: the algebra of characteristic classes of a bundle may be generated by the corresponding Chern classes, *i.e.* any characteristic class of P is a polynomial in the Chern classes.

Much in the same way that the matrix X is eventually replaced by $\frac{i}{2\pi}\Omega$ in the determination of the Chern forms, we may imagine X in a diagonal form, with diagonal elements λ_i given by $\lambda_i = \frac{i}{2\pi}\Omega_i$. Then each of the n terms $(1+\lambda_i)$ in $\det(1+X)$ in (2.5.13) may be interpreted as the Chern class of a one-dimensional (line bundle) L_i, $c(L_i) = c_0(L_i) + c_1(L_i) = 1 + \frac{i}{2\pi}\Omega_i$. As far as the Chern classes are concerned, it is possible to pretend that the n-dimensional complex vector bundle is the Whitney sum (section 1.3) $E = L_1 \oplus ... \oplus L_n$; then its Chern class is given by $c(E) = \det(1 + X) = \prod_{i=1}^{n}(1+\lambda_i) = \prod_{i=1}^{n} c(L_i)$. This is the *splitting principle* for the Chern class. The same way of reasoning shows that, if E and E' are vector bundles,

$$c(E \oplus E') = c(E) \wedge c(E') \tag{2.5.16}$$

since $\det(X \oplus X') = \det X.\det X'$. For line bundles, $c_1(L \otimes L') = c_1(L) + c_1(L')$.

We may find the corresponding property for the Chern character by looking at the invariant polynomial (2.5.14). Replacing X by the direct sum $X \oplus X'$ gives

$$ch(E \oplus E') = ch(E) + ch(E') \quad . \tag{2.5.17}$$

Moreover, and in contrast with the total Chern class which does not behave well for tensor products, the Chern character satisfies

$$ch(E \otimes E') = ch(E) \wedge ch(E') \quad . \tag{2.5.18}$$

2.6 Chern-Simons forms of the Chern characters

Let us now find, using the transgression formula (2.4.17) for $P(\Omega) = ch_l(\Omega)$, the local expression for the Chern-Simons form of the Chern character. The expression for the $\bar{\Omega}_t$ appearing there is given by (2.4.10); setting locally $\bar{\omega}_0 = 0$, $\bar{\omega}_t = t\bar{\omega}$, we find on $U \subset M$

$$\bar{\Omega}_t = d\bar{\omega}_t + \bar{\omega}_t \wedge \bar{\omega}_t = td\bar{\omega} + t^2\bar{\omega} \wedge \bar{\omega} = t\bar{\Omega} + (t^2 - t)\bar{\omega} \wedge \bar{\omega} \qquad (2.6.1)$$

which leads us to the following

(2.6.1) DEFINITION (*Chern-Simons form of a Chern character*)

The local expression for the Chern-Simons form of the Chern character $ch_l(\bar{\Omega})$ *is given by*

$$Q^{(2l-1)}(\bar{\omega}, \bar{\Omega}) = \frac{1}{(l-1)!}\left(\frac{i}{2\pi}\right)^l \int_0^1 dt\, s\mathrm{Tr}(\bar{\omega}, \overset{l-1}{\bar{\Omega}_t \wedge \cdots \wedge \bar{\Omega}_t}) \quad , \qquad (2.6.2)$$

where sTr *indicates symmetrized trace,*

$$s\mathrm{Tr}(X_1, ..., X_l) = \frac{1}{l!} \sum_{\text{perm } \sigma} \mathrm{Tr}(X_{\sigma(1)}, ..., X_{\sigma(l)}) \qquad (2.6.3)$$

(in chapter 10 we shall denote the Chern-Simons form $Q^{(2l-1)}$ of the Chern character $ch_l(\bar{\Omega})$ by ω_0^{2l-1}, but we shall not use this notation here since in this chapter ω is reserved for the connection). For $l = 1$, the application of (2.6.2) trivially leads to

$$Q^{(1)}(\bar{\omega}, \bar{\Omega}) = \frac{i}{2\pi} \int_0^1 dt\, s\mathrm{Tr}\, \bar{\omega} = \frac{i}{2\pi}\mathrm{Tr}\, \bar{\omega} \quad . \qquad (2.6.4)$$

For $l = 2$, we find

$$\begin{aligned}
Q^{(3)}(\bar{\omega}, \bar{\Omega}) &= \left(\frac{i}{2\pi}\right)^2 \int_0^1 dt\, s\mathrm{Tr}(\bar{\omega}, td\bar{\omega} + t^2\bar{\omega} \wedge \bar{\omega}) \\
&= \left(\frac{i}{2\pi}\right)^2 \left\{\frac{1}{2}s\mathrm{Tr}(\bar{\omega}, d\bar{\omega}) + \frac{1}{3}s\mathrm{Tr}(\bar{\omega}, \bar{\omega}^2)\right\} \\
&= \left(\frac{i}{2\pi}\right)^2 \frac{1}{2}\left\{\frac{1}{2}\mathrm{Tr}(\bar{\omega} \wedge d\bar{\omega} + d\bar{\omega} \wedge \bar{\omega}) + \frac{1}{3}\mathrm{Tr}\, 2\bar{\omega}^3\right\} \\
&= \frac{1}{2}\left(\frac{i}{2\pi}\right)^2 \mathrm{Tr}(\bar{\omega} \wedge d\bar{\omega} + \frac{2}{3}\bar{\omega}^3) = \frac{-1}{8\pi^2}\mathrm{Tr}(\bar{\Omega} \wedge \bar{\omega} - \frac{1}{3}\bar{\omega}^3)
\end{aligned} \qquad (2.6.5)$$

and, similarly,

$$Q^{(5)}(\bar\omega,\bar\Omega) = \frac{1}{2}\left(\frac{i}{2\pi}\right)^3 \int_0^1 dt\, s\mathrm{Tr}(\bar\omega,(td\bar\omega + t^2\bar\omega\wedge\bar\omega)^2)$$

$$= \frac{1}{2}\left(\frac{i}{2\pi}\right)^3 [\frac{1}{3}s\mathrm{Tr}(\bar\omega,d\bar\omega,d\bar\omega) + \frac{1}{2}s\mathrm{Tr}(\bar\omega,d\bar\omega,\bar\omega^2)$$

$$+ \frac{1}{5}s\mathrm{Tr}(\bar\omega,\bar\omega^2,\bar\omega^2)]$$

$$= \frac{1}{6}\left(\frac{i}{2\pi}\right)^3 \mathrm{Tr}(\bar\omega\wedge d\bar\omega^2 + \frac{3}{2}\bar\omega^3\wedge d\bar\omega + \frac{3}{5}\bar\omega^5),$$

$$Q^{(7)}(\bar\omega,\bar\Omega) = \frac{1}{4!}\left(\frac{i}{2\pi}\right)^4 \mathrm{Tr}(\bar\omega\wedge d\bar\omega^3 + \frac{8}{5}\bar\omega^3\wedge d\bar\omega^2 + \frac{4}{5}\bar\omega\wedge d\bar\omega\wedge\bar\omega^2\wedge d\bar\omega$$

$$+ 2\bar\omega^5 d\bar\omega + \frac{4}{7}\bar\omega^7),$$

$$(2.6.6)$$

etc. (see section 10.13). In the above expressions the power of a form indicates the appropriate number of wedge products of the form by itself. Clearly, these expressions become simpler by omitting the symbol $\wedge$ completely; we shall adhere to this practice in chapter 10 (and in some expressions in this chapter).

It is illustrative of the manipulations that will be done in chapter 10 to derive the transgression formula (2.6.2) directly from the expression for the Chern character (2.5.12). If we ignore temporarily the factor $(1/l!)(i/2\pi)^l$ and, thinking of the Yang-Mills strength (cf. section 2.2 (b)), write A, F for $\bar\omega$, $\bar\Omega$, we find that the Chern character forms are expressed as

$$ch_l(F) \simeq \mathrm{Tr}\, F^l \; (= s\mathrm{Tr}\, F^l) \;. \qquad (2.6.7)$$

Now, using (2.2.16), we can explicitly check that the Chern characters are closed due to the presence of the trace:

$$d\, ch_l(F) \propto l\, \mathrm{Tr}(dF F^{l-1}) = l[\mathrm{Tr}(\mathscr{D}F F^{l-1}) - \mathrm{Tr}([A,F]F^{l-1})] = 0 \quad (2.6.8)$$

since $\mathrm{Tr}([A,F]F^{l-1})=0$. The gauge invariance of $ch_l(F)$ is obvious since F is gauge-covariant; also eq. (2.2.15) gives $\delta_\zeta ch_l(F) = l\, \mathrm{Tr}([F,\zeta]F^{l-1}) = 0$. Let δA now be any variation of A; then $\delta F = d(\delta A) + \delta A\wedge A + A\wedge\delta A$.

Under this variation, and using that $[A, F] = -dF$, we obtain

$$\delta ch_l(F) \propto l \, \mathrm{Tr}([d(\delta A) + \delta A \wedge A + A \wedge \delta A] \wedge F^{l-1})$$
$$= l \, \mathrm{Tr}(d(\delta A) \wedge F^{l-1} + \delta A \wedge A \wedge F^{l-1} - \delta A \wedge F^{l-1} \wedge A)$$
$$= l \, \mathrm{Tr}(d(\delta A) \wedge F^{l-1} + \delta A \wedge [A, F]F^{l-2} + \delta A \wedge F[A, F]F^{l-3} + \dots$$
$$+ \delta A \wedge F^{l-2} \wedge [A, F]) = l \, d[\mathrm{Tr}(\delta A F^{l-1})] \, .$$
$$(2.6.9)$$

Let now the connection and the curvature be given by $A_t = tA$, $F_t = tdA + t^2 A^2$, and let $\delta A_t \equiv dt \frac{\partial A_t}{\partial t} = dtA$. Then eq. (2.6.9) gives

$$ch_l(F) = dQ^{(2l-1)}(A, F) \, , \qquad (2.6.10)$$

and, restoring the factor omitted in (2.6.7), eq. (2.6.2) is recovered,

$$Q^{(2l-1)}(A, F) = \frac{1}{(l-1)!} \left(\frac{i}{2\pi}\right)^l \int_0^1 dt \, \mathrm{Tr}\{A(tdA + t^2 A^2)^{l-1}\} \, . \qquad (2.6.11)$$

Clearly, this computation is completely equivalent to the application of the homotopy formula (see (2.4.26)) for a local chart on which $\bar{\omega}_0 \equiv A_0 = 0$.

Behaviour of the Chern-Simons forms under a gauge transformation
The gauge invariance $\delta_\zeta ch_l(F) = 0$ gives

$$0 = \delta_\zeta(dQ^{(2l-1)}) = d(\delta_\zeta Q^{(2l-1)}) \, . \qquad (2.6.12)$$

Thus, *the gauge variation of the Chern-Simons form is a closed form* and, again locally, $\delta_\zeta Q^{(2l-1)}(A) = dQ^{(2l-2)}(\zeta, A)$ (this is intimately tied to the cohomological descent procedure that will be discussed in chapter 10). The Chern character forms are gauge-invariant; the Chern-Simons forms, which depend on A, are not. Let us now compute the variation of a Chern-Simons form under a gauge transformation by using the Cartan homotopy formula. Let

$$A_t^g = g^{-1}dg + tg^{-1}Ag \, , \quad F_t^g = dA_t^g + A_t^g \wedge A_t^g = g^{-1}(tdA + t^2 A \wedge A)g \qquad (2.6.13)$$

so that $A_0^g = g^{-1}dg$, $F_0^g = 0$. Then eq. (2.4.24) gives for the present case

$$Q^{(2l-1)}(A^g, F^g) - Q^{(2l-1)}(g^{-1}dg, 0) = (dk_{01} + k_{01}d)Q^{(2l-1)}(A_t^g, F_t^g) \, , \quad (2.6.14)$$

where $Q^{(2l-1)}(A, F)$ is the Chern-Simons form of $ch_l(F)$ so that $dQ^{(2l-1)}(A_t^g, F_t^g) = ch_l(F_t^g) = ch_l(F_t) = ch_l(tdA + t^2 A \wedge A)$. Thus, $k_{01}dQ^{(2l-1)}(A_t^g, F_t^g) = Q^{(2l-1)}(A, F)$, eq. (2.6.11), and we obtain

$$Q^{(2l-1)}(A^g, F^g) - Q^{(2l-1)}(A, F) = d\alpha^{(2l-2)}(A, F, a) + Q^{(2l-1)}(g^{-1}dg, 0) \, , \qquad (2.6.15)$$

where $Q^{(2l-1)}(g^{-1}dg, 0)$ is a closed form which because of its interest is discussed below and

$$\alpha^{(2l-2)}(A, F, a) = k_{01}Q^{(2l-1)}(A_t^g, F_t^g) = k_{01}Q^{(2l-1)}(tA + a, tdA + t^2A \wedge A),$$
(2.6.16)

with $a \equiv dgg^{-1}$. Eq. (2.6.15) gives the expression for the closed form that constitutes the gauge variation (2.6.12) of $Q^{(2l-1)}(A, F)$.

An interesting particular case of (2.6.11) appears when A is pure gauge, $A = g^{-1}dg$. Then $F(A = g^{-1}dg) = 0$ and, as a result, the Chern forms are identically zero. Then, since $d(g^{-1}dg) = -(g^{-1}dg) \wedge (g^{-1}dg)$, $Q^{(2l-1)}(g^{-1}dg, 0) \equiv Q^{(2l-1)}(g^{-1}dg)$ reads

$$Q^{(2l-1)}(g^{-1}dg) = \frac{1}{(l-1)!}\left(\frac{i}{2\pi}\right)^l \int_0^1 dt\,(t^2 - t)^{l-1}\text{Tr}\,(g^{-1}dg)^{2l-1} \quad . \quad (2.6.17)$$

The integral can be worked out by successive integrations by parts,

$$\int_0^1 dt\,(t^2 - t)^{l-1} = (-1)^{l-1}\frac{[(l-1)!]^2}{(2l-1)!} \quad (2.6.18)$$

(it is, in fact, $(-1)^{l-1}$ times the beta function $B(m, n) = \frac{\Gamma(m)\Gamma(n)}{\Gamma(m+n)}$ for $m = n = l$), so that

$$Q^{(2l-1)}(g^{-1}dg) = (-1)^{l-1}\left(\frac{i}{2\pi}\right)^l \frac{(l-1)!}{(2l-1)!}\text{Tr}(g^{-1}dg)^{2l-1} \quad . \quad (2.6.19)$$

Notice that, since the Chern class is zero, $Q^{(2l-1)}(g^{-1}dg)$ is closed (and so locally exact), a fact that can be checked directly since $d[\text{Tr}(g^{-1}dg)^{2l-1}] = -\text{Tr}(g^{-1}dg)^{2l} = 0$ (moving the last $g^{-1}dg$ to the first place under the trace requires the reordering of two odd forms and this introduces a minus sign). Consider eq. (2.6.19), for instance, when $l = 2$, $G = SU(2)$ and $M = S^3$. Then the integral of the above form on S^3 (the functions $g(x)$ appearing in (2.6.19) are in this case defined on S^3) gives the 'winding number' of the mapping $g : S^3 \rightarrow SU(2)$, $g : x \mapsto g(x)$. This is just the instanton number that characterizes (through $\pi_3(SU(2)) = Z$) the $SU(2)$-bundles over S^4 (section 2.8); it is given by

$$k = \frac{1}{24\pi^2}\int_{S^3} \text{Tr}(g^{-1}dg \wedge g^{-1}dg \wedge g^{-1}dg) \quad . \quad (2.6.20)$$

Eq. (2.6.20) is just a particular case of a more general situation, which can be formulated (Bott and Seeley) by means of the following

(2.6.1) THEOREM

Let V_n be an n-dimensional vector space over C, and let $G = \text{Aut } V$. Then $\pi_{2l-1}(G)$ is isomorphic to Z provided that $n \geq l$ and such an isomorphism is obtained by assigning to a map $f : S^{2l-1} \to G$ the integer

$$\int_{S^{2l-1}} f^*\omega \qquad (2.6.21)$$

where $\omega \doteq \left(\frac{i}{2\pi}\right)^l \frac{(l-1)!}{(2l-1)!} \text{Tr}(\theta \wedge \overset{2l-1}{\cdots} \wedge \theta)$ and θ is the $\mathscr{G}$-valued canonical form on G (Definition 1.7.1).

This is a consequence of the fact that $\text{Aut } V$ has the cohomology of its maximal compact subgroup, which may be taken as the subgroup $U(V)$ preserving a hermitian structure, and of the fact that $\pi_{2l-1}(U(n)) = Z$ for $n \geq l$ (cf. Bott's periodicity theorem, section 1.11).

2.7 The magnetic monopole

Ordinary electromagnetism is a $U(1)$-gauge theory. The Maxwell equations,

$$\nabla \times \mathbf{B} - \frac{\partial \mathbf{E}}{\partial x^0} = \mathbf{j} \quad , \quad \nabla \cdot \mathbf{E} = \rho \quad ;$$

$$\nabla \times \mathbf{E} + \frac{\partial \mathbf{B}}{\partial x^0} = 0 \quad , \quad \nabla \cdot \mathbf{B} = 0 \quad , \qquad (2.7.1)$$

are rewritten in terms of the field strength $F_{\mu\nu} = \partial_\mu A_\nu - \partial_\nu A_\mu$ and its dual $\mathscr{F}_{\mu\nu} = \frac{1}{2}\epsilon_{\mu\nu\rho\sigma}F^{\rho\sigma}$, $\mu = 0, 1, 2, 3$, as

$$\partial_\mu F^{\mu\nu} = j^\nu \quad , \quad \partial_\mu \mathscr{F}^{\mu\nu} = 0 \quad (\partial_\mu F_{\rho\sigma} + \partial_\rho F_{\sigma\mu} + \partial_\sigma F_{\mu\rho} = 0) \quad , \qquad (2.7.2)$$

where

$$F_{\mu\nu} = \begin{pmatrix} 0 & E_x & E_y & E_z \\ -E_x & 0 & -B_z & B_y \\ -E_y & B_z & 0 & -B_x \\ -E_z & -B_y & B_x & 0 \end{pmatrix} \quad ,$$

$$\mathscr{F}_{\mu\nu} = \begin{pmatrix} 0 & -B_x & -B_y & -B_z \\ B_x & 0 & -E_z & E_y \\ B_y & E_z & 0 & -E_x \\ B_z & -E_y & E_x & 0 \end{pmatrix} \quad ; \qquad (2.7.3)$$

if the indices are up, the first column and row change signs. The first set of equations follows from the Lagrangian density

$$\mathscr{L} = -\frac{1}{4}F_{\mu\nu}F^{\mu\nu} - j_\mu A^\mu \qquad (2.7.4)$$

$(-\frac{1}{2}F_{\mu\nu}F^{\mu\nu} = (\mathbf{E}^2 - \mathbf{B}^2))$ and the second follows from the definition $F_{\mu\nu} := \partial_\mu A_\nu - \partial_\nu A_\mu$ (or, equivalently, $\mathbf{E} = -\partial \mathbf{A}/\partial x^0 - \nabla\phi$, $\mathbf{B} = \nabla \times \mathbf{A}$, $\mathbf{A} \equiv A^i$). In terms of the field strength two-form $F = dA$ and the current one-form $j = j_\mu dx^\mu$ the action may be written (see eq. (1.5.20)) as

$$I = - \int \left(\frac{1}{2} F \wedge (*F) + A \wedge (*j) \right) \quad . \tag{2.7.5}$$

It is easy to find the equations of motion from (2.7.5). Under a variation $A \to A + \alpha$, the variation of the Lagrangian form comes from the terms linear in α, i.e., is given by $-\frac{1}{2}\{d\alpha\wedge *F + F\wedge *d\alpha\} - \alpha\wedge(*j) = -d\alpha\wedge *F - \alpha\wedge(*j)$ where (1.5.22) has been used. Inserting this into the integral and using that α does not give a contribution at the boundary, we obtain from (1.5.28) that

$$\delta I = - \int \alpha \wedge (*\delta F + *j) \tag{2.7.6}$$

and $\delta I = 0$ gives the Euler-Lagrange equations, $*\delta F + *j = 0$. Since $\delta = *d*$ for the four-dimensional Minkowski space and $*^2 = 1$ for $n = 4$, p odd (eq. (1.5.17)), the two sets in (2.7.2) are, respectively,

$$d * F = - * j \quad , \quad dF = 0 \quad , \tag{2.7.7}$$

where $*j = \frac{1}{3!}j^\mu \epsilon_{\mu\nu\rho\sigma} dx^\nu \wedge dx^\rho \wedge dx^\sigma$. Eqs. (2.7.7) may also be written in the form[*]

$$\delta F = -j \quad , \quad \delta(*F) = 0 \quad , \tag{2.7.8}$$

implying the electric current conservation equation $\delta j = 0$, $\partial_\mu j^\mu = 0$ in coordinates. In differential form the Lorentz gauge $\partial_\mu A^\mu = 0$ reads $\delta A = 0$ (cf. eq. (1.5.27)); thus, in this gauge, $-\delta F = j$ is equivalent to $-\delta dA = -(\delta d + d\delta)A = \partial^\mu \partial_\mu A = j$. An advantage of writing Maxwell's equations in the form (2.7.8) is that, unlike eqs. (2.7.2), they remain valid in any curved space; moving from the flat metric to a general (Minkowski signature) metric is automatically taken into account by the Hodge $*$ operator in δ.

The asymmetry of the two sets of equations suggests studying what would happen if the second set is completed by adding a (magnetic) current k, so that eqs. (2.7.2) become $\partial_\mu F^{\mu\nu} = j^\nu$, $\partial_\mu \mathscr{F}^{\mu\nu} = -k^\nu$ and eqs. (2.7.8) take the form

$$\delta F = -j \quad , \qquad \delta(*F) = k \quad (dF = - * k) \quad . \tag{2.7.9}$$

---

[*] We check, for instance, that $d * F = \frac{1}{2}\partial_\rho \mathscr{F}_{\mu\nu} dx^\mu \wedge dx^\nu \wedge dx^\rho$ and that, using (1.5.12), $\delta F = \frac{1}{4}\partial_\rho \epsilon_{\mu\nu\alpha\beta} F^{\alpha\beta} \epsilon^{\mu\nu\rho\sigma} dx_\sigma = \frac{1}{4}\partial_\rho F^{\alpha\beta}(-2!)(\delta_\alpha^\rho \delta_\beta^\sigma - \delta_\beta^\rho \delta_\alpha^\sigma)dx_\sigma = -\partial_\alpha F^{\alpha\beta} dx_\beta = -\partial^\alpha F_{\alpha\beta} dx^\beta$. Thus, $\delta F = -j$ reproduces the first equation in (2.7.2).

The electric charge is given by $q = \int_{M^3} *j = -\int_{S^2} *F$. The magnetic charge is given by $g = \int_{M^3} *k = -\int_{S^2} F$, and, in contrast with q, which is a Noether charge, its conservation is topological as we shall see.

In the form (2.7.1), the addition of k^μ means that Faraday's law and $\nabla \mathbf{B} = 0$ (which states the absence of magnetic monopoles) are replaced by

$$\nabla \times \mathbf{E} + \frac{\partial \mathbf{B}}{\partial x^0} = -\mathbf{k} \quad , \quad \nabla \mathbf{B} = \rho_m \quad , \qquad (2.7.10)$$

ρ_m corresponding to the density of magnetic charge (such a modification was already suggested by Heaviside in 1893). But, as is well known, a magnetic monopole cannot be described by a non-singular potential since $\nabla \mathbf{B} = \nabla(\nabla \times \mathbf{A}) \equiv 0$ or, equivalently, if $F = dA$, $dF = 0$. This argument can be rephrased: if the field strength F of the monopole were given by dA, applying Stokes' theorem twice to compute the magnetic flux out of the two hemispheres north and south of the common equator S^1 would give zero for the total flux out of the two-sphere S^2 surrounding the monopole, and not the required $4\pi g$. The circulation of A would be equivalently expressed as

$$\oint A = \Omega_+ = \Omega_- \quad , \qquad (2.7.11)$$

and this would give $\Omega_+ - \Omega_- = 0$ for the total flux out of S^2 since the flux *out* of the southern hemisphere is given by $-\Omega_-$. We can conclude in general that if a magnetic field is created ($\mathbf{B} = \nabla \times \mathbf{A}$) by a potential $\mathbf{A}$, the magnetic flux through any closed surface is zero.

It is not difficult, however, to find a *singular* potential for the monopole. For this, it is sufficient to imagine that the magnetic field created by a very thin solenoid looks like the field of a monopole outside of the solenoid itself. Graphically,

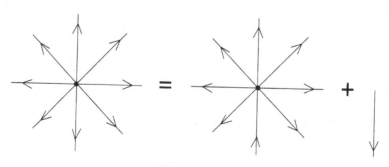

Figure 2.7.1

and

$$g\frac{\mathbf{r}}{r^3} = \nabla \wedge \mathbf{A}_{sol} - 4\pi g\theta(-z)\delta(x)\delta(y)\mathbf{n}_z \quad . \qquad (2.7.12)$$

The solenoid field $\mathbf{B}_{sol} = \nabla \wedge \mathbf{A}_{sol}$ is clearly sourceless, $\nabla \mathbf{B}_{sol} = 0$, even at the origin of R^3 but, as the above figure shows, the potential $\mathbf{A}$ for the monopole magnetic field has to be singular along the negative z-axis: this is the *Dirac string* in Dirac's famous 1931 description of the monopole. This is the content of equation (2.7.9): in terms of *singular* differential forms, it corresponds to writing $F = dA + \omega$, where the first term is necessarily sourceless and the second is the string-like term accounting for the r.h.s. of (2.7.9), $d\omega = -*k$.

The string of singularities of A (which invalidates the argument based on Stokes' theorem since it can no longer be used) is, however, the result of the non-trivial topology implied by the presence of the magnetic monopole. As was shown by Wu and Yang, this singularity can be avoided much in the same way that the singularities of the latitude and longitude parametrization of the earth's surface (the north and south poles) are eliminated by using two charts overlapping over the tropics. Thus, the string of singularities is neither a necessity nor a difficulty and, as is shown below, disappears when local charts are taken.

Let us remove the origin, assumed to be the monopole location. Of course, the electric field created by a point electric charge is also singular at the origin, and outside it satisfies $\nabla \mathbf{E} = 0$ in analogy with $\nabla \mathbf{B} = 0$; the difference between the two cases appears when we try to express the fields in terms of the potentials ϕ, $\mathbf{A}$ since, as we have seen, even if $\nabla \mathbf{B} = 0$ outside the origin, the monopole field cannot be written as $\mathbf{B} = \nabla \times \mathbf{A}$ in that region. In contrast, ϕ is given by the Coulomb potential. Once the singularity at the origin is removed, the three-dimensional space becomes $S^2 \times R$; R is topologically trivial and a sphere requires two charts. Using spherical coordinates (r, θ, ϕ) it is now easy to find the *local* expressions of the potential for the magnetic field $mg(\mathbf{r}/r^3)$ on the charts $(+, -)$ that exclude the (south, north) poles:

$$A_r^+ = 0 \quad , \quad A_\theta^+ = 0 \quad , \quad A_\phi^+ = \frac{m}{2r}\frac{(1 - \cos\theta)}{\sin\theta} \quad ;$$
$$A_r^- = 0 \quad , \quad A_\theta^- = 0 \quad , \quad A_\phi^- = -\frac{m}{2r}\frac{(1 + \cos\theta)}{\sin\theta} \quad , \tag{2.7.13}$$

where m is an integer (the unit g of magnetic charge has been set equal to one). These potentials are singular along the negative $(\theta = \pi)$ and positive $(\theta = 0)$ z-axis, but these regions are not included in the respective chart domains. Nevertheless $F = dA^{\pm} = d(\frac{m}{2}(\pm 1 - \cos\theta)d\phi) = d\left(\frac{m}{2r}\frac{xdy - ydx}{z \pm r}\right) = \frac{m}{2}\frac{1}{r^3}(xdy \wedge dz + ydz \wedge dx + zdx \wedge dy) = \frac{m}{2}\sin\theta d\theta \wedge d\phi$ so that $\mathbf{B} = \frac{m}{2}\frac{\mathbf{r}}{r^3}$; eq. (2.7.9) is $dF = 2\pi m\delta(\mathbf{x})dx \wedge dy \wedge dz$ and zero for $\mathbf{x} \neq 0$. Since both $A_{\pm}$ describe the same magnetic field, they have to be related by a gauge transformation. With $g = e^{im\phi}$ (which is singular along the z axis where ϕ

has no meaning) we check that

$$\frac{im}{r\sin\theta} = i\mathbf{A}^+ - i\mathbf{A}^- = g^{-1}\nabla g = im\nabla\phi \quad, \tag{2.7.14}$$

since $\nabla = (\partial/\partial r, (1/r)\partial/\partial\theta, (1/r\sin\theta)\partial/\partial\phi)$. Equation (2.7.14) is a specific example of eq. (2.2.21). The case $m = 1$ corresponds to A being a connection form on the Hopf bundle $SU(2)(U(1), S^2)$; other values of m correspond to other $U(1)$-fibrations of S^2. Thus, singular potentials are avoided by using non-trivial bundles and the appropriate local charts.

Let us now check with this example that the characteristic classes are topological numbers associated with the bundle itself, and not with the curvature form being integrated. Let us compute the Chern number (there is only one, $C_1(P) = \frac{i}{2\pi}\int_{S^2}\bar{\Omega}$) of the different $P(U(1), S^2)$ fibrations of S^2. Using the previous expression for $\bar{\Omega} = iF$ (we are not including in this simple $U(1)$ case the i in the expressions of A (see eq. (2.7.14)) and F so that they are here real rather than imaginary (cf. section 2.2 (b))) it is found that

$$C_1 = \frac{-1}{2\pi}\int_{S^2} F = \frac{-m}{4\pi}\int_{S^2}(x\,dy\wedge dz + y\,dz\wedge dx + z\,dx\wedge dy) = -m \quad. \tag{2.7.15}$$

In terms of $A^\pm$, the same calculation gives

$$C_1 = \frac{i}{2\pi}(\int_{S^+} idA_+ + \int_{S^-} idA_-) = \frac{i}{2\pi}\oint_{S^1}(iA_+ - iA_-)$$
$$= \frac{i}{2\pi}\oint_{S^1} g^{-1}dg = \frac{-1}{2\pi}\oint_{S^1} m\,d\phi = -m \quad. \tag{2.7.16}$$

This shows explicitly that C_1 is independent of the gauge field (connection). *It depends only on the gauge transformation* $g^{-1}dg$, i.e. on the winding number of the mapping $S^1(\text{equator}) \longrightarrow S^1(\text{structure group})$: if $g = e^{im\phi}$, the group winds around the circle m times; if m is negative, this means that the winding is in the opposite direction; only the transition function enters into (2.7.16). The integer m characterizes the bundle topologically; any connection on the given $P(G, M)$ bundle will lead to the same m. This underlines the fact that the magnetic charge, in contrast with the electric one, is a topological (rather than Noether) charge so that it is identically conserved. In fact, the possible form of a magnetic monopole field strength F is quite determined: if F is the two-form on S^2 describing a monopole, it satisfies $\delta F = 0$ and $dF = 0$. Hence F is an harmonic two-form, $\Delta F = 0$, and as a consequence of Theorem 1.5.3 it is proportional to the surface element on S^2 so that $\int_{S^2} F$ is 4π times a constant. The argument can be extended, in fact, to higher order magnetic antisymmetric tensors[*] where F becomes proportional to the (hyper)surface element of S^{p+2}.

To conclude this discussion of the magnetic monopole let us give Dirac's quantization condition. It can be derived in various equivalent ways. In the original Dirac formulation with the string, the Dirac quantization comes by imposing its unobservability: it results from the condition that the flux along the Dirac string should not give rise to an Aharonov-Bohm effect. In Wu and Yang's fibre bundle description this condition can be obtained by noticing that, since there are two local potentials $A^{\pm}$, the Schrödinger equation should take this fact into account. This is tantamount to saying that the wavefunction becomes a *wavesection* (a section of the associated complex vector line bundle) and that, since the bundle is non-trivial, two local wavesections $\psi^{\pm}$ are needed. There are, thus, two Schrödinger equations for the respective charts

$$\left[\frac{1}{2M} \left(p - \frac{q}{c} A^{\pm} \right) + V \right] \psi^{\pm} = E \psi^{\pm} \ , \qquad (2.7.17)$$

and ψ^{+} and ψ^{-} are related by the exponential of the gauge transformation which relates A^{+} with A^{-}. Including the magnetic charge g explicitly, this means that $\psi^{+} = \psi^{-} exp \, i\frac{qg2\phi}{\hbar c}$, and the consistency of the solution requires that $\frac{2qg}{\hbar c} = m \in Z$. This is Dirac's quantization condition for the product of the electric and monopole charges; it follows from it that if there is a monopole somewhere in the universe, the electric charges have to be quantized, a consequence which constituted the main theoretical argument for introducing magnetic monopoles. This m is, of course, the same m that appeared before and which in the previous analysis also had to be an integer[†].

---

[*] It is interesting to remark that it is possible to introduce an *abelian* 'electrodynamics' for arbitrary order tensors in a D-dimensional space based on the same two equations $\delta F = -j$ and $\delta(*F)=k$. In it, higher order $(p+1)$-forms couple to objects of 'extension' p in much the same way as a one-form potential couples to point $(p=0)$ particles. The form F is taken to be a $(p+2)$-form and, consequently, $*F$ is a $(D-(p+2))$-form. The 'electric' and 'magnetic' currents are forms of order $(p+1)$ (j) and $D-(p+3)$ (k). This scheme describes an 'electric' object of 'spatial dimension' p and a 'magnetic' one of spatial dimension $p^{\#}=D-4-p$; $q_{(p+1)}$ and $g_{(D-(p+3))}$ have dimensions $[q_{(p+1)}]=-((D/2)-p-2)$, $[g_{(D-p-3)}]=((D/2)-p-2)$ in natural units, and the product qg satisfies a Dirac quantization condition. The Minkowski case, $D=4$, is rather unique in the sense that it corresponds to both electric and magnetic dimensionless point-like charges; $p=0=p^{\#}$, and F and $*F$ are forms of equal order.

[†] The previous geometrical discussion of the (undiscovered!) magnetic monopole provides us with a motivation for reproducing here a few opinions of Dirac, Einstein and Yang on the interplay of physics and mathematics, of which gauge theories constitute perhaps the paramount example. In his classic 1931 paper on the monopole, Dirac wrote: 'There are at present fundamental problems in theoretical physics awaiting solution, *e.g.*, the relativistic formulation of quantum mechanics and the nature of atomic nuclei (to be followed by more difficult ones as the problem of life), the solution of which problems will presumably require a more drastic revision of our fundamental concepts than any that have gone before. Quite likely these changes will be so great that it will be beyond the power of human intelligence to get the necessary new ideas by direct attempts

2.8 Yang-Mills instantons

Let us now discuss the case of Yang-Mills *instantons* or *pseudoparticles*. Consider a Yang-Mills theory on a four-dimensional manifold with an $SU(2)$ gauge group. Let $A = A^j T_j$ and $T_j = (\sigma_j/2i)$ where σ_j are the Pauli matrices (1.10.3) so that $[T_i, T_j] = \epsilon_{ijk} T_k$ (cf. eqs. (2.2.6), (2.2.7)). In the absence of matter couplings, the Lagrangian density is expressed by the four-form $Tr(F \wedge *F)$ and the action

$$
\begin{aligned}
I &= \int Tr(F \wedge *F) = -\frac{1}{2} \int F^i \wedge (*F_i) \\
&= \frac{1}{2} \int d^4x \, Tr(F_{\mu\nu} F^{\mu\nu}) = -\frac{1}{4} \int d^4x \, F^i_{\mu\nu} F^{\mu\nu}_i
\end{aligned}
\tag{2.8.1}
$$

(see eqs. (1.5.20) and (2.2.7)) gives the Euler-Lagrange (Yang-Mills) equations,

$$
\mathscr{D}(*F) = d(*F) + A \wedge (*F) - (*F) \wedge A = 0 \quad (\mathscr{D}_\mu F^{\mu\nu} = 0) \quad , \tag{2.8.2}
$$

for the Yang-Mills fields A^i. In contrast with the abelian Maxwell case, where $\mathscr{D}$ reduces to d, solving these equations (with the appropriate boundary conditions) is a difficult problem since they are highly non-linear (the curvature is a quadratic expression in the connection, and thus the equations contain quadratic and cubic terms in the Yang-Mills potentials). However, this can be made much simpler if it is noticed that

---

to formulate the experimental data in mathematical terms. The theoretical worker in the future will therefore have to proceed in a more indirect way. The most powerful method of advance that can be suggested at present is to employ all the resources of pure mathematics in attempts to perfect and generalize the mathematical formalism that forms the existing basis of theoretical physics, and after each success in this direction, to try to interpret the new mathematical features in terms of physical entities.'

On his part, Einstein said in 1934: 'The theoretical scientist is compelled in a increasing degree to be guided by purely mathematical, formal considerations... The theorist who undertakes such a labor should not be carped at as "fanciful"; on the contrary, he should be granted the right to give free rein to his fancy, for there is no other way to the goal' (*The problem of space, ether, and the field in physics*). In 1950, he added: 'On the other hand, it must be conceded that a theory has an important advantage if its basic concepts and fundamental hypotheses are "close to experience", and greater confidence in such a theory is certainly justified. There is less danger of going completely astray, particularly since it takes so much less time and effort to disprove such theories by experience. Yet more and more, as the depth of our knowledge increases, we must give up this advantage in our quest for logical simplicity and uniformity in the foundations of physical theory...' (*Scientific American*, 1950).

C.N. Yang also shares the belief in the relevance of the fundamental mathematical structures in the formulation of the basic theories of physics. But he also said in 1979: 'Deep as the relationship is between mathematics and physics, it would be wrong, however, to think that the two disciplines overlap that much. They do not. And they have their separate aims and tastes. They have distinctly different value judgements, and they have different traditions. At the fundamental conceptual level they *amazingly* share some concepts, but even there, the life force of each discipline runs along its own veins.' (*Einstein and his impact on the physics of the second half of the century*, second Marcel Grossmann meeting).

For other aspects of the relation between mathematics and physics see the essay by Jaffe and Quinn (1993) and the interesting correspondence motivated by this paper.

the *purely geometrical* Bianchi equations (2.2.16), $\mathscr{D}F = 0$, become the field equations if $*F = \lambda F$. For *Euclidean* (in general, elliptic) *space*, this implies $\lambda = \pm 1$ (section 1.5). This means that it is possible to find solutions of the second order equations (2.8.2) by solving the system of first-order equations in A^i

$$*F = \pm F \quad . \tag{2.8.3}$$

For the solutions satisfying (2.8.3) the action becomes $\pm \int \text{Tr}(F \wedge F)$. As a result, the integrand no longer depends on the metric on the four-dimensional manifold (which entered through $*$) and becomes purely topological. It is then not surprising that these solutions are characterized by a topological number which, as the expression of the above integral for (2.8.3) shows, is the second Chern class.

The solutions which make the Euclidean Yang-Mills action finite and which satisfy the self-duality and anti-self-duality conditions are called instanton ($*F=F$) and anti-instanton ($*F=-F$) solutions. They are, in fact, regular solutions which minimize the action. To see this let us write the two-form F_i as the sum of its self-dual (s) and anti-self-dual (a) parts, $F_i = (F_i^s + F_i^a)$, $F_i^{s,a} = \frac{1}{2}(F_i \pm *F_i)$. Then the (positive) Euclidean action may be written (sum over the group index understood)

$$I_E = \frac{1}{2} \int (F_i^s + F_i^a) \wedge (*[F_i^s + F_i^a]) = \frac{1}{2}[(F_i^s, F_i^s) + (F_i^a, F_i^a)] \quad , \tag{2.8.4}$$

where we have used the notation of (1.5.23) and the fact, which easily follows from that expression, that the decomposition $F = F^s + F^a$ of F is an orthogonal decomposition, $(F^s, F^a) = 0$. We may now compare (2.8.4) with the second Chern number (see (2.5.7))

$$C_2 = \frac{-1}{16\pi^2} \int F_i \wedge F_i = \frac{-1}{16\pi^2} \int (F_i^s + F_i^a) \wedge (*[F_i^s - F_i^a])$$
$$= \frac{-1}{16\pi^2}[(F_i^s, F_i^s) - (F_i^a, F_i^a)] \quad . \tag{2.8.5}$$

Since (,) is positive definite for a Riemannian metric, it follows that

$$2I_E \geq |(F_i^s, F_i^s) - (F_i^a, F_i^a)| \quad , \quad I_E \geq 8\pi^2|C_2| \quad . \tag{2.8.6}$$

This bound, which determines the absolute minima of the action for each C_2, is saturated when $F^a = 0$ ($C_2 < 0$) or $F_i^s = 0$ ($C_2 > 0$), i.e. when F_i is self-dual or anti-self-dual[*].

---

[*] It is worth mentioning that the self-duality condition is *conformally invariant* (in fact, the Lagrangian form $F_i \wedge *F_i$, and hence the Lagrange equations (2.8.2), are conformally invariant). This follows from the fact that conformal transformations (see section 9.7) are defined by the property that they re-scale the metric $g'_{ij}(x) = \lambda(x)g_{ij}(x)$, $\lambda(x) > 0$ ($g'^{ij} = \lambda^{-1}g^{ij}$). As a result, if we now compute (1.5.15) for the $*'$ operator (which uses g') we find $(*'\alpha)_{j_{p+1}\cdots j_n} =$

Let us now look at the instanton solutions. We shall not derive their explicit form, and will restrict ourselves to checking that they are classified topologically by the second Chern class which in turn characterizes the (instanton) $SU(2)$-bundles over S^4. At first sight, we might be tempted to say that the finiteness of the action requires that F must fall off sufficiently fast as the radius goes to infinity ($F = O(1/r^3)$ assuming a power behaviour) and that A should produce that behaviour. This is not necessary, however, since the vanishing of F at the boundary may also be achieved by having a potential which becomes pure gauge at infinity sufficiently fast. Assuming a power behaviour, this implies that for large r the potential should be of the form

$$gdg^{-1} + O(1/r^2) \tag{2.8.7}$$

and not simply $O(1/r^2)$.

If the points of fixed large r in Euclidean R^4 space are visualized as a three-sphere, the g in (2.8.7) defines a mapping of S^3 in $SU(2)$. The mappings $g : S^3 \longrightarrow SU(2) \sim S^3$ fall into non-homotopic classes, $\pi_3(SU(2)) = Z$ (section 1.11), and they are easy to construct. For instance, the constant mapping $g_0 : x \in S^3 \mapsto e \in SU(2)$ characterizes the trivial class. The mapping

$$g_1 = g : x \in S^3 \mapsto g_1(x) = \frac{1}{r}(\mathbf{1}x_4 - i\sigma^j x_j) \quad, \tag{2.8.8}$$

where $r = (x_4^2 + \mathbf{x}^2)^{1/2}$ and σ^j are the Pauli matrices, is singular at $r = 0$ and identifies the three-sphere $x_4^2 + \mathbf{x}^2 = r^2$ with $SU(2)$; it is the identity map[*] (it is obvious that $g_1 g_1^\dagger = 1$ and thus $g_1(x) \in SU(2)$). The construction of the general case is straightforward: the mapping g_k,

$$g_k : x \in S^3 \mapsto g_k(x) = \frac{1}{r^k}(\mathbf{1}x_4 - i\sigma^j x_j)^k \quad, \tag{2.8.9}$$

corresponds to the class k. These mappings cannot be deformed into each other, and they characterize accordingly a solution whose asymptotic form is (2.8.7).

$(1/p!)g^{i_1 j_1} \ldots g^{i_p j_p} \epsilon_{j_1 \ldots j_n} \sqrt{g'} \alpha_{i_1 \ldots i_p} = (1/p!)\lambda^{-p} g^{i_1 j_1} \ldots g^{i_p j_p} \epsilon_{j_1 \ldots j_n} \lambda^{n/2} \sqrt{g} \alpha_{i_1 \ldots i_p}$. In other words, $(*'\alpha) = \lambda^{\frac{n}{2}-p}(*\alpha)$ and the $(n-p)$-forms $*\alpha$ and $*'\alpha$ are conformally related. Let now f be an orientation preserving conformal mapping, and let $f^*g = g' = \lambda(x)g$ $((f^*g)_{ij} \equiv g'_{ij})$. Then Theorem 1.5.1 gives $f^*(*\beta) = *'f^*(\beta) = \lambda^{\frac{n}{2}-p} * (f^*\beta)$ using the previous result for $\alpha = f^*\beta$. Thus, if n is even and $p = n/2$ (our case), the pull-back f^* of the conformal mapping commutes with the Hodge operator, $f^*(*\beta) = *(f^*\beta)$, which shows that the duality equations $*F_i = \pm F_i$ are conformally invariant. This has the effect that the instanton solution over R^4 can be extended to the conformal compactification of R^4, i.e. to the sphere S^4 (see below).

[*] In terms of the Euler angles, $x^4 = \cos\frac{\theta}{2}\cos\frac{1}{2}(\psi+\phi)$, $x^3 = \cos\frac{\theta}{2}\sin\frac{1}{2}(\psi+\phi)$, $x^1 = \sin\frac{\theta}{2}\cos\frac{1}{2}(\psi-\phi)$, $x^2 = \sin\frac{\theta}{2}\sin\frac{1}{2}(\psi-\phi)$ (i.e., $x^4 = y^1, x^3 = y^2, x^1 = y^4$ and $x^2 = -y^3$ of (1.3.15)).

Consider now the case for which g is given by (2.8.8), which corresponds to the single instanton solution. It may be shown, by solving $F = *F$ for A with the ansatz $A = f(r)g^{-1}(x)dg(x)$, that the (one-instanton) solution is given by

$$A = \frac{r^2}{r^2 + a^2}g^{-1}dg \quad , \tag{2.8.10}$$

where a is a parameter referred to as the 'instanton size'. In fact, this is the form of an instanton centred at the origin; if not, r is replaced by $(r - r_0)$. This is due to the translation invariance; dilatations modify the parameter a. Clearly, A is well behaved at the origin ($A(r = 0) = 0$) but it cannot be extended to infinity since it does not decay sufficiently fast for $r \longrightarrow \infty$, eq. (2.8.7) ($g^{-1}dg \simeq 1/r$). If we make a gauge transformation by g^{-1} we get

$$A' = gAg^{-1} + gdg^{-1} = \frac{a^2}{r^2 + a^2}gdg^{-1} \quad . \tag{2.8.11}$$

We now have a well-behaved potential at $r \longrightarrow \infty$ ($A' \simeq 1/r^3$), but in contrast with A it is singular at the origin. Thus, we have a situation analogous to that of the monopole: the bad behaviour of A, A' at $r \longrightarrow \infty$ and $r=0$ respectively corresponds to having defined local representatives on the four-sphere S^4 (one-point compactification of the Euclidean space, $S^4 = R^4 \cup \{\infty\}$) in which the north and south poles are identified with $r = 0, \infty$ respectively. Thus, eq. (2.8.11) may be rewritten, in analogy to eqs. (2.2.21) and (2.7.14), as

$$A^+ = g^{-1}A^-g + g^{-1}dg \quad , \tag{2.8.12}$$

where $A^{\pm}$ are defined on charts excluding the south and north poles. It is seen that the instanton solution corresponds to a specific $SU(2)$-fibration of the S^4 sphere; it is, in fact, the Hopf fibration $S^7(SU(2), S^4)$ (the bundle corresponding to the trivial mapping g_0 is the trivial bundle $S^3 \times SU(2)$).

It now remains to check that, as advertised, the transition function g determines the winding number, one, of the instanton solution by computing the second Chern class. We know by the general analysis that $C_2(P)$, where P refers to the specific bundle being considered, does not depend on A. We may, nevertheless, use this example to check this once more. From (2.5.7) we get

$$C_2 = \frac{1}{8\pi^2}\int_{S^4_+} \mathrm{Tr}(F^+ \wedge F^+) + \frac{1}{8\pi^2}\int_{S^4_-} \mathrm{Tr}(F^- \wedge F^-) \quad . \tag{2.8.13}$$

Since $\mathrm{Tr}(F \wedge F) = d\mathrm{Tr}(F \wedge A - \frac{1}{3}A \wedge A \wedge A)$ (this is easily checked by using that $F = dA + A \wedge A$, but see the first equation in (2.6.5) for the

Chern-Simons form ch_2), eq. (2.8.13) gives

$$C_2 = \frac{1}{8\pi^2} \int_{S^3} \{ \mathrm{Tr}(F^+ \wedge A^+ - \frac{1}{3}(A^+)^3) - \mathrm{Tr}(F^- \wedge A^- - \frac{1}{3}(A^-)^3) \} \quad (2.8.14)$$

where the relative minus sign in moving from S^4 to S^3 is due to the opposite orientation of the boundary S^3 with respect to the two charts $S^4_{\pm}$. Using now eq. (2.8.12), C_2 reduces to (minus) eq. (2.6.20). To compute this expression, we notice that, since g is the identity map (2.8.8),

$$\mathrm{Tr}(g^{-1}dg \wedge g^{-1}dg \wedge g^{-1}dg) = \mathrm{Tr}(\epsilon_{ijk} i\sigma^i i\sigma^j i\sigma^k)\mu = 12\mu \quad , \qquad (2.8.15)$$

where μ is the hypersurface element on S^3. Since this surface is given* by $2\pi^2$, we obtain that $C_2 = -1$. For the higher order maps g_k, one finds that $C_2 = -k$, where k is the *instanton number* (the negative values of k appear when the winding maps a triad of tangent vectors into another one of contrary handedness). The reason for this result is that we can look at g_k as the product $g_k = g^k g$, and the winding number of the composite transformation is the sum $1 + \underset{...}{k} + 1 = k$ of the winding numbers. This can be checked by direct computation: if $g = g_a g_b$, its winding number is obtained by integrating $\mathrm{Tr}[(g_a g_b)^{-1}d(g_a g_b) \wedge (g_a g_b)^{-1}d(g_a g_b) \wedge (g_a g_b)^{-1}d(g_a g_b)]$ over S^3. But since for $g = g_a g_b$

$$\mathrm{Tr}(g^{-1}dg \wedge g^{-1}dg \wedge g^{-1}dg) =$$
$$\sum_{i=a,b} \mathrm{Tr}(g_i^{-1}dg_i \wedge g_i^{-1}dg_i \wedge g_i^{-1}dg_i) + d(3\mathrm{Tr}(g_a^{-1}dg_a \wedge g_b^{-1}dg_b)) \quad ,$$

$$(2.8.16)$$

it follows that the winding number of the mapping g is the sum of those of g_a, g_b. This construction also shows that if $g(x)$ is replaced by $g(x)h(x)$, where $h(x)$ is in the zero homotopy class, the winding number is not changed. The topologically trivial gauge transformations $h(x)$, *i.e.* those that are homotopic with the trivial mapping g_0, are called in physics *small gauge transformations*; those altering C_2 are called *large gauge transformations*.

Clearly, the analysis may be repeated for the anti-self-dual solutions by solving $F = - * F$; these are called *anti-instanton* solutions. In general, the solutions A for $F = \pm * F$ and $|k| > 1$ are called *multi-instantons*; they are connections on $SU(2)$-bundles over S^4 characterized by k. They minimize the L^2-norm $\|F\|^2$ of the curvature F; they are classified by the integer k (the instanton number) and correspond to different asymptotic conditions for A. These topologically separated solutions have $\|F\|^2 = I_E = 8\pi^2 |k|$

---

* The (hyper)surface of the n-sphere of radius R is given by $2\pi^{(n+1)/2}R^n / \Gamma((n+1)/2)$.

and are not only of great mathematical interest[†]. They are related with the existence of a countable infinity of different vacua labelled by k, a fact which is relevant for the vacuum structure of gauge theories and their path integral formulation.

2.9 Pontrjagin classes and the Euler class

The Pontrjagin and Euler classes are characteristic classes associated with real vector bundles and their associated principal bundles. Let R^m be the fibre and let $GL(m, R)$ be the group of η_E and of its associated principal bundle; the elements $X \in gl(m, R)$ are $m \times m$ matrices with real entries. Then it is clear that the polynomial (cf. (2.5.1), where X is complex)

$$P(X, \lambda) = \det(1\lambda + X) = \sum_{l=0}^{m} p_l(X)\lambda^{m-l} \tag{2.9.1}$$

is $Ad(GL(m, R))$-invariant. Let now ω be a connection on P and Ω its curvature. By the Chern-Weil theorem, the forms $p_l(\Omega)$ obtained by replacing X by $\Omega/2\pi$ in the polynomial coefficients $p_l(X)$ project as before onto a unique closed form $p_l(\bar{\Omega})$ on M. The Pontrjagin classes are then given by the following

(2.9.1) DEFINITION (*Pontrjagin classes*)

The closed 4k-form $p_{2k}(\bar{\Omega})$ on M represents the k-th Pontrjagin class of the real bundle over M with fibre R^m and structure group $GL(m, R)$; it is an element of $H_{DR}^{4k}(M, R)$. The form $P(\bar{\Omega}, \lambda = 1)$ defines the total Pontrjagin class. The Pontrjagin classes of the tangent bundle of a manifold M are the Pontrjagin classes $p_{2k}(\tau(M))$ of the manifold (frequently denoted $p_{2k}(M)$). Clearly, $p_{2k} = 0$ if $2k > m$ or $2k > n = \dim M$ (Note: p_{2k} is frequently denoted p_k).

We see from their definition that the Pontrjagin classes are the analogue for the real bundles of the Chern classes for complex vector bundles (Definition 2.5.1). The reason to consider only the forms given by the even terms p_{2k} in (2.9.1) is that the forms $p_{2k+1}(\bar{\Omega})$ are cohomologous to zero. This is because, by introducing a metric on the fibres of η_E, the

---

[†] The study of the *moduli space* of all solutions of the self-dual Yang-Mills equations (*i.e.*, the orbit space of these solutions under the action of the group of gauge transformations) on four-dimensional manifolds has played an essential rôle in the modern analysis of the structure of the manifolds themselves. The combination of Donaldson's results for a *smooth*, simply connected four-dimensional manifold (obtained by using the Yang-Mills equations) with the classification by Freedman of simply-connected *topological* four-manifolds performed a dozen years ago implies a striking result: although there is only one smooth structure on R^n for $n \neq 4$, on R^4 there are uncountably many *exotic* or *fake* smooth structures, *i.e.* smooth four-manifolds R_f^4 homeomorphic to, but not diffeomorphic to, R^4 with the standard differentiable structure R_{st}^4.

structure group can be restricted to the m-dimensional orthogonal group. By reducing the structure group from $GL(m, R)$ to $O(m)$ (section 1.3), we can always use a connection which is $o(m)$-valued. As a result, the Lie algebra matrices X in (2.9.1) may be taken to be skew-symmetric, and then the equality $\det(1\lambda + X) = \det[(1\lambda + X)^T] = \det(1\lambda - X)$ implies that $\sum_{l=0}^{m} p_l(X)\lambda^{m-l} = \sum_{l=0}^{m} (-1)^l p_l(X)\lambda^{m-l}$ and hence that $p_l(X) = 0$ for l odd. Thus, we can restrict ourselves to the $o(m)$-valued $4k$-forms $p_{2k}(\bar{\Omega})$; their cohomology class is independent of the metric and of the connection. Notice that, in contrast with the Chern complex case, the characteristic forms are different for $GL(m, R)$ and $O(m)$ vector bundles; nevertheless their characteristic *classes* are the same. This follows from the fact that any $GL(m, R)$-invariant polynomial $P(\bar{\Omega})$ may be written as the sum $P = P_1(p_1, p_2, ..., p_{max}) + P_2(\bar{\Omega})$, where P_1 is a polynomial and P_2 is zero when Ω is $o(m)$-valued and cohomologous to zero in general, *i.e.* even when Ω takes values in $gl(m, R)$, in which case $P_2(\bar{\Omega}) \neq 0$. This implies that the $GL(m, R)$ and $O(m)$ characteristic *classes* are the same even if the characteristic forms differ.

In order to compute the expansion (2.9.1), let us consider polynomials on $o(m)$ and introduce the normalization factor by writing $X = \frac{1}{2\pi} Y$, $Y \in o(m)$. Since X is antisymmetric its diagonal form has complex eigenvalues. By means of a similarity transformation $Y/2\pi$ can be put in the form

$$\frac{1}{2\pi} Y' = \begin{pmatrix} 0 & \lambda_1/2\pi & & & \\ -\lambda_1/2\pi & 0 & & & \\ & & 0 & \lambda_2/2\pi & \\ & & -\lambda_2/2\pi & 0 & \\ & & & & \ddots \end{pmatrix}, \qquad (2.9.2)$$

which means that their diagonal form is given by

$$\frac{1}{2\pi} Y'' = \begin{pmatrix} i\lambda_1/2\pi & & & & \\ & -i\lambda_1/2\pi & & & \\ & & i\lambda_2/2\pi & & \\ & & & -i\lambda_2/2\pi & \\ & & & & \ddots \end{pmatrix}, \qquad (2.9.3)$$

where the last element is zero if m is odd. Clearly, these transformations leave $p_l(X)$ unaltered. Using (2.9.3), we find that

$$p_2 = \left(\frac{1}{2\pi}\right)^2 \sum_{i_1} \lambda_{i_1}^2 \ , \quad p_4 = \left(\frac{1}{2\pi}\right)^4 \sum_{i_1 < i_2} \lambda_{i_1}^2 \lambda_{i_2}^2 \ , \quad \cdots \ ,$$

$$p_{2k} = \left(\frac{1}{2\pi}\right)^{2k} \sum_{i_1 < ... < i_k} \lambda_{i_1}^2 ... \lambda_{i_k}^2 \ , \quad \cdots \ , \quad p_{2[m]} = \left(\frac{1}{2\pi}\right)^{2[m]} \lambda_1^2 \lambda_2^2 ... \lambda_{[m]}^2 \ ,$$

$$(2.9.4)$$

where the indices run from 1 to $[m]$; $[m]$ is given by m itself if m is even and by $(m-1)$ if it is odd. Let us now compute the first few Pontrjagin classes. In order to replace the above products of eigenvalues by products of traces of (even) powers of Y, we first check that, since

$$\text{Tr}\, Y^{2k} = (-1)^k 2 \sum_{i=1}^{[m]} \lambda_i^{2k} \quad , \tag{2.9.5}$$

the first terms in (2.9.4) can be written in the form

$$\sum_i \lambda_i^2 = -\frac{1}{2}\text{Tr}\, Y^2 \quad ,$$

$$\sum_{i_1 < i_2} \lambda_{i_1}^2 \lambda_{i_2}^2 = \frac{1}{2}[(\sum_i \lambda_i^2)^2 - \sum_i \lambda_i^4] = \frac{1}{2}[\frac{1}{4}(\text{Tr}\, Y^2)^2 - \frac{1}{2}\text{Tr}\, Y^4] \quad ,$$

$$\sum_{i_1 < i_2 < i_3} \lambda_{i_1}^2 \lambda_{i_2}^2 \lambda_{i_3}^2 = \frac{1}{6}[(\sum_i \lambda_i^2)^3 + 2\sum_i \lambda_i^6 - 3(\sum_i \lambda_i^2)(\sum_i \lambda_i^4)]$$

$$= \frac{1}{6}[-\frac{1}{2^3}(\text{Tr}\, Y^2)^3 - \text{Tr}\, Y^6 + \frac{3}{4}\text{Tr}\, Y^2\text{Tr}\, Y^4] \quad , \quad \text{etc.} \tag{2.9.6}$$

Thus, replacing Y by the curvature form and using (2.9.4), the first three Pontrjagin classes are characterized by the following closed 4-, 8- and 12-forms on M:

$$p_2(\bar{\Omega}) = -\frac{1}{2}\left(\frac{1}{2\pi}\right)^2 \text{Tr}(\bar{\Omega}^2) \quad ,$$

$$p_4(\bar{\Omega}) = \frac{1}{8}\left(\frac{1}{2\pi}\right)^4 ((\text{Tr}\,\bar{\Omega}^2)^2 - 2\text{Tr}\,\bar{\Omega}^4) \quad , \tag{2.9.7}$$

$$p_6(\bar{\Omega}) = \frac{1}{48}\left(\frac{1}{2\pi}\right)^6 (-(\text{Tr}\,\bar{\Omega}^2)^3 - 8\text{Tr}\,\bar{\Omega}^6 + 6\text{Tr}\,\bar{\Omega}^2\,\text{Tr}\,\bar{\Omega}^4) \quad .$$

The total Pontrjagin class is given by $p(\bar{\Omega}) = 1 + p_2(\bar{\Omega}) + p_4(\bar{\Omega}) + ...$

In general compact form, the Pontrjagin classes are represented by (cf. (2.5.3)) the forms

$$p_{2k}(\bar{\Omega}) = \frac{1}{(2k)!}\left(\frac{1}{2\pi}\right)^{2k} \epsilon_{j_1...j_{2k}}^{i_1...i_{2k}} \bar{\Omega}_{\cdot j_1}^{i_1} \wedge ... \wedge \bar{\Omega}_{\cdot j_{2k}}^{i_{2k}} \quad , \tag{2.9.8}$$

where ϵ is the fully antisymmetric Kronecker symbol. It is not difficult to write the Pontrjagin classes in terms of the Chern classes. The relation follows from the fact that it is always possible to complexify a real vector bundle by the natural inclusion of $GL(m, R)$ into $GL(m, C)$. Then, if X is a real skew-symmetric matrix, $\det(1 + iX) = \sum_k i^{2k} p_{2k}(X)$, $k = 0, 1, 2, ...$ This means that

$$p_{2k}(\bar{\Omega}) = (-1)^k c_{2k}(\bar{\Omega}) \quad , \tag{2.9.9}$$

as clearly follows from comparing (2.5.3) with (2.9.8) with $l = 2k$. Of course, the odd Chern classes c_{2k+1} of this complexified bundle are zero: thus, if one of the odd Chern classes of a complex vector bundle does not vanish, the bundle cannot be reduced to a real one of the *same* dimension, as is the case for the complex $U(1)$ monopole bundle (section 2.7).

To discuss the Euler class is convenient to introduce previously the *Pfaffian* of an antisymmetric matrix X. Its definition stems from the property that the determinant of a skew-symmetric matrix is always a perfect square (zero, in fact, in the odd-dimensional case). Thus, the Pfaffian $Pf(X)$ is defined by being the square root of the determinant $\det X$,

$$\det X := \{Pf(X)\}^2 \quad ; \tag{2.9.10}$$

the sign may be fixed by requiring that the Pfaffian of the diagonal blocks matrix $Pf(\mathrm{diag}(i\sigma_2, \overset{q}{\ldots}, i\sigma_2))$, where $i\sigma_2 = \begin{pmatrix} 0 & 1 \\ -1 & 0 \end{pmatrix}$, is $(-1)^q$. This gives, for instance, $Pf\begin{pmatrix} 0 & a \\ -a & 0 \end{pmatrix} = -a$ (it is common to define $Pf(x)$ by using 1 instead of $(-1)^q$, a factor which is introduced for the Euler form below). Another equivalent way of introducing the Pfaffian is the following: if we have a two-form $\Omega := \frac{1}{2}\Omega_{ij}dx^i \wedge dx^j$ on a $2q$-dimensional manifold, the Pfaffian of the $2q \times 2q$ antisymmetric matrix Ω_{ij} is the coordinate of the $2q$-form

$$Pf(\Omega) := \frac{(-1)^q}{q!}\Omega \wedge \overset{q}{\ldots} \wedge \Omega = Pf(\Omega_{ij})dx^1 \wedge \ldots \wedge dx^{2q} \quad . \tag{2.9.11}$$

Explicitly, the Pfaffian of a $2q \times 2q$ matrix Ω_{ij} is given by

$$Pf(\Omega_{ij}) = \frac{(-1)^q}{2^q q!}\epsilon^{i_1\ldots i_{2q}}_{1\ldots 2q}\Omega_{i_1 i_2}\ldots\Omega_{i_{2q-1}i_{2q}} \quad . \tag{2.9.12}$$

If a matrix X is antisymmetric, so is AXA^t and so we can compute its Pfaffian. The behaviour of the Pfaffian under this transformation is given by the following classical

(2.9.1) LEMMA

$$Pf(AXA^t) = \det A \cdot Pf(X) \quad . \tag{2.9.13}$$

This property indicates that the Pfaffian is an $SO(2q)$-invariant polynomial (it changes sign under an improper transformation (reflection) of $O(2q)$, since then $\det A = -1$). Thus, the polynomial $Pf(X)$ may be used as a characteristic polynomial to represent another characteristic class, the Euler class, which may be defined for any oriented vector bundle. Notice that since $Pf(A)$ is $SO(2q, R)$-invariant, but not invariant under the orientation preserving group $GL_+(2q, R)$ of matrices with positive determinant, the Euler characteristic class represented by the Pfaffian has to be found using a metric connection in contrast with the Pontrjagin case. In fact, it is possible to construct oriented vector bundles E admitting flat non-metric connections for which the Euler class $e(E) \neq 0$. Let then η_E be a $2q$-dimensional real and oriented vector bundle and let Ω be a (pseudo-)Riemannian curvature two-form on its principal bundle. Then the Euler class is given by

(2.9.2) DEFINITION (*Euler class*)

The Euler class of a real oriented vector bundle E with fibre R^{2q} and euclidean metric is represented by the class in $H^{2q}_{DR}(M, R)$ of the closed $2q$-form on M

$$e(\bar{\Omega}) = Pf\left(\frac{\bar{\Omega}}{2\pi}\right) \quad , \tag{2.9.14}$$

where Ω is the curvature of the Riemannian connection. The Euler class of an oriented manifold M is the Euler class $e(\tau(M))$ of the tangent bundle (frequently written $e(M)$). If the fibre vector space is odd-dimensional, $e(\bar{\Omega}) \equiv 0$ since the determinant of an odd-dimensional antisymmetric matrix is zero.

Writing explicitly the Pfaffian (2.9.12), it is found that the *Euler class* is represented by the $2q$-*Euler form*

$$e(\bar{\Omega}) = \frac{(-1)^q}{2^q q!} \frac{1}{(2\pi)^q} \epsilon^{1...2q}_{i_1...i_{2q}} \bar{\Omega}^{i_1}_{i_2} \wedge ... \wedge \bar{\Omega}^{i_{2q-1}}_{i_{2q}} \quad . \tag{2.9.15}$$

Now, since

$$p_{2q}(\bar{\Omega}) = \det\left(\frac{\bar{\Omega}}{2\pi}\right) \quad , \tag{2.9.16}$$

it follows from (2.9.14) that

$$p_{2q}(\bar{\Omega}) = e(\bar{\Omega})e(\bar{\Omega}) \quad , \tag{2.9.17}$$

which defines the Euler class as the 'square root' of the (highest) Pontrjagin class. Of course, the characteristic class $e(M)$ is independent of the Riemannian connection used. Notice that expression (2.9.17) has to be understood in general in a matrix algebraic sense, since the Pontrjagin polynomial may lead to an Euler form which is non-zero even if the differential $4q$-form $p_{2q}(\bar{\Omega})$ on M is identically zero.

Let M^{2q} now be a Riemannian manifold and $\tau(M^{2q})$ its tangent bundle. Then the fundamental Gauss-Bonnet theorem takes the following form:

(2.9.1) THEOREM (*Generalized Gauss-Bonnet theorem*)

The Euler characteristic $\chi(M^{2q})$ *of a compact, oriented (pseudo-)Riemannian manifold* M^{2q} *(eq. (1.5.3)) is given by the integral of the Euler class:*

$$\chi(M^{2q}) = \int_{M^{2q}} e(\tau(M^{2q})) = \int_{M^{2q}} e(\bar{\Omega}) \tag{2.9.18}$$

(if M is odd-dimensional, $\chi(M) = 0 = e(\tau(M))$).

For $q = 1$, this reproduces the classical Gauss-Bonnet theorem, since then $e(\bar{\Omega}) = -\frac{1}{4\pi}\delta^{12}_{i_1 i_2}\bar{\Omega}^{i_1}_{i_2} = \frac{1}{2\pi}\bar{\Omega}^2_1 = \frac{1}{2\pi}K d\mu$ where K is the Gaussian curvature and $d\mu$ is the surface element on $M^2 \subset R^3$; eq. (2.9.18) reads $\int K d\mu = 2\pi\chi(M)$, where $\chi(M)$ is the Euler-Poincaré characteristic of M.* For the sphere, eq. (2.9.18) reads $\int_{S^2} e(\bar{\Omega}) = \chi(S^2)$ and, since $\bar{\Omega}^2_1 = \bar{\Omega}_{121}\theta^1 \wedge \theta^2 \equiv \left(\frac{1}{r^2}\right)(r d\theta) \wedge (r \sin\theta\, d\varphi)$, it gives $\chi(S^2) = 2$. In contrast, T^2 admits a flat connection, and thus the curvature is identically zero (seen as $S^1 \times S^1$ in R^4, the torus has zero K). Then $e(T^2) \equiv 0$ and $\chi(T^2) = 0$ (in general, $\chi(T^n) = 0$, section 1.5). This example provides an illustration of the previous comment: although the two-form $e(S^2)$ does not vanish, the Pontrjagin four-form on S^2 which is obtained by squaring the Euler form is obviously identically zero. The same considerations hold for $SO(4, R)$ and $\tau(M^4)$, where M^4 is an oriented manifold. Since the algebra is determined by four-dimensional matrices, the Euler form is a four-form and the Pontrjagin form on M^4 determined by its square is identically zero. Nevertheless, the reader may wish to check directly that the square of

$$\frac{1}{8}\frac{1}{(2\pi)^2}\epsilon^{1...4}_{i_1...i_4}\bar{\Omega}_{i_1 i_2}\bar{\Omega}_{i_3 i_4} \tag{2.9.19}$$

(which is obtained from (2.9.15) for $q = 2$) is given by $\frac{1}{8}\left(\frac{1}{2\pi}\right)^4[(\mathrm{Tr}\,\bar{\Omega}^2)^2 - 2\mathrm{Tr}\,\bar{\Omega}^4]$ (cf. eq. (2.9.7)) which, as an eight-form on M^4, is identically zero.

---

* As mentioned (section 1.5), all compact orientable two-dimensional surfaces are homeomorphic to spheres S^2_g with g 'handles'; $\chi(S^2_g) = 2 - 2g$ (negative 'beyond' the torus, which has one handle). This can be shown by defining the *connected sum* # of two surfaces N, N', as the surface which is obtained by opening a hole in N and in N' and then joining the two by a cylinder the ends of which are attached to each of the holes. It is then clear that $S^2\#M = M$ for any surface M, since the 'blob' produced by the addition of S^2 can be reabsorbed. It may be shown that $\chi(N\#N') = \chi(N) + \chi(N') - 2$. For instance $(S^2_0 \sim S^2)$, $S^2\#S^2 = S^2$ and $\chi(S^2\#S^2) = 2 + 2 - 2 = \chi(S^2)$. Similarly $(S^2_1 \sim T^2)$, $\chi(S^2_1\#S^2_1) = \chi(T^2) + \chi(T^2) - 2 = -2 = \chi(S^2_2)$ and it is readily seen that S^2_2, the two-sphere with two 'handles', is the connected sum $T_2\#T_2$.

2.10 Index theorems for manifolds without boundary

Index theorems equate a quantity or *index*, which is computed analytically, with another one which is of a topological nature and which is given by a characteristic class. In fact, the Atiyah-Singer theorem can be formulated by saying that the '*analytical index*' and the '*topological index*' are equal. To see how this may be achieved, let us first reformulate the Hodge-de Rham theory contained in Theorems 1.5.2 and 1.5.3 as an example of an index theorem.

Let M be a compact manifold without boundary endowed with a *Riemannian* metric. Then the exterior derivative d and the codifferential δ are adjoints to each other because of (1.5.28), and $(\,,\,)$ is positive definite. Clearly, on the ring of differential forms $\Lambda(M)$ on M the action of d induces a sequence

$$0 \longrightarrow \Lambda^0(M) \xrightarrow{d_0} \Lambda^1(M) \xrightarrow{d_1} \cdots \xrightarrow{d_{i-1}} \Lambda^i(M) \xrightarrow{d_i} \Lambda^{i+1}(M)$$
$$\xrightarrow{} \cdots \xrightarrow{d_{n-1}} \Lambda^n(M) \xrightarrow{d_n} 0 \quad, \tag{2.10.1}$$

where a subscript i has been added to d just to indicate that d_i acts on i-forms. Similarly, the action of the codifferential operator δ generates another sequence the arrows of which now point to the left,

$$\cdots \longleftarrow \Lambda^{i-1}(M) \xleftarrow{\delta_{i-1}} \Lambda^i(M) \xleftarrow{\delta_i} \Lambda^{i+1}(M) \xleftarrow{\delta_{i+1}} \cdots \tag{2.10.2}$$

and where now the subscript i of δ_i refers to the order of the form *resulting* from its action. With this notation, eq. (1.5.28) now reads

$$(d_i\alpha, \beta) = (\alpha, \delta_i\beta) \quad, \quad \alpha \in \Lambda^i(M)\,,\ \beta \in \Lambda^{i+1}(M) \quad; \tag{2.10.3}$$

we may also write $d_i^\dagger = \delta_i$, $\delta_i^\dagger = d_i$, where the dagger indicates adjoint with respect to the product of forms $(\,,\,)$. As we saw in section 1.5, neither of the above two sequences is exact, *i.e.* $\ker d_i \neq \text{range } d_{i-1}$ (similarly for δ) although clearly $\text{range } d_{i-1} \subset \ker d_i$. This property, or, equivalently, the fact that $d^2 \equiv 0$, characterizes the sequence (2.10.1) as a *complex*, in this case the *de Rham complex*. In fact, the de Rham cohomology groups measure the lack of exactness of (2.10.1) since $H^i_{DR}(M, R) := \ker d_i/\text{range } d_{i-1}$. Similarly, $\alpha \in \Lambda^i(M)$ was defined as being co-closed (co-exact) if $\alpha \in \ker \delta_{i-1}$ ($\alpha \in \text{range } \delta_i$). The operator Δ was defined as the (homogeneous) differential operator given by the square of the (inhomogeneous) Hodge-de Rham operator $\displaystyle\not{d} := d + \delta$. With the present notation, eq. (1.5.31) reads, on $\alpha \in \Lambda^i(M)$,

$$\Delta_i := \delta_i d_i + d_{i-1}\delta_{i-1} \quad. \tag{2.10.4}$$

Let us now look at eq. (2.10.3), and assume that the $(i+1)$-form β can be expressed as $\delta_{i+1}\beta'$, $\beta' \in \Lambda^{i+2}(M)$. If this is so, the product $(d_i\alpha, \delta_{i+1}\beta')$ is zero. Proceeding similarly, we conclude that

$$\text{range } d_{i-1} \perp \text{range } \delta_i \perp \ker \Delta_i \qquad (2.10.5)$$

or, in other words, that Theorem 1.5.2 expresses that $\Lambda^i(M)$ has an (in fact, unique having specified (,)) splitting of the form

$$\Lambda^i(M) = \text{range } d_{i-1} \oplus \text{range } \delta_i \oplus \ker \Delta_i \qquad (2.10.6)$$

and, consequently, since $\ker \Delta_i = \text{Harm}^i(M)$,

$$\alpha_i = d_{i-1}\alpha_{i-1} + \delta_i\alpha_{i+1} + h_i \quad , \quad \alpha \in \Lambda^i(M) \quad , \qquad (2.10.7)$$

where h_i is an harmonic i-form, $\Delta_i h = 0$. Theorem 1.5.3 now states that every i-th de Rham cohomology class is represented by one and only one harmonic form:

$$H^i_{DR}(M, R) = \ker \Delta_i = \ker d_i / \text{range } d_{i-1} \quad . \qquad (2.10.8)$$

If the *analytical index of the de Rham complex* (2.10.1) is now the integer defined by the alternating sum

$$index(\Lambda(M), d) = \sum (-)^i \dim(\ker \Delta_i) \qquad (2.10.9)$$

and then (2.10.8), (1.5.3) and (2.9.18) are used, we find

$$index(\Lambda(M), d) = \sum_{i=0}^{N} (-)^i b^i(M) = \chi(M) = \int_M e(\tau(M)) \quad ; \qquad (2.10.10)$$

clearly, the r.h.s. of this expression is of a topological nature: it is a *topological index*. Thus, the Gauss-Bonnet theorem may be looked at as the prototype of an index theorem: *the index of the de Rham complex over the manifold M is the Euler-Poincaré characteristic of M.* Eq. (2.10.10) exhibits the general structure of an index theorem: its l.h.s. is given by an analytical property, which is expressed in terms of the number of zero-frequency solutions of the Laplace equation, *i.e.* the number of harmonic linearly independent forms on the manifold (eq. (2.10.9)); the r.h.s. is the Euler-Poincaré characteristic, which in turn is given in terms of the Betti numbers of the manifold M. The Laplacian depends on the specific Riemannian metric used (and so depends the space of the harmonic forms[*]); the Euler characteristic and the Betti numbers are topological invariants and hence defined regardless of the metric. If M is odd-dimensional, $index(\Lambda(M^{odd}), d) = 0$ since $\chi(M^{odd}) = 0$; this will remain true of index theorems for other differential operators.

---

[*] There is one special case in which the harmonic forms do not change with a change of metric; it is the case of forms of 'middle rank' $k = n/2$ when the metrics are conformally related (see the first footnote in section 2.8).

We may reformulate (2.10.9) by recalling that Δ and $d + \delta$ have the same kernel (harmonic $\Longleftrightarrow$ closed and co-closed). Let us split $\Lambda(M)$ into the sum $\Lambda(M) = \Lambda^{even}(M) \oplus \Lambda^{odd}(M)$ of even and odd forms, $\Lambda^{even}(M) = \sum_{i\oplus} \Lambda^{2i}(M)$, $\Lambda^{odd}(M) = \sum_{i\oplus} \Lambda^{2i+1}(M)$. Let D_+ and D_- be the operators defined by

$$D_+ \equiv D := \sum_i (d_{2i} + \delta_{2i-1}) \quad , \quad D_- := D^\dagger = \sum_i (d_{2i-1} + \delta_{2i}) \quad . \quad (2.10.11)$$

Then D is a mapping $D : \Lambda^{even}(M) \longrightarrow \Lambda^{odd}(M)$ (by this it is meant that $D(\alpha_{(0)}, \alpha_{(2)}, \alpha_{(4)}, ...) = (d_0\alpha_{(0)} + \delta_1\alpha_{(2)}, d_2\alpha_{(2)} + \delta_3\alpha_{(4)}, ...))$, where the subscript indicates that $\alpha_{(i)}$ is an i-form). Its adjoint $D^\dagger$ is a mapping $D^\dagger : \Lambda^{odd}(M) \longrightarrow \Lambda^{even}(M)$. The associated 'Laplacians' are given by

$$\Delta_+ := D^\dagger D = \sum_i \Delta_{2i} \quad , \quad \Delta_- := DD^\dagger = \sum_i \Delta_{2i-1} \quad . \quad (2.10.12)$$

Thus, we may replace the definition (2.10.9) of the analytical index of the de Rham complex by

$$index(\Lambda(M), D) = \dim(\ker \Delta_+) - \dim(\ker \Delta_-) \quad\quad (2.10.13)$$

or, equivalently, by

$$index(\Lambda(M), D) = \dim(\ker D) - \dim(\ker D^\dagger) \quad\quad (2.10.14)$$

since $\ker \Delta_+ \equiv \ker(D^\dagger D) = \ker D$ ($\beta \in \ker \Delta_+ \Rightarrow (\beta, D^\dagger D\beta) = 0 \Rightarrow (D\beta, D\beta) = 0 \Rightarrow D\beta = 0$) and $\ker \Delta_- \equiv \ker(DD^\dagger) = \ker D^\dagger$. In fact, range D (resp. $\ker D$) is the orthogonal complement of $\ker D^\dagger$ (resp. range $D^\dagger$). The analytic index has still one more expression: since

$$\operatorname{coker} D := \Lambda^{odd} / \operatorname{range} D = \ker D^\dagger \quad , \quad\quad (2.10.15)$$

it may be given by

$$index(\Lambda(M), D) = \dim(\ker D) - \dim(\operatorname{coker} D) \quad . \quad\quad (2.10.16)$$

The form (2.10.14) of the analytical index for the de Rham complex is, again, not specific to this case. But, to see what it has in common with the index theorem for other complexes, let us look at its general structure. *First*, we notice that the de Rham sequence could have been written

$$0 \longrightarrow \Gamma(M, E_0) \xrightarrow{D_0} \Gamma(M, E_1) \longrightarrow \cdots \Gamma(M, E_i) \xrightarrow{D_i} \Gamma(M, E_{i+1})$$
$$\longrightarrow \cdots \Gamma(M, E_n) \xrightarrow{D_n} 0 \quad\quad (2.10.17)$$

where $\Gamma(M, E_i)$ is the module of cross sections of the vector bundle $E_i = \wedge^i \tau^*(M)$ since, as we saw in the examples after Definition 1.3.3, the differential i-forms of $\Lambda^i(M)$ may be regarded as sections of the vector bundle $\wedge^i \tau^*(M)$. The writing of (2.10.17) only requires the existence of a differential operator of degree 1 acting on a sequence of sections of vector bundles E_i such that $D_{i+1} \circ D_i \equiv 0$; with this, the sequence (2.10.17) qualifies as a *complex*.

Secondly, the fact that expression (2.10.9) or the difference (2.10.13) was a well-defined one was guaranteed by the nature of the Laplacian operator $\Delta = (d + \delta)^2$, the kernel of which is finite-dimensional. This is a consequence of the fact that Δ is an *elliptic operator*. To formalize this assertion, consider a differential operator $\mathscr{D} : \Gamma(M, E) \longrightarrow \Gamma(M, E')$, $\mathscr{D} : s(x) \longmapsto s'(x)$, acting on the cross sections of a vector bundle E. Let $x^i, i = 1, ..., n$, be local coordinates on $U \subset M$. Then a *general differential operator* $\mathscr{D}$ of degree K over U may be expressed through its action on sections as

$$[\mathscr{D}s(x)]^{l'} = \sum_{\tilde{\alpha}} A^{\alpha l'}{}_{\cdot l}(x) D_\alpha s^l(x) \quad , \quad D_\alpha = \left(\frac{\partial}{\partial x^1}\right)^{\alpha_1} \cdots \left(\frac{\partial}{\partial x^n}\right)^{\alpha_n} \quad ,$$

(2.10.18)

where $l' = 1, ..., m'$, $l = 1, ..., m$ are the respective fibre indices, $(A^\alpha)^{l'}{}_l$ is an $m' \times m$ matrix on U, α is the multi-index $\alpha = (\alpha_1, ..., \alpha_n)$, $\alpha_i \geq 0$, $\tilde{\alpha} \equiv \alpha_1 + ... + \alpha_n$ and the sum over $\tilde{\alpha}$ is a sum over the multi-index α which includes all possibilities for $\tilde{\alpha} \leq K$.

We may now introduce the following

(2.10.1) DEFINITION (*Leading or principal symbol of* $\mathscr{D}$)

Let $\xi = (\xi_1, ..., \xi_n)$ *be a real n-tuple. The leading symbol* $\sigma_x(\xi, \mathscr{D})$ *associated with the local representation of* $\mathscr{D}$ *at the point* x *is the linear map from the fibre* F_x *to* F'_x *defined by the matrix*

$$\sigma_x(\xi, \mathscr{D})^{l'}{}_l = \sum_{\tilde{\alpha}=K} A^{\alpha l'}{}_{\cdot l}(x) \xi_\alpha \quad , \tag{2.10.19}$$

$(\mathscr{D}_\alpha \longrightarrow \xi_\alpha)$ *the entries of which are polynomials in* $\xi_1, ..., \xi_n$ *(notice that only the highest order* K *derivatives in* $\mathscr{D}$ *enter in the definition).*

Let now $E = E'$. The ellipticity of a differential operator is given by

(2.10.2) DEFINITION (*Elliptic operator*)

A differential operator $\mathscr{D} : \Gamma(M, E) \longrightarrow \Gamma(M, E)$ *is said to be elliptic on* M *if* $\forall x \in M$ *the linear mapping given by the* $m \times m$ *matrix* $\sigma_x(\xi, \mathscr{D})$ *is an isomorphism for every* $\xi \neq 0$, *i.e. if the matrix is invertible.*

An elliptic operator defined on a compact manifold has a finite-dimensional kernel and cokernel, and expressions of the type of (2.10.16)

are well defined for them[*]. For instance, if the fibre is one-dimensional and $\mathscr{D}$ is given by

$$\mathscr{D} = A^{ij}\frac{\partial}{\partial x^i}\frac{\partial}{\partial x^j} + B^i\frac{\partial}{\partial x^i} + C \qquad (2.10.20)$$

the leading symbol is given by the one-dimensional matrix $A^{ij}\xi_i\xi_j$, and $\mathscr{D}$ is elliptic if it is non-zero for $\xi \neq 0$. This will be the case if the quadratic form is positive (or negative) definite, i.e., if $A^{ij}\xi_i\xi_j = c$ is an ellipsoid (hence the name of 'elliptic' operator; other possibilities for the quadratic form lead to hyperbolic – maximal rank, mixed signature – or parabolic operators). As a result, the operator $d\delta + \delta d$ is elliptic for a Riemannian metric (it is not for the Minkowski signature). Indeed, at a point x a Riemannian manifold can be made Euclidean up to first order by using geodesic coordinates and then $\Delta = -\sum_{i=1}^{n}(\partial/\partial x^i)^2$ gives $\sigma_x(\xi, \Delta) = -(\xi_1^2 + \cdots + \xi_n^2)$; in contrast, for a pseudo-Riemannian metric of signature (p,q), the quadratic form $\xi_1^2 + \cdots + \xi_p^2 - \xi_{p+1}^2 - \cdots - \xi_{p+q}^2$ is not invertible on the light cone. An important property of elliptic operators is that, since the leading symbol of a composite operator is the composite of symbols, the composites of elliptic operators (and so powers and 'roots') will also be elliptic. Another example is provided by the familiar operators ∇ (gradient), $\nabla\times$ (curl) and $\nabla\cdot$ (divergence) for $M = R^3$; their symbols are given by the matrices

$$\begin{bmatrix} \xi_1 \\ \xi_2 \\ \xi_3 \end{bmatrix}, \quad \begin{bmatrix} 0 & \xi_3 & -\xi_2 \\ -\xi_3 & 0 & \xi_1 \\ \xi_2 & -\xi_1 & 0 \end{bmatrix}, \quad [\xi_1 \ \ \xi_2 \ \ \xi_3] \ ; \qquad (2.10.21)$$

they correspond to mappings $\lambda \mapsto \lambda\xi$, $\vec{\lambda} \mapsto \vec{\xi} \times \vec{\lambda}$, $\vec{\lambda} \mapsto \vec{\xi} \cdot \vec{\lambda}$. As (2.10.21) shows, these operators are not elliptic.

Looking at the de Rham complex on the compact, boundaryless manifold M we conclude that it is an elliptic complex since its associated Laplacians (2.10.4) are elliptic. This suggests a more general

(2.10.3) DEFINITION (*Elliptic complex*)

Let E be a vector bundle. An elliptic complex (E, D) is a finite sequence (2.10.17) of differential operators $D_i : \Gamma(M, E_i) \longrightarrow \Gamma(M, E_{i+1})$ acting on smooth sections such that $D_{i+1} \circ D_i \equiv 0$ and the Laplacians of the complex $\Delta_i := D_i^\dagger D_i + D_{i-1}D_{i-1}^\dagger$, where $D_i^\dagger$ is the adjoint operator with respect to the scalar product on the fibres with a smooth density on M, are elliptic on $\Gamma(M, E_i)$.

---

[*] An operator $\mathscr{D}$ is a *Fredholm* operator if $\ker\mathscr{D}$ and $\mathrm{coker}\,\mathscr{D}$ are finite-dimensional. On a compact manifold, an elliptic operator is a Fredholm operator.

Since $(D_{i+1} \circ D_i)^\dagger = D_i^\dagger \circ D_{i+1}^\dagger$, it follows that if the complex $(\Gamma(M, E_i), D_i)$ is elliptic, so is the complex $(\Gamma(M, E_{i+1}), D_i^\dagger)$ where the arrows point in the opposite direction.

To relate this picture to the form (2.10.13) or (2.10.14) of the index for a de Rham complex, we have to reduce the elliptic complex to a *two-term elliptic complex* (we have to 'roll up the complex', cf. (2.10.17)) and see that the new complex has the same index as the original one $(\Gamma(M, E), D)$. This is where the comparison with eqs. (2.10.11)–(2.10.14) comes in. Defining the even and odd bundles $E^{even} = \sum_{i\oplus} E_{2i}$, $E^{odd} = \sum_{i\oplus} E_{2i+1}$,

$$\Gamma(M, E^{even}) = \sum_{i\oplus} \Gamma(M, E_{2i}) \quad , \quad \Gamma(M, E^{odd}) = \sum_{i\oplus} \Gamma(M, E_{2i+1}) \quad , \tag{2.10.22}$$

$$D = \sum_{i\oplus}(D_{2i} + D_{2i-1}^\dagger) \quad , \quad D^\dagger = \sum_{i\oplus}(D_{2i-1} + D_{2i}^\dagger) \quad , \tag{2.10.23}$$

and the associated Laplacians as in (2.10.4), (2.10.12),

$$\Delta_i := D_i^\dagger D_i + D_{i-1} D_{i-1}^\dagger \; , \; \Delta_+ := \sum_i \Delta_{2i} = D^\dagger D \; , \; \Delta_- := \sum_i \Delta_{2i-1} = DD^\dagger \; ,$$

$$\tag{2.10.24}$$

we are led to the following general

(2.10.4) DEFINITION (*Analytical index of an elliptic complex*)

The analytical index of an elliptic complex $(\Gamma(M, E), D)$ is defined to be the integer

$$index(\Gamma(M, E), D) := \sum_i (-)^i \dim(\ker \Delta_i) = \dim(\ker \Delta_+) - \dim(\ker \Delta_-) \quad .$$

$$\tag{2.10.25}$$

We recall the basic ingredients needed to define (2.10.25) for a compact manifold M: the differential operator defining a complex, the (Riemannian) scalar product defining its adjoint, and the ellipticity property which guarantees that the r.h.s. of (2.10.25) is well defined (an integer). Clearly, in order to have a non-trivial index the operator D cannot be self-adjoint.

The Atiyah-Singer theorem now states that the analytic index (2.10.20) is equal to the topological index of the complex, which is given by the r.h.s. in the formula of

(2.10.1) THEOREM (*Atiyah and Singer*)

Let $(\Gamma(M,E),D)$ be an elliptic complex over a compact boundaryless manifold M of even dimension n. Then the index of the complex is given by

$$index(\Gamma(M,E),D) = (-)^{n(n+1)/2} \int_M ch\left(\sum_{i=0}^{n} \oplus(-)^i E_i\right) \frac{Td(\tau(M)^C)}{e(\tau(M))} \quad,$$

$$(2.10.26)$$

where in the integrand only n-forms are retained. If the manifold is odd-dimensional, the index of the differential operator D is zero.[*]

In eq. (2.10.26), $Td(\tau(M)^C)$ is the Todd class of the complexified tangent bundle $\tau(M)^C$ (see below), and $e(\tau(M))$ is the Euler class; the division is formal (moreover, since the index is zero for M^{odd}, we may assume that M is even-dimensional when writing (2.10.26)). If the elliptic term is a two-term complex, $\Gamma(M,E) \xrightarrow{D} \Gamma(M,E')$, the index is given by

$$index\ D = (-)^{n(n+1)/2} \int_M (chE - chE')\frac{Td(\tau(M)^C)}{e(\tau(M))} \quad. \qquad (2.10.27)$$

It may be instructive to recover (2.10.10) for the de Rham complex by sketching the computation of the topological index for this case. First, we note that E_i is the *complex* tensor bundle $\wedge^i\tau^*(M)^C$ (Chern classes are computed for complex bundles). This complexification, which amounts to considering complex-valued differential forms, will not affect however the l.h.s. (the analytical index) since $\dim H_{DR}^i(M,R)$ and the complex dimension of $H_{DR}^i(M,C)$ are the same. We have then to compute $ch(\sum_{i=0}^{n}(-)^i \wedge^i \tau^*(M)^C)$ or, equivalently, $\sum_i(-)^i ch(\wedge^i\tau^*(M)^C)$, since the Chern character is well behaved with respect to Whitney sums: if F and F' are two complex vector bundles over M, $ch(E \oplus E') = chE + chE'$, eq. (2.5.17). Since for a line bundle $ch(L_i) = e^{\lambda_i}$, $\lambda_i = c_1(L_i)$ (eq. (2.5.11)), we get

$$\sum_{i=0}^{n}(-)^i ch(\wedge^i\tau^*(M)^C)$$

$$= 1 - \sum_{j_1} e^{-\lambda_{j_1}} + \sum_{j_1<j_2} e^{-\lambda_{j_1}}e^{-\lambda_{j_2}} + \ldots + (-)^n \sum_{j_1<\ldots<j_n} e^{-\lambda_1}\ldots e^{-\lambda_{j_n}}$$

$$= \sum_{i=0}^{n}(-)^i \sum_{1\le j_1<\ldots<j_i\le n} e^{-(\lambda_1+\ldots+\lambda_n)} = \prod_{i=1}^{n}(1 - e^{-\lambda_i}) \quad.$$

$$(2.10.28)$$

---

[*] There are elliptic operators, called *pseudo*-differential operators, for which this is not true; they will not be considered here.

In computing (2.10.28) we have looked at $\wedge^i \tau^*(M)^C$ as a product of line bundles and used that $ch(E \otimes E') = chE \wedge chE'$. The fact that we are concerned with dual bundles is reflected in the fact that $\lambda \longrightarrow -\lambda$; the natural curvature for the dual line bundle L^* is the complex conjugate of the curvature for L, and thus $c_1(L^*) = -c_1(L) = -\frac{i}{2\pi}\bar{\Omega}$ (this also follows from the fact that $L \otimes L^*$ is trivial and so $c_1(L \otimes L^*) = 0$; see below eq. (2.5.16)).

We have not yet introduced the *Todd class* associated with a complex vector bundle, but it is sufficient to know that, in much the same way as the Chern class or the Chern characters have $\det(1+X) = \prod_i^n(1+\lambda_i)$ and $\mathrm{Tr}\, e^X = \sum_{i=1}^n e^{\lambda_i}$, respectively, as their generating functions (eqs. (2.5.13) and (2.5.8)), the Todd class is generated by

$$
Td(E) = \prod_{i=1}^n \frac{\lambda_i}{1 - e^{-\lambda_i}}
$$

$$
\left(= 1 + \frac{1}{2}c_1(E) + \frac{1}{12}(c_1^2(E) + c_2(E)) + \frac{1}{24}c_1(E)c_2(E) + \cdots \right) .
$$

(2.10.29)

Multiplying (2.10.29) and (2.10.28) and recalling that the square of the Euler class is given by the highest Pontrjagin class p_n (eq. (2.9.17)) which is $(-)^{n/2}$ times the highest Chern class c_n, and that this last is generated by $\lambda_1 \ldots \lambda_n$, it is seen that the index is indeed given by (2.10.10).

The de Rham complex is one of the four classical elliptic complexes: the *de Rham complex*, the *signature complex*, the *Dolbeault complex* and the *spin complex*. In the next section we describe the spin complex, since its associated index theorem will appear in the description of anomalies. In fact, the computation of the abelian anomaly by using the Fujikawa technique to be described in chapter 10 may be considered as a derivation of the Dirac index theorem by using the heat-kernel regularization in the context of quantum field theory. We shall then close this chapter with a few comments on twisted complexes.

2.11 Index theorem for the spin complex. Twisted complexes

Let M be an even-dimensional compact, oriented Riemannian manifold and let us assume that it admits a spin structure. As mentioned, this happens if certain characteristic classes, the first (orientability) and second (spin) Stiefel-Whitney classes w_1, w_2, are zero (which is true, in particular, of $M = S^n$). This means that it is possible to define a *spinor bundle* the structure group of which is Spin(n), the universal covering group of $SO(n)$ (Spin(n) is the double cover of $SO(n)$). The physical *spinor fields* ψ are cross sections of this spinor bundle; their transformation law is $\psi' = S(A)\psi$ and S is the appropriate spinorial representation of $A \in$ Spin(n).

The n gamma matrices γ^α of the Clifford algebra of Spin(n) are defined by the fact that they satisfy $\{\gamma^\alpha, \gamma^\beta\} = 2\delta^{\alpha\beta}\mathbf{1}_{|2^{n/2}}$, $\alpha, \beta = 0, 1, \ldots, (n-1)$ (recall that the metric is Euclidean, $\eta^{\alpha\beta} = \delta^{\alpha\beta}$). It is well known that the dimension of the Clifford algebra (the number of independent products of gamma matrices) is 2^n, and that for even dimension there is only one irreducible representation of dimension $2^{n/2}$. This representation of the Clifford algebra gives rise to a reducible representation of the spin group, which splits into the direct sum of two representations of dimension $2^{(n/2-1)}$; these representations have their representation spaces made out of spinors of different chirality. For instance, for $n = 4$ (cf. (1.10.2) for the Minkowski signature) the hermitian gamma matrices are given by

$$\gamma^\alpha = \begin{pmatrix} 0 & \alpha^\alpha \\ \tilde{\alpha}^\alpha & 0 \end{pmatrix} \quad , \quad \gamma_5 \equiv -\gamma^0\gamma^1\gamma^2\gamma^3 = \begin{pmatrix} 1 & 0 \\ 0 & -1 \end{pmatrix} \quad , \tag{2.11.1}$$

where $\alpha^\alpha = (1, -i\sigma^j)$, $\tilde{\alpha}^\alpha = (1, i\sigma^j)$, $j = 1, 2, 3$, and σ^j are the Pauli matrices; $\tilde{\alpha}^{\alpha\dagger} = \alpha^\alpha$. The defining Euclidean anticommutation relations $\{\gamma^\alpha, \gamma^\beta\} = 2\delta^{\alpha\beta}$ imply that the 2×2 matrices α satisfy

$$\alpha^\alpha\tilde{\alpha}^\beta + \alpha^\beta\tilde{\alpha}^\alpha = 2\delta^{\alpha\beta}\mathbf{1} \quad , \quad \tilde{\alpha}^\alpha\alpha^\beta + \tilde{\alpha}^\beta\alpha^\alpha = 2\delta^{\alpha\beta}\mathbf{1} \quad , \tag{2.11.2}$$

as may be checked directly.

It is not difficult to find an iterative procedure that gives the gamma matrices for any dimension D. Let $D = 2m$, and assume that the $2(m-1)$ matrices of the $D = 2(m-1)$ case are known; they are $2^{(m-1)} \times 2^{(m-1)}$ matrices defined by the property $\{\gamma^\alpha, \gamma^\beta\} = 2\delta^{\alpha\beta}\mathbf{1}_{|2^{(m-1)}}$. Then we add to this set the unit matrix $\Sigma^0 = \mathbf{1}_{|2^{(m-1)}}$ and the matrix $\gamma^{(2m-1)} \equiv \sigma^3 \otimes \mathbf{1}_{|2^{(m-2)}}$.[*] If this matrix is now called $\Sigma^{(2m-1)}$, and the γ matrices of the $2(m-1)$-dimensional case are called Σ^j, $j = 1, \ldots, 2(m-1)$, the $2^m \times 2^m$ matrices Γ of the $2m$-dimensional case are given by

$$\Gamma^0 \equiv \sigma^1 \otimes \Sigma^0 = \begin{bmatrix} 0 & \Sigma^0 \\ \Sigma^0 & 0 \end{bmatrix},$$

$$\Gamma^j = \sigma^2 \otimes \Sigma^j = \begin{bmatrix} 0 & -i\Sigma^j \\ i\Sigma^j & 0 \end{bmatrix}, \quad j = 1, \ldots, (2m-1) ; \tag{2.11.3}$$

they are all hermitian and 'antidiagonal'. Clearly, (2.11.1) corresponds to starting with the simplest $D = 2$ case with matrices $\gamma^1 = \sigma^1$ and $\gamma^2 = \sigma^2$. Adding $\sigma^3 = \sigma^3 \otimes 1$ we get the matrices of Spin(3) = $SU(2)$, and adding σ^0 we obtain (2.11.1) by the above construction (we note in passing that the same procedure obviously applies to the Minkowski signature $(+, -, \ldots, -)$ just by multiplying the matrices Γ^i of (2.11.3) by

---

[*] The $(2m-1)$ matrices obtained by adding γ^{2m-1} to the previous $2(m-1)$ gammas generate the Clifford algebra corresponding to Spin($2m-1$).

i). In general even dimension n, the matrix γ_{n+1} or 'γ_5' and the chiral projector are given by

$$\gamma_5 = i^{n/2} \prod_{\alpha=0}^{(n-1)} \gamma^\alpha \quad , \quad \gamma_5^2 = 1 \quad , \quad \{\gamma_5, \gamma^\alpha\} = 0 \quad , \quad P_\pm = \frac{1}{2}(1 \pm \gamma_5) \quad (2.11.4)$$

(γ_5 would be $i^{(n/2-1)} \prod_{\alpha=0}^{(n-1)} \gamma^\alpha$ for a Minkowskian signature). The positive (ψ_+) and negative (ψ_-) chirality spinors are eigenvectors of γ_5, $\gamma_5 \psi_\pm = \pm \psi_\pm$ and are orthogonal; the space of Dirac spinors thus has the splitting $\psi = \psi_+ \oplus \psi_-$, $\psi_+(\psi_-)$ being the upper (lower) components of ψ.

Let us now introduce the *Dirac differential operator* in a curved space. It is given by $\not{D} = \gamma^\alpha D_\alpha$, where D_α is the *spinorial covariant derivative* D_α which plays on spinors the same rôle that the ordinary covariant derivative plays on tensors: it guarantees that the covariant derivative of a spinor transforms as a local $SO(n)$-vector and Spin(n)-spinor, $(D_\alpha \Psi)' = \Lambda_\alpha{}^\beta(x) S(\Lambda(x)) D_\beta \Psi$, where Λ is the $SO(n)$ element associated with the Spin(n) element A. It is given by

$$D_\alpha = E_\alpha^\mu (\partial_\mu + \omega_\mu) \quad , \quad (2.11.5)$$

where $E_\alpha^\mu(x)$ is the inverse of the *vielbein* $E_\mu^\alpha(x)$ and ω_μ is the *spin connection*. E_α^μ takes a basis dx^μ of $T_x^*(M)$ to an orthogonal basis $\theta^\alpha = E_\mu^\alpha dx^\mu$ of $T_x^*(M)$; thus

$$\eta_{\alpha\beta} E_\mu^\alpha(x) E_\nu^\beta(x) = g_{\mu\nu}(x), \quad \eta_{\alpha\beta} = E_\alpha^\mu E_\beta^\nu g_{\mu\nu}, \quad E_\alpha^\mu E_\mu^\beta = \delta_\alpha^\beta, \quad E_\alpha^\mu E_\nu^\alpha = \delta_\nu^\mu$$
$$(2.11.6)$$

(clearly, E_α^μ takes the basis ∂_μ of $T_x(M)$ to the orthogonal one D_α of $T_x(M)$, $D_\alpha = E_\alpha^\mu \partial_\mu$). For its part, the spin connection is Spin(n)-valued and transforms under a change of frame as a connection should,

$$\omega_\mu' = S\omega_\mu S^{-1} + S\partial_\mu S^{-1} \quad , \quad (2.11.7)$$

and it is given by

$$\omega_\mu = \frac{1}{2}\omega_\mu^{\alpha\beta}\left(\frac{1}{4}[\gamma_\alpha, \gamma_\beta]\right) = -\omega_\mu^\dagger \quad . \quad (2.11.8)$$

The appearance of the term in the gammas should have been expected since the $\binom{n}{2}$ different commutators $\frac{1}{4}[\gamma_\alpha, \gamma_\beta]$ are nothing but the antihermitian Lie algebra generators of the Spin(n) transformations on the spinors (the antisymmetry property $\omega_\mu^{\alpha\beta} = -\omega_\mu^{\beta\alpha}$, or 'metricity condition' is the equivalent in the 'tetrad' formalism of the metric condition $\nabla_\mu g_{\rho\sigma} = 0$).

Since $\{\gamma_5, \gamma^\alpha\} = 0$, the Dirac operator $\not{D}$ also anticommutes with γ_5. In the chiral (or Weyl) realization of the gamma matrices given above, the operator $i\not{D}$ (where the i has been added for hermiticity reasons) may be written as

$$i\rlap{/}D = \begin{pmatrix} 0 & D_- \\ D_+ & 0 \end{pmatrix} \quad , \qquad (2.11.9)$$

where $D_\pm = \rlap{/}D P_\pm$ and $D_- = D_+^\dagger$ (∂_μ and ω_μ are antihermitian); $D_\pm$ are the *Weyl differential operators*. The operator D_+ takes positive chirality spinors into negative chirality spinors, $D_+ : \psi_+ \longrightarrow \psi_-$; similarly, $D_- :$ $\psi_- \longrightarrow \psi_+$. Such a splitting leads to a two-term complex

$$0 \longrightarrow \Gamma(M, E_+) \overset{D_+}{\underset{D_-}{\rightleftarrows}} \Gamma(M, E_-) \longrightarrow 0 \quad , \qquad (2.11.10)$$

where $\Gamma(M, E_\pm)$ are now sections of definite chirality, *i.e.*, sections of the pair of spin bundles of local coordinates (x, ψ_+) and (x, ψ_-) respectively. Let ξ be a vector. Then we read from (2.11.2) that $(\alpha\xi)(\tilde{\alpha}\xi) = \xi^2 = (\tilde{\alpha}\xi)(\alpha\xi)$. Thus, the leading terms of the Laplacians $\triangle_+ = D_-D_+$, $\triangle_- = D_+D_-$ are elliptic, and the complex is an elliptic complex (we could also have looked at the matrix $\gamma^\alpha E_\alpha^\mu \xi_\mu$ and checked using (2.11.6) that it admits an inverse, $\gamma^\beta E_\beta^\nu \xi_\nu / \xi^2$, which exists for any $\xi \neq 0$ since the metric is Riemannian). The *index of the Dirac operator* is now defined by

$$index(\Gamma(M, E_\pm), \rlap{/}D) = \dim(\ker D_+) - \dim(\ker D_-) \quad . \qquad (2.11.11)$$

It is thus given by the difference $n_+ - n_-$ of the number of zero-frequency (harmonic) normalizable spinors of chirality ± 1. The index theorem for the spin complex equates this quantity to the topological index (cf. (2.10.27))

$$index \, i\rlap{/}D = n_+ - n_- = \int_M (chE_+ - chE_-) \frac{Td(\tau(M)^C)}{e(\tau(M))} \quad . \qquad (2.11.12)$$

The r.h.s. of (2.11.12) turns out to be the integral of the $\hat{A}$-*roof genus* of the manifold M, and (2.11.12) then reads

$$n_+ - n_- = \int_M \hat{A}(\tau(M)) \quad . \qquad (2.11.13)$$

We shall not be concerned with the derivation of the expression for $\hat{A}$ which is the even function of λ_i given by $\hat{A} = \prod_{i=1}^{n/2} [\frac{\lambda_i/2}{\sinh(\lambda_i/2)}]$. We shall just mention that it may be expressed as a polynomial of the Pontrjagin classes and that, accordingly, it is zero unless $\dim M = 4p$ (this is why the sign factor in front of the r.h.s. of (2.11.12) was omitted).

The index theorem (2.11.12) or (2.11.13) cannot be applied yet to non-abelian gauge theories, where spinors have an additional index running over the representation vector space of the group G. A spinor which also has a group representation index is a section of the tensor product (section 1.3) of the spinor bundle and a vector bundle. This situation is not restricted to the spin complex; in general, the complex obtained by moving from the ordinary complex to a complex defined on the sections of a certain tensor product bundle is called a *twisted complex*. For the de Rham complex this would correspond to considering, instead of ordinary q-forms, vector-valued q-forms. This modifies (2.10.26); without discussing it in detail, the change is not difficult to justify. The terms of the complex are now sections of the tensor product $E_i \otimes V = \wedge^i \tau^*(M) \otimes V$. This means that the change affects (2.10.26) by replacing E_i in it by $E_i \otimes V$. But, since the Chern character of the product is the product of the Chern characters, the calculation of the previous section applied to the present case just adds $ch(V)$, with the result that the index theorem for the *twisted de Rham complex* is given by

$$\text{index } D_V = \int_M ch(V) \wedge e(\tau(M)) \quad . \tag{2.11.14}$$

However, since the Euler class is already an even n-form, the only contribution of $ch(V)$ to (2.11.14) must come from the first term in the expansion (2.5.10) since the other terms give vanishing contributions. This finally gives, with $k = \dim V$,

$$\text{index } D_V = k \int_M e(\tau(M)) = k(\text{index} D) \quad . \tag{2.11.15}$$

In the case of the *twisted spin complex*, the two-term complex is replaced by

$$\Gamma(M, E_+ \otimes V) \underset{D_{V,-}}{\overset{D_{V,+}}{\rightleftarrows}} \Gamma(M, E_- \otimes V) \quad . \tag{2.11.16}$$

The Dirac operators D_+, D_- are now obtained as in (2.11.9) although now D involves the Yang-Mills connection A_μ, which is valued in the Lie algebra $\mathscr{G}$ and thus contains its $(\dim V) \times (\dim V)$ representation matrices. Thus,

$$\slashed{D} = \gamma^\alpha D_\alpha \quad , \quad D_\alpha = E_\alpha^\mu (\partial_\mu + \omega_\mu + A_\mu) \quad . \tag{2.11.17}$$

It may be seen that the index theorem now reads

$$n_+ - n_- = (-)^{n/2} \int_M \hat{A}(\tau(M)) \wedge ch(V) \quad , \tag{2.11.18}$$

where, as before, only the form proportional to the volume n-form on M contributes. The twisted spin complex will now have a non-zero index for dimensions which are not multiples of 4 in contrast with eq. (2.11.13). A particularly interesting case corresponds to $M = S^{2p}$. It turns out that $\hat{A}(\tau(S^{2p})) = 1^\dagger$ (in this case, the intrinsic geometry of the manifold does not play a rôle and the spin connection may be ignored) and the index theorem for S^{2p} reads

$$n_+ - n_- = (-)^p \int\limits_{S^{2p}} ch_p(V) \quad . \tag{2.11.19}$$

For the case of the instanton bundle $P(SU(2), S^4)$ the integrand is just $ch_2(F) = -\frac{1}{8\pi^2}\text{Tr}(F \wedge F)$. Thus, the instanton/anti-instanton number k (section 2.8) is the index of the twisted spin complex. Actually, the index theorem may be strengthened by showing that the k of the $F = \pm *$ F instanton solutions is just the number of zero-frequency harmonic solutions of different chirality; when $\ker D_+ = \ker D_- D_+$ (resp. $\ker D_-$) is non-zero, $\ker D_- = \ker D_+ D_-$ (resp. $\ker D_+$) is zero so that $k > 0$ ($k < 0$) implies $n_- = 0$ ($n_+ = 0$). Such a 'vanishing theorem' (n_+ or n_- vanishes) may be derived easily from the fact that we are considering self-dual or anti-self-dual gauge field configurations.

Bibliographical notes for chapter 2

A basic reference on connections on fibre bundles is the first volume of the treatise of Kobayashi and Nomizu (1963); our treatment often follows this reference. For the theory of connections with special emphasis on its physical applications see, *e.g.*, the review of Eguchi *et al.* (1980) and the books of Choquet-Bruhat *et al.* (1982), Nash and Sen (1983), Trautman (1984) and Nakahara (1990). For the theory of invariant connections see Wang (1958) (see also Kobayashi and Nomizu (1963)); a recent reference on the subject is Turner-Laquer (1992).

The fibre bundle approach to Yang-Mills theories starts with the papers of Wu and Yang (1975a,b); see also Greub and Petry (1975) and Yang (1977). For a textbook treatment of their mathematical aspects the reader may look at Bleecker (1981). More details can be found in the review papers of Daniel and Viallet (1980) and Marathe and Martucci (1989, 1992). An introduction to Yang-Mills theories is given, *e.g.*, in Faddeev and Slavnov (1991) and in Jackiw (1980b); see Jackiw and Manton (1980) for some specific properties of the conservation laws in gauge theories. The

---

† This is consistent with the fact that in terms of the Pontrjagin classes $\hat{A}$ may be written as $\hat{A} = 1 - \frac{1}{24}p_2 + \frac{1}{5760}(7p_2^2 - 4p_4) + ...$ and that all the Pontrjagin classes of spheres are zero (recall that our p_{2l} is also denoted p_l).

group-theoretical content of gauge theories is described in O'Raifeartaigh (1986). For a quantum field theory textbook treatment see, *e.g.*, Ramond (1989) and Bailin and Love (1993). More references concerning the topological structure of gauge theories are given in chapter 10.

The theory of characteristic classes is expounded in the second volume of Kobayashi and Nomizu (1969) and in Milnor and Stasheff (1974). See also, *e.g.*, Greub *et al.* (II, 1973), Spivak (1979), the appendix of Chern (1979), Gilkey (1984), Hirzebruch (1966) and Bott and Tu (1986). A comprehensive account of the different characteristic classes with an eye on their applications to physics is given in the review by Eguchi *et al.* (1980); see also the already cited books of Nash and Sen (1983), Choquet-Bruhat and De Witt-Morette (1989), Nakahara (1990) and Nash (1991). In connection with the Euler characteristic see also Witten (1982a). In relation with the transgression and homotopy formulae and the Chern-Simons forms see Chern and Simons (1974), Chern (1979) and Quillen (1989) and, in connection with their applications to anomalies, Zumino *et al.* (1984), Zumino (1985a), Álvarez-Gaumé and Ginsparg (1985) (this paper contains a general mathematical introduction) and Mañes *et al.* (1985). The generalization of the Chern-Simons secondary characteristic classes mentioned in section 2.4 is given in Guo *et al.* (1985). Chern-Simons forms on superfibre bundles are discussed in Bartocci *et al.* (1990).

The magnetic monopole is discussed in the classical papers of Dirac (1931) and Wu and Yang (1975b); see also Greub and Petry (1975) and Ryder (1980). The relation between non-trivial fibre bundles and monopoles is further analysed in Asorey and Boya (1979). For a physics textbook's treatment see Jackson (1975) or any of the books with physical applications cited above. A detailed review containing further references is given by Goddard and Olive (1978); see also Aitchison (1987). For a mathematical treatment see Atiyah and Hitchin (1988). The dynamical symmetry of the magnetic monopole is discussed in Jackiw (1980a). For monopoles in non-abelian gauge theories, and in particular for t'Hooft and Polyakov ideas, see the reviews of Goddard and Olive (1978), Coleman (1982) and Preskill (1984) which contain the original references. For the higher order antisymmetric tensor duality mentioned in the footnote in section 2.7, see Nepomechie (1985), Teitelboim (1986) and Duff and Lu (1992). For a review of the electromagnetic duality, including Seiberg and Witten (1994) duality, see Olive (1996).

Yang-Mills instantons were introduced in Belavin *et al.* (1975); see also the references on the Hopf fibrings given in the bibliographical notes to chapter 1. For a general discussion see Atiyah (1979). The Erice lectures by Coleman (1985), as well as Eguchi *et al.* (1980), Choquet-Bruhat *et al.* (1982), Nash and Sen (1983), Jackiw in Treiman *et al.* (1985) and Göckeler parameters that characterize the instanton solutions, or dimension of the moduli space (not discussed in the text), see Eguchi *et al.* (1980) where the original (mathematical and physical) references can be found; see also Nash (1991) and Marathe and Martucci (1992). This number depends on the Lie group: the k-instanton $SU(n)$ solutions ($SU(n)$ self-dual connections with $C_2 = -k$) depend on $4kn - n^2 + 1$ parameters; see Bernard *et al.* (1977) and Atiyah *et al.* (1978). The relation of Yang-Mills equations with the structure of four-manifolds is described in Atiyah (1984); for a brief discussion see Nash (1991). Detailed mathematical accounts are given in Freed and Uhlenbeck (1984), Freedman and Luo (1989) and in Donaldson and Kronheimer (1990). For topological aspects of quantum field theory in general see Schwarz (1993).

Early references to the Atiyah-Singer theorem are Atiyah and Singer (1968, especially III) and, already in connection with the chiral anomaly, Atiyah and Singer (1984). An excellent review of the various index theorems for manifolds without and with a boundary can be found in the already mentioned review by Eguchi *et al.* (1980). It is also discussed in the notes of Bott's Montreal lectures by Kulkarni (1975). The book of Shanahan (1978) contains detailed calculations; see also Gilkey (1984). A presentation of the index theorem with special emphasis on its applications is given in the books of Nakahara (1990) and Nash (1991); see also the appendix of Ball (1989). For the comment after eq. (2.11.19) see Jackiw and Rebbi (1977). The index theorem for Dirac operators on Euclidean open spaces of odd dimension was found by Callias (1978); see also Bott and Seeley (1978) where the generalized winding numbers are given. It is possible to give a derivation of the Atiyah-Singer theorem for the Dirac operator by using elementary properties of quantum-mechanical supersymmetric systems; see Álvarez-Gaumé (1983), Friedan and Windey (1984) and De Witt (1992). Further references in connection with anomalies will be given in chapter 10.

The properties of Dirac gamma matrices in any dimension may be found in Van Nieuwenhuizen (1984) and in the appendix of Sohnius (1985); see also Álvarez-Gaumé (1986). The subject of n-dimensional spinors goes back to Brauer and Weyl (1935).

3

A first look at cohomology
of groups and related topics

This chapter is devoted to providing a physical motivation for group cohomology and group extensions by analysing some specific features of classical and quantum mechanics of non-relativistic particles. In particular the need of considering projective representations of the Galilei group and Bargmann's superselection rule are discussed. The projective representations lead to the introduction of two-cocycles; the three-cocycle appears as a result of a consistent breach of associativity. The consistency conditions that have to be fulfilled in both cases are analysed, and illustrated in the case of two-cocycles with the examples of the Weyl-Heisenberg group and the Galilei group. All these concepts will be mathematically defined in chapters 4, 5, and 6, and addressed again in a more elaborate way from the point of view of classical physics in chapter 8.

The relation between cohomology and quantization and in particular the topics of symplectic cohomology, dynamical groups and geometric quantization are briefly treated in this chapter. Again, the Galilei and the Weyl-Heisenberg groups will be used as illustrations of the theory.

Finally, it is shown how the group contraction procedure can generate non-trivial group cohomology.

3.1 Some known facts of 'non-relativistic' mechanics: two-cocycles

A theory is called 'non-relativistic' if its formulation is covariant under the Galilei group G, which is the *relativity group* of a 'non-relativistic' theory. It should be noticed, however, that theories covariant under the Poincaré group and the Galilei group are both, strictly speaking, equally *relativistic*; only their relativity group is different. It is only due to the fact that the notion of covariance was introduced first in connection with Einstein's relativity* that, by contrast, Galilean theories are customarily referred to as being 'non-relativistic'.

The elements of the ten-parameter Galilei group are parametrized by

$$g = (B, \mathbf{A}, \mathbf{V}, R) \quad \text{(time, space, boosts, rotations)} \ .$$

Its group law $g'' = g' * g$ is expressed by

$$g'' = (B' + B, \ \mathbf{A}' + R'\mathbf{A} + \mathbf{V}'B, \ \mathbf{V}' + R'\mathbf{V}, \ R'R) \ , \tag{3.1.1}$$

and the identity and inverse elements are given by

$$e = (0, \mathbf{0}, \mathbf{0}, 1), \ g^{-1} = (-B, -R^{-1}(\mathbf{A} - \mathbf{V}B), -R^{-1}\mathbf{V}, R^{-1}) \ . \tag{3.1.2}$$

The action of G on spacetime $(t, \mathbf{x})$ can be given by the 5×5 matrices

$$D(g) = \begin{pmatrix} R & \mathbf{V} & \mathbf{A} \\ 0 & 1 & B \\ 0 & 0 & 1 \end{pmatrix} \quad \text{acting on} \quad \begin{pmatrix} \mathbf{x} \\ t \\ 1 \end{pmatrix} \ , \tag{3.1.3}$$

since it is defined by

$$\mathbf{x}' = R\mathbf{x} + \mathbf{V}t + \mathbf{A} \ , \quad t' = t + B \ . \tag{3.1.4}$$

The additional (and irrelevant) '1' is needed to represent a group of *affine* transformations in matrix (homogeneous) form†. Let us now analyse the Galilean invariance in the two simplest (classical and quantum) cases.

(a) *Classical mechanics of a free Newtonian particle*
Under a Galilean boost, $\dot{\mathbf{x}} \to \dot{\mathbf{x}} + \mathbf{V}$, the Lagrangian $L = \frac{1}{2}m\dot{\mathbf{x}}^2$ is *quasi-invariant*. By this we mean that the transformed Lagrangian L' differs from L by a total derivative instead of being equal to it,

$$L \to L' = L + \frac{d}{dt}m(\mathbf{x} \cdot \mathbf{V} + \frac{1}{2}\mathbf{V}^2 t) \equiv L + \frac{d}{dt}\Delta(t, \mathbf{x}; \mathbf{V}) \ . \tag{3.1.5}$$

There is no way of removing the function $\Delta(t, \mathbf{x}; g)$ (a more precise notation would be $\Delta_m(t, \mathbf{x}; g)$) *for all* the transformations g of the Galilei group

---

* Nevertheless, the Galilean principle of relativity was already established in the description of the experiment in the hold of a ship contained in the *Dialogo sopra i due massimi sistemi del mondo* (1632).
† Since there is no reason to select an origin O in spacetime, any relativity group must include the spacetime translations. In other words, flat spacetime is an *affine* space rather than a *vector* space, and a coordinate change in an affine space is non-homogeneous. Nevertheless, a non-homogeneous transformation may be written in homogeneous matrix form as in eq. (3.1.3). For instance, the analogous expression for the Poincaré group $\mathscr{P}$, which is given by the 5×5 matrices

$$\begin{pmatrix} \Lambda^\mu_{\ \nu} & a^\mu \\ 0 & 1 \end{pmatrix}$$

that act on $(x^\nu, 1)$ where x^ν are the spacetime coordinates, is the so-called Bargmann representation of $\mathscr{P}$. Of course, once an origin O is given, the familiar one-to-one correspondence $P \leftrightarrow \overrightarrow{OP}$ between the points P of the affine space and the vectors of its associated vector space is immediately established.

by adding a total derivative to L (see section 3.5). The variation (3.1.5) looks unimportant in classical mechanics, since the action principle, *i.e.*, the Lagrange equations, is not affected by this change. It might be tempting to discard eq. (3.1.5) as uninteresting, but Δ is however essential when defining conserved quantities, since the quasi-invariance of a Lagrangian requires that the Noether charges include Δ in their definition (see section 8.1).

(b) *Quantum mechanics, free Schrödinger equation*
The Galilean invariance means here that the Schrödinger equation has the same form in any two Galilei-related (primed and unprimed) reference frames, *i.e.*

$$\left(-\frac{\hbar^2}{2m}\nabla^2\right)\psi(x) = i\hbar\frac{\partial\psi(x)}{\partial t} \Leftrightarrow \left(-\frac{\hbar^2}{2m}\nabla'^2\right)\psi'(x') = i\hbar\frac{\partial\psi'(x')}{\partial t'} \quad . \quad (3.1.6)$$

In contrast with the spin zero relativistic case, where the Galilei group is replaced by the Poincaré group and the Schrödinger wavefunction by the Klein-Gordon field, there is no way of implementing Galilei invariance by using a transformation of the form

$$\psi'(\mathbf{x}',t') = \psi(\mathbf{x},t) \tag{3.1.7}$$

for the wavefunction. This is a surprising situation, the more so if, influenced by the case of the relativistic Klein-Gordon equation, (3.1.7) is thought of as the natural transformation law for a spinless wavefunction. However, as was very early stressed by Weyl, pure states are described by *rays*, not by vectors of a Hilbert space $\mathcal{H}$. We have

$$\{ \text{ rays } \} = \mathcal{H}/\mathcal{R} \tag{3.1.8}$$

where $\mathcal{R}$ is the equivalence relation which identifies vectors ψ and ψ' of $\mathcal{H}$ which differ in a phase since we observe probabilities and not wavefunctions, and probabilities depend only on rays, $\|\psi\| = \|\psi'\|$. Thus, let us allow for a spacetime dependent phase factor in (3.1.7) and replace it by

$$\psi'(\mathbf{x}',t') = \exp\left(\frac{i}{\hbar}\Delta(t,\mathbf{x})\right)\psi(t,\mathbf{x}) \quad , \tag{3.1.9}$$

where $\Delta(t,\mathbf{x})$ is to be determined. Using eq. (3.1.4) we find that, for $g = (B,\mathbf{A},\mathbf{V},1)$,

$$\frac{\partial}{\partial t'} = \frac{\partial}{\partial t} - \mathbf{V}\cdot\frac{\partial}{\partial\mathbf{x}} \quad , \quad \frac{\partial}{\partial\mathbf{x}'} = \frac{\partial}{\partial\mathbf{x}} \quad . \tag{3.1.10}$$

Then the terms appearing in the Schrödinger equation in the primed reference frame can be expressed as

$$\frac{\partial \psi'}{\partial t'} = \exp(\frac{i}{\hbar}\Delta)(\frac{\partial \psi}{\partial t} + \frac{i}{\hbar}\frac{\partial \Delta}{\partial t}\psi - \frac{i}{\hbar}\mathbf{V}\cdot\frac{\partial \Delta}{\partial \mathbf{x}}\psi - \mathbf{V}\cdot\frac{\partial \psi}{\partial \mathbf{x}}) \quad ,$$

$$\nabla'^2\psi' = \exp(\frac{i}{\hbar}\Delta)(\nabla^2\psi + [\frac{i}{\hbar}\nabla^2\Delta - \frac{1}{\hbar^2}(\frac{\partial \Delta}{\partial \mathbf{x}})^2]\psi + 2\frac{i}{\hbar}\frac{\partial \Delta}{\partial \mathbf{x}}\cdot\frac{\partial \psi}{\partial \mathbf{x}}) \quad ,$$

(3.1.11)

and thus (3.1.6) implies that

$$-i\mathbf{V} = -\frac{i}{m}\frac{\partial \Delta}{\partial \mathbf{x}} \quad , \qquad -\frac{\partial \Delta}{\partial t} + \mathbf{V}\cdot\frac{\partial \Delta}{\partial \mathbf{x}} = -\frac{i\hbar}{2m}\nabla^2\Delta + \frac{1}{2m}(\frac{\partial \Delta}{\partial \mathbf{x}})^2 \quad , \quad (3.1.12)$$

from which we obtain

$$\frac{\partial \Delta}{\partial \mathbf{x}} = m\mathbf{V} , \qquad \frac{\partial \Delta}{\partial t} = \frac{1}{2}m\mathbf{V}^2 . \qquad (3.1.13)$$

Thus, the invariance of the Schrödinger equation under the Galilei group can be achieved if the transformation law for the Schrödinger wavefunction is taken as in (3.1.9) with

$$\Delta(t,\mathbf{x}) = m(\mathbf{V}\cdot\mathbf{x} + \frac{1}{2}\mathbf{V}^2 t) \equiv \Delta(t,\mathbf{x};g) , \quad g \in G \quad ; \qquad (3.1.14)$$

notice the appearance of the factor m. If spin is included, the same reasoning leads to the general transformation law,

$$\psi'_{\zeta'}(t',\mathbf{x}') = \exp\frac{i}{\hbar}m(\mathbf{V}\cdot R\mathbf{x} + \frac{1}{2}\mathbf{V}^2 t)\, D_{\zeta'\zeta}(R)\psi_\zeta(t,\mathbf{x}) , \qquad (3.1.15)$$

where $D(R)$ is the matrix of the given spin representation ($\zeta,\zeta' = 1, ..., 2s+1$). Notice that the exponential in (3.1.15) is the same Δ that would appear in (3.1.5) including the rotations and, again, the appearance of m; it is obvious that in general (3.1.9) and (3.1.5) have to be the related effects (as we might expect given that $\psi \sim \exp\frac{i}{\hbar}\int L dt$) of a common cause, which will be shown to be the non-trivial cohomology of the Galilei group.

Is the exponential in (3.1.15) important? It is essential, otherwise contradictions occur. Let f be a classical harmonic travelling wave,

$$f(t,x) = A \sin 2\pi \left(\frac{x}{\lambda} - vt\right) \quad . \qquad (3.1.16)$$

Under a Galilean transformation, $x' = x + Vt$, the imposition of the transformation law (3.1.7),

$$f'(t',x') = f(t,x) \Rightarrow f'(t',x') = A \sin 2\pi \left(\frac{x'}{\lambda} - \left(\frac{V}{\lambda}+v\right)t'\right) \quad , \qquad (3.1.17)$$

gives the non-relativistic Doppler effect

$$\lambda = \lambda' \quad , \qquad v' = v + \frac{V}{\lambda} \quad . \qquad (3.1.18)$$

However, these expressions are *inconsistent* with the de Broglie and Einstein relations

$$p = \frac{h}{\lambda} \quad , \quad E = h\nu \quad , \tag{3.1.19}$$

because, under a Galilean boost, momentum and energy transform according to

$$p \rightarrow p' = p + mV \quad , \quad E \rightarrow E' = E + pV + \frac{1}{2}mV^2 \quad ; \tag{3.1.20}$$

this is the old Landé paradox. However, in Galilean quantum mechanics the probability wavefunction is complex and, as we have seen, has a phase which changes under a boost according to (3.1.9), (3.1.14). For a harmonic travelling wave $\psi(t, x) = \exp i2\pi(\frac{x}{\lambda} - \nu t)$ this now implies

$$\frac{1}{\lambda'} = \frac{1}{\lambda} + \frac{mV}{h} \quad , \quad \nu' = \nu + \frac{V}{\lambda} + \frac{1}{2}m\frac{V^2}{h} \quad , \tag{3.1.21}$$

in agreement with (3.1.20).

The transformation law (3.1.9) also allows us to find the composition law for two successive transformations. Let us put $(t, \mathbf{x}) = x$ and write (3.1.9) in the form

$$\psi'(x') \equiv [U(g)\psi](gx) = \exp \frac{i}{\hbar} \Delta(x; g)\, \psi(x) \tag{3.1.22}$$

where $x' = gx$. We can express the above transformation law by saying that, if the transformation is projective, ${}^g\psi \circ g = \exp\frac{i}{\hbar}\Delta(\cdot \ ; g)\psi$ with ${}^g\psi \equiv \psi'$; then $({}^g\psi \circ g)(x) = (\exp\frac{i}{\hbar}\Delta(\cdot \ ; g)\psi)(x)$ reproduces (3.1.22) (for an ordinary representation we would be able to remove the phase factor to have ${}^g\psi \circ g = \psi$ in the spinless case and $D(g) \circ \psi = {}^g\psi \circ g$ in the general one where $D(g)$ represents the rotation group acting on the vector space of the components of ψ). If $x'' = g'x' = g'gx$, we may similarly write

$$[U(g'g)\psi](x'') = \exp \frac{i}{\hbar} \Delta(x; g'g)\, \psi(x) \,. \tag{3.1.23}$$

To compare $U(g'g)$ with $U(g')U(g)$ we first notice that ${}^{g'g}\psi \circ g'g = \exp\frac{i}{\hbar}\Delta(\cdot \ ; g'g)\psi$ and that ${}^{g'}({}^g\psi) \circ g'g = \exp\frac{i}{\hbar}\Delta(g\cdot \ ; g'){}^g\psi \circ g = \exp\frac{i}{\hbar}\Delta(g\cdot \ ; g')\exp\frac{i}{\hbar}\Delta(\cdot \ ; g)\psi$ or, in other words, that

$$\begin{aligned}
[U(g')U(g)\psi](x'') &= [U(g')(U(g)\psi)](g'x') \\
&= \exp\frac{i}{\hbar}\Delta(x'; g').(U(g)\psi)(x') \\
&= \exp\frac{i}{\hbar}\Delta(gx; g').\exp\frac{i}{\hbar}\Delta(x; g)\,\psi(x) \quad .
\end{aligned} \tag{3.1.24}$$

Then, from (3.1.23) and (3.1.24), we obtain

$$U(g')U(g) = U(g'g)\exp\frac{i}{\hbar}\{\Delta(gx;g') + \Delta(x;g) - \Delta(x;g'g)\} \quad . \quad (3.1.25)$$

Since the l.h.s. of (3.1.25) cannot depend here on the spacetime argument x (we shall come back to this point in section 8.4) we may define

$$\xi(g',g) := \Delta(gx;g') + \Delta(x;g) - \Delta(x;g'g) \qquad (3.1.26)$$

and rewrite (3.1.25) in the form

$$U(g')U(g) = \exp\frac{i}{\hbar}\xi(g',g)\,U(g'g) \equiv \omega(g',g)U(g'g) \qquad (3.1.27)$$

introducing the unimodular factors $\omega(g',g)$ defined by $\omega(g',g)= exp\frac{i}{\hbar}\xi(g',g)$. We shall see in the next section that (3.1.27) defines a *projective* or *ray* representation of G, and that (3.1.26) defines a *two-cocycle ξ* on G. The fact that ξ cannot be made zero *for all* the group elements g',g of the Galilei group (*i.e.*, that the projective representation of the Galilei group used in quantum mechanics cannot be transformed into an ordinary one) will be expressed by saying that ξ is a non-trivial two-cocycle (it is not a two-coboundary) on the Galilei group. Notice that this statement is not contradicted by eq. (3.1.26) above since Δ depends on x and accordingly it does not define a mapping $\Delta : G \rightarrow R$ (cf. section 5.4) as would be needed to conclude that ξ is an R-valued two-coboundary on G.

(c) *Bargmann superselection rule*
The phase factor in (3.1.15) has another consequence: it induces a *superselection rule*. Superselection rules appear in quantum mechanics due to the presence of non-trivial operators that commute with all the observables and thus belong to any complete set of commuting observables. As a result, these *superselection operators* decompose the Hilbert space of all possible states of a system into *coherent subspaces* characterized by their eigenvalues (it is usually assumed that all the observables defining superselection rules commute among themselves). The superposition principle holds only inside these superselection subspaces, and no observable may have non-zero matrix elements between states of different superselection eigenvalues. Well-known examples are the charge or the spin (or *univalence*) superselection rules which separate the spaces with different charge or $(-1)^{2J}$ values. As a result, there are no pure states which are the superposition of two states with different charges or of states with spin integer and half-integer; these states are not *physically realizable*.

Consider now the unit transformation written as the product of Galilean transformations

$$e = (0,0,-\mathbf{V},1)(0,-\mathbf{A},0,1)(0,0,\mathbf{V},1)(0,\mathbf{A},0,1) \quad . \qquad (3.1.28)$$

Taking (3.1.15) into account, it is immediately seen that the right hand side is represented on ψ by the phase factor

$$\exp\left(\frac{i}{\hbar}m\mathbf{A}\cdot\mathbf{V}\right)\ ,\tag{3.1.29}$$

where m is the mass of the system considered. Consider now the possibility of defining a state Ψ as the sum of two states ψ_1 and ψ_2 of masses m_1 and m_2. Clearly, under the transformations of the r.h.s. of (3.1.28),

$$\Psi\mapsto\Psi'=\exp\left(\frac{i}{\hbar}m_1\mathbf{A}\cdot\mathbf{V}\right)\psi_1+\exp\left(\frac{i}{\hbar}m_2\mathbf{A}\cdot\mathbf{V}\right)\psi_2\neq(\text{phase})\Psi$$
$$\tag{3.1.30}$$

and thus Ψ and $U(e)\Psi$ cannot represent the same physical state. Physical subspaces corresponding to states of different mass are incoherent, and the superposition principle does not hold among them; this is the result of the *Bargmann mass superselection rule*. In other words, states of two representations of a group corresponding to two inequivalent two-cocycles (here, different mass) cannot be coherently mixed. In fact, it may be seen (see section 3.4 and the remark in (3.2.18) below) that different masses in the Schrödinger equation define different (although isomorphic) extended Galilei groups. Thus, and within the domain of application of non-relativistic quantum mechanics (the superselection rule is obviously approximate since Nature is relativistic) there are no states with a mass spectrum and therefore unstable particles.

(d) *Associativity of* $U(g'')U(g')U(g)$

We have seen that to represent the Galilei group on the Schrödinger wavefunctions the customary *representation* property $U(g')U(g)=U(g'g)$ has to be replaced by the projective representation condition (3.1.27). Nevertheless, the associative property of the $U(g)$'s still holds; in fact, this is a *necessary property* of linear operators, and imposes a condition on the phase factors ω, the *two-cocycle condition*. To find it, let us compute $U(g'')U(g')U(g)$ in two different ways:

$$[U(g'')U(g')]U(g)=U(g''g')\omega(g'',g')U(g)=\omega(g'',g')\omega(g''g',g)U(g''g'g),$$
$$U(g'')[U(g')U(g)]=U(g'')U(g'g)\omega(g',g)=U(g''g'g)\omega(g'',g'g)\omega(g',g).$$
$$\tag{3.1.31}$$

The equality of both expressions implies that the phase factors ω satisfy $\omega(g'',g')\,\omega(g''g',g)=\omega(g'',g'g)\omega(g',g)$. In terms of the exponents $\xi(g',g)$ the two-cocycle condition reads

$$\xi(g'',g')+\xi(g''g',g)-\xi(g'',g'g)-\xi(g',g)=0\ ,\tag{3.1.32}$$

where the zero can be replaced by nh in $\exp\frac{i}{\hbar}\xi$. Eq. (3.1.32) has to be satisfied for any local exponent $\xi(g',g)$.

3.2 Projective representations of a Lie group: a review of Bargmann's theory

As already mentioned, the physical motivation to consider projective representations in physics is that, since pure states are represented in quantum mechanics by rays, symmetry operators may be realized by unitary *ray operators* $\bar{U}$,

$$\bar{U}(g')\bar{U}(g) = \bar{U}(g'g), \quad g',g \in G \ , \tag{3.2.1}$$

where the bar indicates the *class* of equivalent operators which differ in a phase:

$$U, \ U' \in \bar{U} \Leftrightarrow U \mathcal{R} U' \Leftrightarrow U = \gamma U' \ , \quad |\gamma| = 1 \ .$$

We shall assume that the group G is connected, so that the possibility of having anti-unitary ray representations does not arise: since the product of two anti-unitary operators is unitary, the elements of the identity component (and, in particular, the identity itself) cannot be represented by anti-unitary operators.

Let us now consider briefly the abstract mathematical problem of projective or ray representations (hence the physical Planck constant $\hbar$ of section 3.1 will be absent here). Let $g, g' \in \Pi$, Π being a neighbourhood of $e \in G$. If elements $U(g')$, $U(g)$ are chosen inside the classes $\bar{U}(g')$, $\bar{U}(g)$, all that is known is that the product $U(g')U(g)$ belongs by eq. (3.2.1) to the class $\bar{U}(g'g)$. If a representative $U(g'g)$ is selected in the class $\bar{U}(g'g)$, (3.2.1) will be replaced by

$$U(g')U(g) = \omega(g',g)U(g'g) , \quad |\omega(g',g)| = 1 \ . \tag{3.2.2}$$

Note that the above considerations have a local character (they refer to Π above) so that the ω's should be called *local factors*. To study the ω's it is often convenient to use the *local exponents* $\xi(g',g)$ already introduced, which are defined by

$$\omega(g',g) := \exp i\xi(g',g) \ . \tag{3.2.3}$$

If different representatives U' are selected in each class, a new $\omega'(g',g)$ will also be selected,

$$U'(g')U'(g) = \omega'(g',g)U'(g'g) \ . \tag{3.2.4}$$

But since $U'(g)$ and $U(g)$ belong to the same class, they are related by a phase for each g, $U'(g) = \gamma(g)U(g)$, $|\gamma(g)| = 1$. This in turn relates ω and ω',

$$\omega'(g',g) = \omega(g',g)\gamma^{-1}(g'g)\gamma(g')\gamma(g) \ . \tag{3.2.5}$$

When it is possible to select a $\gamma(g)$ such that $\omega' = 1$ ($\xi' = 0$) it is said that the local factors (exponents) are equivalent to 1 (0).

Given a group G, it is not always possible to extend the continuous set of representatives on the neighbourhood Π to the whole group G. However, the following theorem due to Bargmann, which will be given without proof, holds:

(3.2.1) THEOREM

Let G be a connected and simply connected group, ω a factor on the whole group G, and $\bar{U}(g)$ a continuous ray representation such that the local factor defined by a chosen continuous set of representatives $U(g)$ coincides with ω on a neighbourhood of the identity element e of G. Then there exists a uniquely determined continuous set of representatives for the whole group which has ω as its system of factors.

Thus, it is convenient to substitute the covering group for the group G if G is not simply connected (of course since G and its universal covering group are locally isomorphic, their local exponents coincide: it suffices to identify their elements locally). If all local exponents of G are zero, it is possible to transform a ray representation of the covering group into an ordinary representation. This does not mean, however, that this can also be done for a non-simply connected group, as the example of the double-valued (spinorial) representations of the rotation group $SO(3)$ to be discussed below shows. We shall assume here that G is (connected and) simply connected (see also section 4.3).

As has been mentioned, if the factor $\omega(g', g)$ can be eliminated through (3.2.5) (ω is 'equivalent to one') the ray representation becomes an ordinary (vector) representation. Clearly, eq. (3.2.5) establishes an equivalence relation among factors. Thus, *a ray representation determines a class of equivalent factors*.

By using the associative property we have already obtained from (3.1.31) that

$$\omega(g'', g')\omega(g''g', g) = \omega(g'', g'g)\omega(g', g) . \tag{3.2.6}$$

Clearly, $U(e) = 1$ (this normalization condition will always be assumed) gives $\omega(e, e) = 1$. Setting successively $g'' = g' = e$, $g' = g = e$ and $g'' = g'^{-1}$, $g = g''$ in (3.2.6) we obtain

$$\omega(e, g) = 1 = \omega(g, e) \quad , \quad \omega(g, g^{-1}) = \omega(g^{-1}, g) \quad . \tag{3.2.7}$$

We may look at the factors ω as (in general, local) mappings

$$\omega : G \times G \longrightarrow U(1) \quad \text{(phase group)} \tag{3.2.8}$$

such that $\omega(e, e) = 1$ and eq. (3.2.6) is satisfied. They are normalized (because $\omega(e, e) = 1$ is assumed) *two-cocycles*, since (3.2.6) is the two-cocycle condition which will be discussed at length in chapter 5. Two-cocycles of the form

$$\omega_{cob}(g',g) = \gamma^{-1}(g'g)\gamma(g')\gamma(g) \tag{3.2.9}$$

generated by mappings

$$\gamma : G \longrightarrow U(1) \tag{3.2.10}$$

are trivial two-cocycles or *two-coboundaries*; inequivalent ray representations are associated with non-equivalent two-cocycles. Two two-cocycles ω and ω' are equivalent, $\omega' \mathcal{R} \omega$, when there exists a two-coboundary ω_{cob} such that $\omega'(g',g) = \omega(g',g)\,\omega_{cob}(g',g)$; the classes of inequivalent two-cocycles define the *second cohomology group* $H^2(G,U(1))$.

The multiplicative notation can be transformed into an additive one by using the local exponents $\xi(g',g)$; they satisfy the corresponding two-cocycle and normalization equations[*]

$$\xi(g'',g') + \xi(g''g',g) = \xi(g'',g'g) + \xi(g',g) \quad,$$

$$\xi(g,e) = 0 = \xi(e,g) \quad, \quad \xi(g,g^{-1}) = \xi(g^{-1},g) \quad, \tag{3.2.11}$$

and a local exponent is a two-coboundary $\xi_{cob}(g',g)$ if there exists a function $\eta(g)$ such that

$$\xi_{cob}(g',g) = \eta(g') + \eta(g) - \eta(g'g) \quad; \tag{3.2.12}$$

two local exponents are equivalent, $\xi \sim \xi'$, if they differ in a coboundary, $\xi'(g',g) = \xi(g',g) + \xi_{cob}(g',g)$. As may be expected, the equivalence classes are stable under the inner automorphisms of G, *i.e.* if $\xi_{g_0}(g',g) := \xi(g_0 g' g_0^{-1}, g_0 g g_0^{-1})$ is the exponent obtained by the action of $g_0 \in G$, $\xi_{g_0}(g',g) \sim \xi(g',g)$ and the coboundary realizing the equivalence is generated by $\eta_{g_0}(g) = \xi(g_0^{-1}, g_0 g g_0^{-1}) - \xi(g, g_0^{-1})$. The following lemma (Bargmann), establishing the differentiability of local exponents, holds:

(3.2.1) LEMMA (*Differentiable exponents*)

Every local exponent ξ of a Lie group is equivalent to a differentiable local exponent, i.e. to an exponent $\xi(g',g)$ which is differentiable in a neighbourhood of the identity. If two differentiable local exponents ξ_1, ξ_2 are equivalent, $\xi_1 = \xi_2 + \xi_{cob}$, ξ_{cob} is generated by a differentiable function $\eta(g)$.

---

[*] When $\xi(g',g)$ is the local exponent of a $U(1)$-factor $\omega(g',g)$, as is the case here, we may take $\xi(g,e) = 0 \pmod{2\pi}$.

If the Lie group G is connected and simply connected, every differentiable local exponent may be extended to a differentiable exponent on the whole group G.

Let us now write the different operators inside the class $\bar{U}(g)$ in the form

$$e^{i\theta} U(g) \quad , \tag{3.2.13}$$

by introducing a new variable θ. Then

$$e^{i\theta'} U(g') e^{i\theta} U(g) = e^{i(\theta'+\theta)} e^{i\xi(g',g)} U(g'g) \equiv e^{i\theta''} U(g'') \quad . \tag{3.2.14}$$

Thus, given a group G and a local exponent $\xi : G \times G \to R$, a new group $\bar{G}$ (the *local group*) with elements $\bar{g} = (\theta, g)$ may be defined by the group law

$$(\theta'', g'') = (\theta', g') * (\theta, g) = (\theta' + \theta + \xi(g', g), g'g) \quad . \tag{3.2.15}$$

One finds

$$(0, e) = \bar{e} \in \bar{G} \ , \quad (\theta, g)^{-1} = (-\theta - \xi(g, g^{-1}), g^{-1}), \quad \xi(g, g^{-1}) = \xi(g^{-1}, g) \ ; \tag{3.2.16}$$

(3.2.15) may also be written as $(\zeta = e^{i\theta}, \omega(g', g) = \exp i\xi(g', g))$

$$(\zeta', g')(\zeta, g) = (\zeta'\zeta \,\omega(g', g), g'g) \quad . \tag{3.2.17}$$

In the form (3.2.17) the new group $\tilde{G}$ of parameters (ζ, g) is such that:

a) it contains $U(1)$ as an invariant subgroup and $\tilde{G}/U(1) = G$. In fact the $U(1)$ invariant subgroup is a central one: it is immediately verifiable that the elements (ζ, e) commute with any element (ζ', g') of $\tilde{G}$. In other words, $\tilde{G}$ is a *central extension* of G by $U(1)$ (see section 5.2, case (a)).

b) G is *not* a subgroup of $\tilde{G}$, but it is possible to represent G projectively by representing $\tilde{G}$ and then looking at G as $\tilde{G}/U(1)$.

Remarks

(a) Central extensions defined by proportional exponents $\xi_2 = \lambda\xi_1$ are given by isomorphic groups: the mapping $(\theta_1, g) \to (\theta_2 = \lambda\theta_1, g)$ is an isomorphism, because

$$(\theta'_1 + \theta_1 + \xi_1(g', g), g'g) \mapsto (\lambda\theta'_1 + \lambda\theta_1 + \lambda\xi_1(g', g), g'g)$$
$$= (\theta'_2 + \theta_2 + \xi_2(g', g), g'g) . \tag{3.2.18}$$

Nevertheless, they are different as group extensions as will be shown in section 4.2.

(b) Eqs. (3.2.17) and (3.2.15) define, by themselves, central extensions $\tilde{G}$ (resp. $\bar{G}$) of G by $U(1)$ (resp. by R). Both groups are locally isomorphic and lead to the same Lie algebra in the same way as $U(1)$ and R do.

(c) Finite-dimensional ray representations for G simply connected are always equivalent to ordinary representations. If the Hilbert space of the representation (3.2.2) is finite-dimensional, the operators $U(g)$ are given by $n \times n$ unitary matrices $D(g)$, and $|\det D(g)| = 1$. Then (3.2.2) implies, with $\omega = exp\, i\xi$, that

$$\det D(g') \det D(g) = e^{in\xi(g',g)} \det D(g'g) \qquad (3.2.19)$$

or $[\det D(g')]^{\frac{1}{n}}[\det D(g)]^{\frac{1}{n}} = \omega(g',g)[\det D(g'g)]^{\frac{1}{n}}$. Thus, it is sufficient to introduce new representatives $D'(g)=D(g)[\det D(g)]^{-\frac{1}{n}}$ to find $D'(g')D'(g) = D'(g'g)$, q.e.d. Notice, however, that in defining $D'(g)$ we have selected a particular root $[\det D(g)]^{\frac{1}{n}}$. The result is thus valid in the neighbourhood of the identity, but not globally unless the group is simply connected, since only then can the root be defined uniquely along a closed path. For instance, the rotation group $SO(3)$ is doubly connected and thus for n even (or, since $n = 2s + 1$, for s half integer) we obtain $D_s(g')D_s(g)=\pm D_s(g'g)$.

(d) Finally we stress that, as mentioned, the two-cocycle relation (3.2.11) is essential for (3.2.15) to be a group law; in fact, the associativity of (3.2.15) implies the two-cocycle condition in (3.2.11).

3.3 The Weyl-Heisenberg group and quantization

The Weyl-Heisenberg Lie group (WH) is the simplest example of a group with a central extension structure; moreover it is the underlying group of basic quantum commutation relations such as $[q^i, p_j] = i\hbar\delta^i_j$, or $[a^i, a^\dagger_j] = \delta^i_j$, and thus worth discussing as a first example. To simplify the notation, we shall simply write (q,p) for (q^i, p_i) in most places since the two- and $2n$-dimensional cases are structurally identical.

The WH group is a three-dimensional (in general, $(2n+1)$-dimensional) manifold (q, p, ζ) with a group law given by

$$q'' = q' + q \; , \; p'' = p' + p \; , \; \zeta'' = \zeta'\zeta exp\frac{i}{2\hbar}(q'p - p'q) \; , \qquad (3.3.1)$$

$$(\zeta; q, p)^{-1} = (\zeta^{-1}; -q, -p) \, ,$$

where, thinking of the simplest physical application, we look at q and p as coordinates and momenta respectively. With the usual assignments of dimensions, the exponent in (3.3.1) requires the presence of a constant with the dimensions of an action (the presence of a constant accompanying the cocycle will be further discussed later on). With the notation of the previous section, the two-cocycle $\xi(g', g)$ is given by

$$\xi(g',g) = \frac{1}{2\hbar}(q'p - p'q) \; . \qquad (3.3.2)$$

This two-cocycle is only a representative of its class. For instance, adding $\xi_{cob}(g',g) = \frac{1}{2\hbar}(q'p + p'q)$, which is the two-coboundary generated by $\eta(g) = -\frac{1}{2\hbar}pq$ (see eq. (3.2.12)), the new form of the two-cocycle is

$$\xi(g',g) = \frac{1}{\hbar}q'p \quad . \tag{3.3.3}$$

The RIVFs on the WH group are easily computed from (1.1.13),

$$\tilde{X}^R_{(q)} = \frac{\partial}{\partial q} + \frac{i}{2\hbar}p\Xi \quad , \quad \tilde{X}^R_{(p)} = \frac{\partial}{\partial p} - \frac{i}{2\hbar}q\Xi \quad , \quad \tilde{X}_{(\zeta)} = \zeta\frac{\partial}{\partial\zeta} \equiv \Xi \quad . \tag{3.3.4}$$

They have zero commutation relations among themselves, with the exception of

$$[\tilde{X}^R_{(q)}, \tilde{X}^R_{(p)}] = -\frac{i}{\hbar}\Xi \quad , \tag{3.3.5}$$

Ξ being the central generator. We see in the presence of the two-cocycle in (3.3.1) (and in the non-zero commutator (3.3.5)) the central extension character of the Weyl–Heisenberg group. As mentioned in the previous general discussion, once the third expression in (3.3.1) containing the two-cocycle is added to the $2n$-dimensional abelian group (q,p) to define the group law of the $(2n+1)$-dimensional WH group, the $g = (p,q)$ parameters do *not* close into a subgroup since with $(\zeta;g) \equiv (\zeta;q,p)$ it is found that $(1;q',p')(1;q,p) = (\zeta'' = exp\frac{i}{2\hbar}(q'p - p'q); q' + q, p' + p)$. This is reflected in the corresponding Weyl–Heisenberg Lie algebra wh: the even-dimensional abelian algebra generated by $X^R_{(q)} = \frac{\partial}{\partial q}, X^R_{(p)} = \frac{\partial}{\partial p}$ is centrally extended by Ξ to give the odd-dimensional algebra wh generated by $\tilde{X}_{(q)}, \tilde{X}_{(p)}$ and $\tilde{X}_\zeta \equiv \Xi$, of which the former is *not* a subalgebra due to the commutator (3.3.5) (Lie algebra extensions will be studied in chapter 6).

If we now define the operators $P = -i\hbar\tilde{X}^R_{(q)}$ and $Q = i\hbar\tilde{X}^R_{(p)}$, the algebra wh may be rewritten as

$$[Q,Q] = 0 = [P,P] \quad , \quad [Q,P] = i\hbar\Xi \quad . \tag{3.3.6}$$

This is clearly isomorphic to the Poisson bracket algebra

$$\{q,q\} = 0 = \{q,q\} \quad , \quad \{q,p\} = 1 \quad . \tag{3.3.7}$$

The LIFs on the WH group are found, *e.g.* from eq. (1.6.10),

$$\tilde{\omega}^{L(q)} = dq \quad , \quad \tilde{\omega}^{L(p)} = dp \quad , \quad \tilde{\omega}^{L(\zeta)} = \frac{i}{2\hbar}(pdq - qdp) + \frac{d\zeta}{\zeta} \quad . \tag{3.3.8}$$

They are preserved by the generators of the left translations, $L_{\tilde{X}^R}\tilde{\omega}^L = 0$ (eq. (1.8.4)); their invariance can also be checked directly in (3.3.8) by considering the primed variables in (3.3.1) as the parameters of the WH group transformation $L_{g'}$. Defining the *quantization form* by

$$\Theta \equiv \frac{\hbar}{i}\tilde{\omega}^{L(\zeta)} = \frac{1}{2}(pdq - \dot{q}dp) + \hbar\frac{d\zeta}{i\zeta}, \qquad (3.3.9)$$

we see that the 'vertical' one-form $(\tilde{\omega}^{L(\zeta)}$ is dual to the central $U(1)$ generator Ξ, $\tilde{\omega}^{L(\zeta)}(\tilde{X}_{(\zeta)}) = 1)$ allows us to define a symplectic form* Ω

$$d\Theta = \Omega = dp \wedge dq \qquad (dp_i \wedge dq^i) \qquad (3.3.10)$$

on the base manifold $S = WH/U(1) \sim R^2 (R^{2n})$ (the WH group has a principal bundle structure $WH(U(1), S)$). The form Θ is a form of *constant class $2n + 1$*[†] and the pair (WH, Θ) defines the so-called *quantum manifold* which is a basic ingredient of the geometric approaches to quantization; on the $2n$-dimensional manifold $S = WH/U(1)$ there is a symplectic structure defined by (3.3.10). Since this manifold is the result of taking the quotient by $U(1)$, and the symplectic 2-form ω is associated with the definition of the Poisson bracket, it is not surprising that Poisson brackets realize through eq. (3.3.7) a projective representation of the abelian algebra generated by $\partial/\partial q^i, \partial/\partial p_i$.

To see the implications of this fact more explicitly, let us recall first how in general a symplectic structure (S, Ω) on a $2n$-dimensional manifold S determines the Poisson brackets. Let us still denote the local coordinates on S by (q, p) $((q^i, p_i))$. Given a function $f(q, p)$, the mapping

$$i_{X_f}\Omega = -df \longmapsto X_f = \left(\frac{\partial f}{\partial p_i}\frac{\partial}{\partial q^i} - \frac{\partial f}{\partial q^i}\frac{\partial}{\partial p_i}\right) \qquad (3.3.11)$$

defines an isomorphism between the module of the vector fields X_f and that of the one-forms df on S. Then the Poisson bracket is defined by

$$\{f, g\} := -\Omega(X_f, X_g) = \left(\frac{\partial f}{\partial q^i}\frac{\partial g}{\partial p_i} - \frac{\partial f}{\partial p_i}\frac{\partial g}{\partial q^i}\right) \qquad (3.3.12)$$

---

* A *symplectic form* Ω on an even-dimensional manifold S^{2n} is a closed two-form of maximal $(2n)$ rank (the *rank* of a form α is the minimal number of independent one-forms needed to express α); the pair (S, Ω) is called a *symplectic manifold*. In general, Ω is a symplectic form iff $\Omega \wedge \overset{n}{\cdots} \wedge \Omega \equiv \Omega^n$ is a volume form on S^{2n}. Darboux' theorem guarantees that Ω may be written locally as $\Omega = dx^i \wedge dy_i$ on the symplectic chart (x^i, y_i), $i = 1, ..., n$; we see that Ω is expressed using $2n$ one-forms, dx^i and dy_i. For instance, on the cotangent space $T^*(M)$ parametrized by (q^i, p_i) the local expression of Ω is given by $\Omega = dq^i \wedge dp_i$. Given two symplectic manifolds (S, Ω), (S', Ω'), $\phi : M \to M'$ is a *symplectic transformation* if $\phi^*(\Omega') = \Omega$.

† Given a differential form α on a manifold M, a *characteristic vector field* X on M of α is a vector field that belongs to the radicals of α and $d\alpha$, i.e. such that $i_X\alpha = 0$, $i_Xd\alpha = 0$. Since $i_{fX} = f i_X$, the characteristic vector fields determine a submodule $\mathscr{C}(\alpha)$ of $\mathscr{X}(M)$; the constant class is the codimension of $\mathscr{C}(\alpha)$. In the case of Θ above, $\dim(\mathrm{rad}\Theta \cap \mathrm{rad}\,d\Theta) = 0$. The *class of a form* α of constant class is the minimal number of functions that are needed to express it.

or, equivalently, by

$$\{f,g\} := i_{X_f} i_{X_g} \Omega = L_{X_g} f = -L_{X_f} g = df(X_g) = -dg(X_f) \quad . \tag{3.3.13}$$

As is well known, the Poisson bracket is linear and antisymmetric in both arguments,

$$\{f,g\} = -\{g,f\} \quad , \quad \{\lambda_1 f_1 + \lambda_2 f_2, g\} = \lambda_1 \{f_1, g\} + \lambda_2 \{f_2, g\} \quad , \tag{3.3.14}$$

and satisfies the Jacobi identity

$$\{f, \{g, h\}\} + \{g, \{h, f\}\} + \{h, \{f, g\}\} = 0 \quad . \tag{3.3.15}$$

Properties (3.3.14) and (3.3.15) endow the associative algebra $\mathscr{F}(S)$ of the smooth functions on S with a Lie algebra structure. Moreover, the Poisson bracket also satisfies the Leibniz rule,

$$\{f, g_1 g_2\} = \{f, g_1\} g_2 + g_1 \{f, g_2\} \quad . \tag{3.3.16}$$

Eq. (3.3.16), together with (3.3.14), tells us that the mapping $f \longmapsto \{f, \cdot\}$ is the derivation on $\mathscr{F}(S)$ given by X_f in (3.3.11)*. The symplectic form Ω also defines the *Lagrange bracket* of two vector fields $X, Y \in \mathscr{X}(M)$ as the *function* $[X, Y]_{\mathscr{L}} = -\Omega(X, Y)$; $[X_f, X_g]_{\mathscr{L}} = \{f, g\}$.

Clearly, if $f = q, p$ the resulting vector fields X_q, X_p commute despite the fact that $\{q, p\} = 1$. Moreover, if $f = \text{const}$, $X_f = 0$. But it is possible to define a mapping $f \longmapsto \tilde{X}_f$, $\{f, g\} \longmapsto \tilde{X}_{\{f,g\}}$, which is a Lie algebra isomorphism. To this end, let us extend (in fact, *lift*) the vector fields X_f on S to vector fields $\tilde{X}_f$ on a $(2n+1)$-dimensional contact manifold (P, Θ) by defining

$$\tilde{X}_f = X_f + i[f - \Theta(X_f)]\Xi \quad . \tag{3.3.17}$$

The expression (3.3.17) may be justified by noticing that it is the unique solution of the equations

$$i_{\tilde{X}_f} \Theta = f \quad , \quad i_{\tilde{X}_f} d\Theta = -df \quad , \tag{3.3.18}$$

which are the necessary and sufficient conditions for $\tilde{X}_f$ to preserve the contact form Θ,

$$L_{\tilde{X}_f} \Theta = 0 \quad . \tag{3.3.19}$$

---

* In general, a mapping of $\mathscr{F}(M) \times \mathscr{F}(M) \longrightarrow \mathscr{F}(M)$, $(f, g) \longmapsto \{f, g\}$, satisfying the above properties (antisymmetry, Leibniz rule and Jacobi identity) characterizes M as a *Poisson manifold*. This means that $\mathscr{F}(M)$ has also the structure of an infinite-dimensional Lie algebra with a commutator given by the Poisson bracket. This Lie structure on $\mathscr{F}(M)$ is compatible with its associative algebra structure in the sense that the Lie (Poisson) bracket is a derivation with respect to the multiplication in $\mathscr{F}(M)$.

Using (3.3.17) it is now simple to check that the mapping $f \longmapsto \tilde{X}_f$ is a Lie algebra isomorphism because

$$[\tilde{X}_f, \tilde{X}_g] = -\tilde{X}_{\{f,g\}} \tag{3.3.20}$$

(the sign is irrelevant for this purpose). In particular, if we take $f, g = q, p$ it is found that, in contrast with $X_{\{p,q\}} = 0$, $\tilde{X}_{\{q,p\}} = i\Xi$.

Let (WH, Θ) now be the contact manifold defined by the WH group so that Θ is given by (3.3.9). This leads to

$$\tilde{X}_f = \left(\frac{\partial f}{\partial p_j} \frac{\partial}{\partial q^j} - \frac{\partial f}{\partial q^j} \frac{\partial}{\partial p_j} \right) + i \left[f - \frac{1}{2}(q\frac{\partial f}{\partial q} + p\frac{\partial f}{\partial p}) \right] \Xi \tag{3.3.21}$$

and, as a result, we find

$$\tilde{X}_p = \frac{\partial}{\partial q} + \frac{i}{2}p\Xi \equiv \tilde{X}^R_{(q)} \quad , \quad \tilde{X}_q = -\frac{\partial}{\partial p} + \frac{i}{2}q\Xi \equiv -\tilde{X}^R_{(p)} \quad , \tag{3.3.22}$$

and it is seen that eq. (3.3.5) is satisfied ($\hbar = 1$) or, equivalently, eq. (3.3.6) with the identifications $P = -i\hbar\tilde{X}_p$ and $Q = -i\hbar\tilde{X}_q = i\hbar\tilde{X}^R_{(p)}$. Thus, the Poisson brackets on $(WH/U(1), \Theta)$ have a Lie algebra structure isomorphic to the $(2n+1)$-dimensional WH algebra.

The isomorphism between the (q, p) Poisson brackets and the Lie brackets of the associated $(\tilde{X}_q, \tilde{X}_p)$ vector fields on the WH group is just the old Dirac prescription $\{\ ,\ \} \longrightarrow \frac{1}{i\hbar}[\ ,\ ]$ for the quantization of a dynamical system in its simplest version (we are ignoring here all the difficulties which can make this replacement inadequate and in particular the presence of constraints which require Dirac's starred brackets). This relation between Poisson and Lie brackets and the search for a symplectic manifold is encountered, one way or another, in all geometric quantization schemes. The manifold (WH, Θ) is the simplest example of a *quantum manifold*. Within the terminology of geometric quantization, eq. (3.3.17) determines the *quantum lifting* $\tilde{X}_f$ of the Hamiltonian vector field X_f, which is also defined by (3.3.19) as being a *quantomorphism*. The correspondence $f \mapsto \hat{f} \equiv -i\hbar\tilde{X}_f$ leading to the quantum operators $\hat{f}$ does not finish the quantization process, however, since for instance the operators P, Q obtained do not have their usual form $-i\hbar\partial/\partial q$, q (in the coordinate realization), and it is accordingly called *prequantization*. To obtain their customary (irreducible) form, it is necessary to restrict the contact manifold to what is called by Souriau the *Planck manifold*. This requires the introduction of the *polarizations*, the aim of which is to reduce the number of variables (q's or p's, say). The introduction of suitable polarizations in general is the stumbling block on the path of geometric quantization, and we shall not treat this difficult problem here. Let us mention instead that an essential feature in the above discussion of the connection among Poisson brackets and quantization was the existence of a $WH(U(1), WH/U(1))$

principal bundle structure for the WH group. This suggests the possibility of introducing a group approach to geometric quantization by requiring that the starting point should be a group $\tilde{G}$, or *quantum group*, possessing a $U(1)$-principal bundle structure[*] (the quantum or centrally $U(1)$-extended Galilei group below also has such a principal bundle structure). In this way, instead of quantizing a classical system, the quantum group $\tilde{G}$ is viewed as a quantum system from the start; after all, physical systems are always quantum systems since $\hbar$ is small but never zero. In fact, it might well be that, instead of looking for geometrical ways of quantizing a classical system, one should consider quantum systems from the start and *then* worry about the converse mathematical problem, *i.e.* the classical limit. The above group approach requires finding the appropriate $\tilde{G}$ as the starting point, and we shall not discuss it further.

3.4 The extended Galilei group

Another example of the application of the results of section 3.2 is the case where G is the Galilei group. In fact, a two-cocycle defining the extension is given by

$$\xi_m(g', g) = m(\mathbf{V}' \cdot R'\mathbf{A} + \tfrac{1}{2}\mathbf{V}'^2 B) \qquad (3.4.1)$$

(the dependence of ξ_m on m has been stressed by the subscript) to be compared with Δ in (3.1.5),

$$\Delta(t, \mathbf{x}; g) = m(\mathbf{V} \cdot R\mathbf{x} + \tfrac{1}{2}\mathbf{V}^2 t) \ . \qquad (3.4.2)$$

Indeed, (3.4.1) may be easily derived from (3.1.26) since this expression reads, using (3.1.4),

$$m[\mathbf{V}' \cdot R'(R\mathbf{x} + \mathbf{V}t + \mathbf{A}) + \tfrac{1}{2}\mathbf{V}'^2(t + B)]$$
$$+ m[\mathbf{V} \cdot R\mathbf{x} + \tfrac{1}{2}\mathbf{V}^2 t] - m[(\mathbf{V}' + R'\mathbf{V}) \cdot R'R\mathbf{x} + \tfrac{1}{2}(\mathbf{V}' + R'\mathbf{V})^2 t] \ , \qquad (3.4.3)$$

which gives (3.4.1). This shows that (3.4.1) and (3.4.2) imply each other. It is important to notice that the non-trivial two-cocycle (3.4.1) is the

---

[*] This quantum group concept should not be confused with the *quantum groups* recently introduced (and so named by V.G. Drinfel'd), the possible applications of which to physics are at present being explored. These are not really groups; rather, they correspond to q-deformations of universal enveloping Lie algebras or of algebras of functions on Lie group manifolds. Since the generators ('group parameters') of these algebras are non-commutative, their study is closely related to that of non-commutative geometry. Only in the $q \to 1$ limit of the deformation parameter is the 'classical' or Lie structure recovered.

result of the interplay of the translations and the boosts: if, for instance, $g = (0, 0, \mathbf{V}, 1)$ and $g' = (0, 0, \mathbf{V}', 1)$, then (3.1.26) gives zero for $\xi_m(\mathbf{V}', \mathbf{V})$ as is seen in (3.4.1). This will be reflected in the commutation relations of the generators of the translations and boosts of the extended group (see below).

The group law for the extended group $\bar{G}_{(m)}$ is then given by

$$
\left.
\begin{aligned}
B'' &= B' + B \quad, \\
\mathbf{A}'' &= \mathbf{A}' + R'\mathbf{A} + \mathbf{V}'B \quad, \\
\mathbf{V}'' &= \mathbf{V}' + R'\mathbf{V} \quad, \\
R'' &= R'R \quad, \\
\theta'' &= \theta' + \theta + m(\mathbf{V}' \cdot R'\mathbf{A} + \tfrac{1}{2}\mathbf{V}'^2 B) \quad.
\end{aligned}
\right\}
\begin{aligned}
&not \ \ a \ \ subgroup \ \ of \ \ \bar{G}_{(m)}
\end{aligned}
\Biggr\} \ \bar{G}_{(m)}
$$

$$(3.4.4)$$

m characterizes the extension. We may look at $\bar{G}_{(m)}$ as a (central) extension of the Galilei group G by R; we see that the central parameter θ has here the dimensions of an action. (The group law for $\tilde{G}_{(m)}$, which is an extension of G by $U(1)$, is given by replacing the last line of (3.4.4) by $\zeta'' = \zeta'\zeta \exp \frac{i}{\hbar} m(\mathbf{V}' \cdot R'\mathbf{A} + \frac{1}{2}\mathbf{V}'^2 B)$.) Ignoring rotations to simplify, it is now easy to compute the LI and RI vector fields and forms:

LIVFs:

$$\bar{X}^L_{(B)} = \frac{\partial}{\partial B} + \mathbf{V}\frac{\partial}{\partial \mathbf{A}} + \tfrac{1}{2}m\mathbf{V}^2\frac{\partial}{\partial \theta} \quad,$$

$$\bar{X}^L_{(A)} = \frac{\partial}{\partial \mathbf{A}} + m\mathbf{V}\frac{\partial}{\partial \theta} \quad,$$

$$\bar{X}^L_{(V)} = \frac{\partial}{\partial \mathbf{V}} \quad,$$

$$\bar{X}^L_{(\theta)} = \frac{\partial}{\partial \theta} \quad;$$

LIFs:

$$\bar{\omega}^{L(B)} = dB \quad,$$

$$\bar{\omega}^{L(A)} = d\mathbf{A} - \mathbf{V}dB \quad,$$

$$\bar{\omega}^{L(V)} = d\mathbf{V} \quad,$$

$$\bar{\omega}^{L(\theta)} = d\theta + m(\tfrac{1}{2}\mathbf{V}^2 dB - \mathbf{V}d\mathbf{A}) \quad.$$

$$(3.4.5)$$

RIVFs:

$$\bar{X}^R_{(B)} = \frac{\partial}{\partial B} \quad,$$

$$\bar{X}^R_{(A)} = \frac{\partial}{\partial \mathbf{A}} \quad,$$

$$\bar{X}^R_{(V)} = \frac{\partial}{\partial \mathbf{V}} + B\frac{\partial}{\partial \mathbf{A}} + m\mathbf{A}\frac{\partial}{\partial \theta} \quad,$$

$$\bar{X}^R_{(\theta)} = \bar{X}^L_{(\theta)} = \bar{X}_{(\theta)} = \frac{\partial}{\partial \theta} \quad;$$

RIFs:

$$\bar{\omega}^{R(B)} = dB \quad,$$

$$\bar{\omega}^{R(A)} = d\mathbf{A} - Bd\mathbf{V} \quad,$$

$$\bar{\omega}^{R(V)} = d\mathbf{V} \quad,$$

$$\bar{\omega}^{R(\theta)} = d\theta - m\mathbf{A}d\mathbf{V} \quad.$$

$$(3.4.6)$$

One may easily check that $\bar{\omega}^{L,R(\bar{g}^i)}(\bar{X}^{L,R}_{(\bar{g}^j)}) = \delta^{i*}_j$ where $\bar{g}^i = (B, \mathbf{A}, \mathbf{V}, \theta)$, and find

$$[\bar{X}^L_{(B)}, \bar{X}^L_{(A)}] = 0 \, , \quad [\bar{X}^L_{(B)}, \bar{X}^L_{(V)}] = -\bar{X}^L_{(A)} \, , \quad [\bar{X}^L_{(A^i)}, \bar{X}^L_{(V^j)}] = -m\delta_{ij}\bar{X}_{(\theta)} \, ,$$

$$(3.4.7)$$

the other commutators are zero. By computing the Lie brackets of the LIVFs of the extended group, we have found the defining relations of the *extended Galilei algebra*; $[\bar{X}^L_{\bar{g}^i}, \bar{X}_{(\theta)}] = 0 \; \forall \bar{g}^i$ exhibits the central character of the extension. The commutators for the RIVFs are the same but for a global minus sign and $[\bar{X}^L, \bar{X}^R] = 0$ [(1.2.11), (1.2.12)]. As a vector space, $\bar{\mathcal{G}}$ is the sum $\mathcal{G} \oplus R$, but the fact that (3.4.7) is an extension $\bar{\mathcal{G}}$ of the Galilei algebra different from the trivial product of $\mathcal{G}$ and R is exhibited by the new, non-zero commutator between the translation and boost generators; this also shows that $\mathcal{G}$ is not a subalgebra of $\bar{\mathcal{G}}$.

3.5 The two-cocycle ambiguity and the Bargmann cocycle for the Galilei group

The preceding discussion of $\bar{G}_{(m)}$ has been based on the group law (3.4.4), which in turn is determined by the cocycle (3.4.1) obtained from (3.4.2). But since from the point of view of the action principle Lagrangians are not uniquely defined, the question arises as to what happens to the quasi-invariance of a Lagrangian when a total derivative is added to it. Let us use this freedom for the simplest example $L = \frac{1}{2}m\dot{\mathbf{x}}^2$ and define a new Lagrangian $\tilde{L}$ by

$$\tilde{L} = L + \frac{d}{dt}\phi = L + \frac{d}{dt}\left(-\frac{m\mathbf{x}^2}{2t}\right) = \frac{1}{2}m(\dot{\mathbf{x}} - \frac{\mathbf{x}}{t})^2 \, . \tag{3.5.1}$$

We may ask ourselves about the general quasi-invariance properties of $\tilde{L}$ since, in contrast with L, it is obviously invariant under the Galilean boosts. Under a transformation $g = (B, \mathbf{A}, \mathbf{V}, \mathbf{1})$ of the Galilei group (where again the rotations are ignored) we may write, e.g., (cf. (3.1.4))

$$\delta_{(g)}\tilde{L} = \frac{d}{dt}m\left(\frac{1}{2}\left[\frac{\mathbf{x}^2}{t} - \frac{(\mathbf{x} + \mathbf{A} - \mathbf{V}B)^2}{t + B}\right]\right) \equiv \frac{d}{dt}\tilde{\Delta}(t, \mathbf{x}; g) \, . \tag{3.5.2}$$

Thus, although $\delta_{(V)}\tilde{L} = 0$, the quasi-invariance under a *general* transformation still remains since $\tilde{L}$ is now quasi-invariant under translations. Using (3.1.26) again to compute the two-cocycle $\tilde{\xi}$ associated with the new $\tilde{\Delta}$, we find

$$\tilde{\xi}_m(g', g) = m[\mathbf{V} \cdot \mathbf{V}'B' + \frac{1}{2}\mathbf{V}^2B' - \mathbf{V} \cdot \mathbf{A}'] \, . \tag{3.5.3}$$

---

[*] In fact, once the IVFs are known, the corresponding invariant one-forms can be found by solving the system of r equations which results for each $\bar{g}^i$.

This is the Bargmann form of the Galilei two-cocycle; it is equivalent to ξ_m and accordingly defines the same extension $\bar{G}_{(m)}$.

Looking at the above result we might be tempted to say that a change in L ($L \to \tilde{L}$) leads to a new expression for the two-cocycle, $\xi_m \to \tilde{\xi}_m$. However, this is not so as will be shown explicitly below. The important point to notice is that Δ *itself* is not uniquely defined: it is possible to obtain a new one by adding a function $\eta(g)$ on the group parameters since $\frac{d}{dt}\eta(g) = 0$. Thus, what the new $\tilde{L}$ illustrates is that the existence of a non-trivial two-cocycle is an *intrinsic* property of the invariance group G, and that its manifestation – the quasi-invariance of the Lagrangian under a general transformation of G – cannot be removed by adding a total derivative.

Consider a Lagrangian of the form $\tilde{L} = L + \frac{d}{dt}\phi(x)$, where x again represents the spacetime coordinates $(t, \mathbf{x})$, and assume that under a group transformation the original Lagrangian L varies according to $L \to L + \frac{d}{dt}\Delta(x;g)$. Since the group transformations commute with the time derivative, the variation of $\tilde{L}$, $\tilde{L}' - \tilde{L}$ is given by

$$\frac{d}{dt}\tilde{\Delta}(x;g) \equiv \frac{d}{dt}[\Delta(x;g) + (\phi(gx) - \phi(x))] . \tag{3.5.4}$$

This means that, in general,

$$\tilde{\Delta}(x;g) = \Delta(x;g) + \phi(gx) - \phi(x) + \eta(g) ; \tag{3.5.5}$$

eq. (3.5.5) shows that the new $\tilde{\Delta}$ depends on ϕ. However, if we now use eq. (3.1.26) to compute[*] the new form of the two-cocycle, we find

$$\tilde{\xi}(g',g) = \xi(g',g) + \eta(g') + \eta(g) - \eta(g'g) = \xi(g',g) + \xi_{cob}(g',g) , \tag{3.5.6}$$

where

$$\xi_{cob}(g',g) = \eta(g') + \eta(g) - \eta(g'g) \tag{3.5.7}$$

is the two-coboundary generated by $\eta(g)$ (eq. (3.5.7) simply repeats (3.2.5) in additive rather than multiplicative notation). Thus, the function $\phi(x)$ does not play a rôle in the definition of the two-cocycle; in fact, the operation (3.1.26), applied to $\phi(gx) - \phi(x)$, gives zero for any ϕ[†]. It is

---

[*] Although eq. (3.1.26) was obtained in a quantum mechanical context, it will be shown in section 8.4 that classical considerations based on the Lagrangian lead to the same formula for ξ.

[†] This fact has a simple reinterpretation in terms of the cohomology of G *with values in the space of functions on a manifold*, section 5.4 (this cohomology is different from the one defining $\bar{G}$; the cohomology relevant for this abelian extension (see chapter 5) takes values in R and arguments in G only). It may be seen from the first set of examples given there that $\phi(gx) - \phi(x)$ is just the one-cocycle generated by the zero-cochain $\phi(x)$ (hence $\phi(gx) - \phi(x)$ is a trivial one-cocycle or one-coboundary). Since the r.h.s. of (3.1.26) is just the result of the action of the *coboundary operator* δ on a one-cochain, it has to be zero since δ gives zero on any cochain which is a cocycle, either trivial or not (in fact, as we shall see in chapter 5, δ-cocycles are *defined* in general by this property).

not difficult to find the function $\eta(g)$ which generates $\tilde{\xi}$; once $\Delta(x;g)$ and $\tilde{\Delta}(x;g)$ are fixed, $\eta(g)$ is the only unknown in eq. (3.5.5) since, by (3.5.1), $\phi(x) = -\frac{mx^2}{2t}$. The result is

$$\eta(g) = m(\mathbf{A} \cdot \mathbf{V} - \frac{1}{2}\mathbf{V}^2 B) . \qquad (3.5.8)$$

The function $\eta(g)$ generates [(3.5.7)] the coboundary

$$\xi_{cob}(g',g) = -m(\mathbf{A}' \cdot \mathbf{V} + \mathbf{A} \cdot \mathbf{V}' + \frac{1}{2}(\mathbf{V}'^2 B - \mathbf{V}^2 B') - \mathbf{V} \cdot \mathbf{V}'B') \qquad (3.5.9)$$

and it is immediately verifiable that ξ_{cob}, when added to (3.4.1), gives (3.5.3). We shall come back to this point in section 8.4.

The group law for $\bar{G}_{(m)}$, as defined by the Bargmann cocycle (3.5.3), is as in (3.4.4) provided the last line is replaced by

$$\theta'' = \theta' + \theta + m[\mathbf{V} \cdot \mathbf{V}'B' + \frac{1}{2}\mathbf{V}^2 B' - \mathbf{V}\mathbf{A}'] \qquad (3.5.10)$$

(here we have ignored the rotation subgroup). For completeness, we give below the left- and right-invariant vector fields and one-forms (cf. (3.4.5), (3.4.6)) that are obtained when (3.5.10) is used:

LIVFs:

$$\bar{X}^L_{(B)} = \frac{\partial}{\partial B} + \mathbf{V}\frac{\partial}{\partial \mathbf{A}} ,$$

$$\bar{X}^L_{(A)} = \frac{\partial}{\partial \mathbf{A}} ,$$

$$\bar{X}^L_{(V)} = \frac{\partial}{\partial \mathbf{V}} - m(\mathbf{A} - B\mathbf{V})\frac{\partial}{\partial \theta} ,$$

$$\bar{X}^L_{(\theta)} = \frac{\partial}{\partial \theta} .$$

LIFs:

$$\bar{\omega}^{L(B)} = dB ,$$

$$\bar{\omega}^{L(A)} = d\mathbf{A} - \mathbf{V}dB ,$$

$$\bar{\omega}^{L(V)} = d\mathbf{V} ,$$

$$\bar{\omega}^{L(\theta)} = d\theta + m(\mathbf{A} - \mathbf{V}B)d\mathbf{V} .$$

$$(3.5.11)$$

RIVFs:

$$\bar{X}^R_{(B)} = \frac{\partial}{\partial B} + \frac{1}{2}m\mathbf{V}^2\frac{\partial}{\partial \theta} ,$$

$$\bar{X}^R_{(A)} = \frac{\partial}{\partial \mathbf{A}} - m\mathbf{V}\frac{\partial}{\partial \theta} ,$$

$$\bar{X}^R_{(V)} = \frac{\partial}{\partial \mathbf{V}} + B\frac{\partial}{\partial \mathbf{A}} ,$$

$$\bar{X}^R_{(\theta)} = \frac{\partial}{\partial \theta} .$$

RIFs:

$$\bar{\omega}^{R(B)} = dB ,$$

$$\bar{\omega}^{R(A)} = d\mathbf{A} - Bd\mathbf{V} ,$$

$$\bar{\omega}^{R(V)} = d\mathbf{V} ,$$

$$\bar{\omega}^{R(\theta)} = d\theta + m(\mathbf{V}d\mathbf{A} - d(\frac{1}{2}m\mathbf{V}^2 B)) .$$

$$(3.5.12)$$

Notice that if the expressions for the IVF's and IF's for a certain two-cocycle are known, it is sufficient to know the function $\eta(g)$ generating the two-coboundary to obtain the new expressions for the invariant vector fields and forms corresponding to the new equivalent cocycle. Thus, it is sufficient to know eqs. (3.4.5), (3.4.6) and (3.5.8) to obtain the IVFs and IFs above which, of course, can also be computed directly from (3.5.10).

3.6 Dynamical groups and symplectic cohomology: an introduction

We have already encountered in section 3.3 the notion of a symplectic manifold. When the rank of the closed two-form Ω is even but not maximal, rank $\Omega = 2s < \dim M$, the pair (M, Ω) is called a *presymplectic manifold* (the fact that rank Ω is not maximal can be checked by observing that the module of vector fields $\{X | i_X \Omega = 0\}$ has non-null dimension).

Let $f : G \times M \to M$ be a left action of the Lie group G on M, and let us realize the Lie algebra $\mathscr{G}$ by means of the vector fields X_M on M induced by this action (we shall use here the subscript in X_M to distinguish the vector fields on M from $X \in \mathscr{G} \sim T_e(G)$). Then G is called a dynamical group if the condition of the following definition is met:

(3.6.1) DEFINITION (*Dynamical group*)

A Lie group G is a dynamical group with respect to the presymplectic manifold (M, Ω) if

$$L_{X_M} \Omega = 0 \qquad \forall X \in \mathscr{G} \quad , \tag{3.6.1}$$

where X_M is the generator of the action on M. Then M is called a G-presymplectic manifold.

Since Ω is closed, eq. (3.6.1) also implies $d i_{X_M} \Omega = 0$ or, in other words, that the vector field X_M is *locally Hamiltonian*. The locally Hamiltonian vector fields form a Lie subalgebra of $\mathscr{X}(M)$. If $i_{X_M} \Omega$ is also exact,

$$i_{X_M} \Omega = d\beta \quad , \tag{3.6.2}$$

then X_M is said to be a (globally) *Hamiltonian vector field*.

(3.6.2) DEFINITION (*Momentum mapping*)

Let M be a symplectic or presymplectic manifold. A momentum mapping μ of a dynamical group G is a $\mathscr{G}^*$-valued function $\mu : M \to \mathscr{G}^*$ on M, where $\mathscr{G}^*$ is the dual algebra (or coalgebra) such that

$$i_{X_M} \Omega = -d(\mu(X)) \equiv d(f_X) \qquad \forall X \in \mathscr{G} \quad . \tag{3.6.3}$$

It is convenient to note, however, that eq. (3.6.3) does not determine a momentum mapping μ *uniquely* since it will also be satisfied by another momentum $\mu' = \mu + \mu_0$ where $\mu_0 : M \to \mathscr{G}^*$ is a constant mapping $(d(\mu_0(X)) = 0)$. For a connected manifold, $H^0_{DR}(M) = R$; thus $d[(\mu' - \mu)(X)] = 0$ tells us that on a connected manifold all momentum mappings associated with a given action (X_M) differ by a constant one.

In view of the previous discussion, it is clear that a dynamical group admits a momentum mapping whenever M is simply connected, since this is what is required for the Poincaré lemma to hold on M $(H^1_{DR}(M) = 0$ for M simply connected). But if $\mathscr{G}$ is equal to its derived algebra, $\mathscr{G} = [\mathscr{G}, \mathscr{G}]$ (which is in particular true for $\mathscr{G}$ semisimple), then the dynamical group G always admits a momentum mapping. This is because in this case it is possible to express any basis $X_{(i)}$ of $\mathscr{G}$ in terms of commutators, $X_{(i)} = [Y_{(i)}, Z_{(i)}]$ say, and then

$$i_{X_{M(i)}}\Omega = i_{[Y_{M(i)}, Z_{M(i)}]}\,\Omega = -d[\Omega(Y_{M(i)}, Z_{M(i)})] = -d(\mu(X_{(i)})) \qquad (3.6.4)$$

(omitting the subscripts, the relation $[L_X, i_Y] = i_{[X,Y]}$ and the fact that $i_X\Omega$ is closed immediately give $i_{[Y,Z]}\Omega = (L_Y i_Z - i_Z L_Y)\Omega = d(i_Y i_Z\Omega) = -d(\Omega(Y,Z))$). Eq. (3.6.4) indicates that the coordinates $\mu_i(x) = \mu(x)(X_{(i)})$ of a momentum mapping satisfying (3.6.4) may be given by the functions $\mu_i(x) = \Omega(Y_{M(i)}, Z_{M(i)})$.

A special example occurs when Ω admits a potential Λ, $\Omega = d\Lambda$ (this is the case, for instance, for the (symplectic) cotangent manifold $S = T^*(N)$ to a manifold N, which may be parametrized by the coordinates (q^i, p_i), Λ being the canonical *Liouville form* $\Lambda = -p_i dq^i$ and $\Omega = d\Lambda = dq^i \wedge dp_i$). Then $L_{X_S}\Lambda = d(i_{X_S}\Lambda - \mu(X))$. Thus, if Λ is invariant (respectively quasi-invariant) the equation $L_{X_S}\Lambda = 0$ (resp. $L_{X_S}\Lambda = du_X$) shows that $\mu(X) = i_{X_S}\Lambda$ (resp. $i_{X_S}\Lambda - u_X$) where u_X is a function.

The transformation properties of the momentum mapping are reflected by the following

(3.6.1) THEOREM

Let (M, Ω) be a (pre)symplectic connected manifold. Then the $\mathscr{G}^*$-valued function

$$\mu - (Ad^*g^{-1}) \circ \mu \circ f_g \ , \qquad (3.6.5)$$

where Ad^*G is the coadjoint representation, is constant on M.

Proof: The proof relies on the assumed G-invariance of the (pre)-symplectic form Ω. First, we notice that an action defines for each $x \in M$ a mapping $\varphi_x : G \to M$ by means of the identification $\varphi_x(g) = f(g;x)$ $(\equiv f_g(x))$ (section 1.1). The tangent mapping at $e \in G$, $\varphi_x^T(e)$, associates a tangent vector $X_M(x) \in T_x(M)$ to each $X \in \mathscr{G}$. Thus, the mappings φ_x^T

$(x \in M)$ determine a vector field $X_M \in \mathscr{X}(M)$ associated with each $X \in \mathscr{G}$. The mapping is a homomorphism of algebras, and it moreover satisfies $f_g^T \circ X_M \circ f_{g^{-1}} = (AdgX)_M$, i.e. that the images X_M and $(AdgX)_M$ in $\mathscr{X}(M)$ of X and $AdgX$ by $\varphi_x^T(e)$ are f_g-related[*]. Now, proving that (3.6.5) is a constant mapping is tantamount to saying that $d(\mu(X)) = d(\mu \circ f_g(AdgX))$ since $Ad^* g^{-1} = (Adg)^*$ (section 1.9). Using the definition of μ, eq. (3.6.3), we find

$$-d(\mu \circ f_g(AdgX)) = i_{(AdgX)_M \circ f_g} \Omega \circ f_g = i_{f_g^T X_M} \Omega \circ f_g \tag{3.6.6}$$
$$= i_{X_M}(f_g)^* \circ \Omega \circ f_g = i_{X_M}\Omega$$

since by hypothesis Ω is f_g-invariant, q.e.d.

Since (3.6.5) takes the same value irrespective of its argument x, it may be used to define a mapping $\gamma : G \to \mathscr{G}^*$. More precisely,

(3.6.3) DEFINITION (Symplectic cocycle on G)

The smooth mapping $\gamma : G \to \mathscr{G}^*$ defined by (cf. eq. (3.6.5))

$$\gamma(g) = \mu \circ f_g - (Ad^* g) \circ \mu \tag{3.6.7}$$

is called a symplectic cocycle. Clearly, $\gamma(e) = 0$.

The mapping $\gamma(g)$ is a (one)-cocycle because it satisfies the following (one)-cocycle condition:

$$(\delta\gamma)(g', g)) \equiv \gamma(g') + (Ad^* g')\gamma(g) - \gamma(g'g) = 0 \ . \tag{3.6.8}$$

Proof : Since Ad^* is a representation, $Ad^* g'g = (Ad^* g')(Ad^* g)$ and

$$\gamma(g'g) = \mu(g'gx) - (Ad^* g')(Ad^* g)\mu(x)$$
$$\equiv \mu(g'x') - (Ad^* g')(Ad^* g)\mu(x)$$
$$= \gamma(g') + (Ad^* g')[\mu(x') - (Ad^* g)\mu(x)] \tag{3.6.9}$$
$$= \gamma(g') + (Ad^* g')\gamma(g) \ ,$$

q.e.d.

---

[*] This means that (eq. (1.2.3)) $f_g^T \circ X_M \circ f_g^{-1} = (AdgX)_M$. The left hand side gives $f_g^T(g^{-1}x)X_M(g^{-1}x) = f_g^T(g^{-1}x) \circ \varphi_{g^{-1}x}^T(e)X = (f_g \circ \varphi_{g^{-1}x})^T(e)X = (f_g \circ \varphi_x \circ R_{g^{-1}})^T(e)X = (\varphi_x \circ L_g \circ R_{g^{-1}})^T(e)X = \varphi_x^T(e) \circ (L_g \circ R_{g^{-1}})^T(e)X = \varphi_x^T(e)(AdgX) = (AdgX)(x) ((AdgX)_M$; notice that, as follows from their definitions, $f_g \circ \varphi_x = \varphi_x \circ L_g$ and, since $\varphi_{g^{-1}x}(g') = g'g^{-1}x = \varphi_x(g'g^{-1}) = \varphi_x \circ R_{g^{-1}}(g'), \varphi_{g^{-1}x} = \varphi_x \circ R_{g^{-1}})$. The fact that the vector field f_g-related to X_M is the vector field $(AdgX)_M$ on M associated with $AdgX$ is a natural result already encountered. For instance, if M is a Lie group G, $x = g'$, then $\varphi_{g'} : g = gg'$ $(\varphi_{g'} = R_{g'})$ and, accordingly, X_G is an RI vector field on G. Thus, the above formula reads $L_g^T \circ X_G \circ L_{g^{-1}} = (AdgX)_G$. Also, by selecting a certain $X_{(j)}$ and taking $g = exp \, g^i X_{(i)}$ we recover from (1.4.44) that $L_{X_{(i)}} X_{(j)} = [X_{(i)}, X_{(j)}]$.

Let μ_0 be a constant mapping $\mu_0 : M \to \mathscr{G}^*$ (hence $\mu_0 \in \mathscr{G}^*$). Then

$$\delta\mu_0(g) \equiv (Ad^*g)\mu_0 - \mu_0 = \gamma_{cob}(g) \qquad (3.6.10)$$

automatically satisfies condition (3.6.8). The one-cocycle $\delta\mu_0$ is generated by μ_0, and hence trivial: it is a *symplectic coboundary*.

This construction defines a cohomology group, the *symplectic cohomology group* $H_s(G, \mathscr{G}^*)$ associated with a dynamical group G. We shall see in section 5.1 that eqs. (3.6.8) and (3.6.10) are particular examples of more general n-cocycle conditions defined *e.g.* on functions which depend on n group arguments and take values on the carrier vector space V of a linear representation D of G (here, $V = \mathscr{G}^*$ and $D = Ad^*$). The present particular case, however, is specially interesting since it allows us to construct a family of representations $(Ad^*)_\gamma$ of G modifying the coadjoint one Ad^*. These representations are characterized by the γ's and permit the construction of a family of symplectic orbits in $\mathscr{G}^*$ which generalize the coadjoint orbits introduced by Kirillov. To motivate these symplectic orbits in the coalgebra of any Lie group G let us go back to eq. (3.6.7), which will be rewritten as

$$\gamma(g) = \gamma_1(g, x) - \gamma_2(g, x) \qquad (3.6.11)$$

with

$$\gamma_1(g, x) \equiv \mu(gx) \quad , \quad \gamma_2(g, x) \equiv (Ad^*g)\mu(x) \quad . \qquad (3.6.12)$$

We may look at γ_1 as the composite mapping $\gamma_1 = \mu \circ \varphi_x : G \xrightarrow{\varphi_x} M \xrightarrow{\mu} \mathscr{G}^*$ so that

$$D\gamma_1(g)X = D(\mu \circ \varphi_x)(g)X = D\mu(gx) \circ D\varphi_x(g)X \quad . \qquad (3.6.13)$$

Thus, setting $g = e$ and using again that $\varphi_x^T(e)X = X_M(x)$, we get

$$D\gamma_1(e, x)X = D\mu(x)X_M \quad . \qquad (3.6.14)$$

Since $(Ad^*)^T(e)X = ad^*X$, the derivative $D\gamma_2$ with respect to g at e is a mapping $\mathscr{G} \to \mathscr{G}^*$ which on $X \in \mathscr{G}$ is simply the coadjoint representative ad^*X acting on $\mu(x)$,

$$D\gamma_2(e, x)X = (ad^*X)\mu(x) \quad . \qquad (3.6.15)$$

Then eq. (3.6.11) gives that

$$D\gamma(e) \cdot X \equiv \Gamma(X) = D\mu(x)X_M - (ad^*X)\mu(x) \quad . \qquad (3.6.16)$$

(3.6.4) DEFINITION (*Symplectic cocycle on $\mathscr{G}$*)

Given a symplectic cocycle γ on G, its associated symplectic cocycle on $\mathscr{G}$ is the mapping $\Gamma : \mathscr{G} \to \mathscr{G}^*$ defined by $\Gamma := D\gamma(e)$. Γ may also be viewed as a skew-symmetric mapping $\Gamma : \mathscr{G} \times \mathscr{G} \to R$ by setting

$$\Gamma(X)X' := \Gamma(X, X') \quad . \tag{3.6.17}$$

If Γ is determined by a coboundary $\gamma_{cob}(g)$ [(3.6.10)], $\Gamma_{cob}(X) = D\gamma_{cob}(e)X = (ad^* X)\mu_0$ and then

$$\Gamma_{cob}(X, X') = \mu_0([X, X']) \quad . \tag{3.6.18}$$

The following theorem now holds:

(3.6.2) THEOREM

Let $\Gamma : \mathscr{G} \times \mathscr{G} \to R$ be the symplectic $\mathscr{G}$-cocycle defined by the group symplectic cocycle γ of a dynamical group G. Then

a) the (pre)symplectic form Ω may be rewritten as

$$\Omega(X_M, Y_M) = \mu([X, Y]) + \Gamma(X, Y) \quad , \tag{3.6.19}$$

i.e., as the (Kirillov) term μ plus the symplectic $\mathscr{G}$-cocycle Γ, and

b)

$$\Gamma(X, [Y, Z]) + \Gamma(Y, [Z, X]) + \Gamma(Z, [X, Y]) = 0 \quad . \tag{3.6.20}$$

Proof:

(a) First, we rewrite eq. (3.6.16) as $D\mu(x)X_M = \Gamma(X) + ad^* X \mu(x)$. On $Y \in \mathscr{G}$ the first term gives

$$\begin{aligned}
(D\mu(x)X_M)(Y) &= Y \cdot (D\mu(x)X_M) \\
&= (DY(\mu(x)) \circ D\mu(x)X_M = D(Y \circ \mu)(x) \cdot X_M \\
&= d[Y \circ \mu](X_M(x)) = d[\mu(Y)](X_M(x)) = [-i_{Y_M}(\Omega)](X_M) \\
&= -\Omega(Y_M, X_M) = \Omega(X_M, Y_M) \quad ,
\end{aligned} \tag{3.6.21}$$

where in the first equality we have looked at Y as a mapping $\mathscr{G}^* \to R$ since $\mathscr{G} \sim \mathscr{G}^{**}$, used that the Jacobian of Y coincides with itself in the second equality (Y is linear), identified the differential in the fourth and, finally, used definition (3.6.3) in the sixth equality. Then, eq. (3.6.19) follows from (3.6.21).

(b) Eq. (3.6.20) may be proven directly from

$$\Gamma([X, Y]) = (ad^*X)\Gamma(Y) - (ad^*Y)\Gamma(X) \ , \tag{3.6.22}$$

which itself may be derived from (3.6.9). In fact, we shall see in section 6.1 that this expression just states that Γ is an algebra (ad^*)-one-cocycle (as we might expect since it is obtained from the group (Ad^*)-one-cocycle), q.e.d.[*]

The r.h.s. of expression (3.6.19) suggests the possibility of realizing the G-symplectic manifolds associated with physical problems as *orbits in $\mathscr{G}^*$* of the *modified coadjoint action* $(Ad^*G)_\gamma$ of G on $\mathscr{G}^*$ defined by

$$(Ad^*g)_\gamma\mu := (Ad^*g)\mu + \gamma(g) \ , \quad \mu \in \mathscr{G}^* \ . \tag{3.6.23}$$

$(Ad^*G)_\gamma$ is clearly an action since, using eq. (3.6.8),

$$\begin{aligned}(Ad^*g')_\gamma[(Ad^*g)_\gamma\mu] &= (Ad^*g')(Ad^*g)\mu + (Ad^*g')\gamma(g) + \gamma(g')\\ &= (Ad^*g'g)\mu + \gamma(g'g) \ . \end{aligned} \tag{3.6.24}$$

Indeed, the following theorem generalizing the theory of coadjoint orbits is a consequence of the above results:

(3.6.3) THEOREM (*Symplectic orbits on the coalgebra*)

Let G be a Lie group, $\mathscr{G}$ its Lie algebra and γ a symplectic cocycle. If S is the $(Ad^*G)_\gamma$-orbit $S = \{\mu|\mu = (Ad^*g)\mu_0 + \gamma(g), g \in G\}$ (eq. (3.6.23)), then

---

[*] The proof is as follows. First, we rewrite (3.6.9) as $(\gamma \circ R_g)(g') = \gamma(g') + \sigma(g')\gamma(g)$, where the representation σ is $\sigma \equiv Ad^*$ in our case. Taking the derivative D with respect to g' and using the chain rule (1.1.20) we find

$$(D\gamma)(R_g g') \circ (DR_g)(g') = (D\gamma)(g') + (D\sigma)(g')\gamma(g) \ . \tag{1}$$

Acting on $X \in \mathscr{G}$ at $g' = e$, this expression gives $(D\gamma)(g) \circ (DR_g)(e) \cdot X = (D\gamma)(e) \cdot X + [(D\sigma)(e) \cdot X]\gamma(g)$, i.e. (cf. (1.2.9), (3.6.16))

$$(D\gamma)(g).X^R(g) = \Gamma(X) + \rho(X)\gamma(g) \ , \tag{2}$$

where here the algebra representation $\rho(X)=(D\sigma)(e) \cdot X$ is $\rho \equiv ad^*$. Now, much in the same way $(Df)( )Z=Z.f$ (cf. (1.4.28)), $(D\gamma)(g) \cdot Z(g)$ may be written as $Z(g).\gamma(g)$. Let us then compute $(D\gamma)(g) \cdot [X^R(g), Y^R(g)]$. Omitting the arguments of the vector fields, it may be written as

$$\begin{aligned}X_G^R.Y_G^R.\gamma(g) - Y_G^R.X_G^R.\gamma(g) &= X_G^R.(D\gamma)(g) \cdot Y_G^R - Y_G^R.(D\gamma)(g) \cdot X_G^R\\ &= X_G^R.(\Gamma(Y) + \rho(Y)\gamma(g)) - Y_G^R.(\Gamma(X) + \rho(X)\gamma(g)) = X_G^R.\rho(Y)\gamma(g) - Y_G^R.\rho(X)\gamma(g) \ ,\end{aligned} \tag{3}$$

since $\Gamma(X)$ and $\Gamma(Y)$ are constant (do not depend on the group variables). Eq. (3) may be rewritten as

$$\begin{aligned}&D(\rho(Y) \circ \gamma)(g) \cdot X_G^R - D(\rho(X)\gamma)(g) \cdot Y_G^R\\ &= D(\rho(Y))\gamma(g) \circ (D\gamma)(g) \cdot X_G^R - D(\rho(X))\gamma(g) \circ (D\gamma)(g) \cdot Y_G^R\end{aligned} \tag{4}$$

using agin the chain rule. Now, since $D(\rho(Z))=\rho(Z)$ ($\rho(Z)$ is a linear mapping), we find at $g=e$ $\Gamma([X, Y]) = \rho(X)\Gamma(Y) - \rho(Y)\Gamma(X)$ i.e., eq. (3.6.22) (in the l.h.s. a minus sign has been added because if $[X_G^R, Y_G^R] = Z_G^R$, $[X, Y] = -Z$ since the Lie algebra is defined by the LIVFs. As mentioned, eqs. (3.6.9) and (3.6.22) will be seen to be (in sections 5.1 and 6.1) the conditions expressing that γ and Γ are one-cocycles in the group and algebra cohomologies respectively.

a) a tangent vector $Z_\mu \in T_\mu(\mathscr{G}^*)$ belongs to $T_\mu(S)$ iff there is a $Z \in \mathscr{G}$ such that

$$Z_\mu = (ad^* Z)\mu + \Gamma(Z) \qquad (3.6.25)$$

(notice that eq. (3.6.25) defines a vector field pointwise, and that not every vector field Z_S on S is necessarily of this form since the same Z may not be valid for all μ),

b) the orbit S is endowed with a symplectic structure (S, Ω_S) given by

$$\Omega_\mu(Z_\mu, Z'_\mu) = \mu([Z, Z']) + \Gamma(Z, Z') . \qquad (3.6.26)$$

Proof: Since S is an orbit, G acts transitively on S and $\dim S \le \dim G$. Moreover, eq. (3.6.25) follows from eq. (3.6.23) by derivation.

Let $Z_\mu, Z'_\mu \in T_\mu(S)$. Then, by (a), there exist $Z, Z' \in \mathscr{G}$ such that $Z_\mu = (ad^* Z)\mu + \Gamma(Z)$ and $Z'_\mu = (ad^* Z')\mu + \Gamma(Z')$. Now, with $\Omega_\mu(Z_\mu, Z'_\mu) = (Z_\mu)(Z')$ we see that (3.6.26) is reproduced, since $(Z_\mu)(Z') = ((ad^* Z)\mu)(Z') + \Gamma(Z)(Z') = \mu((adZ)Z') + \Gamma(Z, Z') = \mu([Z, Z']) + \Gamma(Z, Z')$. Since Ω_S is closed, proving that it is symplectic means showing that it is of maximal rank (hence even). But if $i_{Z_\mu}\Omega_\mu = 0$ then it follows that $Z_\mu(Z') = 0 \; \forall Z' \in \mathscr{G}$ which implies that the linear form Z_μ of $\mathscr{G}^*$ is zero, q.e.d.

To conclude this outlook on symplectic cohomology, let us recall that the classes of inequivalent symplectic one-cocycles γ ($\gamma \mathscr{R} \gamma' \iff \gamma - \gamma' = \gamma_{cob}$, eq. (3.6.10)) define the symplectic cohomology group $H_s(G, \mathscr{G}^*)$. The cohomology classes are a characteristic property of the dynamical group, and parametrize the possible (physical) symplectic orbits associated with G. The dual (algebra) version of this is provided by the symplectic cohomology on $\mathscr{G}$; the non-equivalent classes of Lie algebra one-cocycles, the mappings $\Gamma : \mathscr{G} \to \mathscr{G}^*$, determine the symplectic cohomology group $H_s(\mathscr{G}, \mathscr{G}^*)$. We have also seen that this may also be understood as a mapping $\Gamma : \mathscr{G} \times \mathscr{G} \to R$. Thus, the question naturally arises as to whether we can relate the Lie algebra *two*-cocycles Γ with the dual version of the cohomology $H^2(G, R)$ defined by the two-cocycles $\xi(g', g)$, i.e. with the second Lie algebra cohomology group $H^2(\mathscr{G}, R)$. It is not difficult to show that, as it might be expected, $H_s(\mathscr{G}, \mathscr{G}^*) \approx H^2(\mathscr{G}, R)$ (see the footnote in section 6.4). This isomorphism permits us to express the symplectic form (3.6.26) for the central extension $\tilde{G}_\gamma$ of G associated with the cocycle $\gamma \in H_s(\mathscr{G}, \mathscr{G}^*)$ by just the (Kirillov) term in μ of the extended group, since the extended group will have trivial symplectic cohomology. In this form, the symplectic manifold appears as an orbit of the *coadjoint* action of the central extension $\tilde{G}_\gamma$ of the dynamical group $\tilde{G}$. We shall prove in section 6.5 that, if G is semisimple, $H^2(\mathscr{G}, R)$ is zero (Whitehead's lemma). Thus, if G is semisimple, the expression of Ω is also reduced to the Kirillov-Kostant term. This is the case of $SU(2)$, for which $H_s(SU(2), su(2)) = 0$ and μ determines the symplectic form on S^2 (orbit of the coadjoint action).

A first example: the Weyl-Heisenberg group
The simplest illustration of the above theory is the Weyl-Heisenberg group already discussed, for which the symplectic structure on the orbits of the coadjoint action of WH on wh^* is evident from the start. To see this explicitly, it is convenient to write the WH group law by using the cocycle form (3.3.3). With $\zeta = e^{\frac{i}{\hbar}\theta}$, the elements g of the WH group and wh algebra may be written as

$$g = \begin{pmatrix} 1 & q & \theta \\ 0 & 1 & p \\ 0 & 0 & 1 \end{pmatrix}, \; g^{-1} = \begin{pmatrix} 1 & -q & -\theta + pq \\ 0 & 1 & -p \\ 0 & 0 & 1 \end{pmatrix},$$

$$X = \begin{pmatrix} 0 & X_{(q)} & X_{(\theta)} \\ 0 & 0 & X_{(p)} \\ 0 & 0 & 0 \end{pmatrix}. \tag{3.6.27}$$

In this way, the group law ($g'' = g'g : q'' = q' + q$, $p'' = p' + p$, $\theta'' = \theta' + \theta + q'p$) and (left) algebra commutators (all zero but $[X_{(q)}, X_{(p)}] = X_{(\theta)}$) are given, respectively, by products and commutators of matrices. Moreover, since in this way we are dealing with matrices, the adjoint representation is simply given by

$$Adg\, X = gXg^{-1} \tag{3.6.28}$$

so that, with respect to the basis $\{X_{(q)}, X_{(p)}, X_{(\theta)}\}$ of wh, $g \in G$ is represented by the matrix (see section 1.9(a))

$$Adg = \begin{pmatrix} 1 & 0 & 0 \\ 0 & 1 & 0 \\ -p & q & 1 \end{pmatrix}. \tag{3.6.29}$$

Similarly, on the coalgebra wh^* with basis $\{\omega^{(q)}, \omega^{(p)}, \omega^{(\theta)}\}$ (which may also be realized by the LI one-forms $\{\omega^{L(q)}, \omega^{L(p)}, \omega^{L\theta}\}$ on WH dual to the vector fields $\{X_{(q)}^L, X_{(p)}^L, X_{(\theta)}^L\}$) the coadjoint representation is given by

$$Ad^*g = \begin{pmatrix} 1 & 0 & p \\ 0 & 1 & -q \\ 0 & 0 & 1 \end{pmatrix} \tag{3.6.30}$$

since $Ad^*g = (Adg^{-1})^*$; here, the star in the r.h.s. indicates transposition since we assume that the matrices act on column vectors and, as we saw in section 1.9, the form of the mapping $Ad^*g : \mu \mapsto \mu'$, $\mu, \mu' \in \mathcal{G}^*$, is determined by the condition $\mu'(X') = \mu(X)$ where $Adg : X \mapsto X'$, $X, X' \in$

$\mathcal{G}$. This means that an arbitrary element μ of wh with coordinates P, Q, λ,

$$\mu = P\omega^{(q)} + Q\omega^{(p)} + \lambda\omega^{(\theta)} \quad, \tag{3.6.31}$$

is transformed according to

$$P' = P - p\lambda \quad, \quad Q' = Q + q\lambda \quad, \quad \lambda' = \lambda \quad. \tag{3.6.32}$$

There are no relations here among the different coordinates of eq. (3.6.32) that give the points of an orbit, and thus a set of three fixed values (P, Q, λ) determine the orbit which contains this point. Hence the generators of the action on the orbit that goes through $(P, Q, \lambda = \text{const.})$ are directly given by

$$Z_{(q)}(\mu) = \lambda\frac{\partial}{\partial Q} \quad, \quad Z_{(p)}(\mu) = -\lambda\frac{\partial}{\partial P} \quad, \quad Z_{(\theta)}(\mu) = 0 \quad. \tag{3.6.33}$$

Clearly, the value $\lambda = 0$ has to be excluded, since then the orbit is reduced to a single point. The symplectic form on a coadjoint orbit is given by

$$\Omega_S(\mu)(Z_{(q)}(\mu), Z_{(p)}(\mu)) = \mu([X_{(q)}, X_{(p)}]) \tag{3.6.34}$$

and, setting $\lambda = 1$, we obtain the expected result

$$\Omega_S = dP \wedge dQ \quad. \tag{3.6.35}$$

Expression (3.6.34) illustrates the previous discussion of eq. (3.6.26). Since the Weyl-Heisenberg group has trivial cohomology, the term Γ in (3.6.26) does not contribute, and Ω_S is just given by the Kirillov term in μ. But we could have proceeded similarly for the *abelian* group of the (p, q) translations. In that case, in contrast, the term in μ would have been identically zero, and Ω_S would have been simply given by the Γ term. The symplectic manifold (p, q) obtained in this case corresponds to the previous $\lambda = 1$ symplectic orbit in the coalgebra of the WH group; the non-triviality of Γ for the abelian (p, q)-group just reflects that this group can be centrally extended, its extension being precisely the WH group.

3.7 The adjoint and the coadjoint representations of the extended Galilei group and its algebra

Let us now describe the adjoint and coadjoint representations of $\bar{G}_{(m)}$ and $\bar{\mathcal{G}}_{(m)}$ and construct the symplectic form on a particular orbit of the coadjoint action of $\bar{G}_{(m)}$ on its coalgebra $\bar{\mathcal{G}}_{(m)}^*$. The orbit itself can be identified with the manifold of the constants of the (uniform) free particle motion and the symplectic form gives the canonical structure of the free particle theory described by the Galilei group understood as a dynamical group. This will provide a non-trivial example of the method of the *coadjoint orbits*, which has already been applied to the Weyl-Heisenberg group and can be applied to other groups.

Let us first consider the adjoint representation $Ad\bar{G}_{(m)}$ on $\bar{\mathscr{G}}_{(m)}$. Let b, $\mathbf{a}$, $\vec{v}$, $\vec{\omega}$, σ be the parameters that characterize a general element of $\bar{\mathscr{G}}_{(m)}$,

$$X = bX_{(B)} + \mathbf{a} \cdot X_{(\mathbf{A})} + \vec{v} \cdot X_{(\mathbf{V})} + \vec{\omega} \cdot X_{(\vec{\epsilon})} + \sigma X_{(\theta)} \quad , \tag{3.7.1}$$

expressed in terms of the basis elements of the extended Galilei Lie algebra. In matrix form, the elements of $\bar{G}_{(m)}$ can be written as

$$g = \begin{pmatrix} R & \mathbf{V} & 0 & \mathbf{A} \\ 0 & 1 & 0 & B \\ m\mathbf{V}^t R & \frac{m}{2}\mathbf{V}^2 & 1 & \theta \\ 0 & 0 & 0 & 1 \end{pmatrix} \quad , \tag{3.7.2}$$

$$g^{-1} = \begin{pmatrix} R^{-1} & -R^{-1}\mathbf{V} & 0 & -R^{-1}(\mathbf{A}-\mathbf{V}B) \\ 0 & 1 & 0 & -B \\ -m\mathbf{V}^t & \frac{m}{2}\mathbf{V}^2 & 1 & -\theta - \frac{m}{2}B\mathbf{V}^2 + m\mathbf{V}\mathbf{A} \\ 0 & 0 & 0 & 1 \end{pmatrix} \quad ,$$

where $R = R(\vec{\epsilon})$. The 6×6 matrices act on the column vector $(\mathbf{x}, t, \chi, 1)$ where χ is an additional variable (its rôle will be discussed in section 8.3), and the product $g'g$ reproduces (3.4.4). The expression of the generators of $\bar{\mathscr{G}}_{(m)}$ acting on this enlarged spacetime is given by

$$\bar{X}_{(B)} = \frac{\partial}{\partial t} \quad , \quad \bar{X}_{(\mathbf{A})} = \frac{\partial}{\partial \mathbf{x}} \quad ,$$

$$\bar{X}_{(\mathbf{V})} = t\frac{\partial}{\partial \mathbf{x}} + m\mathbf{x}\frac{\partial}{\partial \chi} \quad , \quad \bar{X}_{(\epsilon^i)} = -\left(\mathbf{x} \times \frac{\partial}{\partial \mathbf{x}}\right)_i \equiv -\eta_{ij.}{}^k x^j \frac{\partial}{\partial x^k} \quad , \tag{3.7.3}$$

$$\bar{X}_{(\theta)} = \frac{\partial}{\partial \chi} \quad ;$$

they satisfy the (left group action, right-type algebra) commutation relations (cf. (3.4.7), (1.9.16))

$$[\bar{X}_{(B)}, \bar{X}_{(\mathbf{V})}] = \bar{X}_{(\mathbf{A})} \quad , \quad [\bar{X}_{(A^i)}, \bar{X}_{(V^j)}] = m\delta_{ij}\bar{X}_{(\theta)} \quad ,$$

$$[\bar{X}_{(\epsilon^i)}, \bar{X}_{(\epsilon^j)}] = \eta_{ij.}{}^k \bar{X}_{(\epsilon^k)} \quad , \quad [\bar{X}_{(\epsilon^i)}, \bar{X}_{(A^j)}] = \eta_{ij.}{}^k \bar{X}_{(A^k)} \quad , \tag{3.7.4}$$

$$[\bar{X}_{(\epsilon^i)}, \bar{X}_{(V^j)}] = \eta_{ij.}{}^k \bar{X}_{(V^k)} \quad ,$$

where the vanishing commutators have been omitted. In matrix form, the Lie algebra element X is written in terms of the coefficients in (3.7.1) as

$$\begin{pmatrix} -j(\vec{\omega}) & \vec{v} & 0 & \mathbf{a} \\ 0 & 0 & 0 & b \\ m\vec{v}^t & 0 & 0 & \sigma \\ 0 & 0 & 0 & 0 \end{pmatrix} \quad , \tag{3.7.5}$$

where, using Souriau's notation, $j(\vec{\omega})$ is the cross product operator, $j(\vec{\omega})\mathbf{x} = \vec{\omega} \times \mathbf{x}$. Thus, $\delta_{\vec{\omega}}\mathbf{x} = (\vec{\omega} \cdot X_{(\vec{\epsilon})})\mathbf{x} = -\vec{\omega} \times \mathbf{x} = -j(\vec{\omega})\mathbf{x}$. As a matrix, $j(\mathbf{u})^i{}_j \equiv$

$u^k \eta^i_{kj}$; $j(\mathbf{u})^T = -j(\mathbf{u})$. Noticing that $Rr(\vec{\omega})R^{-1} = r(\overrightarrow{R\omega})$, a calculation using (3.6.28) gives the following expressions for the adjoint representation of the extended Galilei group on its algebra:

$$b' = b \quad ,$$

$$\mathbf{a}' = R\mathbf{a} + b\mathbf{V} - (\overrightarrow{R\omega}) \times (\mathbf{A} - B\mathbf{V}) - B(\overrightarrow{R\vec{v}}) \quad ,$$

$$\vec{v}\,' = \overrightarrow{R\vec{v}} - (\overrightarrow{R\omega}) \times \mathbf{V} \quad ,$$

$$\vec{\omega}' = \overrightarrow{R\omega} \quad ,$$

$$\sigma' = \sigma + b\frac{m}{2}\mathbf{V}^2 + m\mathbf{V} \cdot (R\mathbf{a}) - m(\overrightarrow{R\vec{v}}) \cdot \mathbf{A} - m\mathbf{V} \cdot [(\overrightarrow{R\omega}) \times (\mathbf{A} - B\mathbf{V})] \quad .$$

$$(3.7.6)$$

Equations (3.7.6) can now be used to compute the adjoint representation $ad\bar{\mathcal{G}}_{(m)}$ of $\mathcal{G}_{(m)}$ on itself. In matrix form, and acting on $(b, \mathbf{a}, \vec{v}, \vec{\omega}, \sigma) \in \mathcal{G}_{(m)}$, it is given by

$$adX = \begin{pmatrix} 0 & 0 & 0 & 0 & 0 \\ \vec{v}^t & -j(\vec{\omega}) & -b & j(\mathbf{a}) & 0 \\ 0 & 0 & -j(\vec{\omega}) & j(\vec{v}) & 0 \\ 0 & 0 & 0 & -j(\vec{\omega}) & 0 \\ 0 & m\vec{v}^t & -m\mathbf{a}^t & 0 & 0 \end{pmatrix} \quad . \qquad (3.7.7)$$

Let us now compute the *coadjoint* representation. Let μ be the generic element of $\bar{\mathcal{G}}^*_{(m)}$ characterized by the parameters E, $\mathbf{J}$, $\mathbf{P}$, $\mathbf{K}$, λ,

$$\mu = E\omega^{(B)} + \mathbf{P} \cdot \omega^{(A)} + \mathbf{K} \cdot \omega^{(V)} + \mathbf{J} \cdot \omega^{(\vec{e})} + \lambda\omega^{(\theta)} \quad . \qquad (3.7.8)$$

Using the expression for X' obtained from (3.7.6) and identifying the coefficients in $\vec{\omega}$, $\vec{v}$, $\mathbf{a}$, b and τ, it is easily seen that the coadjoint representation $Ad^*\bar{G}_{(m)}$ is given by

$$E' = E - (R\mathbf{P}) \cdot \mathbf{V} + \frac{m}{2}\mathbf{V}^2\lambda \quad ,$$

$$\mathbf{P}' = R\mathbf{P} - m\mathbf{V}\lambda \quad ,$$

$$\mathbf{K}' = R\mathbf{K} + BR\mathbf{P} + m(\mathbf{A} - \mathbf{V}B)\lambda \quad , \qquad (3.7.9)$$

$$\mathbf{J}' = R\mathbf{J} + \mathbf{A} \times (R\mathbf{P}) + \mathbf{V} \times (R\mathbf{K}) + m\mathbf{V} \times \mathbf{A}\lambda \quad ,$$

$$\lambda' = \lambda \quad .$$

For $\lambda = 1$ these transformations are, we note in passing, the values of the kinetic energy $\frac{1}{2}m\dot{\mathbf{x}}^2$, momentum $m\dot{\mathbf{x}}$, 'centre of mass' $m(t\dot{\mathbf{x}} - \mathbf{x})$ and angular momentum $m\dot{\mathbf{x}} \times \mathbf{x}$ for the transformation of parameters $-\mathbf{A}$, $-\mathbf{V}$ (instead of $\mathbf{A}$, $\mathbf{V}$). From (3.7.9) it is not difficult to find the vector fields that generate the action of $\bar{G}_{(m)}$ on its coalgebra $\bar{\mathscr{G}}^*_{(m)}$, which have the expression

$$\bar{Y}_{(B)}(\mu) = \mathbf{P}\frac{\partial}{\partial\mathbf{K}} \quad , \quad \bar{Y}_{(A)}(\mu) = -\mathbf{P}\times\frac{\partial}{\partial\mathbf{J}} + m\lambda\frac{\partial}{\partial\mathbf{K}} \quad ,$$

$$\bar{Y}_{(V)}(\mu) = -\mathbf{P}\frac{\partial}{\partial E} - m\lambda\frac{\partial}{\partial\mathbf{P}} - \mathbf{K}\times\frac{\partial}{\partial\mathbf{J}} \quad ,$$

$$\bar{Y}_{(\bar{\varepsilon})}(\mu) = -\mathbf{P}\times\frac{\partial}{\partial\mathbf{P}} - \mathbf{K}\times\frac{\partial}{\partial\mathbf{K}} - \mathbf{J}\times\frac{\partial}{\partial\mathbf{J}} \quad ,$$

$$\bar{Y}_{(\theta)}(\mu) = 0 \quad ,$$

(3.7.10)

and satisfy the commutation relations of (3.7.4). The matrix form of the infinitesimal action of $\bar{G}_{(m)}$ on $\bar{\mathscr{G}}^*_{(m)}$ can be obtained directly from (3.7.9) or by writing (3.7.10) in matrix form. This gives the adjoint representation $ad^*\bar{\mathscr{G}}_{(m)}$ of the extended Galilei algebra on its coalgebra $\bar{\mathscr{G}}^*_{(m)}$,

$$ad^*(X) = \begin{pmatrix} 0 & -\vec{v} & 0 & 0 & 0 \\ 0 & -j(\vec{\omega}) & 0 & 0 & -m\vec{v} \\ 0 & b & -j(\vec{\omega}) & 0 & m\mathbf{a} \\ 0 & j(\mathbf{a}) & j(\vec{v}) & -j(\vec{\omega}) & 0 \\ 0 & 0 & 0 & 0 & 0 \end{pmatrix} \quad .$$

(3.7.11)

It is easily verified, looking at (3.7.7) and (3.7.11), that the general defining relation $ad^*X = -(adX)^t$ between the adjoint and coadjoint representations of any Lie algebra holds.

Let us now take the orbit of the coadjoint action (3.7.9) that goes through $\mathbf{P} = \mathbf{K} = \mathbf{J} = \mathbf{0}$, $E = 0$, $\lambda = 1$. The elements of this orbit are determined by

$$E = \frac{m}{2}\mathbf{V}^2, \quad \mathbf{P} = -m\mathbf{V}, \quad \mathbf{K} = m(\mathbf{A}-\mathbf{B}V), \quad \mathbf{J} = \mathbf{A}\times(-m\mathbf{V}) \quad , \quad (3.7.12)$$

from which the following constraints are deduced:

$$E = \frac{\mathbf{P}^2}{2m} \quad , \quad \mathbf{J} = \frac{\mathbf{K}}{m}\times\mathbf{P} \quad .$$

(3.7.13)

Clearly, E, $\mathbf{P}$, $\mathbf{J}$, $\mathbf{K}$ can be identified with the constants of motion of a free Galilean particle of mass m. Thus, we have obtained the manifold of solutions (possible trajectories) of the free particle motion, which is characterized by the values of $\mathbf{P}$, $\mathbf{K}$, as an orbit of the coadjoint representation of the extended Galilei group; by construction, $\bar{G}_{(m)}$ operates transitively on the orbit. The canonical structure defined on it can be given by means of the symplectic two-form Ω_S of eq. (3.6.26) which is reduced to the Kirillov term

$$\Omega_S(\mu)(\bar{Z}_{(i)}(\mu), \bar{Z}_{(j)}(\mu)) := \mu([\bar{X}_{(i)}, \bar{X}_{(j)}]) \quad , \tag{3.7.14}$$

where $\bar{Z}_{(i)}(\mu)$ is the value of the vector field that generates the action along the parameter (i) of $\bar{G}_{(m)}$ at the point μ of the orbit (*i.e.*, the projection of the generators (3.7.10) on the orbit), and $\bar{X}_{(i)}$ is the corresponding element of the Lie algebra $\bar{\mathscr{G}}_{(m)}$.

It is instructive to check explicitly that the generators $\bar{Y}_{(i)}(\mu)$ project to $\bar{Z}_{(i)}(\mu)$ on the orbit. If we define new variables by means of the mapping $\phi : (E, \mathbf{P}, \mathbf{K}, \mathbf{J}) \mapsto (\hat{E}, \hat{\mathbf{P}}, \hat{\mathbf{K}}, \hat{\mathbf{J}})$,

$$\hat{E} = E - \frac{\mathbf{P}^2}{2m} \quad , \quad \hat{\mathbf{P}} = \mathbf{P} \quad , \quad \hat{\mathbf{K}} = \mathbf{K} \quad , \quad \hat{\mathbf{J}} = \mathbf{J} - \frac{\mathbf{K}}{m} \times \mathbf{P} \,, \tag{3.7.15}$$

it is clear that the above orbit is characterized by the conditions $\hat{E} = 0$, $\hat{\mathbf{J}} = 0$. We can now write the generators (3.7.10) in terms of the new variables (in other words, we can compute the action of the tangent mapping of the mapping ϕ, eq. (3.7.15), on the vector fields of (3.7.10)), which gives

$$\bar{Y}_{(B)}(\mu) = \hat{\mathbf{P}} \cdot \frac{\partial}{\partial \hat{\mathbf{K}}} \quad , \quad \bar{Y}_{(A)}(\mu) = m\lambda \frac{\partial}{\partial \hat{\mathbf{K}}} + (\lambda - 1)\hat{\mathbf{P}} \times \frac{\partial}{\partial \hat{\mathbf{J}}} \quad ,$$

$$\bar{Y}_{(v)}(\mu) = -m\lambda \frac{\partial}{\partial \hat{\mathbf{P}}} + (\lambda - 1)\left(\hat{\mathbf{P}} \frac{\partial}{\partial \hat{E}} + \hat{\mathbf{K}} \times \frac{\partial}{\partial \hat{\mathbf{J}}} \right) \quad , \tag{3.7.16}$$

$$\bar{Y}_{(\tilde{\epsilon})}(\mu) = -\hat{\mathbf{P}} \times \frac{\partial}{\partial \hat{\mathbf{P}}} - \hat{\mathbf{K}} \times \frac{\partial}{\partial \hat{\mathbf{K}}} - \hat{\mathbf{J}} \times \frac{\partial}{\partial \hat{\mathbf{J}}} \quad ,$$

and the restriction to the orbit gives, dropping the hats on $\hat{\mathbf{P}}$, $\hat{\mathbf{K}}$,

$$\bar{Z}_{(v)} = -m \frac{\partial}{\partial \mathbf{P}} \,, \quad \bar{Z}_{(A)} = m \frac{\partial}{\partial \mathbf{K}} \,, \quad \bar{Z}_{(B)} = \mathbf{P} \cdot \frac{\partial}{\partial \mathbf{K}} \,,$$

$$\bar{Z}_{(\tilde{\epsilon})} = -\mathbf{P} \times \frac{\partial}{\partial \mathbf{P}} - \mathbf{K} \times \frac{\partial}{\partial \mathbf{K}} \quad . \tag{3.7.17}$$

Clearly, the same result is obtained if we substitute directly the constraints

(3.7.13) in (3.7.9) to express everything in terms of **P** and **K**, so that

$$\mathbf{P}' = R\mathbf{P} - m\mathbf{V} \quad ,$$
$$\mathbf{K}' = R\mathbf{K} + B R\mathbf{P} - m B\mathbf{V} + m\mathbf{A} \quad , \tag{3.7.18}$$

and then compute the generators $\bar{Z}$ of the action on the orbit using (3.7.18).

Coming back to (3.7.14), we see that

$$\Omega_S(\bar{Z}_{(V^i)}, \bar{Z}_{(A^j)}) = \mu([\bar{X}_{(V^i)}, \bar{X}_{(A^j)}]) = -\mu(m\delta_{ij}\bar{X}_{(\theta)}) = -m\delta_{ij} \quad , \tag{3.7.19}$$

which gives the familiar symplectic form

$$\Omega_S = \frac{dK^i}{m} \wedge dP_i \quad . \tag{3.7.20}$$

It is now simple to check that the symplectic form is preserved under the action of (3.7.18), *i.e.*, that $L_{\bar{Z}_{(i)}}\Omega_S = 0$: thus, the orbit is a $\bar{G}_{(m)}$-*symplectic manifold* and the $\bar{Z}_{(i)}$ generate symplectic diffeomorphisms or *symplecto-morphisms*. Moreover, $L_{\bar{Z}_{(i)}}\Omega_S = 0$ implies that $i_{\bar{Z}_{(i)}}\Omega_S$ is a closed (in fact exact) form, the differential of a function, the momentum associated with the i-th generator. As expected, these 'momenta' are given by $\frac{1}{2m}\mathbf{P}^2$, **P**, **K**, $\mathbf{P} \times \mathbf{K}/m$ on the manifold of the solutions of the free particle motion.

3.8 On the possible failure of the associative property: three-cocycles

We saw at the beginning of this chapter that the two-cocycle condition for ξ, eq. (3.1.32), automatically guarantees the associativity property for $U(g'')U(g')U(g)$ as is checked, *e.g.*, for the case of the free particle (eq. (3.4.1)). Nevertheless, eq. (3.1.32) (or (3.2.6)), looked at as an abstract relation, prompts us to go further and consider what would happen in general if the associative property were to fail for certain operators $O(g)$. Although a non-associative composition law cannot be realized in terms of well-defined linear operators acting on a quantum Hilbert space, let us consider nevertheless the consequences of the presence of a non-trivial exponent ξ_3 in

$$\{[O(g'')O(g')]O(g)\}\psi(g''g'gx) = \exp\frac{i}{\hbar}\xi_3(x; g'', g', g) \cdot$$
$$\{O(g'')[O(g')O(g)]\}\psi(g''g'gx) \quad . \tag{3.8.1}$$

Again, there is a consistency condition which has to be satisfied by the exponent ξ_3. Consider $\{[O(g''')O(g'')][O(g')O(g)]\}\psi(g'''g''g'gx)$. Using

(3.8.1) repeatedly, we find

$$\{[O(g''')O(g'')][O(g')O(g)]\}\psi(g'''g''g'gx)$$

$$= exp\frac{i}{\hbar}\xi_3(x;g''',g'',g'g)\{O(g''')[O(g'')[O(g')O(g)]]\}\psi(g'''g''g'gx)$$

$$= exp\frac{i}{\hbar}[\xi_3(x;g''',g'',g'g)$$

$$- \xi_3(x;g'',g',g)]\{O(g''')[[O(g'')O(g')]O(g)]\}\psi(g'''g''g'gx)$$

$$= exp\frac{i}{\hbar}[\xi_3(x;g''',g'',g'g) - \xi_3(x;g'',g',g) - \xi_3(x;g''',g''g',g)].$$

$$\{[O(g''')[O(g'')O(g')]]O(g)\}\psi(g'''g''g'gx)$$

$$= exp\frac{i}{\hbar}[\xi_3(x;g''',g'',g'g) - \xi_3(x;g'',g',g) - \xi_3(x;g''',g''g',g)$$

$$- \xi_3(gx;g''',g'',g')]\{[[O(g''')O(g'')]O(g')]O(g)\}\psi(g'''g''g'gx)$$

$$= exp\frac{i}{\hbar}[\xi_3(x;g''',g'',g'g) - \xi_3(x;g'',g',g)$$

$$- \xi_3(x;g''',g''g',g) - \xi_3(gx;g''',g'',g')$$

$$+ \xi_3(x;g'''g'',g',g)]\{[O(g''')O(g'')][O(g')O(g)]\}\psi(g'''g''g'gx) .$$

$$\tag{3.8.2}$$

Notice that in the fourth step the re-association of $O(g''')[O(g'')O(g')]$ to $[O(g''')O(g'')]O(g')$ is done without affecting $O(g)$, so that the argument of the exponent is gx rather than x. Clearly, (3.8.2) implies the consistency condition

$$\xi_3(x;g'',g',g) - \xi_3(x;g'''g'',g',g) + \xi_3(x;g''',g''g',g)$$
$$- \xi_3(x;g''',g'',g'g) + \xi_3(gx;g''',g'',g') = 0 \tag{3.8.3}$$

(or nh). We shall see in section 5.4 that this is the condition for ξ_3 being a *three-cocycle*: thus, in general, *the breaking of the associativity property in a consistent manner implies the presence of a (non-trivial) three-cocycle.* We stress, nevertheless, that eq. (3.8.2) should be considered as a formal one inasmuch as ψ is identified with a wavefunction since the Hilbert space operators form an associative algebra.

Clearly, the breakdown of the associativity of the group law implies at the algebra level the failure of the Jacobi identity. Thus, we should consider the breaking of the Jacobi identity as the signature of the presence of a three-cocycle. To see this, consider the relation among group operators

$$O(g'')(O(g')O(g)) = exp\,i\xi(g'',g',g)\,(O(g'')O(g'))O(g) \tag{3.8.4}$$

where $O(g) = exp\ i\zeta^j \hat{X}_{(j)}$. Using the Baker-Campbell-Hausdorff formula*
for the products of exponentials to compute both sides of (3.8.4) and
antisymmetrizing with respect to ζ^i, ζ'^i, ζ''^i, we find that the term linear
in each ζ is

$$i\zeta''^{[i}\zeta'^j\zeta^{k]}\frac{1}{4}[\hat{X}_{(i)},[\hat{X}_{(j)},\hat{X}_{(k)}]]$$

$$= -i\zeta''^{[i}\zeta'^j\zeta^{k]}\frac{1}{4}[\hat{X}_{(i)},[\hat{X}_{(j)},\hat{X}_{(k)}]] \qquad (3.8.5)$$

$$+ i\zeta''^{[i}\zeta'^j\zeta^{k]}\partial_{g''^i}\partial_{g'^j}\partial_{g^k}\xi_3(g'',g',g)\big|_{g''=g'=e} \quad,$$

where one has to remember that ξ_3 is a normalized cocycle, so that
the first term in its expansion is proportional to $\zeta''^i\zeta'^j\zeta^k$. Then we
have

$$\partial_{g''^i}\partial_{g'^j}\partial_{g^k}\xi_3(g'',g',g)\big|_{g''=g'=e} = ([\hat{X}_{(i)},[\hat{X}_{(j)},\hat{X}_{(k)}]] + \text{cycl. perm.}) \quad ; (3.8.6)$$

this expression ties the three-cocycle to the failure of the Jacobi identity.

Is there a place in physics where we could observe the presence of
a three-cocycle? Consider a charged point particle in a magnetic field.
To avoid any reference to wavefunctions (moreover, as we mentioned in
section 2.7, the wavefunctions must be replaced by *wavesections*) let us
look to the operator formulation of the equations of motion. These are
given by $m\ddot{x} = q\dot{x} \times \mathbf{B}$, where q stands for the particle's electric charge
and the velocity of light c has been set equal to one. The Hamiltonian is
simply given by $H = \frac{1}{2}m\dot{x}^2$ because the magnetic field does no work and
does not contribute to the energy of the particle. If we were to construct
operators for $\mathbf{x}$ and $\dot{x}$ ($\hat{x}$ and $\hat{\dot{x}}$) in a quantum theory, they would have to
satisfy

$$\hat{\dot{x}} = \frac{i}{\hbar}[\hat{H},\hat{x}] \quad, \quad m\hat{\ddot{x}} = \frac{i}{\hbar}[\hat{H},m\hat{\dot{x}}] = q\hat{\dot{x}} \times \mathbf{B} \quad. \qquad (3.8.7)$$

Thus, the condition that the canonical evolution equations reproduce the
equations of motion requires the fundamental relations

$$[\hat{x}^i,\hat{x}^j] = 0 \quad, \quad [\hat{x}^i,m\hat{\dot{x}}^j] = i\hbar\delta^{ij} \quad, \quad [m\hat{\dot{x}}^i,m\hat{\dot{x}}^j] = iq\hbar e^{ij}{}_k B^k \quad. \qquad (3.8.8)$$

---

* *i.e.*, that $e^A e^B = e^C$, $C \simeq A + B + \frac{1}{2}[A,B] + \frac{1}{12}[A,[A,B]] + \frac{1}{12}[B,[B,A]]$. In this case this means
that $O(g')O(g) \simeq 1 + i\zeta'^i\hat{X}_{(i)} + i\zeta^i\hat{X}_{(i)} - \frac{1}{2}\zeta'^i\zeta^j[\hat{X}_{(i)},\hat{X}_{(j)}]$ plus terms that do not contribute below
or are of higher order.

The consequence of (3.8.8) is now that the Jacobi identity is given by

$$[m\hat{x}^i, [m\hat{x}^j, m\hat{x}^k]] + [m\hat{x}^k, [m\hat{x}^i, m\hat{x}^j]] + [m\hat{x}^j, [m\hat{x}^k, m\hat{x}^i]] \tag{3.8.9}$$
$$= q\hbar^2\{\epsilon^{jk}{}_l\partial^i B^l + \epsilon^{ij}{}_l\partial^k B^l + \epsilon^{ki}{}_l\partial^j B^l\} = \epsilon^{ijk}q\hbar^2\nabla\cdot\mathbf{B} \quad.$$

Eq. (3.8.9) measures a potential failure of the Jacobi identity for the operators $m\hat{x}$, which can be identified with the translation generators on account of the second commutator in (3.8.8). If $\mathbf{B}$ is the magnetic field of a monopole, then $\nabla\cdot\mathbf{B} = 4\pi g\delta(\mathbf{x} - \mathbf{x}_0)$ and the right hand side is not zero. Of course, it is still zero and the Jacobi identities fulfilled once the singular point $\mathbf{x}_0$ is excluded.

The methods of section 6.9 (cf. Theorem 6.9.2) will provide a way for calculating the finite (group) cocycle starting from the infinitesimal (algebra) one $q\epsilon_{ijk}\hbar^2\nabla\cdot\mathbf{B}$. Assuming that they can be transcribed to the present case, these methods imply that, for three translations $\mathbf{a}''$, $\mathbf{a}'$, $\mathbf{a}$,

$$\xi_3(\mathbf{x};\mathbf{a}'',\mathbf{a}',\mathbf{a}) = \int_T d^3b\,\nabla\cdot\mathbf{B}(\mathbf{x}+\mathbf{b}) \quad, \tag{3.8.10}$$

where T is the tetrahedron in *group* space that joins the four elements $\mathbf{0}$, $\mathbf{a}$, $\mathbf{a}+\mathbf{a}'$ and $\mathbf{a}+\mathbf{a}'+\mathbf{a}''$. This is precisely the magnetic flux of $\mathbf{B}$ through a tetrahedron in space with vertices $\mathbf{x}$, $\mathbf{x}+\mathbf{a}$, $\mathbf{x}+\mathbf{a}+\mathbf{a}'$ and $\mathbf{x}+\mathbf{a}+\mathbf{a}'+\mathbf{a}''$. When there are no magnetic sources, this is always zero (the factor $\hbar^2$ is absent because the analogue to (3.8.6) contains a factor $\hbar^{-2}$ in the r.h.s. if $i/\hbar$ replaces i in the exponent of the operators O, and this cancels the $\hbar^2$ in (3.8.9)). But, in the presence of monopoles, the result is non-zero if the tetrahedron contains a magnetic charge in its interior. For these values of the vertices, $e^{\frac{i}{\hbar}\xi_3}$ gives $exp(\frac{i}{\hbar}q4\pi g)$, with g the magnetic charge of the monopole (the total flux of $\mathbf{B}$ is then $4\pi g$). Dirac's quantization condition, however, which is a consistency condition for the fibre bundle description of the monopole (in which there is no string), or a condition of unobservability of the string in the original Dirac theory of the monopole in quantum mechanics, *already* requires that $qg = k\frac{\hbar}{2}$. Thus, there is no break of associativity and no three-cocycle is present in this case.

3.9 Some remarks on central extensions

Let us now make some observations concerning the central extensions of the Galilei and the Poincaré groups.

(a) The function Δ in (3.1.5) clearly has the dimensions of an action, and so have ξ_m and θ (θ 'goes' with the Lagrangian or with the differential form associated with the Lagrangian; see chapter 8).

(b) Exponents are dimensionless; thus ξ_m, when appearing in an exponent, *has to* be multiplied by a *universal* constant with the appropriate dimensions, *i.e.*, by $\frac{1}{\hbar}$. Thus, if we write

$$\zeta = e^{\frac{i}{\hbar}\theta} , \quad \omega_m(g',g) = \exp \frac{i}{\hbar}\xi(g',g) \tag{3.9.1}$$

(ζ 'goes' with the phase of the wavefunction ψ), the last line of the group law reads

$$\zeta'' = \zeta' \zeta \, \omega_m(g',g) \quad . \tag{3.9.2}$$

Thus, for $\tilde{G}$, the vector fields $\tilde{X}$ are obtained by replacing $\frac{\partial}{\partial\theta}$ by $\tilde{X}_{(\zeta)} = \frac{i}{\hbar}\zeta\frac{\partial}{\partial\zeta}$, and the forms $\tilde{\omega}$ by substituting $\hbar\frac{d\zeta}{i\zeta}$ for $d\theta$. For instance (cf. (3.4.5)),

$$\tilde{\omega}^{L(\zeta)} = \frac{1}{2}m\mathbf{V}^2 dB - m\mathbf{V}\cdot d\mathbf{A} + \hbar\frac{d\zeta}{i\zeta} \quad ; \tag{3.9.3}$$

$\tilde{\omega}^{L(\zeta)}$ may be shown to be the quantization form of the group approach to geometric quantization mentioned at the end of section 3.3. In fact, the extended Galilei group may be considered as the $U(1)$-bundle $\tilde{G}_{(m)}(U(1), G)$ over the Galilei group G. This bundle may be endowed with a left-invariant connection (see section 2.2 (d)), which is provided by the $U(1)$-component of the canonical LI form on $\tilde{G}_{(m)}$ above; the curvature of the connection is non-zero since $\tilde{G}$ is a non-trivial extension[*]. Explicitly,

$$d\tilde{\omega}^{L(\zeta)} = -C^{\zeta}_{A^iV^j}\tilde{\omega}^{L(A^i)} \wedge \tilde{\omega}^{L(V^j)} \quad (C^{\zeta}_{A^iV^j} = -m\delta_{ij}) \quad . \tag{3.9.4}$$

(c) The commutator $[\bar{X}^R_{(V^i)}, \bar{X}^R_{(A^j)}] = -\delta_{ij}m\bar{X}_{(\theta)}$ is non-zero for $\bar{G}_{(m)}$ (it is zero for Galilei); it is like the Poisson bracket

$$\{K_i, P_j\} = m\delta_{ij}$$

which is encountered when the Galilei group is realized in terms of canonical (contact) transformations of the dynamical system (free particle of mass m); the parameter m characterizes the extensions of the Galilei group, hence the subscript m. The appearance of m in $\{ \, , \, \}_{PB}$ when realizing the symmetries of G indicates that the Poisson brackets realize the algebra of $\bar{G}_{(m)}$, instead. This is another example of the rule already mentioned when discussing the Weyl-Heisenberg group according to which

---

[*] As mentioned, the approach uses $\tilde{G}$ as the basic manifold; the more familiar $(2n+1)$-dimensional *quantum manifold* can be derived from it, but it is not a necessary step and, for instance, the time evolution can be maintained as (3.9.3) shows. The case of $\tilde{G}_{(m)}$, which is a central extension, is a simple possibility (and the WH group the simplest one among this class), but other examples exist. For instance, the group $SU(2)$ (which contains $U(1)$ but which, being simple, does not have an extension structure) can also be considered as a quantum group, in fact a quantum group without a classical limit ($U(1)$ cannot be opened to R without losing the group principal bundle structure).

Poisson brackets realize $\bar{G}$, not G. In view of the above properties, the $U(1)$-extended Galilei group, $\tilde{G}_{(m)}$, may rightly be called the group of non-relativistic quantum mechanics or *quantum Galilei group* for short.

The quasi-invariance of L, the existence of ξ, and (c) above are all consequences of the fact that G may be extended to $\tilde{G}_{(m)}$. The quantum system seems not to preserve the classical symmetry. It has become customary to refer to this fact as an 'anomaly'. However, and as the naïve quantization rule $\{\,,\,\} \leftrightarrow \frac{1}{i\hbar}[\,,\,]$ shows, the Poisson and the Lie (quantum) Galilei algebras are isomorphic and so there is nothing anomalous; this point will be discussed further in chapter 8. The *extension $\bar{G}$ of G by R* (central variable θ) is related to the transformation properties of the classical system (Lagrangian) which still retains the original symmetry (determined by $G^{\dagger}$); the *extension by $U(1)$* (central variable ζ, which *requires $\hbar$*) characterizes the *quantum* system. The groups $\bar{G}$ and $\tilde{G}$ are locally isomorphic. The mathematical fact behind all this is the existence of some non-trivial cohomology; this leads us naturally to the general study of the cohomology of groups which will be started in the next chapter.

(d) The previous discussion in (c) refers to the Galilei group. What can be said about the Poincaré group, $\bar{\mathscr{P}} = Tr_4 \circ SL(2, C)$? ($\bar{\mathscr{P}}$ denotes the universal covering group of $\mathscr{P} = Tr_4 \circ L_+^\uparrow$ and $\circ$ here indicates the semi-direct product of the four-dimensional translation group and the restricted Lorentz group). It is known that in this case Lagrangians L are invariant rather than quasi-invariant, that the transformation laws of the relativistic fields are not projective (for instance, for a scalar field $\phi'(x') = \phi(x)$), etc. The above reasonings then imply that $\bar{\mathscr{P}}$ cannot be extended by $U(1)$ in a non-trivial way (or, in the terminology of section 5.2, that the second cohomology group of $\bar{\mathscr{P}}$ is trivial, $H_0^2(\bar{\mathscr{P}}, U(1)) = 0$). That this is indeed the case was proved by Wigner in 1939; later on, Wigner called $\bar{\mathscr{P}}$ the 'quantum mechanical Poincaré group'.

The (strict) invariance of Lagrangians and the non-appearance of phases is thus due to the trivial cohomology of the Poincaré group. But the cause of this simple behaviour is also the origin of complications in other respects: the quantization of a system as simple as the relativistic particle obliges us to take the mass-shell constraint properly into account.

(e) *Every local exponent of a semisimple group is equivalent to zero* (the proof will be given in chapter 6, Theorem 6.5.2). Thus, every local exponent of an n-dimensional pseudo-orthogonal group $SO(p, n - p)$, $n \geq 2$, is equivalent to zero (the groups $SO(p, n - p)$, $n > 2$, are all semisimple).

---

† It is also possible, by adding the new variable χ (see the previous section) to spacetime, $[\chi] =$ action, to achieve $\bar{G}_{(m)}$ invariance, as we shall see in chapter 8.

(f) The case (d) of the Poincaré group (which is the inhomogeneous group I_4^1 of a four-dimensional pseudo-orthogonal space of signature $(1, 3)$) can be generalized to arbitrary dimensions and signature. It was proved by Bargmann that *every local exponent of I_n^p (n > 2) is equivalent to zero.* However, the Poincaré group in *two*-dimensional $(1, 1)$ spacetime still may be centrally extended as will be shown below. It is interesting to observe that, with the usual assignments, the parameter F which determines the extension has the dimensions of a force (in the terminology of section 5.2, $H_0^2(I_2^1, U(1))$ is one-dimensional and its elements are then characterized by a parameter). In the Galilean contraction $c \rightarrow \infty$ limit, it corresponds to the gravity force on a lift which is eliminated by its free fall (the vector space which characterizes the extension of the two-dimensional Galilei group G_2, $H_0^2(G_2, U(1))$, is two-dimensional and characterized by the parameters (m, F)).

As for the universal coverings of the inhomogeneous spacetime groups I_n^p $(n > 2)$, the above result implies that every representation can be reduced to an ordinary representation.

3.10 Contractions and group cohomology

The previous comments in section 3.9(d) immediately pose a question. The Galilei group is a ten-parameter group which can be obtained by a Wigner-İnönü contraction from the Poincaré group. The Poincaré group may be extended only trivially by $U(1)$, i.e. as the direct product* $\mathscr{P} \otimes U(1)$. The Galilei group, however, has non-trivial central extensions $\tilde{G}_{(m)}$. If the $c \rightarrow \infty$ contraction limit, where c is the velocity of light, brings the Poincaré group to the Galilei group, how can a *direct* product extension like $\mathscr{P} \otimes U(1)$ lead to a central one, $\tilde{G}_{(m)}$, which is not? The reason for this behaviour is that a direct product extension $\mathscr{P} \otimes U(1)$ may be given in terms of a two-coboundary that, in the contraction limit, leads to a non-trivial two-cocycle of the contracted group. In other words, *the contraction process may generate group cohomology.*

It is not difficult to characterize in the general case the two-coboundaries on the original group G which will lead to non-trivial two-cocycles for the contracted group: they are generated by functions $\beta(g)$ which are *not* defined in the contraction limit but that generate two-coboundaries $\xi_{cob}(g', g) = \beta(g') + \beta(g) - \beta(g'g)$ with a well-defined contraction limit. The result of this limit will be a non-trivial two-cocycle since it cannot be obtained from a one-cochain. If we call G_c the group which is obtained by contraction from a group G, the above is summarized by saying that if the group law of $G \otimes U(1)$ is expressed in such a way that it leads to

---

* In the mathematical literature, the symbol $\times$ is commonly used for the direct product of groups.

$G_c \otimes U(1)$ in the contraction limit, a change of coboundary (which still defines $G \otimes U(1)$ before the contraction) may lead after the contraction to a central extension $\tilde{G}_c$ of G_c. In other words, the diagram

$$
\begin{array}{ccc}
& \xrightarrow{\;\text{contraction}\;} & \\
G \otimes U(1) & & G_c \otimes U(1) \\
\Big\downarrow & & \;\;\Big\uparrow\!\!\!\!\diagup \\
& \xrightarrow{\;\text{contraction}\;} & \\
G \otimes U(1) & & \tilde{G}_c
\end{array}
$$

change of coboundary

Diagram 3.9.1

is not a commutative one. The direct product extension will be called *pseudoextension* when it is written in terms of a coboundary leading to a non-trivial extension in the contraction limit.

Let us illustrate this explicitly with the example of the four-dimensional Poincaré group. Consider the direct product extension $\mathscr{P} \otimes U(1)$ as given by the two-coboundary generated by the one-cochain $\eta(g) = -mcA^0$ on $\mathscr{P}$, where A^0 is the time component of the four-translation A^μ, $\mu = 0, 1, 2, 3$. Clearly, the $c \to \infty$ limit of $\eta(g)$ does not exist. Using (3.5.7), this two-coboundary is given by

$$
\begin{aligned}
\xi_{cob}(g', g) &\equiv \xi_{cob}(A'^\mu, \Lambda'; A^\mu, \Lambda) = \eta(g') + \eta(g) - \eta(g'g) \\
&= -mc(A'^0 + A^0 - \Lambda'^0_{\cdot\nu}A^\nu - A'^0),
\end{aligned} \tag{3.10.1}
$$

since the Poincaré group law is given by $g' * g \equiv (\Lambda', A') * (\Lambda, A) = (\Lambda'\Lambda, \Lambda'A + A')$ where $\Lambda'A$ means $\Lambda'^\mu_{\cdot\nu}A^\nu$, etc. As a result, the group law of $\mathscr{P} \otimes U(1)$ is given by

$$
(\theta', A', \Lambda') * (\theta, A, \Lambda) = (\theta' + \theta + \xi_{cob}(A', \Lambda'; A, \Lambda), \Lambda'A + A', \Lambda'\Lambda) \tag{3.10.2}
$$

rather than by $(\theta', A', \Lambda') * (\theta, A, \Lambda) = (\theta' + \theta, \Lambda'A + A', \Lambda'\Lambda)$. The Lorentz matrices Λ are given by

$$
\Lambda = \begin{pmatrix} \gamma & \frac{1}{c}\gamma V^t R \\ \frac{1}{c}\gamma V & R + \frac{\gamma^2}{c^2(1+\gamma)}(V \otimes V^t)R \end{pmatrix} \tag{3.10.3}
$$

where $\gamma = 1/\sqrt{1 - (V^2/c^2)}$, V is the vector of the Lorentz boost parameters, V^t is its transpose and R is the 3×3 rotation matrix contained in the decomposition $\Lambda = (\text{boost})(\text{rotation})$. Setting $A^0 = cB$, where B is the time translation, the above two-coboundary is given by

$$
\xi_{cob}(g', g) = mc \left(-cB + \frac{cB}{\sqrt{1 - \dfrac{V'^2}{c^2}}} + \frac{V' \cdot R'A}{c\sqrt{1 - \dfrac{V'^2}{c^2}}} \right) \tag{3.10.4}
$$

and in the $c \to \infty$ limit ξ_{cob} reproduces the non-trivial Galilean two-cocycle (3.4.1).

As a second example[*] of group cohomology generated by the contraction process, consider the two-dimensional (anti-) de Sitter group. Its $so(2,1)$ algebra is given by

$$[P_\mu, J] = \epsilon_\mu{}^\nu P_\nu \quad , \quad [P_\mu, P_\nu] = -\frac{\Lambda_\mathscr{S}}{2}\epsilon_{\mu\nu}J \quad (\mu, \nu = 0, 1) \quad , \qquad (3.10.5)$$

where $\epsilon_{01} = 1$ and, thinking of P_μ as the translations of the two-dimensional contracted Poincaré group that may be obtained from (3.10.5), the constant $\Lambda_\mathscr{S}$ has been introduced with dimensions $[\Lambda_\mathscr{S}] = L^{-2}$ in natural units (thus, $[P_\mu] = L^{-1}$ and $[J]$ is dimensionless). An element of the $SO(2,1)$ group is obtained by exponentiation of the algebra,

$$U(g) = \exp(\alpha J + \alpha^\mu P_\mu) \quad , \qquad (3.10.6)$$

(α, α^μ) being the three group parameters. The generators are antihermitian, $P_\mu^\dagger = -P_\mu$, $J^\dagger = -J$; indices are raised and lowered with $\eta_{\mu\nu} = (1, -1)$.

It is not difficult to obtain the group law for $SO(2,1)$. A judicious use of i's allows us to introduce a new set of generators J_i,

$$J_1 \equiv iJ \quad , \quad J_2 \equiv i\sqrt{\frac{2}{\Lambda_\mathscr{S}}}P_0 \quad , \quad J_3 \equiv \sqrt{\frac{2}{\Lambda_\mathscr{S}}}P_1 \quad . \qquad (3.10.7)$$

Using this familiar trick, the algebra satisfied by them is the $so(3)$ algebra,

$$[J_i, J_j] = -\epsilon_{ijk}J_k \quad i, j, k = 1, 2, 3 \quad . \qquad (3.10.8)$$

The $SO(3)$ group law was given in (1.9.15). Since an element of the $so(2,1)$ algebra may be now written as $X = \alpha J + \alpha^\mu P_\mu = \vec{\epsilon} \cdot \vec{J}$ with

$$\epsilon^1 = -i\alpha \quad , \quad \epsilon^2 = -i\sqrt{\frac{\Lambda_\mathscr{S}}{2}}\alpha^0 \quad , \quad \epsilon^3 = \sqrt{\frac{\Lambda_\mathscr{S}}{2}}\alpha^1 \quad , \qquad (3.10.9)$$

we may use expression (1.9.15) to find the $SO(2,1)$ group law as

$$\alpha'' = \alpha'\sqrt{1 + \frac{\alpha^2}{4} + \frac{\Lambda_\mathscr{S}}{8}\alpha^\nu\alpha_\nu} + \alpha\sqrt{1 + \frac{\alpha'^2}{4} + \frac{\Lambda_\mathscr{S}}{8}\alpha'^\nu\alpha'_\nu} - \frac{1}{4}\Lambda_\mathscr{S}\epsilon_{\nu\rho}\alpha'^\nu\alpha^\rho \quad ,$$

$$\alpha''^\mu = \alpha'^\mu\sqrt{1 + \frac{\alpha^2}{4} + \frac{\Lambda_\mathscr{S}}{8}\alpha^\nu\alpha_\nu}$$

$$+ \alpha^\mu\sqrt{1 + \frac{\alpha'^2}{4} + \frac{\Lambda_\mathscr{S}}{8}\alpha'^\nu\alpha'_\nu} - \frac{1}{2}\epsilon^\mu{}_\nu(\alpha'^\nu\alpha - \alpha^\nu\alpha') \quad .$$

$$(3.10.10)$$

---

[*] Contributed by César F. Talavera.

As is obvious for the algebra (3.10.5), which in the contraction limit gives the two-dimensional Poincaré algebra, the $\Lambda_{\mathscr{G}} \to 0$ contraction of the $SO(2,1)$ group in (3.10.10) gives the two-dimensional Poincaré group. To see this, we make the coordinate change

$$\sinh(\tfrac{\gamma}{2}) = \tfrac{\alpha}{2} \quad , \quad x^\mu = \mathcal{M}(\tfrac{\gamma}{2})^\mu{}_\nu \alpha^\nu \tag{3.10.11}$$

(where $\mathcal{M}(\gamma)^\mu{}_\nu = \cosh\gamma\,\delta^\mu{}_\nu + \sinh\gamma\,\epsilon^\mu{}_\nu$), after taking the limit in (3.10.10). A small calculation leads to the $(1+1)$-dimensional Poincaré group law

$$\gamma'' = \gamma' + \gamma \quad , \quad x''^\mu = x'^\mu + \mathcal{M}(\gamma')^\mu{}_\nu x^\nu \quad . \tag{3.10.12}$$

The $SO(2,1)$ group, being simple, cannot be extended by $U(1)$ by the remark (e) above. Let us nevertheless add a two-coboundary (eq. (3.5.7)) to its group law generated by

$$\eta(g) = \frac{1}{\Lambda_{\mathscr{G}}}\eta_1(g) + \eta_2(g) \equiv \frac{1}{\Lambda_{\mathscr{G}}}2\Lambda_{\mathscr{G}}\operatorname{arcsinh}(\tfrac{\alpha}{2}) - \frac{1}{16}\Lambda_{\mathscr{G}}\alpha^\mu\alpha_\mu\frac{\alpha}{\sqrt{1+\frac{\alpha^2}{4}}} \quad , \tag{3.10.13}$$

where $g = (\alpha, \alpha^\mu)$ and, since $[\alpha^\mu] = L$, the additional constant $\Lambda_{\mathscr{G}}$ has dimensions $[\Lambda_{\mathscr{G}}] = L^{-2}$. With the two coboundary $\xi_{cob}(g',g) \equiv \frac{1}{\Lambda_{\mathscr{G}}}\xi_{1,cob}(g',g) + \xi_{2,cob}(g',g)$ generated by (3.10.13) the group law is modified to (3.2.15), where θ is the $U(1)$ coordinate. The extension of $SO(2,1)$ by $U(1)$ (equivalent to the direct product) has an algebra with an additional central generator Ξ. The pseudoextension (3.10.13) corresponds to making the change of basis $(\Xi, J) \to (\Xi, J')$ where

$$J \longmapsto J' = J + \frac{\Lambda_{\mathscr{G}}}{\Lambda_{\mathscr{G}}}\Xi \quad . \tag{3.10.14}$$

With this change the Lie algebra $so(2,1) \otimes R$ takes the form, dropping the prime in J',

$$[P_\mu, J] = \epsilon_\mu{}^\nu P_\nu \quad ,$$
$$[P_\mu, P_\nu] = -\frac{1}{2}\Lambda_{\mathscr{G}}\epsilon_{\mu\nu}J + \frac{1}{2}\Lambda_{\mathscr{G}}\epsilon_{\mu\nu}\Xi \quad , \tag{3.10.15}$$

as may be verified by explicit commutation between the vector fields obtained from the group law (3.10.10) complemented with the part for the central parameter θ given by $\xi_{cob}(g',g)$.

The contraction of this pseudoextended algebra is carried out by taking the limit $\Lambda_{\mathscr{G}} \to 0$. Clearly, this limit cannot be taken in (3.10.13) (or in (3.10.14)), so that if the cocycle generated by (3.10.13) has a limit it is again non-trivial since it cannot be generated by the limit of (3.10.13). To find the group law defining the contracted group we write

$$\xi_{cob} = \frac{1}{\Lambda_{\mathscr{S}}}\left(\eta_1(g') + \eta_1(g) - \eta_1(g'g)\right) + \left(\eta_2(g') + \eta_2(g) - \eta_2(g'g)\right) \quad.$$

(3.10.16)

Looking at (3.10.10) we see that only $g'g$ contains $\Lambda_{\mathscr{S}}$. The contributions to the limit come from the term linear in $\Lambda_{\mathscr{S}}$ from the first bracket ($\xi_{1,cob}(\Lambda_{\mathscr{S}} = 0) = 0$) and by setting $\Lambda_{\mathscr{S}} = 0$ in the second. Thus

$$\lim_{\Lambda_{\mathscr{S}} \to 0} \xi_{cob} = \left.\frac{\partial \xi_{1,cob}(g',g)}{\partial \Lambda_{\mathscr{S}}}\right|_{\Lambda_{\mathscr{S}}=0} + \xi_{2,cob}(g',g)|_{\Lambda_{\mathscr{S}}=0} \quad.$$

(3.10.17)

The first term of the sum is, after making the coordinate change $(\alpha, \alpha^\mu) \to (\gamma, x^\mu)$ of (3.10.11),

$$\left.\frac{\partial \xi_{1,cob}(g'g)}{\partial \Lambda_{\mathscr{S}}}\right|_{\Lambda_{\mathscr{S}}=0} =$$

$$-\Lambda_{\mathscr{P}}\frac{1}{\sqrt{1 + \frac{\alpha''(\Lambda_{\mathscr{S}}=0)^2}{4}}}\left(\frac{1}{16}\frac{\alpha^\mu \alpha_\mu}{\sqrt{1 + \frac{\alpha}{4}}}\alpha' + \frac{1}{16}\frac{\alpha'^\mu \alpha'_\mu}{\sqrt{1 + \frac{\alpha'}{4}}}\alpha - \frac{1}{4}\epsilon^{\mu\nu}\alpha'_\mu \alpha_\nu\right) =$$

$$-\frac{1}{8}\Lambda_{\mathscr{P}}\left(x^\mu x_\mu \frac{\sinh\frac{\gamma'}{2}}{\cosh\frac{\gamma''}{2}\cosh\frac{\gamma}{2}} + x'^\mu x'_\mu \frac{\sinh\frac{\gamma}{2}}{\cosh\frac{\gamma''}{2}\cosh\frac{\gamma'}{2}}\right.$$

$$\left.- \frac{2}{\cosh\frac{\gamma''}{2}}x'^\mu \epsilon_{\mu\nu}\mathscr{M}\left(\frac{\gamma'-\gamma}{2}\right)^\nu_{\ \rho} x^\rho\right) \quad,$$

(3.10.18)

where $\alpha''(\Lambda_{\mathscr{S}} = 0) = \alpha'\sqrt{1 + \frac{\alpha^2}{4}} + \alpha\sqrt{1 + \frac{\alpha'^2}{4}}$, and $\gamma'' = \gamma' + \gamma$. The second term of the sum in (3.10.17) is, after making the coordinate change of (3.10.11),

$$\xi_{2,cob}(g'g)|_{\Lambda_{\mathscr{S}}=0}$$

$$= \frac{1}{8}\Lambda_{\mathscr{P}}\left[x^\mu x_\mu\left(\tanh\frac{\gamma''}{2} - \tanh\frac{\gamma}{2}\right)\right.$$

$$\left. + x'^\mu x'_\mu\left(\tanh\frac{\gamma''}{2} - \tanh\frac{\gamma'}{2}\right) + 2\tanh\frac{\gamma''}{2}x'_\mu \mathscr{M}(\gamma')^\mu_{\ \nu}x^\nu\right] \quad.$$

(3.10.19)

Using various relations between hyperbolic functions like

$$\frac{\sinh\frac{\gamma'}{2}}{\cosh\frac{\gamma''}{2}\cosh\frac{\gamma}{2}} = \tanh\frac{\gamma''}{2} - \tanh\frac{\gamma}{2} \quad,$$

$$\sinh\gamma'\tanh\frac{\gamma''}{2} = \cosh\gamma' - \frac{\cosh\frac{\gamma'-\gamma}{2}}{\cosh\frac{\gamma'}{2}} \quad,$$

the contracted group is found to be the centrally extended two-dimensional Poincaré group

$$\gamma'' = \gamma' + \gamma \quad ,$$
$$x''^{\mu} = x'^{\mu} + M(\gamma)^{\mu}{}_{\nu}x^{\nu} \quad , \tag{3.10.20}$$
$$\theta'' = \theta' + \theta + \frac{1}{4}\Lambda_{\mathscr{P}}x'^{\mu}\epsilon_{\mu\nu}\mathscr{M}(\gamma')^{\nu}{}_{\rho}x^{\rho}$$

(see (f) above). Its central extension is defined by the two-cocycle

$$\xi_{\Lambda_{\mathscr{P}}}(\gamma', x'^{\mu}; \gamma, x^{\mu}) = \frac{1}{4}\Lambda_{\mathscr{P}}x'^{\mu}\epsilon_{\mu\nu}\mathscr{M}(\gamma')^{\nu}{}_{\rho}x^{\rho} \quad . \tag{3.10.21}$$

The corresponding extended algebra may be obtained from (3.10.20) (or by taking the $\Lambda_{\mathscr{S}} \to 0$ limit in (3.10.15)):

$$[P_{\mu}, J] = \epsilon_{\mu}{}^{\nu}P_{\nu} \quad , \quad [P_{\mu}, P_{\nu}] = \frac{1}{2}\Lambda_{\mathscr{P}}\epsilon_{\mu\nu}\Xi \quad . \tag{3.10.22}$$

The above two groups (pseudoextended (anti-) de Sitter group and extended Poincaré group) have been recently used in the gauge formulation of two-dimensional gravity models.

Bibliographical notes for chapter 3

Rays were introduced in Weyl (1931). The classical reference on multipliers and ray representations is Bargmann (1954); see also Hamermesh (1962), Parthasarathy (1969a,b) and Varadarajan (1984). For the relevance of these representations in mechanics see Pauri and Prosperi (1966). Other aspects are treated in García Prada *et al.* (1988) and in Boya and Sudarshan (1991).

The Galilei group is discussed in Bargmann (1954) and in the review of Lévy-Leblond (1971; see also 1963), where more details can be found and which contains further references. It is also treated, *e.g.*, in the books of Souriau (1969), Gilmore (1974) and Guillemin and Sternberg (1984). Further aspects of its projective representations are given in Cariñena and Santander (1981). For dynamical symmetry in mechanics see, *e.g.*, Mariwalla (1975). The paper of Dadashev (1986) is devoted to Galilean field theory. For the $(1 + 1)$-dimensional Poincaré covariant field theory see Baumann (1992).

The reader who is interested in geometric quantization may read the book of Souriau (1969), which is a classic reference despite the rather personal notation used; his approach (and notes by V. Aldaya) has been used in section 3.6. More mathematical in spirit are the paper by Kostant (1970) and Kirillov's book (1976); see also Chu (1974). The book by Abraham and Marsden (1978) also contains a short résumé on geometric quantization. Other references are Śniaticki (1980), Woodhouse (1992),

Tuynman and Wiegerinck (1987) and Guillemin and Sternberg (1984) for various physical applications. For modern aspects of the momentum mapping see Duistermaat and Heckman (1982) and Atiyah and Bott (1984); see also Wu (1993) which includes an application to the WZW model of Chapter 10. For the *deformation* approach to quantization (not treated in the text) see Bayen *et al.* (1978) and Lichnerowicz (1988). The group manifold approach to quantization mentioned in the text is given in Aldaya and de Azcárraga (1982, 1985a, 1987); for further extensions see Aldaya *et al.* (1992). For other aspects of geometric quantization see Isham (1984).

The coadjoint orbits and the quasi-invariance are also discussed in Inamoto (1993). For recent applications of the method of coadjoint orbits in connection with two-dimensional gravity (see section 9.6) see, *e.g.*, Alekseev and Shatashvili (1989) and Delius *et al.* (1990).

A collection of reprints on quantum (*deformed*) groups and algebras (not treated in the text), including the pioneering papers on the subject by Drinfel'd, Faddeev, Reshetikhin and Takhtajan, Jimbo, and others, can be found in Jimbo (1990). For an introduction to quantum groups see, *e.g.*, Takhtajan (1989), or Gómez *et al.* (1995) and Chari and Pressley (1994) for textbook presentations. For an introduction to Woronowicz differential calculus on quantum groups see *e.g.* Aschieri and Castellani (1993); for an introduction to non-commutative geometry in general see Connes (1990, 1994).

Wigner-İnönü contractions were introduced in İnönü and Wigner (1953); see also İnönü (1964) and the book by Gilmore (1974). For the problem of how the contraction may generate cohomology see Saletan (1961) and Aldaya and de Azcárraga (1985a).

The possible appearance of a three-cocycle in quantum mechanics in the presence of a monopole and the consequent violation of the Jacobi identity was put forward in the papers of Grossmann (1985, 1986), Jackiw (1985a,b, 1987), Wu and Zee (1985), and subsequently discussed by Boulware *et al.* (1985), Mickelsson (1985a); see also Hou *et al.* (1986). For discussions on the three-cocycle in field theory and non-abelian anomalies see, *e.g.*, Jo (1985), Wang (1989) and Carey *et al.* (1995), from which further references may be traced.

The (1 + 1)-gravity models mentioned in section 3.10 have been put forward by Teitelboim (1983) and Jackiw (see Cangemi and Jackiw (1992) for their group contents); see also the lectures by Jackiw (1993).

4

An introduction to abstract
group extension theory

This is the first of a set of four chapters that constitute a self-contained introduction to Lie group and Lie algebra cohomology and extensions. In the present one, the general theory of group extensions is considered by first introducing the formalism of exact sequences of homomorphisms and by using the fibre bundle terminology of chapter 1.

The central concept underlying a group extension is the factor system, which will be shown to determine and be determined by the extension.

4.1 Exact sequences of group homomorphisms

An essential tool in the discussion of group extensions is the notion of an exact sequence of group homomorphisms. Since it will be extensively discussed in what follows, we shall briefly introduce it here. It is expressed by the following

(4.1.1) DEFINITION (*Exact sequence of group homomorphisms*)

Let $f_i : G_i \to G_{i+1}$ be a collection of group homomorphisms,

$$... \to G_i \overset{f_i}{\to} G_{i+1} \overset{f_{i+1}}{\to} G_{i+2} \to ... \quad . \tag{4.1.1}$$

The sequence f_i is called exact if

$$\mathrm{range} f_i = \ker f_{i+1} \quad . \tag{4.1.2}$$

As a result $f_i \circ f_{i-1} = 0$ for an exact sequence, *i.e.* the composition of two adjacent homomorphisms is the trivial homomorphism.

Let G be a group and let S be an invariant or normal subgroup, $S < G$, so that the group law of G induces a group law on the quotient set G/S which then becomes the quotient group. The canonical surjective homomorphism $f : G \to G/S$ maps $g \in G$ onto the element of G/S defined by the class to which g belongs in G. Clearly, $\ker f = S$ since the subgroup S, viewed as an element of G/S, is the identity of G/S. Thus, if $S < G$, the sequence of homomorphisms

$$1 \to S \to G \xrightarrow{f} G/S \to 1 \quad , \tag{4.1.3}$$

where f is the above canonical homomorphism, is exact since $S \to G$ is an injection and $G/S \to 1$ is the trivial homomorphism. Moreover, the exact sequence

$$1 \to S \to G \to B \to 1 \tag{4.1.4}$$

implies $B \approx G/S$.

Many group properties can be expressed in terms of exact sequences. For instance, the exact sequences

$$1 \to Z_2 \to SU(2) \to SO(3, R) \to 1 \quad ,$$

$$1 \to Z \to R \to U(1) \to 1 \quad , \tag{4.1.5}$$

constitute another way of expressing the familiar relations of the three-dimensional rotation group and the one-dimensional unitary group with their respective universal covering groups $SU(2)$ and R; the kernels of the corresponding covering homomorphisms are the first (or fundamental) homotopy groups, $\pi_1(SO(3, R)) = Z_2$ and $\pi_1(U(1)) = Z$. Similarly,

$$1 \to G' \to G \to Ab\, G \to 1 \tag{4.1.6}$$

expresses that the quotient of a group G by its commutator normal subgroup G' (the group *generated* by all the elements of the form $g_1 g_2 g_1^{-1} g_2^{-1}$ $\forall g_1, g_2 \in G$, or commutator of g_1, g_2) is the (abelian) group denoted $Ab\, G$. As we shall see extensively in the next sections, several exact sequences may give rise to a diagram. Consider, for example, the exact sequences

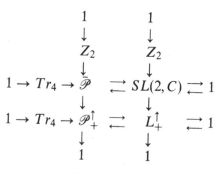

Diagram 4.1.1

The horizontal homomorphisms (left to right) state simply that $SL(2,C)$ $(L_+^\uparrow)$ is the quotient of the quantum mechanical Poincaré group $\bar{\mathscr{P}}$ (restricted Poincaré group $\mathscr{P}_+^\uparrow$) by the four-dimensional translation group Tr_4. Moreover, the additional reversed arrows allow us to conclude that $\bar{\mathscr{P}}$ and $\mathscr{P}_+^\uparrow$ have a semidirect structure (see section 5.2). The arrows do not go further to the one-element group at the left because this would imply that $\bar{\mathscr{P}}$ ($\mathscr{P}_+^\uparrow$) would be the direct product of $SL(2,C)$ $(L_+^\uparrow)$ and the translation group, which obviously is not the case: $SL(2,C)$ and $L_+^\uparrow$ are subgroups of $\bar{\mathscr{P}}$ and $\mathscr{P}_+^\uparrow$ respectively, but not invariant subgroups. As for the vertical arrows, they constitute another way of expressing that the homomorphism relating $\mathscr{P}_+^\uparrow$ and $L_+^\uparrow$ to their respective universal covering groups $\bar{\mathscr{P}}$ and $SL(2,C)$ is two-to-one.

As another example concerning infinite-dimensional Lie groups we may rewrite (1.3.26) as

$$1 \to \text{Aut}_v(P) \to \text{Diff}_M(P) \to \text{Diff } M \to 1 \quad . \tag{4.1.7}$$

With these simple examples in mind, let us now turn to discuss in general the structure of a group extension.

4.2 Group extensions: statement of the problem in the general case

Let us start by describing the structure of an extension and by analysing the information which is contained in the statement that the *abstract group $\tilde{G}$ is an extension of G by K* (the tilde will be used here to denote the extension nature of the group $\tilde{G}$). In the previous chapter we have considered in some detail the case in which K is the group $U(1)$ (or R). In this chapter we shall consider the more general case and, in particular, it will not be assumed that K is abelian.

(4.2.1) DEFINITION (*Group extension*)

Let G, K be abstract groups. A group $\tilde{G}$ is said to be an extension of G by K if K is an invariant subgroup of $\tilde{G}$, $K \prec \tilde{G}$, and $\tilde{G}/K = G$.

In terms of exact sequences, the above is equivalent to saying that

$$1 \to K \to \tilde{G} \to G \to 1$$

is exact; thus K is injected into $\tilde{G}$ and $\tilde{G}$ projected onto G by the canonical homomorphism so that $G = \tilde{G}/K$. The reader has already encountered a few examples of extensions, such as the central extension of the Galilei group. Also, the universal covering group $\bar{G}$ of a Lie group G can be

viewed as a central extension of G by its (discrete) first homotopy group[*]
$\pi_1(G)$; for instance, R is a central extension of $SO(2)$ by Z ($R/Z = SO(2)$).
Clearly, a group which is the *direct product* of two groups can be viewed
as an extension of any of the factors by the other; for instance, $O(3)$ is
an extension of $SO(3)$ (Z_2) by Z_2 ($SO(3)$) since $O(3) = SO(3) \otimes Z_2$. If
$\tilde{G} = K \otimes G$, then $\tilde{G}/G = K$ and $\tilde{G}/K = G$, which means that in the direct
product the arrows of the exact sequence above can be inverted.

It must be said immediately that the above exact sequence states that
$\tilde{G}$ is an extension of G by K, but that the mere knowledge of K and G
does not define $\tilde{G}$ uniquely. The simplest example of all is provided by
the case $K = G = Z_2$. In it, $\tilde{G}$ can be given by the cyclic group Z_4 or
by V (the group consisting of the identity and three involutive elements,
or *Vierergruppe*, which for instance may be realized on spacetime by the
discrete spacetime reflections e, P, T, PT). Obviously, $V = Z_2 \otimes Z_2 \neq Z_4$
but $Z_4/Z_2 = Z_2$ and $V/Z_2 = Z_2$.

Assume that a known group $\tilde{G}$ is an extension of G by K. Then:

a) Since K is a normal subgroup of $\tilde{G}$, the elements of $\tilde{G}$ act on K by
conjugation. Let $\text{Aut}\,K$ be the group of all automorphisms of K. Then
$\exists f, f : \tilde{G} \to \text{Aut}\,K$, $f : \tilde{g} \in \tilde{G} \mapsto [\tilde{g}] \in \text{Aut}\,K$, where $[\tilde{g}]$ is defined by

$$[\tilde{g}] : k \in K \mapsto \tilde{g}k\tilde{g}^{-1} \in K \quad . \tag{4.2.1}$$

The kernel of f is the centralizer $\phi_{\tilde{G}}(K)$ of K in $\tilde{G}$ since, by definition,
the centralizer is composed of the elements of $\tilde{G}$ that commute with
all elements of K,

$$\phi_{\tilde{G}}(K) = \{\tilde{h} \in \tilde{G} | \ k\tilde{h} = \tilde{h}k \ \forall k \in K\} \quad .$$

Thus, the following exact sequence of group homomorphisms exists:

$$1 \to \phi_{\tilde{G}}(K) \to \tilde{G} \xrightarrow{f} \text{Aut}\,K \quad . \tag{4.2.2}$$

b) The centre C_K of K is invariant in $\phi_{\tilde{G}}(K)$; if H denotes the quotient
group $H = \phi_{\tilde{G}}(K)/C_K$,

$$1 \to C_K \to \phi_{\tilde{G}}(K) \to H \to 1 \tag{4.2.3}$$

is exact.

c) Given K, we know $\text{Aut}\,K$ (in principle; in practice, this may be a
difficult problem). Let $\text{Int}\,K$ be the group of *inner* automorphisms $[k]$,

$$\forall k \in K, \ [k] : k' \mapsto kk'k^{-1}.$$

---

[*] Any discrete normal subgroup of a connected group belongs to its centre.

By definition, the *external* (*outer*) automorphisms are those which are not internal*. The classes of outer automorphisms are given by Out K = Aut K/ Int K, *i.e.*

$$1 \to \text{Int}\,K \to \text{Aut}\,K \to \text{Out}\,K \to 1 \quad ; \qquad (4.2.4)$$

by an abuse of language, one may also refer to Out K as the group of outer automorphisms.

d) Elements of $\tilde{G}$ which belong to the same equivalence class by K (*i.e.*, to the same coset $\tilde{g} * K$) produce automorphisms which differ by an inner automorphism of K. Thus we have

$$1 \to H \to G \overset{\sigma}{\to} \text{Out}\,K \quad . \qquad (4.2.5)$$

The mapping σ need not be surjective, and this is why the exact sequence (4.2.5) ends in Out K.

Thus, *given* an extension $\tilde{G}$ of G by K, all the information is summarized in the following commutative diagram of exact sequences:

$$
\begin{array}{ccccccccc}
& & 1 & & 1 & & 1 & & \\
& & \downarrow & & \downarrow & & \downarrow & & \\
1 & \to & C_K & \to & \phi_{\tilde{G}}(K) & \overset{\pi}{\to} & H & \to & 1 \\
& & \downarrow i & & \downarrow i & & \downarrow i & & \\
1 & \to & K & \to & \tilde{G} & \overset{\pi}{\to} & G & \to & 1 \\
& & \downarrow \pi & & \downarrow f & & \downarrow \sigma & & \\
1 & \to & \text{Int}\,K & \to & \text{Aut}\,K & \to & \text{Out}\,K & \to & 1 \\
& & \downarrow & & & & & & \\
& & 1 & & & & & &
\end{array}
$$

i = injection; π = projection; f, σ = hom

Diagram 4.2.1

Let us check further some parts of Diagram 4.2.1:

i) The elements of H are the *cosets* $h = \{\tilde{h} * C_K\}$, where $\tilde{h} \in \phi_{\tilde{G}}(K)$. H is characterized as a subgroup of G by means of the injection i, $i : \{\tilde{h} * C_K\} \mapsto \{\tilde{h} * K\} \in G$. The homomorphism i is in fact one-to-one since if $i(\{\tilde{h} * C_K\}) = i(\{\tilde{h}' * C_K\})$ then $\tilde{h} * K = \tilde{h}' * K$ which implies that

---

* For instance, the complex conjugation is an outer automorphism of $SL(2, C)$. However, since $\forall A \in SL(2, C)$ it is found that $\epsilon A \epsilon^{-1} = (A^{-1})^T$ where $\epsilon = \begin{pmatrix} 0 & 1 \\ -1 & 0 \end{pmatrix} = i\sigma^2$, the complex conjugation is an inner automorphism of its $SU(2)$ subgroup since for a unitary matrix $(A^{-1})^T = A^*$.

there is a $k \in K$ such that $\tilde{h}' = \tilde{h}k$, i.e. $\tilde{h}^{-1}\tilde{h}' = k$. This in turn implies that $k \in K \cap \phi_{\tilde{G}}(K) = C_K$ and that $\tilde{h} * C_K = \tilde{h}' * C_K$. Thus, the diagram

$$
\begin{array}{ccc}
\tilde{h} * C_K & \xrightarrow{\;\;\pi\;\;} & h \\
\downarrow i & & \downarrow i \\
\tilde{h} * K & \xrightarrow{\;\;\pi\;\;} & h
\end{array}
$$

is commutative.

Now, H is a normal subgroup of G since

$$(\tilde{g} * K) * (\tilde{h} * K) * (\tilde{g}^{-1} * K) = (\tilde{g} * \tilde{h} * \tilde{g}^{-1}) * K$$

and $\phi_{\tilde{G}}(K)$ is a normal subgroup of $\tilde{G}$.

ii) ker $f = \phi_{\tilde{G}}(K)$, because if $\tilde{h} \in \phi_{\tilde{G}}(K)$, $[\tilde{h}]k = \tilde{h}k\tilde{h}^{-1} = k$. Because $G = \tilde{G}/K$, all elements $g \in G$ induce external automorphisms of K. However, those of H induce the trivial automorphism; and thus ker $\sigma = H$, as implied by the diagram.

In view of the above discussion, the problem of extending G by K may be formulated as

'Given two groups G and K, and a homomorphism $\sigma : G \to \mathrm{Out}\,K$, find $\tilde{G}$ and $\cdots >$ in Diagram 4.2.2':

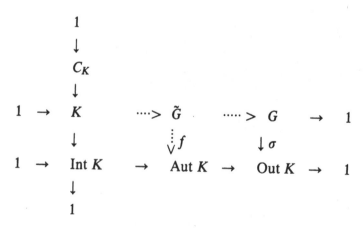

Diagram 4.2.2

(4.2.2) DEFINITION (*G-kernel* (K,σ)*, extendible G-kernel*)

A G-kernel (K,σ) *is a group K and a homomorphism* $\sigma : G \to \mathrm{Out}\,K$. *A G-kernel is extendible if there is a* $\tilde{G}$ *which is a solution of the problem above.*

In general, a G-kernel (K, σ) is not extendible (but if K is abelian there is always a solution as we shall see in section 5.2(b), the semidirect product $\tilde{G} = K \circledS G$).

Given a G-kernel (K, σ), there is another G-kernel associated with it, the G-kernel (C_K, σ_0), obtained by restricting the action of any element $\sigma(g) \in \mathrm{Out}\, K$ to C_K, i.e. by defining $\sigma_0 = \sigma' \circ \sigma$ in

$$G \xrightarrow{\sigma} \mathrm{Out}\, K$$
$$\searrow \sigma'$$
$$\mathrm{Out}\, C_K \qquad (= \mathrm{Aut}\, C_K)$$

Diagram 4.2.3

where σ' is the natural restriction of the elements of $\mathrm{Out}\, K$ to $\mathrm{Out}\, C_K$. The existence of σ' is guaranteed by the fact that the centre of any group is *characteristic*, i.e. is mapped onto itself by all the automorphisms of the group. This follows from the fact that $ck = kc$ if $c \in C_K$ and so if $a \in \mathrm{Aut}\, K$ this commutativity is also true for their images by a, $a(c)a(k) = a(k)a(c)\ \forall k$. Thus, $a(C_K) = C_K$. Since the internal automorphisms of $\mathrm{Int}\, K$ act trivially on C_K, there is a natural homomorphism σ' between $\mathrm{Out}\, K$ and $\mathrm{Out}\, C_K$. Thus, (C_K, σ_0) is indeed a G-kernel:

(4.2.3) DEFINITION (*Centre of a G-kernel*)

Given a G-kernel (K, σ), the G-kernel (C_K, σ_0) where $\sigma_0 = \sigma' \circ \sigma$ and σ' is the natural homomorphism $\sigma' : \mathrm{Out}\, K \to \mathrm{Out}\, C_K$, is called the centre of the G-kernel (K, σ)

(4.2.4) DEFINITION (*Equivalence of extensions*)

Let $\tilde{G}_1, \tilde{G}_2$ be two extensions of the G-kernel (K, σ). They are equivalent if there is an isomorphism $\tilde{f} : \tilde{G}_1 \to \tilde{G}_2$ such that

$$
\begin{array}{ccccc}
 & i_1 & \tilde{G}_1 & \pi_1 & \\
 & \nearrow & & \searrow & \\
1 \to K & & \downarrow \tilde{f} & & G \to 1 \\
 & \searrow & & \nearrow & \\
 & i_2 & \tilde{G}_2 & \pi_2 &
\end{array}
\qquad
\begin{array}{l}
i_2 = \tilde{f} \circ i_1 \\
\pi_1 = \pi_2 \circ \tilde{f}
\end{array}
$$

Diagram 4.2.4

is commutative.

Example

In section 3.2 we discussed the central extensions $\bar{G}$ of a group G defined by the two-cocycles $\xi(g',g)$. Let $\bar{G}_1$ and $\bar{G}_2$ be two central extensions of G defined by two different two-cocycles ξ_1, ξ_2, and let their group laws be

$$(\theta',g')(\theta,g) = (\theta'+\theta+\xi_1(g',g),g'g)\,, \quad [\theta',g'][\theta,g] = [\theta'+\theta+\xi_2(g',g),g'g]\,,$$
(4.2.6)

where the round [square] brackets denote the elements of $\bar{G}_1$ [$\bar{G}_2$]. Let us assume that there exists an isomorphism $\bar{f} : \bar{G}_1 \to \bar{G}_2$ which makes Diagram 4.2.4 commutative. Since $(\theta,g) = (\theta,e)(0,g)$ and $\bar{f}$ is a homomorphism, it is clear that $\bar{f}$ is completely determined once the images of the elements (θ,e) and $(0,g)$ are given. On the other hand, the commutativity of the diagram introduces some conditions for $\bar{f}$:

$$\bar{f} \circ i_1 = i_2 \quad \Rightarrow \quad \bar{f}(\theta,e) = [\theta,e]\quad,$$
$$\pi_2 \circ \bar{f} = \pi_1 \quad \Rightarrow \quad \bar{f}(0,g) = [\eta(g),g]\,.$$
(4.2.7)

Thus, $\bar{f}$ has to be of the generic form $\bar{f}(\theta,g) = [\theta+\eta(g),g]$; once $\eta(g)$ is known, $\bar{f}$ is determined. But, since $\bar{f}$ is a homomorphism,

$$\bar{f}(\theta'+\theta+\xi_1(g',g),g'g) = [\theta'+\theta+\xi_1(g',g)+\eta(g'g),g'g]$$
(4.2.8)

must be equal to

$$\bar{f}(\theta',g')\bar{f}(\theta,g) = [\theta'+\eta(g'),g'][\theta+\eta(g),g]$$
$$= [\theta'+\theta+\xi_2(g',g)+\eta(g')+\eta(g),g'g]\,,$$
(4.2.9)

and hence $\eta(g)$ must satisfy the relation

$$\xi_1(g',g) = \xi_2(g',g)+\eta(g')+\eta(g)-\eta(g'g) \equiv \xi_2(g',g)+\xi_{cob}(g',g)\,, \quad (4.2.10)$$

where $\xi_{cob}(g',g)$ is the two-coboundary generated by $\eta(g)$ [(3.2.9), (3.5.7)]. Thus, if Diagram 4.2.4 is commutative, ξ_1 and ξ_2 define the same extension. Proportional exponents $\xi_2 = \lambda\xi_1$ cannot satisfy (4.2.10) for all g', g; thus they define isomorphic groups $\bar{G}_1$, $\bar{G}_2$ that are different as *extensions*, as remarked in section 3.2. A physical example of this situation is provided by the extended Galilei groups $\bar{G}_{(m)}$ and $\bar{G}_{(m')}$ associated with two particles of masses m and m'; they are clearly isomorphic as groups, although they correspond to different central extensions. In the terminology of section 5.2(a) they define different elements of the second cohomology group H_0^2 of the Galilei group, *i.e.* different real numbers m, m'.

4.3 Principal bundle description of an extension $\tilde{G}(K,G)$

Let G, K and $\tilde{G}$ be *abstract* groups. It is often convenient to look at the extension $\tilde{G}$ of G by K using the terminology of a principal bundle with structure group* K.

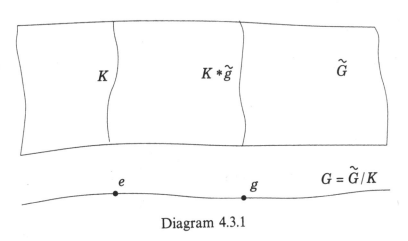

Diagram 4.3.1

The cosets of K in $\tilde{G}$, each one defining an element $g \in G$, are the fibres over g; their elements are

$$K * \tilde{g} \ (= \tilde{g} * K \text{ as a } coset \text{ since } K \text{ is normal in } \tilde{G}) \ .$$

Thus, the fibres are generated by the *left* action (see footnote)

$$k : \tilde{g} \mapsto k\tilde{g} \quad \forall k \in K$$

of K on a given $\tilde{g}$; $\pi^{-1}(\pi(\tilde{g}_0)) = \{\tilde{g}|\tilde{g} = k\tilde{g}_0, k \in K\}$ is the fibre through $\tilde{g}_0 \in \tilde{G}$.

A section of $\tilde{G}(K, \tilde{G}/K = G)$,

$$s : G \to \tilde{G} \ , \quad s : (g) \mapsto s(g) \ ,$$

selects an element of $\tilde{G}$ in each coset (= each fibre) $K * \tilde{g}$. Given a section, the elements of $\tilde{G}$ can be referred to it because, if (a) $s(g) = \tilde{g}_1$ and (b) $\pi(\tilde{g}_2) = \pi(\tilde{g}_1) = g$, then there exists a *unique* k such that $\tilde{g}_2 = k\tilde{g}_1 = ks(g)$. We may then refer to $s(g)$ as a *trivializing section*; notice that the set $\{s(g) \in \tilde{G} \mid g \in G\}$ is *not*, in general, a subgroup of $\tilde{G}$. In other words, s is one-to-one but not necessarily a homomorphism and, in general, G is not a subgroup of the (total group space) $\tilde{G}$. In the notation of Definition

---

* For this purpose, the action of the structure group on the group $\tilde{G}$ will be taken here to be a *left* action. See also the remark before Theorem 4.3.1 below.

1.3.1 with K acting *on the left* we would write

$$\Phi_s^{-1}(k, g) = ks(g) \quad , \tag{4.3.1}$$

where Φ_s is a global trivialization of the bundle $\tilde{G}(K, G)$; since $\pi \circ s$ is the identity mapping on G, the diagram

$$
\begin{array}{ccc}
\tilde{G} & \xrightarrow{\ \Phi_s\ } & K \times G \\
\pi \downarrow & & \downarrow p_2 \\
G & =\!=\!=\!=\!= & G
\end{array}
$$

Diagram 4.3.2

where p_2 is the Cartesian projection onto the second factor, is commutative. If G is a subgroup of $\tilde{G}$, then the injection of G into $\tilde{G}$ $s : g \mapsto (0, g) \in \tilde{G}$, $g \in G$, determines a section which is also a monomorphism.

Remark

As will be shown in the next section, the *abstract* group $\tilde{G}$ will be characterized in terms of the system of factors defined through the trivializing section, and we shall make liberal use of this fibre bundle language in what follows. However, this description of $\tilde{G}$ may not be feasible for *topological* or *Lie* groups since, in general, it is not possible to find a *continuous* mapping $s : G \to \tilde{G}$ such that $\pi \circ s = 1_G$. Such a construction can still be carried out under the restrictions of the following theorem which refers to group extensions of *Lie* groups:

(4.3.1) THEOREM (*Hochschild*)

Let K be a connected finite-dimensional Lie group, G a simply connected Lie group, and $\tilde{G}$ an extension of G by K. Then there exists an analytic mapping s of G into $\tilde{G}$ such that $\pi \circ s$ is the identity mapping.

Also, we have the following:

(4.3.2) THEOREM

Let G and K be connected Lie groups, and let G also be simply connected. Then the correspondence which associates with each extension of G by K the induced extension of the corresponding Lie algebras $\mathcal{G}$ and $\mathcal{K}$ gives rise to a one-to-one mapping of the respective sets of equivalence classes of extensions.

4.4 Characterization of an extension through the factor system ω

Let $\tilde{G}$ be an extension of the G-kernel (K, σ), and let $s : G \rightarrow \tilde{G}$ be a trivializing section (see Diagram 4.4.1).

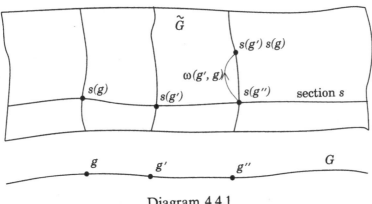

Diagram 4.4.1

Necessarily,

$$\pi(s(g'')) = \pi(s(g')\, s(g)) \quad ;$$

thus, there exists a factor $\omega(g', g) \in K$ such that

$$s(g')s(g) = \omega(g', g)s(g'g) \quad ; \tag{4.4.1}$$

in fact, (4.4.1) defines a factor $\omega(g', g)$. By taking $g' = e$ or $g = e$ we obtain

$$\omega(g', e) = \omega(e, g) = s(e) \quad . \tag{4.4.2}$$

It is convenient to take s in such a way that $s(e) = \tilde{e} \in \tilde{G}$, which is tantamount to saying that the diagram

$$\begin{array}{ccc}
K & \xrightarrow{\ i_1\ } & K \times G \\
{\scriptstyle i}\big\downarrow & & \big\downarrow{\scriptstyle \Phi_s^{-1}} \\
\tilde{G} & =\!=\!=\!= & \tilde{G}
\end{array}$$

Diagram 4.4.2

is commutative since then (eq. (4.3.1)) $\Phi_s^{-1}(k, e) = ks(e) = k$. A trivializing section satisfying $s(e) = \tilde{e} \in \tilde{G}$ will be called a *normalized section*. Since this can always be achieved, we shall consider only normalized sections from now on. For a normalized s, we obtain

$$\omega(g, e) = \omega(e, g) = \omega(e, e) = e \in K \quad ; \tag{4.4.3}$$

these factors are called *normalized factors*.

On $\tilde{G}$, the associativity of the group law,

$$(s(g'')s(g'))s(g) = s(g'')(s(g')s(g)) \quad , \tag{4.4.4}$$

implies that $\omega : G \times G \to K$ satisfies

$$\omega(g'', g')\omega(g''g', g) = ([s(g'')]\omega(g', g))\omega(g'', g'g) \tag{4.4.5}$$

where [   ] indicates (see (4.2.1)) the automorphism of K defined by $s(g) \in \tilde{G}$:

$$[s(g)]k := s(g)ks(g)^{-1} \quad \forall k \in K \quad . \tag{4.4.6}$$

(We may omit the bracket (   ) in the r.h.s. of (4.4.5) if we adopt the convention that [   ] acts only on the first element of K appearing at the right.) The previous discussion is summarized in the following

(4.4.1) DEFINITION (*Factor system relative to a section s*)

Relative to the (normalized) trivializing section $s : G \to \tilde{G}$, *a factor system is a mapping* $\omega : G \times G \to K$ *which satisfies* (4.4.3), (4.4.5).

If K is abelian, Aut K = Out K and $[s(g)] = \sigma(g)$ is independent of the trivializing section and the condition (4.4.5) does not make reference to s.

In the above construction there is a degree of arbitrariness: the choice of the trivializing section. Assume now that another one is selected, $u : G \to \tilde{G}$, $u : g \mapsto u(g)$, $u(e) = e$. But then a mapping

$$\gamma : G \to K \quad , \quad \gamma(e) = e \quad , \tag{4.4.7}$$

exists such that

$$u(g) = \gamma(g)s(g) \quad . \tag{4.4.8}$$

The section u now defines a different factor system ω',

$$u(g')u(g) = \omega'(g', g)u(g'g) \quad , \tag{4.4.9}$$

which satisfies (4.4.5) where s is replaced by u. Now, (4.4.8) and (4.4.1) give

$$\gamma(g')([s(g')]\gamma(g))\omega(g', g) = \omega'(g', g)\gamma(g'g) \quad , \tag{4.4.10}$$

i.e.,

$$\omega'(g', g) = \gamma(g')([s(g')]\gamma(g))\omega(g', g)\gamma^{-1}(g'g) \quad . \tag{4.4.11}$$

We thus see that an extension determines a factor system $\omega : G \times G \to K$ up to the equivalence defined by (4.4.11). This expression can be compared with the simpler one (3.2.5), where K was the abelian group $U(1)$ (and thus the ordering of the factors $\gamma(g) \in U(1)$ did not matter) and where $[s(g)] = \sigma(g)$ was the identity automorphism.

4.5 Group law for $\tilde{G}$ in terms of the factor system

Let $\tilde{G}$ be an extension of G by K. As stated, any $\tilde{g} \in \tilde{G}$ can be related to $s(g)$ by writing $\tilde{g} = ks(g)$. We may now represent

$$s(g) \quad \text{by} \quad (1,g) \quad \text{and} \quad ks(g) \quad \text{by} \quad (k,g) \qquad (4.5.1)$$

(i.e., $\Phi^{-1}(1,g) = s(g), \Phi^{-1}(k,g) = ks(g))$; as a set, $\tilde{G} = K \times G$. Then the group law of $\tilde{G}$ is given by

$$k's(g')ks(g) = k'[s(g')]ks(g')s(g) = k'[s(g')]k\omega(g',g)s(g'g) \quad . \qquad (4.5.2)$$

Using the trivialization Φ, the above group law corresponds to writing

$$(k'',g'') = (k',g') *_s (k,g) = (k'[s(g')]k\omega(g',g),g'g) \qquad (4.5.3)$$

where in *this* formula we have written $*_s$ to stress the dependence of the expression (4.5.3) on the trivializing section s (or associated trivialization Φ_s). It is slightly cumbersome to refer to $s(g)$ all the time; it is simpler to omit s in $*_s$ and to write

(a) $s(g) = (1,g)$, provided we do not forget that the elements $\{(1,g) \in \tilde{G} | g \in G\}$ do *not* define a group in general, and that $(1,g)$ is defined through a particular section,

(b) $[s(g)] = (f \circ s)(g)$ (or simply $[s(g)] = f(g)$, for short) for the element of Aut K defined by $g \in G$ again through the section $s : G \to \tilde{G}$ (*not* a homomorphism in general),

$$1 \to K \to \tilde{G} \xrightarrow[\substack{\pi \\ \longleftarrow \\ s}]{} G \to 1 \quad .$$

Then the *group law* of $\tilde{G}$ (4.5.2) reads

$$\tilde{g}'' = \tilde{g}' * \tilde{g} \quad \text{or} \quad (k',g') * (k,g) = (k'f(g')k\,\omega(g',g),g'g) \quad . \qquad (4.5.4)$$

Clearly, the extended group $\tilde{G}$ must be well defined with independence of the particular trivializing section selected. Let now $u(g) = \gamma(g)s(g)$ be another section, and $\tilde{g} = \bar{k}u(g)$. Then, using (4.5.1), (4.5.4), we may write, since $\bar{k}\gamma(g) \in K$,

$$\tilde{g} = \bar{k}u(g) = \bar{k}\gamma(g)s(g) = (\bar{k}\gamma(g),g) \quad , \quad \tilde{g}' = (\bar{k}'\gamma(g'),g') \quad , \qquad (4.5.5)$$

$$\tilde{g}' * \tilde{g} = (\bar{k}'\gamma(g')f(g')(\bar{k}\gamma(g))\omega(g',g) \,, \, g'g) \quad . \qquad (4.5.6)$$

The first part can be rewritten as

$$\bar{k}'\gamma(g')f(g')(\bar{k}\gamma(g))\omega(g',g) = \bar{k}'\gamma(g')(f(g')\bar{k})(f(g')\gamma(g))\omega(g',g)$$
$$= \bar{k}'\gamma(g')(f(g')\bar{k})\gamma(g')^{-1}\gamma(g')(f(g')\gamma(g))\omega(g',g)$$
$$= \bar{k}'(\bar{f}(g')\bar{k})\gamma(g')(f(g')\gamma(g))\omega(g',g)$$

and, using (4.4.10), (4.4.11), as equal to

$$\bar{k}'(\bar{f}(g')\bar{k})\omega'(g',g)\gamma(g'g) \, . \tag{4.5.7}$$

Thus,

$$\tilde{g}'' = \tilde{g}' * \tilde{g} \equiv (\bar{k}''\gamma(g'g), \; g'g) \tag{4.5.8}$$

with

$$\bar{k}'' = \bar{k}'[\bar{f}(g')\bar{k}]\omega'(g',g) \, , \quad \bar{f}(g')\bar{k} = \gamma(g')(f(g')\bar{k})\gamma(g')^{-1} \, . \tag{4.5.9}$$

This means that if we define

$$\bar{f}(g) = [u(g)] = [\gamma(g)s(g)] \tag{4.5.10}$$

or, more precisely, if $\bar{f}(g)$ is $(f \circ u)(g)$ and $u(g)$ is used as a trivializing section to write

$$\tilde{g} = \bar{k}u(g) \quad \text{as} \quad (\bar{k}, g) \quad , \tag{4.5.11}$$

the group law is now given by

$$(\bar{k}', g')(\bar{k}, g) = (\bar{k}'(\bar{f}(g')\bar{k})\omega'(g',g) \, , \; g'g) \tag{4.5.12}$$

where the new ω' is related to ω by (4.4.11). To move from (4.5.12) to (4.5.4) it is sufficient to make the redefinition

$$\bar{k}\gamma(g) = k \quad . \tag{4.5.13}$$

Although the group laws (4.5.4), (4.5.12) use different trivializing sections, they are nevertheless equivalent. Their explicit form differs in the use of ω, ω' (equivalent by (4.4.11)) and in $f(g) = [s(g)]$, $\bar{f}(g) = [u(g)]$, (4.5.10). It is clear that $f(g)$ and $\bar{f}(g)$ differ in an internal automorphism of K ($[\gamma(g)]$); thus, both belong to the same class $\sigma(g)$ in Out K [(4.2.5)].

In the same way, it is not difficult to see that, conversely, a system of automorphisms $F(g)$ of K and a factor system $\omega(g',g) \in K$, $g',g \in G$, define an extension $\tilde{G}$ of G by K. We can state these results in the following form:

(4.5.1) THEOREM

Every extension $\tilde{G}$ of G by K determines a class [(4.4.11)] of factor systems ω. Conversely, a system of automorphisms $F : G \to \text{Aut}K$ and a mapping $\omega : G \times G \to K$ satisfying the properties $F(g')F(g) = [\omega(g',g)]F(g'g)$ and $\omega(g'',g')\omega(g''g',g) = (F(g'')\omega(g',g))\omega(g'',g'g)$ define a unique extension $\tilde{G}$ of G by K. (In the previous discussion, where $\tilde{G}$ was assumed known, $F(g)$ was written $[s(g)]$ (see (b) on preceding page); cf. (4.4.1) and (4.4.5).)

The group law of $\tilde{G}$ is defined by (4.5.4) where, as mentioned, $f(g)$ is an abbreviation for $(f \circ s)(g)$:

$$((f \circ s)g)k \equiv [s(g)]k = s(g)ks(g)^{-1} \quad . \tag{4.5.14}$$

The identity element of $\tilde{G}$ is given by

$$\tilde{e} = (1, e) \, , \tag{4.5.15}$$

where 1 is the unit of K, and the inverse of $\tilde{g} = (k, g)$ by

$$\tilde{g}^{-1} = ((f(g)^{-1}k^{-1})\omega^{-1}(g^{-1}, g), g^{-1}) = (f(g)^{-1}(k^{-1}\omega^{-1}(g, g^{-1})) \, , \, g^{-1}) \quad . \tag{4.5.16}$$

It is not difficult to check that $\tilde{g}^{-1}$ is given by (4.5.16). In so doing, some of the following expressions are useful:

$$\omega(g, g^{-1}) = [s(g)]\omega(g^{-1}, g) \quad , \quad [s(g)]^{-1}\omega(g, g^{-1}) = \omega(g^{-1}, g) \quad , \tag{4.5.17}$$

$$\omega^{-1}(g, g^{-1}) = [s(g)]\omega^{-1}(g^{-1}, g) \quad , \tag{4.5.18}$$

$$\omega(g^{-1}, g) = [s(g^{-1})]\omega(g, g^{-1}) \quad , \quad \omega^{-1}(g^{-1}, g) = [s(g^{-1})]\omega^{-1}(g, g^{-1}) \quad , \tag{4.5.19}$$

$$s(g^{-1})s(g) = \omega(g^{-1}, g) \quad . \tag{4.5.20}$$

They are obtained from (4.4.5) by setting $g'' = g, g' = g^{-1}, g = g$ [(4.5.17)]; $g'' = g^{-1}, g' = g, g = g^{-1}$ [(4.5.19)]. Eq. (4.5.18) follows from (4.5.17) ([$s(g)$] is a homomorphism) and (4.5.20) from (4.4.1) for $g' = g^{-1}$.

Observation. As already remarked $s(g)^{-1} \neq s(g^{-1})$, s being a section, *not* a homomorphism in general.

We now face the problem of finding *when* a G-kernel (K, σ) is extendible. In section 5.2 we shall see that, if $K = A$ is abelian, there is always a solution, the semidirect product $\tilde{G} = A \circledS G$. In the general case there may be an obstruction and a solution may not exist. We shall give in section 7.2 a necessary and sufficient condition for a G-kernel (K, σ) to be extendible, but firstly we shall discuss in the next chapter the notion of cohomology of groups and the problem of extending G by an *abelian* kernel A and in chapter 6 the cohomology of Lie algebras.

Bibliographical notes for chapter 4

The group extension problem was first considered by Schreier and by Baer in the 1920s. A curious early reference is due to the computer science pioneer Turing (1938). General book references on group cohomology and extensions are Hall (1959), Kurosh (1960), Michel (1964), Weiss (1969), and Brown (1982). The basic reference is Eilenberg and Mac Lane (1947a).

For extensions of topological and Lie groups see, in particular, Calabi (1951) and Hochschild (1951), which contains the proof of Theorem 4.3.1.

The problem of extending the Poincaré group and its (impossible) mixing with internal symmetries was extensively discussed in the 1960s. See, *e.g.*, O'Raifeartaigh (1965), Dyson (1966) (which contains a collection of relevant papers with an introduction), Coleman and Mandula (1967), and also Haag *et al.* (1975) in the context of supersymmetry.

The most general setting to discuss the cohomology theory of Lie groups and Lie algebras is that of homology theory. Two classical (and advanced) books in this subject are those of Cartan and Eilenberg (1956) and Mac Lane (1963); see also the texts of Hilton and Stammbach (1970) and of Dubrovin *et al.* (1990).

5

Cohomology groups of a group G and extensions by an abelian kernel

In this chapter the concepts of cocycles and coboundaries for the group cohomology with values in an abelian group A are introduced. The relation between the second cohomology groups and abelian extensions is given, with an analysis of the different possibilities among which the semidirect and central extensions are the most common in physics. This general definition is applied to the case in which the abelian group is the space of functions $\mathscr{F}(M)$ on a manifold M. The reason for considering this construction is that it will be relevant in chapters 8 and 10. The chapter ends by exhibiting the extension nature of superspace considered as the supertranslation group; again, this section may be omitted if so desired.

5.1 Cohomology of groups

Let G be a group, and A an *abelian* group, written additively, on which G operates through the homomorphism $\sigma : G \to \mathrm{Aut}\, A = \mathrm{Out}\, A$ (since A is abelian, $\mathrm{Int}\, A$ is reduced to the trivial automorphism).

(5.1.1) DEFINITION (*n-cochains on* G)

A mapping $\alpha_n : G \times ... \times G \to A$ *is an n-cochain,*

$$\alpha_n : (g_1, ..., g_n) \mapsto \alpha_n(g_1, ..., g_n) \in A \quad . \tag{5.1.1}$$

The cochains (5.1.1) are sometimes called 'non-homogeneous' *cochains.*

The n-cochains form an abelian group, which will be denoted $C^n(G, A)$, the addition being defined by the pointwise addition of functions, *i.e.* by

$$(\alpha'_n + \alpha_n)(g_1, ..., g_n) = \alpha'_n(g_1, ..., g_n) + \alpha_n(g_1, ..., g_n) \quad . \tag{5.1.2}$$

$C^0(G, A)$ is defined by the zero-dimensional cochains, *i.e.* the constant mappings α_0; thus $C^0(G, A) = A$.

The operator $\delta : C^n \to C^{n+1}$ (the *coboundary* operator) can be defined in two ways, depending on the way the action $\sigma(g) \in \mathrm{Aut}\, A$ of the elements g of G is defined on A. When the action is a *left action*, $\sigma(g_1)\sigma(g_2)=\sigma(g_1 g_2)$, δ is given by

(5.1.2a) DEFINITION (*Coboundary operator for the left action*)

$$(\delta\alpha_n)(g_1, ..., g_n, g_{n+1}) := \sigma(g_1)\alpha_n(g_2, ..., g_n, g_{n+1})+$$

$$+ \sum_{i=1}^{n}(-1)^i\alpha_n(g_1, ..., g_{i-1}, g_i g_{i+1}, g_{i+2}, ..., g_{n+1}) \quad (5.1.3)$$

$$+ (-1)^{n+1}\alpha_n(g_1, ..., g_n) \quad .$$

When the action σ is a *right action*, δ is given by

(5.1.2b) DEFINITION (*Coboundary operator for the right action*)

$$(\delta\alpha_n)(g_1, ..., g_n, g_{n+1}) := (-1)^{n+1}\alpha_n(g_2, ..., g_n, g_{n+1})$$

$$+ \sum_{i=1}^{n}(-1)^{i+n+1}\alpha_n(g_1, ..., g_{i-1}, g_i g_{i+1}, g_{i+2}, ..., g_{n+1})$$

$$+ \alpha_n(g_1, ..., g_n)\sigma(g_{n+1}) .$$

$$(5.1.4)$$

It is also possible to write $\sigma(g)$ at the left in (5.1.4), but on the condition that it is remembered that $\sigma(g')(\sigma(g)\alpha_n(...)) = \sigma(gg')\alpha_n(...)$.

The first few cases of the left action (5.1.3) are

$$(\delta\alpha_0)(g) = \sigma(g)\alpha_0 - \alpha_0 \quad ;$$

$$(\delta\alpha_1)(g_1, g_2) = \sigma(g_1)\alpha_1(g_2) - \alpha_1(g_1 g_2) + \alpha_1(g_1) \quad ;$$

$$(\delta\alpha_2)(g_1, g_2, g_3) = \sigma(g_1)\alpha_2(g_2, g_3) + \alpha_2(g_1, g_2 g_3) - \alpha_2(g_1 g_2, g_3)$$

$$- \alpha_2(g_1, g_2) \quad ;$$

$$(\delta\alpha_3)(g_1, g_2, g_3, g_4) = \sigma(g_1)\alpha_3(g_2, g_3, g_4) - \alpha_3(g_1 g_2, g_3, g_4) + \alpha_3(g_1, g_2 g_3, g_4)$$

$$- \alpha_3(g_1, g_2, g_3 g_4) + \alpha_3(g_1, g_2, g_3) \quad .$$

$$(5.1.5)$$

The first few cases of the right action (5.1.4) are

$$(\delta\alpha_0)(g) = \alpha_0\sigma(g) - \alpha_0 \;;$$
$$(\delta\alpha_1)(g_1, g_2) = \alpha_1(g_2) - \alpha_1(g_1 g_2) + \alpha_1(g_1)\sigma(g_2) \;;$$
$$(\delta\alpha_2)(g_1, g_2, g_3) = -\alpha_2(g_2, g_3) + \alpha_2(g_1 g_2, g_3) - \alpha(g_1, g_2 g_3)$$
$$+ \alpha_2(g_1, g_2)\sigma(g_3) \;;$$
$$(\delta\alpha_3)(g_1, g_2, g_3, g_4) = \alpha_3(g_2, g_3, g_4) - \alpha_3(g_1 g_2, g_3, g_4) + \alpha_3(g_1, g_2 g_3, g_4)$$
$$- \alpha_3(g_1, g_2, g_3 g_4) + \alpha_3(g_1, g_2, g_3)\sigma(g_4) \;.$$

$$(5.1.6)$$

From either definition of δ, it follows trivially that

$$\delta(\alpha + \alpha') = \delta\alpha + \delta\alpha' \quad . \tag{5.1.7}$$

It also follows from the above definitions, and it is easily checked on the above examples, that $\delta^2 = 0$. We shall give below the general proof for the operator δ in (5.1.3); the proof for (5.1.4) is analogous.

(5.1.1) THEOREM

The operator δ satisfies

$$\delta \circ \delta = 0 \,. \tag{5.1.8}$$

Proof : Let δ be as in (5.1.3) and let α_n be an n-cochain. The action of δ on the $(n+1)$-cochain $\delta\alpha_n$ can be written, using the definition, as

$$(\delta(\delta\alpha_n))(g_1, ..., g_{n+2}) = \sigma(g_1)(\delta\alpha_n)(g_2, ..., g_{n+2}) +$$

$$\sum_{i=1}^{n+1}(-1)^i(\delta\alpha_n)(g_1, ..., g_{i-1}, g_i g_{i+1}, g_{i+2}, ..., g_{n+2}) + (-1)^{n+2}(\delta\alpha_n)(g_1, ..., g_{n+1}) \,.$$

$$(5.1.9)$$

Then, again using (5.1.3) in (5.1.9) we obtain

$$(\delta(\delta\alpha_n))(g_1, ..., g_{n+2}) = \sigma(g_1)\sigma(g_2)\alpha_n(g_3, ..., g_{n+2})$$

$$+ \sigma(g_1)\sum_{i=2}^{n+1}(-1)^{i-1}\alpha_n(g_2, ..., g_{i-1}, g_i g_{i+1}, g_{i+2}, ..., g_{n+2})$$

$$+ (-1)^{n+1}\sigma(g_1)\alpha_n(g_2, ..., g_{n+1}) - \sigma(g_1 g_2)\alpha_n(g_3, ..., g_{n+2})$$

$$+ \sum_{i=2}^{n+1}(-1)^i\sigma(g_1)\alpha_n(g_2, ..., g_{i-1}, g_i g_{i+1}, g_{i+2}, ..., g_{n+2})$$

$$+ \sum_{i=1}^{n}(-1)^i(-1)^{n+1}\alpha_n(g_1, ..., g_{i-1}, g_i g_{i+1}, g_{i+2}, ..., g_{n+1}) + \alpha_n(g_1, ..., g_n)$$

$$+ \sum_{i=3}^{n+1}\sum_{l=1}^{i-2}(-1)^{i+l}\alpha_n(g_1,...,g_{l-1},g_lg_{l+1},g_{l+2},...,g_{i-1},g_ig_{i+1},g_{i+2},...,g_{n+2})$$

$$+ \sum_{i=2}^{n+1}(-1)^i(-1)^{i-1}\alpha_n(g_1,...,g_{i-2},g_{i-1}g_ig_{i+1},g_{i+2},...,g_{n+2})$$

$$+ \sum_{i=1}^{n}(-1)^i(-1)^i\alpha_n(g_1,...,g_{i-1},g_ig_{i+1}g_{i+2},g_{i+3},...,g_{n+2})$$

$$+ \sum_{i=1}^{n-1}\sum_{l=i+2}^{n+1}(-1)^{i+l-1}\alpha_n(g_1,...,g_{i-1},g_ig_{i+1},g_{i+2},...,g_{l-1},g_lg_{l+1},g_{l+2},...,g_{n+2})$$

$$+ (-1)^{n+2}\sigma(g_1)\alpha_n(g_2,...,g_{n+1})$$

$$+ (-1)^{n+2}\sum_{i=1}^{n}(-1)^i\alpha_n(g_1,...,g_{i-1},g_ig_{i+1},g_{i+2},...,g_{n+1})$$

$$+ (-1)^{n+2}(-1)^{n+1}\alpha_n(g_1,...,g_n) \ .$$

(5.1.10)

There are fourteen terms in (5.1.10); they all add to zero in pairs. The cancellations of the first with the fourth, the second with the fifth, the third with the twelfth, the sixth with the thirteenth, the seventh with the fourteenth, and the ninth with the tenth are trivial; that of the eighth with the eleventh follows from the equality

$$\sum_{i=3}^{n+1}\sum_{l=1}^{i-2}(-1)^{i+l}\alpha_n(g_1,...,g_{l-1},g_lg_{l+1},g_{l+2},...,g_{i-1},g_ig_{i+1},g_{i+2},...,g_{n+2})$$

$$= \sum_{l=1}^{n-1}\sum_{i=l+2}^{n+1}(-1)^{i+l}\alpha_n(g_1,...,g_{l-1},g_lg_{l+1},g_{l+2},...,g_{i-1},g_ig_{i+1},g_{i+2},...,g_{n+2}) \ ,$$

which is the result of the change of limits in the sums when their order is changed. In this way, l goes from 1 to the maximum value of $i-2$, which is $n+1-2 = n-1$ and, given l, the index i (with an upper limit $n+1$) is always greater than or equal to $l+2$ because from the r.h.s. $l \le i-2$, q.e.d.

Due to the property (5.1.7), the following sequence (*not* an exact one) of abelian groups may be defined:

$$0 \xrightarrow{\delta_{-1}} C^0 \xrightarrow{\delta_0} C^1 \xrightarrow{\delta_1} C^2 \to \cdots \xrightarrow{\delta_n} C^{n+1} \to \cdots \ . \qquad (5.1.11)$$

Since $\delta^2 = 0$,

$$\text{range}\,(\delta_{n+1}\circ\delta_n) = 0, \quad \text{range}\,\delta_n \subset \ker\delta_{n+1} \ , \qquad (5.1.12)$$

which again motivates the definitions

$$Z_\sigma^n := \ker \delta_n \equiv \{\text{cocycles}\} \quad, \tag{5.1.13}$$
$$B_\sigma^n := \text{range } \delta_{n-1} \equiv \{\text{coboundaries}\} \quad.$$

Pictorially, we have

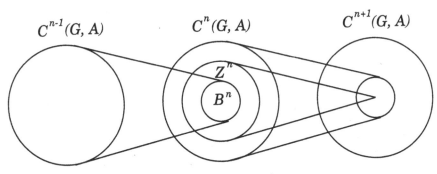

$C^{n-1}(G, A)$ $C^n(G, A)$ $C^{n+1}(G, A)$

Z^n

B^n

Figure 5.1.1

Both B_σ^n and Z_σ^n are subgroups of $C^n(G, A)$ (for $n = 0$, $B^0 = 0$ and the above is still true); the general picture is similar (in fact, the structure is identical) to the de Rham sequence (eq. (2.10.1)) and cohomology which were discussed in section 1.5. Obviously $B_\sigma^n \subset Z_\sigma^n$. This leads to the following

(5.1.3) DEFINITION (*Cohomology groups on G*)

The quotient group

$$H_\sigma^n(G, A) := Z_\sigma^n(G, A) / B_\sigma^n(G, A) \tag{5.1.14}$$

is called the n-th cohomology group of G with values on the abelian group A.

By construction, the different cohomology groups measure the lack of exactness of the sequence (5.1.11): $Z_\sigma^n = B_\sigma^n \Rightarrow H_\sigma^n = 0 \Rightarrow \ker \delta_n = \text{range } \delta_{n-1}$.

Simple cases for Definition 5.1.2a

(1) $H_\sigma^0(G, A)$
A zero-cochain $\alpha_0 \in C^0(G, A)$ is an element a of A. By definition, $B^0(C, A) = 0$. The cochain α_0 is a zero-cocycle, $\alpha_0 \in Z_\sigma^0$, if $\delta \alpha_0 = 0$. This implies

$$(\delta \alpha_0)(g) = \sigma(g)\alpha_0 - \alpha_0 = 0 \Rightarrow \sigma(g)a - a = 0 \quad \forall g \in G \tag{5.1.15}$$

by identifying the constant mapping α_0 with the corresponding element in A. Thus, $H^0_\sigma(G, A)$ is the subgroup of A determined by the fixed invariant points under the action of G defined by σ.

(2) $H^1_\sigma(G, A)$

Let $\alpha_1 \in C^1(G, A)$. The cochain α_1 will belong to $Z^1(G, A)$ if

$$(\delta\alpha_1)(g_1, g_2) = \sigma(g_1)\alpha_1(g_2) - \alpha_1(g_1 g_2) + \alpha_1(g_1) = 0 \quad \forall g_1, g_2 \; ,$$

i.e.,

$$\alpha_1(g_1 g_2) = \sigma(g_1)\alpha_1(g_2) + \alpha_1(g_1) \quad . \tag{5.1.16}$$

Such mappings are known in algebra as *crossed homomorphisms*. Thus,

$$Z^1_\sigma(G, A) = \{\text{crossed homomorphisms of } G \text{ into } A\} \quad . \tag{5.1.17}$$

α_1 will be a 1-coboundary if there exists an element $\alpha_0 = a \in A$ such that

$$\alpha_1(g) = (\delta a)(g) = \sigma(g)a - a \quad . \tag{5.1.18}$$

These particular crossed homomorphisms are called *principal homomorphisms*. Thus,

$$B^1_\sigma(G, A) = \{\text{principal homomorphisms of } G \text{ into } A\} \quad , \tag{5.1.19}$$

and $H^1(G, A)$ *is the group of crossed homomorphisms of* G *into* A *modulo the principal homomorphisms*.

(3) $H^2_\sigma(G, A)$

$\alpha_2 \in C^2(G, A)$ is a two-cocycle if $(\delta\alpha_2)(g_1, g_2, g_3) = 0 \quad \forall g_1, g_2, g_3 \in G$, i.e.,

$$\sigma(g_1)\alpha_2(g_2, g_3) + \alpha_2(g_1, g_2 g_3) = \alpha_2(g_1 g_2, g_3) + \alpha_2(g_1, g_2) \quad . \tag{5.1.20}$$

α_2 will be a two-coboundary if it is generated by a one-cochain,

$$\alpha_2(g_1, g_2) = (\delta\alpha_1)(g_1, g_2) = \sigma(g_1)\alpha_1(g_2) - \alpha_1(g_1 g_2) + \alpha_1(g_1) \quad . \tag{5.1.21}$$

Thus, comparing (5.1.20), (5.1.21) with (4.4.5), (4.4.11) for $(K, *) = (A, +)$, we see that the elements of the second cohomology group $H^2_\sigma(G, A)$ *characterize the extensions* $\tilde{G}$ *of the group* G *by the abelian group* A *for the given action* σ *of* G *on* A (written $[s(g)]$ in section 4.4 and in this section simply denoted by $\sigma(g)$; since A is abelian, $[s(g)] = [u(g)]$ although $s(g) \neq u(g)$).

Is there a significance to the higher cohomology groups? We already saw in section 3.8 that a three-cocycle could be associated with a breaking of the associativity property. We shall see in chapter 7 that the three-cocycle also appears as an obstruction to the extensions by non-abelian kernels (see Theorem 7.2.1). We shall also mention in section 7.5 another situation where the three- and higher order cocycles appear.

5.2 Extensions $\tilde{G}$ of G by an abelian group A

Group extensions by an abelian group have been classified in the previous section by the elements of $H^2_\sigma(G, A)$ where the action $\sigma : G \to$ Out $A =$ Aut A has been indicated explicitly by the subscript σ in H^2_σ. The group elements $\tilde{g} \in \tilde{G}$ are written $\tilde{g} = (a, g)$, $a \in A$, and the group law, which will be written additively for the A part, is given by

$$(a', g')(a, g) = (a' + \sigma(g')a + \xi(g', g), \ g'g) \ , \tag{5.2.1}$$

$$(a, g)^{-1} = (-\sigma(g^{-1})a - \sigma(g^{-1})\xi(g, g^{-1}), \ g^{-1})$$
$$= (-\sigma(g^{-1})a - \xi(g^{-1}, g), \ g^{-1}) \ , \tag{5.2.2}$$

$$(0, e) = \tilde{e} \ , \quad \xi(g, g^{-1}) = \sigma(g)\xi(g^{-1}, g) \ . \tag{5.2.3}$$

Two two-cocycles ξ' and ξ are equivalent (and define the same extension) if

$$\xi'(g', g) = \xi(g', g) + \xi_{cob}(g', g) \tag{5.2.4}$$

where ξ_{cob} is the two-coboundary generated by an $\eta : G \to A$,

$$\xi_{cob}(g', g) = (\delta\eta)(g', g) = \sigma(g')\eta(g) - \eta(g'g) + \eta(g') \ . \tag{5.2.5}$$

A two-coboundary is also called a *trivial* two-cocycle.

Three cases are of particular interest:

(a) *The action σ is trivial* ($\sigma(g) = e \in$ Aut A), *but ξ is not.*
Then the group law of $\tilde{G}$ is

$$(a', g')(a, g) = (a' + a + \xi(g', g), \ g'g) \tag{5.2.6}$$

and the extension is central (A belongs to the centre of $\tilde{G}$). Central extensions are given by $H^2_0(G, A)$; this is the case in section 3.2. It is simple to see the rôle of the coboundary in (5.2.6); it corresponds to a different characterization of the central parameter (see section 4.5). If we take $\tilde{a} = a - \eta(g)$, the group law takes the form

$$(\tilde{a}'' + \eta(g''), g'') = (\tilde{a}' + \eta(g') + \tilde{a} + \eta(g) + \xi(g', g), g'g) \ . \tag{5.2.7}$$

Hence,

$$(\tilde{a}'', g'') = (\tilde{a}', g')(\tilde{a}, g) = (\tilde{a}' + \tilde{a} + \tilde{\xi}(g', g), g'g) \tag{5.2.8}$$

where $\tilde{\xi}(g', g) = \xi(g', g) + \eta(g') + \eta(g) - \eta(g'g) = \xi(g', g) + \xi_{cob}(g', g)$. If it is possible to find an $\eta(g)$ such that the group law reduces to $(\tilde{a}', g')(\tilde{a}, g) = (\tilde{a}' + \tilde{a}, g'g)$, the group $\tilde{G}$ has the structure of a direct product, $\tilde{G} = A \otimes G$ (see (c) below).

(b) *The two-cocycle is trivial, but σ is not.*

Given the G-kernel (A, σ) there is always an extension, the semidirect product, and then G is also a subgroup of $\tilde{G}$,

$$(a', g')\,(a, g) = (a' + \sigma(g')a\,,\ g'g). \tag{5.2.9}$$

In terms of exact sequences,

$$1 \to A \to \tilde{G} \underset{s}{\overset{\pi}{\rightleftarrows}} G \overset{\rightarrow}{\leftarrow} 1, \quad \tilde{G} = A \circledS G \quad . \tag{5.2.10}$$

Such an extension is called a *splitting* extension; s $(\pi \circ s = 1_{|G})$ is now a homomorphism which injects G into $\tilde{G}$. As we show below, $H_\sigma^1(G, A)$ characterizes the different splittings.

Let us assume that (5.2.9) provides the canonical splitting, $s(g) = (0, g)$, where $\{s(g)|g \in G\}$ is the image of G as a *subgroup* of $\tilde{G}$. To look for the possible splittings, we look for all the injections

$$s' : G \to \tilde{G} \quad, \quad s' : g \mapsto (\alpha(g)\,,\ g) \quad, \tag{5.2.11}$$

where $\alpha \in C^1(G, A)$, which are group homomorphisms. This implies

$$s'(g'g) = (\alpha(g'g),\ g'g) = (\alpha(g') + \sigma(g')\alpha(g),\ g'g) \tag{5.2.12}$$

and, by (5.1.16), that α is a crossed homomorphism (also called a *derivation*), $\alpha \in H_\sigma^1(G, A)$. Two splittings s and s' are conjugate to each other if there is an element $a \in A$ such that

$$as'(g)a^{-1} = (a, e)\,(\alpha'(g),\ g)\,(-a,\ e) = s(g) = (\alpha(g),\ g) \tag{5.2.13}$$

or

$$\alpha'(g) - \alpha(g) = \sigma(g)a - a \quad . \tag{5.2.14}$$

This means that α' and α differ in a principal homomorphism (*principal derivation*). Thus, the A-classes of splittings of a splitting extension $\tilde{G} = A \circledS G$ are characterized by the elements of $H_\sigma^1(G, A)$.

The discussion of this case 5.2(b) contains the proof of the following

(5.2.1) PROPOSITION

The following statements are equivalent:

1) *the exact sequence* $1 \to A \overset{i}{\to} \tilde{G} \overset{\pi}{\to} G \to 1$ *splits;*
2) $\tilde{G}$ *contains a subgroup G′ which is mapped by π isomorphically onto G, i.e. G can be injected into $\tilde{G}$ and* $G' \cap i(A) = e;$

3) $\tilde{G}$ contains a subgroup G_0 such that any element $\tilde{g} \in \tilde{G}$ has a unique expression in the form $i(a)g_0$, $g_0 \in G_0$ [*]; and

4) the sequence 1) is equivalent to the semidirect extension $1 \to A \xrightarrow{i'} A \circledS G \xrightarrow{\pi'} G \to 1$ and i', π' are the canonical injection and projection homomorphisms respectively.

Finally, the simplest case is where

(c) *Both ξ and σ are trivial.*
In this case, the lower sequence in (5.2.10) also ends,

$$1 \underset{\leftarrow}{\overset{\rightarrow}{\rightleftarrows}} A \underset{\leftarrow}{\overset{\rightarrow}{\rightleftarrows}} \tilde{G} \underset{\leftarrow}{\overset{\rightarrow}{\rightleftarrows}} G \underset{\leftarrow}{\overset{\rightarrow}{\rightleftarrows}} 1 \quad,$$

and $\tilde{G}$ is the *direct product* $\tilde{G} = A \otimes G$.

5.3 An example from supergroup theory: superspace as a group extension

Let us consider here (with the remarks made at the beginning of section 1.10) the (abelian) group τ given by the group law

$$\theta''^{\alpha} = \theta'^{\alpha} + \theta^{\alpha} , \quad \alpha = 1, 2, 3, 4 , \quad (5.3.1)$$

where the θ's are four-dimensional *anticommuting* Majorana spinors. It is trivial to check that the four-vector

$$\xi^{\mu}(\theta', \theta) = i\bar{\theta}'\gamma^{\mu}\theta , \quad \mu = 0, 1, 2, 3 \quad (5.3.2)$$

(see section 1.10) satisfies the two-cocycle condition (5.1.20) for a trivial action σ of τ on Tr_4 since

$$i\bar{\theta}_2\gamma^{\mu}\theta_3 + i\bar{\theta}_1\gamma^{\mu}(\theta_2 + \theta_3) = i(\bar{\theta}_1 + \bar{\theta}_2)\gamma^{\mu}\theta_3 + i\bar{\theta}_1\gamma^{\mu}\theta_2 .$$

It can also be seen that $\xi^{\mu}(\theta', \theta)$ is real[†], since $(i\bar{\theta}'\gamma^{\mu}\theta)^{\dagger} = -i\theta^{\dagger}\gamma^{\mu\dagger}\gamma^0\theta' = -i\bar{\theta}\gamma^{\mu}\theta'$ and it is a property of the vector bilinear constructed out of two Grassmann Majorana spinors that $\bar{\theta}'\gamma^{\mu}\theta = -\bar{\theta}\gamma^{\mu}\theta'$. Thus, the mapping $\xi^{\mu} : \tau \times \tau \to Tr_4$, $\xi^{\mu}(\theta', \theta) = i\bar{\theta}'\gamma^{\mu}\theta$ is a 2-cocycle with values on the four-dimensional translation group ($\approx R^4$). Since for an anticommuting Majorana spinor $\bar{\theta}\gamma^{\mu}\theta$ is zero, ξ^{μ} cannot be generated by a one-cochain $\eta^{\mu}(\theta)$ and thus ξ^{μ} is not a two-coboundary. As a result, ξ^{μ} determines

---

[*] e.g., if $g_0 = (\alpha(g), g)$, (5.2.12) leads to $\{(a', e)(\alpha(g'), g')\} * \{(a, e)(\alpha(g), g)\} = \{(a' + \sigma(g')a, e)(\alpha(g'g), g'g)\}$ as it should.
[†] Strictly speaking the components of $\xi^{\mu}(\theta', \theta)$ are not real numbers, since for a quantity made out of Grassmann components there is always a finite power which gives zero. Rather, they are real even (commuting) elements (supernumbers) of a Grassmann algebra.

a central extension [(5.2.6)] of the group of the 'odd' translations by the abelian group of the ordinary translations. This extension is the super-translation group, the group law of which was given in (1.10.1) and which now appears as a consequence of (5.2.6). It also exhibits in another context the fundamental character of fermions: the odd translations 'come first'. We can summarize the above and generalize it to any D-dimensional spacetime by saying that superspace is tied to the existence of an *anti-symmetric* vector bilinear which determines a Tr_D-valued non-trivial two-cocycle: *supersymmetry is the result of the non-trivial cohomology of the odd supertranslation group.*

The extension of the odd supertranslation group may be defined by any of the equivalent non-trivial two-cocycles; it is always possible to add a two-coboundary to the particular ξ^μ given in (5.3.2). In so doing, the form of the group law and that of the generators of the algebra are modified. The 'chiral' and 'antichiral' form of the generators and covariant derivatives in supersymmetry can be obtained by selecting a suitable two-cocycle ξ^μ in its equivalence *class*. Notice, by the way, that a multiplicative parameter could be introduced in (5.3.2) and that accordingly the cocycle ξ^μ given above also *defines* the *dimensions* of θ. Taking $c = 1$, eq. (1.10.1) implies the customary assignment $[\theta] = L^{1/2}$, but one might consider, in principle, introducing a dimensional constant multiplying the two-cocycle $\xi^\mu(\theta', \theta)$.

The presence of a *central charge* in supersymmetry is also the result of a central extension. It is simple to see, in the present context, why the group (1.10.1) (*one* Majorana spinor, or $N = 1$ ($D = 4$) supersymmetry) does not admit a central charge. Due to Lorentz covariance, $\xi(\theta', \theta)$ has to be of the form $\bar\theta'\theta$ or $i\bar\theta'\gamma^5\theta$ but, since both bilinears are symmetric, they turn out to be the two-coboundaries generated by the one-cochains $\frac{1}{2}(\bar\theta\theta)$ and $\frac{i}{2}(\bar\theta\gamma^5\theta)$ respectively. At least *two* Majorana spinors ($N=2$-extended[*] superspace, Σ_2) or, alternatively, one Dirac (not selfconjugate, complex) spinor is needed to have a central charge. In this four-dimensional spacetime case, *and with θ now representing a complex spinor*, the extended group law $g'' = g'g$, $g = (x^\mu, \theta^\alpha, \bar\theta^\alpha, \phi)$ is given by (cf. (1.10.1))

$$x''^\mu = x'^\mu + x^\mu + i(\bar\theta'\gamma^\mu\theta - \bar\theta\gamma^\mu\theta') \ ,$$
$$\theta'' = \theta' + \theta \ , \qquad \bar\theta'' = \bar\theta' + \bar\theta \ , \qquad (5.3.3)$$
$$\phi'' = \phi' + \phi + im(\bar\theta'\theta - \bar\theta\theta') \ .$$

---

[*] The expression '*N-extended superspace*' is used in supersymmetry to indicate that the Grassmann sector contains N independent Majorana spinors. In the present context, the name '*N-enlarged* superspace' would be less prone to confusion.

Once (5.3.3) is added to (1.10.1), (1.10.1) is no longer a (sub)group. The relative minus sign in (5.3.3) is crucial; $\xi = \bar{\theta}'\theta + \bar{\theta}\theta'$ is the two-coboundary generated by $\eta(g) = \bar{\theta}\theta$ and, as such, it can be removed as was shown in section 5.2(a). The reader will notice the appearance of the parameter m (mass) in the last line of (5.3.3); m characterizes the elements of $H_0^2(\Sigma_2, U(1))$ much in the same way as it labelled those of $H_0^2(G, U(1))$ for the Galilei group (section 3.4). The reader may check this parallel further by computing the generators $Q_{(\alpha)}^i$, $i = 1, 2$, and finding the corresponding anticommutation relations.

Finally, to conclude this section, we mention that the super-Poincaré group can be described in two ways. The first emphasizes its similarity to the ordinary Poincaré group, which is the semidirect extension of the Lorentz group by the Minkowski space translations. In it, the super-Poincaré group is described as the semidirect extension [(5.2.9)] of the Lorentz group by the supertranslation group Σ. Let $(x^\mu, \theta^\alpha) \in \Sigma$ and $A \in SL(2, C)$, the universal covering group of the Lorentz group. Clearly, A determines an automorphism of Σ since, if $\Lambda(A)^\mu_{.\nu}$ and $S(A)^\alpha_{.\beta}$ are the matrices associated with A corresponding to the vector $(\frac{1}{2}, \frac{1}{2})$ and Dirac $(0, \frac{1}{2}) \oplus (\frac{1}{2}, 0)$ four-dimensional representations of $SL(2, C)$, we can define

$$\sigma(A) : (x^\mu, \theta^\alpha) \mapsto (\Lambda(A)^\mu_{.\nu} x^\nu, S(A)^\alpha_{.\beta} \theta^\beta). \tag{5.3.4}$$

Then the super-Poincaré group law is given by

$$[(x', \theta'), A'][(x, \theta), A] = [(x' + \Lambda(A')x + i\bar{\theta}'\gamma^\mu S(A')\theta, \theta' + S(A')\theta), A'A]. \tag{5.3.5}$$

The other way of obtaining the super-Poincaré group is by extending the group composed of $SL(2, C)$ and the odd translations, itself a semidirect product with group law $(\theta', A')(\theta, A) = (\theta' + S(A')\theta, A'A)$, by the ordinary spacetime translations. The super-Poincaré group law (5.3.5) appears in this case as an example of the general case (5.2.1), where $a = x \in Tr_4$, $g = (\theta, A)$ and $\sigma(g) = \Lambda(A)$. The two-cocycle is given by $\xi^\mu(g', g) = i\bar{\theta}'\gamma^\mu S(A')\theta$ and satisfies

$$\xi^\mu(g', g) + \xi^\mu(g'g, g'') = \sigma(g')^\mu_{.\nu}\xi^\nu(g, g'') + \xi^\mu(g', gg'') \tag{5.3.6}$$

by virtue of the well-known relations $\gamma^0 S^\dagger(A)\gamma^0 = S(A)^{-1}$, $S(A)^{-1}\gamma^\mu S(A) = \Lambda(A)^\mu_{.\nu}\gamma^\nu$, i.e. the two-cocycle condition $\delta\xi = 0$ (third equation in (5.1.5)).

5.4 $\mathcal{F}(M)$-valued cochains: cohomology induced by the action of G on a manifold M^m

In the previous sections we have considered G-cochains valued in a finite-dimensional abelian group A. To conclude this chapter we shall consider a different situation and discuss the cohomology which arises when the group acts on a manifold M – not necessarily a group itself – and the cochains are functions on M, i.e., elements[*] of $\mathcal{F}(M)$. With the natural addition of functions,

$$(f_1 + f_2) : x \mapsto (f_1 + f_2)(x) = f_1(x) + f_2(x) , \quad f_1, f_2 \in \mathcal{F}(M) , \quad (5.4.1)$$

$\mathcal{F}(M)$ is endowed with an abelian group structure. Thus, we have

(5.4.1) DEFINITION ($\mathcal{F}(M)$-*valued n-cochains on G*)

The $\mathcal{F}(M)$-*valued n-cochains are mappings*

$$\Omega_n : G \times \overset{n}{\cdots} \times G \to \mathcal{F}(M) \quad , \quad \Omega_n : (g_1, ..., g_n) \mapsto \Omega_n(\cdot \; ; g_1, ..., g_n) \quad , \quad (5.4.2)$$

where $\Omega_n(\cdot \; ; g_1, ..., g_n)$ *is the function*

$$\Omega_n(\cdot \; ; g_1, ..., g_n) : M \to R \quad , \quad \Omega_n(\cdot \; ; g_1, ..., g_n) : x \mapsto \Omega_n(x; g_1, ..., g_n) \quad .$$
$$(5.4.3)$$

The sum of n-cochains is defined in the natural way (cf. (5.1.2)*),*

$$(\Omega'_n + \Omega_n)(\cdot \; ; g_1, ..., g_n) = \Omega'_n(\cdot \; ; g_1, ..., g_n) + \Omega_n(\cdot \; ; g_1, ..., g_n) , \quad (5.4.4)$$

since the addition of elements of $\mathcal{F}(M)$ *defines the sum* $\Omega'_n + \Omega_n$. *Thus,* $C^n(G, \mathcal{F}(M))$ *is an abelian group;* $C^0(G, \mathcal{F}(M)) = \mathcal{F}(M)$.

To define a cohomology structure a coboundary operator δ : $C^n(G, \mathcal{F}(M)) \to C^{n+1}(G, \mathcal{F}(M))$ is needed. This, in turn, requires defining an action Φ of G on $\mathcal{F}(M)$, i.e. a mapping $\Phi : G \times \mathcal{F}(M) \to \mathcal{F}(M)$ or $\Phi(g) : \mathcal{F}(M) \to \mathcal{F}(M)$ with a certain composition law. The natural action of G on $f \in \mathcal{F}(M)$, $\Phi(g) : f \mapsto (\Phi(g)f) \equiv \Phi(g, f) = f'$ is the one defined by any of the equivalent conditions

$$(\Phi(g)f)(gx) = f(x), \quad (\Phi(g)f)(x) = f(g^{-1}x), \quad (\Phi(g^{-1})f)(x) = f(gx) .$$
$$(5.4.5)$$

From this it follows that $\Phi(g')(\Phi(g)f)(x) = (\Phi(g)f)(g'^{-1}x) = f(g^{-1}g'^{-1}x) = f((g'g)^{-1}x) = (\Phi(g'g)f)(x)$ and thus Φ defined as above is a *left action*.

---

[*] Although here M^m or M for short will be a finite-dimensional or m-manifold and G a finite-dimensional Lie group, this assumption may be relaxed; it is then possible to follow the same steps and develop, at least formally, the corresponding cohomology theory for the infinite-dimensional case. This will be used in chapter 10, where the r-dimensional Lie group G will be replaced by the infinite-dimensional gauge group, the m-dimensional manifold M by the set $\mathscr{A}$ of gauge fields $A_\mu^i(x)$ ($\mu = 0, 1, ..., m-1$, $i = 1, ..., r$) and the module $\mathcal{F}(M)$ of functions $f(x)$ on Minkowski space by the space $\mathcal{F}[\mathscr{A}]$ of functionals $F[A]$ of the gauge fields.

Alternatively we may use the notation $(\sigma(g)f)(x) = f(x^g)^*$. With this notation, σ defines a *left action* on the cochains since $\sigma(g')\sigma(g)=\sigma(g'g)$. Explicitly, we have

$$[\sigma(g)\Omega_n(\cdot\ ;g_1,...,g_n)](x) = \Omega_n(x^g;g_1,...,g_n) \quad,$$
$$[\sigma(g')(\sigma(g)\Omega_n(\cdot\ ;g_1,...,g_n))](x) = [\sigma(g'g)\Omega_n(\cdot\ ;g_1,...,g_n)](x) \qquad (5.4.6)$$
$$= \Omega_n(x^{g'g};g_1,...,g_n) \quad.$$

Then a rewriting of formula (5.1.3) defines δ, $\delta^2 \equiv 0$:

(5.4.1a) DEFINITION (*Coboundary operator for the left action*)

It is defined by (*cf.* (5.1.3))

$$[(\delta\Omega_n)(\cdot\ ;g_1,...,g_{n+1})](x) = (\delta\Omega_n)(x;g_1,...,g_{n+1})$$
$$:= (-1)^{n+1}\Omega_n(x;g_1,...,g_n)$$
$$+ \sum_{i=1}^{n}(-1)^i\Omega_n(x;g_1,...,g_{i-1},g_ig_{i+1},g_{i+2},...,g_{n+1})$$
$$+ \Omega_n(x^{g_1};g_2,...,g_{n+1}) \,, \qquad (5.4.7)$$

where the first group argument in the last term acts on x.

The first four examples of this formula are

$$(\delta\Omega_0)(x;g_1) = -\Omega_0(x) + \Omega_0(x^{g_1}) \quad,$$
$$(\delta\Omega_1)(x;g_1,g_2) = \Omega_1(x;g_1) - \Omega_1(x;g_1g_2) + \Omega_1(x^{g_1},g_2) \quad,$$
$$(\delta\Omega_2)(x;g_1,g_2,g_3) = -\Omega_2(x;g_1,g_2) - \Omega_2(x;g_1g_2,g_3) + \Omega_2(x;g_1,g_2g_3)$$
$$+ \Omega_2(x^{g_1};g_2,g_3) \quad,$$
$$(\delta\Omega_3)(x;g_1,g_2,g_3,g_4) = \Omega_3(x;g_1,g_2,g_3) - \Omega_3(x;g_1g_2,g_3,g_4)$$
$$+ \Omega_3(x;g_1,g_2g_3,g_4) - \Omega_3(x;g_1,g_2,g_3g_4)$$
$$+ \Omega_3(x^{g_1};g_2,g_3,g_4) \quad. \qquad (5.4.8)$$

Eq. (5.4.5) also implies that $(\Phi(g^{-1})(\Phi(g'^{-1})f)(x) = (\Phi(g^{-1}g'^{-1})f)(x) = f(g'gx)$. If we now write $\sigma(g)$ for $\Phi(g^{-1})$, we have $(\sigma(g)f)(x) = f(gx)$,

---

* We could have started by directly defining the *left action* on the elements of $\mathcal{F}(M)$ by

$$(\sigma(g)f)(x) := f(x^g) \Rightarrow (\sigma(g')[\sigma(g)f])(x) = (\sigma(g)f)(x^{g'}) = f(x^{g'g}) = (\sigma(g'g)f)(x) \,,$$

which is a left action since $\sigma(g')(\sigma(g)f) = (\sigma(g'g)f)$. Similarly, a *right action* on $f \in \mathcal{F}(M)$ is defined by

$$(\sigma(g)f)(x) := f(gx) \Rightarrow (\sigma(g')[\sigma(g)f])(x) = (\sigma(g)f)(g'x) = f(gg'x) = (\sigma(gg')f)(x) \,;$$

it is indeed a right action since $\sigma(g')(\sigma(g)f) = (\sigma(gg')f)$.

$(\sigma(g')(\sigma(g)f))(x) = f(gg'x) = (\sigma(gg')f)(x)$ and σ is now a *right action* since g has been replaced by its inverse. The action on cochains is given by

$$[\sigma(g)\Omega_n(\cdot\;;g_1,...,g_n)](x) = \Omega_n(gx;g_1,...,g_n) \quad,$$

$$[\sigma(g')(\sigma(g)\Omega_n(\cdot\;;g_1,...,g_n))](x) = [\sigma(gg')\Omega_n(\cdot\;;g_1,...,g_n)](x) \qquad (5.4.9)$$

$$= \Omega_n(gg'x;g_1,...,g_n) \quad.$$

and the coboundary operator is defined by

(5.4.1b) DEFINITION (*Coboundary operator for the right action*)

The coboundary operator is given by

$$[(\delta\Omega_n)(\cdot\;;g_1,...,g_{n+1})](x)$$

$$= (\delta\Omega_n)(x;g_1,...,g_{n+1})$$

$$:= (-1)^{n+1}\Omega_n(x;g_2,...,g_{n+1})$$

$$+ \sum_{i=1}^{n}(-1)^{i+n+1}\Omega_n(x;g_1,...,g_{i-1},g_ig_{i+1},g_{i+2},...,g_{n+1}) \qquad (5.4.10)$$

$$+ \Omega_n(g_{n+1}x;g_1,...,g_n) \quad.$$

In contrast with (5.4.7), the element that acts on x in the last term is now the last one, g_{n+1}.

For the lowest order cochains (5.4.10) gives

$$(\delta\Omega_0)(x;g_1) = -\Omega_0(x) + \Omega_0(g_1x) \quad,$$

$$(\delta\Omega_1)(x;g_1,g_2) = \Omega_1(x;g_2) - \Omega_1(x;g_1g_2) + \Omega_1(g_2x;g_1) \quad,$$

$$(\delta\Omega_2)(x;g_1,g_2,g_3) = -\Omega_2(x;g_2,g_3) + \Omega_2(x;g_1g_2,g_3) - \Omega_2(x;g_1,g_2g_3)$$

$$+ \Omega_2(g_3x;g_1,g_2) \quad,$$

$$(\delta\Omega_3)(x;g_1,g_2,g_3,g_4) = \Omega_3(x;g_2,g_3,g_4) - \Omega_3(x;g_1g_2,g_3,g_4)$$

$$+ \Omega_3(x;g_1,g_2g_3,g_4) - \Omega_3(x;g_1,g_2,g_3g_4)$$

$$+ \Omega_3(g_4x;g_1,g_2,g_3) \quad.$$

$$(5.4.11)$$

The reader may now identify condition (3.8.3) with the three-cocycle condition $(\delta\xi_3) = 0$ above for $g''' = g_1$, $g'' = g_2$, $g' = g_3$ and $g = g_4$.

Although $\sigma(g)$ was written at the left in (5.4.9) (cf. (5.1.4)), the fact that σ is a right action is visible in the reordering of the $g'g$ arguments. Also, it is the last group element in the argument, g_{n+1}, the one which appears in the last line of eq. (5.4.10). For instance, if we check that $\delta^2 = 0$ by computing $\delta\Omega_2$ with $\Omega_2 = \delta\Omega_1$ in (5.4.11) we immediately notice that the last term in $\delta\Omega_2$ will produce another one, coming from the last one in $\delta\Omega_1$, with argument g_2g_3x. It is the ordering g_2g_3 that allows the cancellation of this contribution to $\delta\Omega_2$ with the term coming from the third one in $\delta\Omega_2$ originated by the first term in $\delta\Omega_1$.

It is easily seen that to the left action on cochains corresponds the right action on the manifold variables x (section 1.1) and vice versa (see the last footnote). If the action σ is the identity, the expressions for both coboundary operators (as was the case for eqs. (5.1.3), (5.1.4)) coincide up to a $(-1)^{n+1}$ sign.

In an analogous way, it is possible to define cochains that take values in $\Lambda^p(M)$ instead of $\Lambda^0(M) = \mathcal{F}(M)$. In this case, the n-cochains are p-forms, and will be denoted Ω_n^p, where the subscript refers to the order of the cochain and the superscript refers to the order of the differential form. Clearly, this change does not affect the definition of the coboundary operator δ, which affects the order of the cochain and not the order of the form.

Bibliographical notes for chapter 5

To the general references already quoted in the bibliographical notes for chapter 4, the mathematically minded reader may add Eilenberg (1949), Hochschild and Serre (1953a) and Hochschild and Mostow (1962).

For an unusual application of cohomology, see Penrose (1991).

For an introduction to supersymmetry see the bibliographical notes for chapter 1. More details on the subject of section 5.3 can be found in Aldaya and de Azcárraga (1985b).

6

Cohomology of Lie algebras

This chapter is devoted to studying some concepts that will be extensively used in the last chapters, namely the cohomology of Lie algebras with values in a vector space, the Whitehead lemmas and Lie algebra extensions (which are related to second cohomology groups). The same three different cases of extensions of chapter 5 as well as the $\mathscr{F}(M)$-valued version of cohomology will be considered. In fact, the relation between Lie group and Lie algebra cohomology will be explored here, first with the simple example of central extensions of groups and algebras (governed by two-cocycles), and then in the higher order case, providing explicit formulae for obtaining Lie algebra cocycles from Lie group ones and vice versa.

The Lie algebra cohomology *à la* Chevalley-Eilenberg, which uses invariant forms on a Lie group, is also presented in this chapter. This will turn out to be specially useful in the construction of physical actions (chapter 8), *i.e.* in the process of relating cohomology and mechanics. The BRST formulation of Lie algebra cohomology will also be discussed here due to its importance in gauge theories.

6.1 Cohomology of Lie algebras: general definitions

We now discuss the cohomology of Lie algebras with values on a vector space V following the same pattern as used in chapter 5 for the cohomology of groups with values in an abelian group. Let $\mathscr{G}$ be a Lie algebra, and let the vector space V over the field K be a $\rho(\mathscr{G})$-module[*]. Although in principle it might be infinite-dimensional (in the last section, the case in which $V = \mathscr{F}(M)$, the module of smooth functions on a manifold, will

---

[*] This means that there is an external operation on V, $\mathscr{G} \times V \to V$, $(X,v) \mapsto \rho(X)v$, which satisfies the properties (a) $\rho(X)(v + v') = \rho(X)v + \rho(X)v'$; (b) $\rho(X + X')v = \rho(X)v + \rho(X')v$; (c) $\rho([X,X'])v = [\rho(X),\rho(X')]v$ (for a left $\rho(\mathscr{G})$-module).

be considered) we shall take V here to be the representation space of a finite-dimensional representation ρ of $\mathcal{G}$.

A few definitions are now necessary.

(6.1.1) DEFINITION (*n-dimensional V-cochains on $\mathcal{G}$*)

An n-dimensional V-cochain ω_n for $\mathcal{G}$ is a skew-symmetric n-linear mapping

$$\omega_n : \underbrace{\mathcal{G} \times \mathcal{G} \times ... \times \mathcal{G}}_{n} \longrightarrow V \quad . \tag{6.1.1}$$

The (abelian) group of all n-cochains will be denoted by $C^n(\mathcal{G}, V)$.

(6.1.2a) DEFINITION (*Coboundary operator for the left action*)

Let V be a left $\rho(\mathcal{G})$-module. The coboundary operator $s_n : C^n(\mathcal{G}, V) \to C^{n+1}(\mathcal{G}, V)$ is defined by its action on the cochains:

$$(s\omega_n)(X_1, ..., X_{n+1}) := \sum_{i=1}^{n+1} (-)^{i+1} \rho(X_i) (\omega(X_1, ..., \hat{X}_i, ..., X_{n+1}))$$

$$+ \sum_{\substack{j,k=1 \\ j<k}}^{n+1} (-)^{j+k} \omega([X_j, X_k], X_1, ..., \hat{X}_j, ..., \hat{X}_k, ..., X_{n+1}) \quad , \tag{6.1.2}$$

where $X_1, ..., X_{n+1} \in \mathcal{G}$.

We give below the first simple examples:

$$(s\omega_0)(X_1) = \rho(X_1)\omega_0 \;;$$
$$(s\omega_1)(X_1, X_2) = \rho(X_1)\omega_1(X_2) - \rho(X_2)\omega_1(X_1) - \omega_1([X_1, X_2]) \;;$$
$$(s\omega_2)(X_1, X_2, X_3) = \rho(X_1)\omega_2(X_2, X_3) - \rho(X_2)\omega_2(X_1, X_3)$$
$$+ \rho(X_3)\omega_2(X_1, X_2) - \omega_2([X_1, X_2], X_3)$$
$$+ \omega_2([X_1, X_3], X_2) - \omega_2([X_2, X_3], X_1) \;;$$
$$(s\omega_3)(X_1, X_2, X_3, X_4) = \rho(X_1)\omega_3(X_2, X_3, X_4) - \rho(X_2)\omega_3(X_1, X_3, X_4)$$
$$+ \rho(X_3)\omega_3(X_1, X_2, X_4) - \rho(X_4)\omega_3(X_1, X_2, X_3)$$
$$- \omega_3([X_1, X_2], X_3, X_4) + \omega_3([X_1, X_3], X_2, X_4) - \omega_3([X_1, X_4], X_2, X_3)$$
$$- \omega_3([X_2, X_3], X_1, X_4) + \omega_3([X_2, X_4], X_1, X_3) - \omega_3([X_3, X_4], X_1, X_2) . \tag{6.1.3}$$

Using the fact that ρ is a Lie algebra homomorphism, $[\rho(X_1), \rho(X_2)] = \rho([X_1, X_2])$, it may be verified that $s^2 = 0$. The proof of this statement is straightforward but cumbersome, although the first examples are easy to check using (6.1.3). Notice, nevertheless, that if ω is taken to be a q-form on a Lie group G and the X's to be vector fields on G, then ω takes values on $\mathscr{F}(G)$, and replacing $\rho(X_i)$ by the action of the vector field X_i on $\mathscr{F}(G)$, eq. (6.1.2) is just the definition of the exterior differential d [(1.4.21)] and $d^2 = 0$. Cocycles and coboundaries are now defined as usual. Consider

$$\cdots \rightarrow C^n(\mathscr{G}, V) \overset{s_n}{\rightarrow} C^{n+1}(\mathscr{G}, V) \rightarrow \cdots \quad . \tag{6.1.4}$$

Then we have the usual definitions:

(6.1.3) DEFINITIONS (*Lie algebra n-cocycle, n-coboundary, cohomology*)

The group $Z_\rho^n(\mathscr{G}, V) = \ker s_n$ is the group of n-cocycles, and $B_\rho^n(\mathscr{G}, V) =$ range s_{n-1} is the group of n-coboundaries. The quotient vector space

$$H_\rho^n(\mathscr{G}, V) := Z_\rho^n(\mathscr{G}, V) / B_\rho^n(\mathscr{G}, V) \tag{6.1.5}$$

is called the n-th cohomology group of $\mathscr{G}$ with coefficients in V corresponding to the representation ρ. Of course,

$$H_\rho^n(\mathscr{G}, V) = 0 \quad \text{for} \quad n > \dim \mathscr{G} \ .$$

It is also possible to define an algebra *cohomology for the right action* in much the same way as we did in the group cohomology case (section 5.1), where we introduced two cohomology operators associated with the left and right actions. Then Definition 6.1.2a is replaced by

(6.1.2b) DEFINITION (*Coboundary operator for the right action*)

This is given by

$$(s\omega)(X_1, ..., X_{n+1}) := \sum_{i=1}^{n+1} (-)^{i+1} \rho(X_i) \left(\omega(X_1, ..., \hat{X}_i, ..., X_{n+1})\right)$$

$$+ \sum_{\substack{j,k=1 \\ j<k}}^{n+1} (-)^{j+k+1} \omega([X_j, X_k], X_1, ..., \hat{X}_j, ..., \hat{X}_k, ..., X_{n+1}) \ .$$

$$\tag{6.1.6}$$

The additional minus sign in the second term guarantees that $s^2 = 0$ when ρ is an algebra anti-homomorphism, i.e., $[\rho(X_1), \rho(X_2)] = -\rho([X_1, X_2])$.

The first examples of (6.1.6) are

$$(s\omega_0)(X_1) = \rho(X_1)\omega_0 ;$$
$$(s\omega_1)(X_1, X_2) = \rho(X_1)\omega_1(X_2) - \rho(X_2)\omega_1(X_1) + \omega_1([X_1, X_2]) ;$$
$$(s\omega_2)(X_1, X_2, X_3) = \rho(X_1)\omega_2(X_2, X_3) - \rho(X_2)\omega_2(X_1, X_3)$$
$$+ \rho(X_3)\omega_2(X_1, X_2) + \omega_2([X_1, X_2], X_3)$$
$$- \omega_2([X_1, X_3], X_2) + \omega_2([X_2, X_3], X_1) ;$$
$$(s\omega_3)(X_1, X_2, X_3, X_4) = \rho(X_1)\omega_3(X_2, X_3, X_4) - \rho(X_2)\omega_3(X_1, X_3, X_4)$$
$$+ \rho(X_3)\omega_3(X_1, X_2, X_4) - \rho(X_4)\omega_3(X_1, X_2, X_3)$$
$$+ \omega_3([X_1, X_2], X_3, X_4) - \omega_3([X_1, X_3], X_2, X_4) + \omega_3([X_1, X_4], X_2, X_3)$$
$$+ \omega_3([X_2, X_3], X_1, X_4) - \omega_3([X_2, X_4], X_1, X_3) + \omega_3([X_3, X_4], X_1, X_2) .$$
$$(6.1.7)$$

It is worth noticing that the differences between (6.1.2) [(6.1.3)] and (6.1.6) [(6.1.7)] are reduced to a sign every time a commutator appears. This is due to the fact that $\mathscr{G}$ is assumed to be defined in both cases by the commutators of the 'left' algebra and thus in (6.1.6) ρ is an anti-homomorphism of this algebra. Clearly, if the defining relations of $\mathscr{G}$ are those of the 'right' copy of $\mathscr{G}$, the ρ in (6.1.6) becomes a homomorphism of the right algebra, a sign appears in the commutators and (6.1.6) becomes (6.1.2).

Simple cases (for Definition 6.1.2a)

(1) $H_\rho^0(\mathscr{G}, V)$

By convention, $B_\rho^0(\mathscr{G}, V) = 0$. The zero-dimensional V-cochains are defined as constant mappings from $\mathscr{G}$ to V. Hence

$$C^0(\mathscr{G}, V) = V . \qquad (6.1.8)$$

Thus, a zero-cochain is a vector $v \in V$. If it is a zero-cocycle, then using (6.1.3),

$$(sv)(X) = \rho(X)v = 0 \quad \forall X \in \mathscr{G} , \qquad (6.1.9)$$

i.e., v is an 'invariant' element (cf. (5.1.15)). Thus,

$$H_\rho^0(\mathscr{G}, V) = V^\mathscr{G} , \qquad (6.1.10)$$

where $V^\mathscr{G}$ denotes the *subspace of invariants* in V. If $V = K$, ρ trivial, we have $Z^0 = C^0 = K$, $B^0 = 0$, $H^0(\mathscr{G}, K) = K$.

(2) $H^1_\rho(\mathcal{G}, V)$

If a one-cochain ω is a one-cocycle,

$$(s\omega)(X_1, X_2) = \rho(X_1)\omega(X_2) - \rho(X_2)\omega(X_1) - \omega([X_1, X_2]) = 0 \quad \forall X_1, X_2 \in \mathcal{G} \quad . \tag{6.1.11}$$

A one-cochain ω_{cob} is a coboundary if there is a zero-cochain v such that $sv = \omega_{cob}$, i.e., if

$$\omega_{cob}(X) = \rho(X)v \qquad \forall X \in \mathcal{G} \quad . \tag{6.1.12}$$

A case of special interest corresponds to taking $V = C$ and $\rho = 0$ (trivial action). Then ω_{cob} is a coboundary if $\omega_{cob} = 0$, and the cocycle condition reads $\omega([X', X]) = 0 \quad \forall X, X' \in \mathcal{G}$. This implies that the one-cocycles are linear mappings vanishing on $[\mathcal{G}, \mathcal{G}]$. Thus

$$H^1_0(\mathcal{G}, V) = (\mathcal{G}/[\mathcal{G}, \mathcal{G}])^* \quad . \tag{6.1.13}$$

If $\mathcal{G}$ is semisimple, $[\mathcal{G}, \mathcal{G}] = \mathcal{G}$ and thus $H^1_0(\mathcal{G}, V) = 0$. (It is also true that $H^1_\rho(\mathcal{G}, V) = 0$, see Theorem 6.5.1, a).)

6.2 Extensions of a Lie algebra $\mathcal{G}$ by an abelian algebra $\mathcal{A}$: $H^2_\rho(\mathcal{G}, \mathcal{A})$

We discussed in section 5.2 the problem of extension of a group G by an abelian group A. Similarly, we shall show now that the elements of the second cohomology group on $\mathcal{G}$ are in one-to-one correspondence with the inequivalent extensions of $\mathcal{G}$ by the abelian algebra $\mathcal{A}$.

(6.2.1) DEFINITION (*Lie algebra extensions*)

A Lie algebra $\tilde{\mathcal{G}}$ is an extension of $\mathcal{G}$ by $\mathcal{A}$ if there is an exact sequence of Lie algebras

$$0 \to \mathcal{A} \xrightarrow{i} \tilde{\mathcal{G}} \underset{\tau}{\overset{\pi}{\underset{\longleftarrow}{\longrightarrow}}} \mathcal{G} \to 0 \quad ; \tag{6.2.1}$$

$\mathcal{A}$ is an ideal of $\tilde{\mathcal{G}}$ and $\tilde{\mathcal{G}}/\mathcal{A} = \mathcal{G}$; we shall assume $\mathcal{A}$ abelian.

Suppose that a solution to (6.2.1) is known. It is clear that $\mathcal{A}$ is a representation space of $\mathcal{G}$ if we define, $\forall X \in \mathcal{G}, \quad \forall Y \in \mathcal{A}$,

$$\rho(X)Y = [\pi^{-1}(X), \ Y] = [\tau(X), \ Y] \quad , \tag{6.2.2}$$

where τ is a (trivializing) linear mapping ($\pi \circ \tau = 1_{|\mathcal{G}}$). $\rho(X)$ is well defined, because all the elements belonging to the same class of $\tau(X)$ in $\tilde{\mathcal{G}}$ (i.e., all the elements in $\pi^{-1}(X)$) differ in an element of $\mathcal{A}$ and, since $\mathcal{A}$ is abelian, define the same action $\rho(X)$ by (6.2.2). In (6.2.2) and below we identify the elements of $\mathcal{A}$ and their images in $\tilde{\mathcal{G}}$ by the injection i in (6.2.1).

Let us now define a two-cochain $\omega \in C^2(\mathcal{G}, \mathcal{A})$ by

$$\omega(X_1, X_2) = [\tau(X_1),\ \tau(X_2)] - \tau([X_1,\ X_2]) \quad \forall X_1,\ X_2 \in \mathcal{G} \ . \qquad (6.2.3)$$

It is easy to see that $\omega \in Z_\rho^2(\mathcal{G}, \mathcal{A})$, since by (6.1.3)

$$\begin{aligned}
(s\omega)\,(X_1, X_2, X_3) = &\ [\tau(X_1), \omega(X_2, X_3)] - [\tau(X_2), \omega(X_1, X_3)] \\
&+ [\tau(X_3), \omega(X_1, X_2)] - [\tau([X_1, X_2]),\ \tau(X_3)] \\
&+ [\tau([X_1, X_3]),\ \tau(X_2)] - [\tau([X_2, X_3]),\ \tau(X_1)] \quad ,
\end{aligned}$$
$$(6.2.4)$$

where the three terms missing in (6.2.4) after using (6.2.3) cancel out because of the Jacobi identity in $\mathcal{G}$. Again using (6.2.3) in (6.2.4) six terms cancel out, and the remaining three add up to zero because of the Jacobi identity in $\tilde{\mathcal{G}}$.

Suppose now that we use another τ', $\tau' \neq \tau$. τ' defines the same ρ [(6.2.2)], so that $v = \tau' - \tau$ $(v(X) = \tau'(X) - \tau(X))$ necessarily defines the trivial representation or, in other words, v is a linear mapping:

$$\tau' - \tau = v : \mathcal{G} \to \mathcal{A}, \quad v \in C^1(\mathcal{G}, \mathcal{A}) \ . \qquad (6.2.5)$$

The cochain v generates a two-coboundary. On the other hand, from the two-cocycles ω and ω' associated with the linear mappings τ and τ' respectively, we obtain

$$\omega'(X_1, X_2) - \omega(X_1, X_2) = [v(X_1) + \tau(X_1),\ v(X_2) + \tau(X_2)]$$

$$-[\tau(X_1),\ \tau(X_2)] - v([X_1, X_2])$$

$$= \rho(X_1)v(X_2) - \rho(X_2)v(X_1) - v([X_1, X_2]) \quad . \qquad (6.2.6)$$

Thus, ω and ω' differ in a two-coboundary and since

$$H_\rho^2(\mathcal{G}, \mathcal{A}) = Z_\rho^2(\mathcal{G}, \mathcal{A})/B_\rho^2(\mathcal{G}, \mathcal{A}) \quad , \qquad (6.2.7)$$

the extension of $\mathcal{G}$ by $\mathcal{A}$ defines an element of $H_\rho^2(\mathcal{G}, \mathcal{A})$; the coboundaries just reflect the freedom of choice of trivializing sections, *i.e.*, the freedom of representing the classes of $\tilde{\mathcal{G}}$ by $\mathcal{A}$ (the elements of $\mathcal{G}$) by different elements in them.

As in Definition 4.2.4, we have now the following:

(6.2.2) DEFINITION (*Equivalence of extensions*)

Two extensions $\tilde{\mathcal{G}}$ and $\tilde{\mathcal{G}}'$ are equivalent if there exists a Lie algebra iso-morphism f such that Diagram 6.2.1 is commutative:

$$
\begin{array}{ccccc}
 & & \overset{i}{\nearrow} \overset{\widetilde{\mathcal{G}}}{} \overset{\pi}{\searrow} & & \\
0 & \to & \mathcal{A} \quad \downarrow f \quad \mathcal{G} & \to & 0 \\
 & & \underset{i'}{\searrow} \underset{\widetilde{\mathcal{G}}' \quad \pi'}{} \overset{\nearrow}{} & &
\end{array}
$$

Diagram 6.2.1

Equivalent extensions lead necessarily to the same ρ in (6.2.2). Given $\rho : \mathcal{G} \to \mathrm{End}\mathcal{A}$, the extension $\widetilde{\mathcal{G}}$ is determined from $\omega \in H_\rho^2(\mathcal{G}, \mathcal{A})$ by the following steps:

a) As a vector space, $\widetilde{\mathcal{G}} = \mathcal{A} \oplus \mathcal{G}$, and its elements may be denoted (Y, X), $Y \in \mathcal{A}$ and X characterized by its coordinates in $\mathcal{G}$ (note, however, that the elements $\{(0, X) = \tau(X)\}$ do not, in general, define a subalgebra of $\widetilde{\mathcal{G}}$).

b) The commutator of $\tilde{X}_1 = (Y_1, X_1)$, $\tilde{X}_2 = (Y_2, X_2)$ is given by

$$[(Y_1, X_1), (Y_2, X_2)] = (\rho(X_1)Y_2 - \rho(X_2)Y_1 + \omega(X_1, X_2), [X_1, X_2]) \quad , \quad (6.2.8)$$

which corresponds to defining
- in $\mathcal{A}$, zero Lie bracket (abelian);
- in $\tau(\mathcal{G})$, $[\tau(X_1), \tau(X_2)] = \omega(X_1, X_2) + \tau([X_1, X_2])$;
- and, between $\mathcal{A}$ and $\tau(\mathcal{G})$,

$$[Y, \tau(X)] = -\rho(X)Y \quad . \tag{6.2.9}$$

Since $[[(Y_1, X_1), (Y_2, X_2)], (Y_3, X_3)] + [[(Y_2, X_2), (Y_3, X_3)], (Y_1, X_1)] + [[(Y_3, X_3), (Y_1, X_1)], (Y_2, X_2)] = 0$ (Jacobi identity) and the extended algebra bracket (6.2.8) implies the two-cocycle property for ω (see (6.1.3)), the above completes the proof of the following

(6.2.1) THEOREM

For a given ρ, the classes of equivalent extensions $\widetilde{\mathcal{G}}$ of $\mathcal{G}$ by the abelian algebra $\mathcal{A}$ are in one-to-one correspondence with the elements of the second cohomology group $H_\rho^2(\mathcal{G}, \mathcal{A})$.

6.3 Three cases of special interest

Let us now give the Lie algebra version of the three cases considered in section 5.2.

(a) $\omega = 0$, $\rho \neq 0$

In this case $\tilde{\mathcal{G}}$ is a *semidirect* extension*, the semidirect product of $\mathcal{G}$ and $\mathcal{A}$ with respect to ρ. As can be read from (6.2.8), (6.2.9), if ω is zero τ is not only a linear application, but a Lie algebra homomorphism. $\tau(\mathcal{G})$ is then a subalgebra of $\tilde{\mathcal{G}}$, which is written

$$\tilde{\mathcal{G}} = \mathcal{A} \oslash \mathcal{G} \ . \tag{6.3.1}$$

where the semidirect symbol $\oslash$ above (a frequent notation is the open direct product symbol $\otimes$) indicates that $\tilde{\mathcal{G}}$ is not a direct product of Lie *algebras*, since $\mathcal{G}$ operates on $\mathcal{A}$ through ρ:

$$[(Y', X'), (Y, X)] = \tag{6.3.2}$$
$$(\rho(X')Y - \rho(X)Y', [X', X]) \quad \forall X, X' \in \mathcal{G} , \ \forall Y, Y' \in \mathcal{A} \ ,$$

and, of course, $\rho(X)Y \in \mathcal{A}$. The underlying vector space of $\tilde{\mathcal{G}}$ is the direct *sum* of the vector spaces, $\mathcal{A} \oplus \mathcal{G}$.

The above analysis shows that given $\mathcal{G}$, $\mathcal{A}$ and an action $\rho : \mathcal{G} \to \text{End } \mathcal{A}$ there is not a unique (splitting) extension. As in section 5.2(b) we may classify the different splittings of $\tilde{\mathcal{G}}$, *i.e.* all possible inclusions τ of $\mathcal{G}$ into $\tilde{\mathcal{G}}$ which preserve the Lie bracket structure of $\mathcal{G}$ and $\tilde{\mathcal{G}}$. Let τ' be a Lie algebra splitting, different from the τ which originally defined (6.3.1) by $\tau(X) = (0, X)$ $\forall X \in \mathcal{G}$. Let τ' be defined by

$$\tau'(X) = (\omega(X), \ X) \tag{6.3.3}$$

where $\omega(X) \in \mathcal{A}$ and, since τ' is a linear mapping, $\omega \in C^1(\mathcal{G}, \mathcal{A})$. If τ' is a Lie algebra homomorphism,

$$\tau'([X_1, X_2]) = (\omega([X_1, X_2]), \ [X_1, X_2]) \tag{6.3.4}$$

must be equal to

$$[\tau'(X_1), \ \tau'(X_2)] = (\rho(X_1)\omega(X_2) - \rho(X_2)\omega(X_1), \ [X_1, X_2]) \ , \tag{6.3.5}$$

where (6.3.2) and (6.3.3) have been used. Thus, equality implies

$$\rho(X_1)\omega(X_2) - \rho(X_2)\omega(X_1) - \omega([X_1, X_2]) \equiv (s\omega) \ (X_1, X_2) = 0 \ , \tag{6.3.6}$$

i.e., $\omega \in Z^1_\rho(\mathcal{G}, \mathcal{A})$. Thus, τ' is a splitting only if the corresponding ω is a cocycle. Among the possible ω's, those given by

$$\omega_Y(X) = \rho(X)Y \quad \forall X \in \mathcal{G} \tag{6.3.7}$$

are defined by a fixed $Y \in \mathcal{A}$. They are coboundaries [(6.1.12)]; thus $H^1_\rho(\mathcal{G}, \mathcal{A})$ classifies the splittings of $\tilde{\mathcal{G}}$ modulo those given in terms of linear mappings ω_Y of the form (6.3.7). The zero element of $H^1_\rho(\mathcal{G}, \mathcal{A})$, represented by $Y = 0$ in (6.3.7), defines the splitting τ, $\tau(X) = (0, X)$ $\forall X \in \mathcal{G}$.

---

* The name is borrowed from the Lie group case; for algebras we could also say *semidirect sum*.

(b) $\rho = 0$ *(trivial action)*, ω *non-trivial*
The extended Lie algebra bracket is given by

$$[\tilde{X}_1, \tilde{X}_2] \equiv [(Y_1, X_1), (Y_2, X_2)] = (\omega(X_1, X_2), [X_1, X_2]) \quad . \tag{6.3.8}$$

The ideal $\mathscr{A} = \{\tilde{Y} | \tilde{Y} = (Y, 0)\}$ is central, and $\tilde{\mathscr{G}}$ is a central extension. Clearly, the two-cocycle condition reduces in this case to

$$\omega([X_1, X_2], X_3) + \omega([X_2, X_3], X_1) + \omega([X_3, X_1], X_2) = 0 \quad . \tag{6.3.9}$$

(c) $\rho = 0$, $\omega = 0$
In this case both $\mathscr{A}$ and $\mathscr{G}$ are ideals of the extended algebra, which splits into the direct product of algebras, $\tilde{\mathscr{G}} = \mathscr{A} \otimes \mathscr{G}$.

6.4 Local exponents and the isomorphism between $H_0^2(G, R)$ and $H_0^2(\mathscr{G}, R)$

As we have seen, case (b) in the previous section characterizes the central extensions through $H_0^2(\mathscr{G}, \mathscr{A})$. It is interesting to analyse now the particular case that corresponds to the Lie algebra associated with the extended local group G which appears when the projective representations of an r-dimensional Lie group G are considered. The group law of $\bar{G}$ was given by (3.2.15) (see also section 5.2 (a)) and was determined by the local exponent $\xi(g', g)$ characterizing the extension; as a two-cocycle, $\xi(g', g)$ satisfied the condition (3.2.11). The problem now is to reformulate this from the point of view of the associated extended algebra $\bar{\mathscr{G}}$. This requires us to look for the Lie algebra two-cocycle associated with ξ and to find how two-cocycles ξ, ξ' in the same group cohomology class determine two-cocycles in the same algebra cohomology class.

Consider the central extension $\bar{G}$ as defined by the central exponent $\xi(g', g)$ in the group law $\bar{g}'' = \bar{g}' * \bar{g} = (\theta' + \theta + \xi(g', g), g'g)$. The LI generators of the $(r + 1)$-dimensional extended algebra $\bar{\mathscr{G}}$ are given by

$$\bar{X}_{(i)}^L = X_{(i)}^L + \left. \frac{\partial \xi(g', g)}{\partial g^i} \right|_{\substack{g'=g \\ g=e}} \frac{\partial}{\partial \theta} \quad (i = 1, ..., r), \quad \bar{X}_{(\theta)}^L = \frac{\partial}{\partial \theta}, \tag{6.4.1}$$

where $X_{(i)}^L$ are the LIVFs of the original algebra $\mathscr{G}$; $[X_{(i)}^L, X_{(j)}^L] = C_{ij}^k X_{(k)}^L$. Consider now the commutators of $\bar{G}$. Clearly, the commutator $[\bar{X}_{(\theta)}^L, \bar{X}_{(i)}^L] = 0$ exhibits the central character of the extension and shows that the structure constants $C_{i\theta}^j = C_{i\theta}^\theta = 0$. The complete set of the structure constants of $\bar{\mathscr{G}}$ can be obtained from eq. (1.2.16). Using it for $\bar{g}''$ given by the above group law, we find

$$
C_{ij}^k = \left(\frac{\partial^2 \bar{g}''^k(\bar{g}',\bar{g})}{\partial g'^i \partial g^j} - \frac{\partial^2 \bar{g}''^k(\bar{g}',\bar{g})}{\partial g'^j \partial g^i} \right)\Bigg|_{\substack{g'=e\\g=e}}
$$

$$
= \left(\frac{\partial^2 g''^k}{\partial g'^i \partial g^j} - \frac{\partial^2 g''^k}{\partial g'^j \partial g^i} \right)\Bigg|_{\substack{g'=e\\g=e}} \qquad (i,j,k = 1,...,r = \dim \mathcal{G}) ,
$$

(6.4.2)

which shows that the structure constants for $\bar{\mathcal{G}}$ are the same as those for $\mathcal{G}$ when their indices take values between 1 and r. When the upper index corresponds to the $(r+1)$-th new parameter θ, they are modified following

(6.4.1) PROPOSITION

The additional structure constants of the centrally extended algebra $\bar{\mathcal{G}}$ are determined from the two-cocycle defining the central extension $\bar{G}$ by

$$
C_{ij}^\theta = \left(\frac{\partial^2 \bar{g}''^\theta}{\partial g'^i \partial g^j} - \frac{\partial^2 \bar{g}''^\theta}{\partial g'^j \partial g^i} \right)\Bigg|_{\substack{g'=e\\g=e}} = \left(\frac{\partial^2 \xi(g',g)}{\partial g'^i \partial g^j} - \frac{\partial^2 \xi(g',g)}{\partial g'^j \partial g^i} \right)\Bigg|_{\substack{g'=e\\g=e}} . \qquad (6.4.3)
$$

Thus, the commutators of $\bar{\mathcal{G}}$ have the form

$$
[\bar{X}_{(\theta)}^L, \bar{X}_{(i)}^L] = 0 , \qquad [\bar{X}_{(i)}^L, \bar{X}_{(j)}^L] = C_{ij}^k \bar{X}_{(k)}^L + C_{ij}^\theta \bar{X}_{(\theta)}^L . \qquad (6.4.4)
$$

For instance, for the Galilei group eqs. (6.4.3) and (3.4.1) give $C_{A^i V^j}^\theta = -m\delta_{ij}$, eq. (3.4.7).

If $\xi(g',g)$ is modified by the addition of a coboundary $\xi_{cob}(g',g)$ generated by $\eta(g)$, the new two-cocycle ξ' is given by $\xi'(g',g) = \xi(g',g) + \eta(g') + \eta(g) - \eta(g'g)$. The LI generators $\bar{X}'_{(i)}^L$ of the extended algebra for the two-cocycle ξ' coincide with the previous ones in their $\partial/\partial g^i$ components, but the derivative of $\xi_{cob}(g',g)$ with respect to g^i at $g = e$ gives additional contributions to their $\partial/\partial\theta$ term. Specifically,

$$
\bar{X}'_{(i)}^L = \bar{X}_{(i)}^L + \left[-X_{(i)}^L.\eta(g) + \frac{\partial\eta(g)}{\partial g^i}\Big|_{g=e} \right] \frac{\partial}{\partial\theta} ; \qquad (6.4.5)
$$

clearly, $\bar{X}'_{(\theta)}^L = \bar{X}_{(\theta)}^L$. If we now compute the structure constants for the coboundary-modified generators $\bar{X}'_{(i)}^L$, we find

$$
[\bar{X}'_{(i)}^L, \bar{X}'_{(j)}^L] = [\bar{X}_{(i)}^L, \bar{X}_{(j)}^L] - [X_{(i)}^L, X_{(j)}^L].\eta(g) \frac{\partial}{\partial\theta}
$$

$$
= C_{ij}^k \bar{X}'_{(k)}^L + \left[C_{ij}^\theta - C_{ij}^k \frac{\partial\eta(g)}{\partial g^k}\Big|_{g=e} \right] \frac{\partial}{\partial\theta} ; \qquad (6.4.6)
$$

notice that only the $(\partial\eta/\partial g^k)|_{g=e}$ term in (6.4.5) modifies C_{ij}^θ. Denoting with a prime the structure constants for the $\bar{X}'^L_{(i)}$, we may write

$$C'^k_{ij} = C^k_{ij} \quad \text{but} \quad C'^\theta_{ij} = C^\theta_{ij} - \mu_{ij}, \quad \mu_{ij} \equiv \eta_k C^k_{ij}, \quad \eta_k \equiv \left.\frac{\partial\eta(g)}{\partial g^k}\right|_{g=e}.$$

$$(6.4.7)$$

Thus, from the point of view of the Lie algebra commutators, the search for a non-trivial central extension can be reduced to the problem of finding certain C_{ij}^θ's which cannot be removed by μ_{ij}'s for which $C_{ij}^\theta - \eta_k C_{ij}^k = 0$. The associativity of $\bar{G}$ implies the Jacobi identity for $\mathcal{G}$; hence the structure constants C_{ij}^θ as well as μ_{ij} satisfy the consistency condition

$$C_{ij}^m C_{mk}^\theta + C_{jk}^m C_{mi}^\theta + C_{ki}^m C_{mj}^\theta = 0. \qquad (6.4.8)$$

This condition for C_{ij}^θ is the two-cocycle condition, and is the result of the analogous condition satisfied by $\xi(g',g)$. It was seen in section 5.2 (a) that the two-coboundary generating function $\eta(g)$ was related to a different characterization of the central parameter of the extended group $\bar{G}$. Similarly, the coboundary of the extended algebra $\mathcal{G}$ is associated with a new basis of generators.

Clearly, we could have computed the modification μ_{ij} of C_{ij}^θ directly from the two-coboundary $\xi_{cob}(g',g) = \eta(g') + \eta(g) - \eta(g'g)$ with the result

$$C'^\theta_{ij} = C^\theta_{ij} - \left(\frac{\partial^2\eta(g'g)}{\partial g''^i\partial g^j} - \frac{\partial^2\eta(g'g)}{\partial g''^j\partial g^i}\right)\Bigg|_{\substack{g'=e\\g=e}} \equiv C^\theta_{ij} - \mu_{ij}. \qquad (6.4.9)$$

From this expression it also follows that

$$\mu_{ij} = \frac{\partial\eta(g'')}{\partial g''^k}\left(\frac{\partial^2 g''^k}{\partial g'^i\partial g^j} - i\leftrightarrow j\right)\Bigg|_{\substack{g'=e\\g=e}} = \left.\frac{\partial\eta(g'')}{\partial g''^k}\right|_{g''=e} C_{ij}^k \qquad (6.4.10)$$

as before, since the term in the second derivative of $\eta(g'')$ gives zero due to its symmetry in (i,j). All this discussion may be summarized by means of

(6.4.2) PROPOSITION

If the two-cocycle of a central extension is modified by the addition of a two-coboundary $\xi_{cob}(g',g) = \eta(g') + \eta(g) - \eta(g'g)$ generated by the function $\eta(g)$ on G, the new generators $\bar{X}'^L_{(i)}(g)$ of the extended Lie algebra $\mathcal{G}$ are given by (6.4.5), and the new structure constants C'^θ_{ij} are given by (6.4.7).

Since the Lie bracket is linear with respect to each of its arguments, the C_{ij}^θ of (6.4.4) define a skew-symmetric bilinear mapping

$$\omega_2 : \mathcal{G} \times \mathcal{G} \to R, \quad \omega_2(X_{(i)}, X_{(j)}) = C_{ij}^\theta, \quad i, j = 1, ..., r . \tag{6.4.11}$$

The two-cocycle condition (6.4.8) can be rewritten as

$$\omega_2([X_{(i)}, X_{(j)}], X_{(k)}) + \omega_2([X_{(j)}, X_{(k)}], X_{(i)}) + \omega_2([X_{(k)}, X_{(i)}], X_{(j)}) = 0 \tag{6.4.12}$$

(or, equivalently, $\omega_2([X, Y], Z) + \omega_2([Y, Z], X) + \omega_2([Z, X], Y) = 0$ $\forall X, Y, Z \in \mathcal{G}$). Eq. (6.4.12) just expresses that $s\omega_2 = 0$ (see eq. (6.1.3) for $\rho = 0$) and that accordingly ω_2 is a two-cocycle for the cohomology of $\mathcal{G}$ with values in R. Similarly, μ_{ij} defines a mapping

$$\omega_{cob} : \mathcal{G} \times \mathcal{G} \to R, \quad \omega_{cob}(X_{(i)}, X_{(j)}) = \omega_1([X_{(i)}, X_{(j)}]) , \tag{6.4.13}$$

where $\omega_1(X_{(k)}) = -\eta_k$ (eq. (6.4.7)). This mapping is again a two-cocycle since $s\omega_{cob}(X, Y, Z) = \omega_1([[X, Y], Z]) + \omega_1([[Y, Z], X]) + \omega_1([[Z, X], Y]) = 0$ [(6.3.9)], and it is a trivial one because $(s\omega_1)(X, Y) = \omega_1([X, Y])$ (see (6.1.3) for ρ trivial). Thus, the structure constants C_{ij}^θ and μ_{ij} define skew-symmetric mappings which correspond to a two-cocycle and a two-coboundary, respectively, in the cohomology of $\mathcal{G}^\dagger$.

The dimension d of the vector space made out of the skew-symmetric mappings $\mathcal{G} \times \mathcal{G} \to R$ is $d = \binom{r}{2}$. The two-cocycles ω_2 satisfy (6.4.12) and constitute a subspace of dimension d_c; the two-coboundaries ω_{cob} define in turn a subspace of the previous one of dimension d_{cob}. Since the coboundaries are generated by mappings $\omega_1 : \mathcal{G} \to R$ through (6.4.13), $d_{cob} \leq r$. The dimension of the quotient space that characterizes the non-trivial two-cocycles is then $d_c - d_{cob}$, and thus the dimension of $H_0^2(\mathcal{G}, R)$ is given by

$$\dim H_0^2(\mathcal{G}, R) = d_c - d_{cob} \leq d - d_{cob} \leq d = \frac{1}{2}r(r-1) . \tag{6.4.14}$$

The maximal dimension $d = \binom{r}{2}$ occurs iff $\mathcal{G}$ (and hence G if it is connected) is abelian. This is so because then eq. (6.4.13) gives $\omega_{cob} = 0$ for any ω_1. Conversely, in this case condition (6.4.12) becomes empty so that every antisymmetric two-form defines a non-trivial local exponent and then $\dim H_0^2(\mathcal{G}, R) = \binom{r}{2}$. This number can be easily visualized: it is the number of *pairs* of elements of a basis of $\mathcal{G}$ for which their abelian commutator can be replaced by another one with a central element in the r.h.s. The dimension of $H_0^2(\mathcal{G}, R)$ is then the number of independent parameters

---

† We note in passing that, with $\omega_1 \equiv \mu_0$ and $\omega_2 \equiv \Gamma$, the relation $\Gamma'(X, Y) = \Gamma(X, Y) + \mu_0([X, Y])$ between two cohomologous cocycles is the same as that for two cohomologous symplectic cocycles as follows from eq. (3.6.18). Thus, and as mentioned in section 3.6, $H_s(\mathcal{G}, \mathcal{G}^*) \approx H_0^2(\mathcal{G}, R)$ (and is isomorphic to $H_0^2(G, R)$).

('*central charges*') the values of which characterize the extension of the original algebra.

6.5 Cohomology groups for semisimple Lie algebras. The Whitehead lemma

(6.5.1) THEOREM

Let $\mathcal{G}$ be a semisimple algebra over a field K of characteristic zero, and let V be a finite-dimensional $\rho(\mathcal{G})$-module. Then

$$a) \ H^1_\rho(\mathcal{G}, V) = 0 \quad ; \quad b) \ H^2_\rho(\mathcal{G}, V) = 0 \quad . \tag{6.5.1}$$

Proof: a) $H^1_\rho(\mathcal{G}, V) = 0$, i.e. $Z^1_\rho(\mathcal{G}, V) = B^1_\rho(\mathcal{G}, V)$.

Let $\omega \in Z^1_\rho(\mathcal{G}, V)$, and let $h(X_1)$, $X_1 \in \mathcal{G}$, be the endormorphism of $Z^1(\mathcal{G}, V)$

$$h(X_1) : \omega \mapsto h(X_1)\omega \in Z^1_\rho(\mathcal{G}, V) \tag{6.5.2}$$

defined by

$$[h(X_1)\omega](X_2) = \rho(X_1)\omega(X_2) - \omega([X_1, X_2]) \quad ; \tag{6.5.3}$$

$h : X \mapsto h(X)$ is a representation of $\mathcal{G}$. Since ω is a one-cocycle, (6.5.3) and (6.1.11) imply

$$[h(X_1)\omega](X_2) = \rho(X_2)\omega(X_1) \tag{6.5.4}$$

and then (6.1.12) shows that the image of ω under $h(X_1)$ is the coboundary generated by the vector (zero-cochain) $\omega(X_1) \in V$. Thus, $h(X_1)$ is a mapping

$$h(X_1) : Z^1_\rho(\mathcal{G}, V) \to B^1_\rho(\mathcal{G}, V) \quad . \tag{6.5.5}$$

Now let Z^1_I be the space of the images $h(X)(Z^1_\rho)$, $X \in \mathcal{G}$, of h. By (6.5.5), $Z^1_I \subset B^1_\rho$. To conclude the proof, it is sufficient to prove that the space

$$Z^1_{ker} = \{\omega \in Z^1_\rho | h(X)\omega = 0 \quad \forall X \in \mathcal{G}\} \quad , \tag{6.5.6}$$

which (h is a representation of $\mathcal{G}$) may be shown to be the complementary space to Z^1_I in Z^1_ρ, $Z^1_\rho = Z^1_I \oplus Z^1_{ker}$, is zero. Let $\omega' \in Z^1_{ker}$. By (6.5.4)

$$\rho(X_2)\omega'(X_1) = 0 \quad \forall X_1, X_2 \tag{6.5.7}$$

and then (6.5.3) implies $\omega'([X_1, X_2]) = 0 \ \forall X_1, X_2$. But, if $\mathcal{G}$ is semisimple, $[\mathcal{G}, \mathcal{G}] = \mathcal{G}$; then $\omega'(\mathcal{G}) = 0$ implies $\omega' = 0$, q.e.d. (For the particular case of ρ trivial, the proof is also trivial: see (6.1.13).)

Part b) of the theorem is lengthier, but can be proven in a similar way, although that will not be done here. We shall present here, instead, the proof that all local exponents of semisimple groups are equivalent to zero:

(6.5.2) THEOREM (*Whitehead's lemma*)

$H_0^2(\mathscr{G}, R) = 0$ if $\mathscr{G}$ is semisimple.

Proof: Let C_{ij}^θ be the additional structure constants of $\mathscr{G}$; they satisfy (6.4.8). Since $\mathscr{G}$ is semisimple, $C_{i\cdot j}^m C_{\cdot l}^{j\cdot i} = \delta_l^m$ (see (1.8.12)). Thus, multiplying (6.4.8) by $C_{\cdot l}^{j\cdot i}$, we obtain

$$C_{k\cdot l}^\theta = C_{\cdot l}^{j\cdot i} C_{j\cdot k}^m C_{m\cdot i}^\theta + C_{\cdot l}^{j\cdot i} C_{k\cdot i}^m C_{m\cdot j}^\theta . \tag{6.5.8}$$

Denoting by B the matrix of elements $B_{\cdot i}^m = C^{m\theta}{}_i$, the above expression gives

$$C_{k\cdot l}^\theta = 2\mathrm{Tr}(B\, adX_{(l)}\, adX_{(k)}) = \mathrm{Tr}(B(adX_{(l)}\, adX_{(k)} - adX_{(k)}\, adX_{(l)})) \tag{6.5.9}$$
$$= C_{l\cdot k}^m \mathrm{Tr}(B\, adX_{(m)}) \equiv C_{l\cdot k}^m \lambda_m \equiv \mu_{lk}$$

where the antisymmetry of $C_{k\cdot l}^\theta$ in (kl) has been used. But, by (6.4.7), this means that the μ_{lk} can be removed by a suitable definition of the basis of $\mathscr{G}$ i.e., they define a two-coboundary, q.e.d.

In contrast, if $\mathscr{G}$ is nilpotent, the second cohomology group is not equal to zero; this is trivially the case of the two-dimensional abelian algebra, which may be extended to give the Weyl-Heisenberg algebra of section 3.3. To conclude this section, we shall now show that Theorem 6.5.1 b) is closely related to the Levi-Mal'čev decomposition of Lie algebras:

(6.5.3) THEOREM (*Levi-Mal'čev*)

Every (finite-dimensional) Lie algebra $\mathscr{G}$ is the split extension of a semisimple Lie algebra by the radical $\mathscr{R}$ of $\mathscr{G}$.

---

* Given a Lie algebra $\mathscr{G}$, the sequence $\mathscr{G}^{(0)} = \mathscr{G}$, $\mathscr{G}^{(1)} = [\mathscr{G}^{(0)}, \mathscr{G}^{(0)}],..., \mathscr{G}^{(i)} = [\mathscr{G}^{(i-1)}, \mathscr{G}^{(i-1)}]$ of ideals of $\mathscr{G}$ is called its *derived series*, and $\mathscr{G}$ is *solvable* if there exists an integer n such that $\mathscr{G}^{(n)} = \{0\}$; the first such n is then called the (derived) length of $\mathscr{G}$. The maximal solvable ideal of $\mathscr{G}$ is called the *radical* $\mathscr{R}$ of $\mathscr{G}$, and $\mathscr{G}/\mathscr{R}$ is semisimple. $\mathscr{G}$ is called *nilpotent* when the sequence $\mathscr{G}^0 = \mathscr{G}$, $\mathscr{G}^1 = [\mathscr{G}, \mathscr{G}]$, $\mathscr{G}^2 = [\mathscr{G}, \mathscr{G}^1],..., \mathscr{G}^i = [\mathscr{G}, \mathscr{G}^{i-1}]$ of ideals of $\mathscr{G}$ (the *descending central series*) ends, $\mathscr{G}^n = 0$. Clearly, $\mathscr{G}^{(i)} \subset \mathscr{G}^i \; \forall i$: nilpotent algebras are solvable. For instance, the harmonic oscillator algebra $[a, a^\dagger] = 1$, $[a^\dagger a, a^\dagger] = a^\dagger$, $[a^\dagger a, a] = -a$ is solvable, but it is not nilpotent since $\mathscr{G}^i$ is the Weyl-Heisenberg algebra for all i. Connected Lie groups with solvable (nilpotent) Lie algebras are themselves called solvable (nilpotent).

Proof (outline):

This proceeds by induction on the derived length of $\mathscr{R}$. Assume first $n = 1$. $\mathscr{R}$ is a $(\mathscr{G}/\mathscr{R})$-module and $H^2(\mathscr{G}/\mathscr{R}, \mathscr{R})$, which classifies the extensions of $\mathscr{G}/\mathscr{R}$ by $\mathscr{R}$ by Theorem 6.2.1, is zero by Theorem 6.5.1 b). Thus, the sequence

$$0 \rightarrow \mathscr{R} \rightarrow \mathscr{G} \underset{\rightarrow}{\overset{\leftarrow}{\rightleftarrows}} \mathscr{G}/\mathscr{R} \rightleftarrows 0$$

splits and as a vector space $\mathscr{G} = \mathscr{R} \oplus \mathscr{G}/\mathscr{R}$ (section 6.3 (a)). Let now $n = 2$ and let us construct Diagram 6.5.1,

$$
\begin{array}{ccccccc}
0 & \longrightarrow & \mathscr{R} \longrightarrow & & \mathscr{G} \longrightarrow & & \mathscr{G}/\mathscr{R} \rightleftarrows & 0 \\
& & & & & & \parallel & \\
0 & \rightarrow & \mathscr{R}/[\mathscr{R},\mathscr{R}] \rightarrow & & \mathscr{G}/[\mathscr{R},\mathscr{R}] \overset{t}{\underset{\rightarrow}{\leftarrow}} & & \mathscr{G}/\mathscr{R} \underset{\rightarrow}{\leftarrow} & 0
\end{array}
$$

Diagram 6.5.1

The previous argument now applies to the lower sequence, which splits. Let $t : \mathscr{G}/\mathscr{R} \rightarrow \mathscr{G}/[\mathscr{R},\mathscr{R}]$ be a splitting homomorphism, and let $t(\mathscr{G}/\mathscr{R}) = \mathscr{H}/[\mathscr{R},\mathscr{R}] \subset \mathscr{G}/[\mathscr{R},\mathscr{R}]$ be the image of $\mathscr{G}/\mathscr{R}$. It is clear that $\mathscr{H}$ is the extension

$$0 \longrightarrow [\mathscr{R},\mathscr{R}] \rightarrow \mathscr{H} \overset{t'}{\underset{\rightarrow}{\leftarrow}} \mathscr{H}/[\mathscr{R},\mathscr{R}] \underset{\rightarrow}{\leftarrow} 0$$

and that it splits, say, by t'. Then $s = t' \circ t : \mathscr{G}/\mathscr{R} \rightarrow \mathscr{H} \subset \mathscr{G}$ is a homomorphism and the top sequence also splits.

The proof is completed by extending the above reasonings to any n, q.e.d.

6.6 Higher cohomology groups

Theorem 6.5.1 extends to the higher cohomology groups in the following way:

(6.6.1) THEOREM (*Whitehead*)

Let $\mathscr{G}$ be a finite-dimensional semisimple Lie algebra over a field of characteristic zero and let V be a finite-dimensional irreducible $\rho(\mathscr{G})$-module such that $\rho(\mathscr{G})V \neq 0$ (ρ non-trivial). Then

$$H_\rho^q(\mathscr{G}, V) = 0 \quad \forall q \geq 0 \quad . \tag{6.6.1}$$

If $q = 0$, the non-triviality of ρ and the irreducibility imply that $\rho(\mathscr{G}) \cdot v = 0$ ($v \in V$) holds only for $v = 0$. This means [(6.1.10)] that $H_\rho^0(\mathscr{G}, V) = 0$. The cases $q = 1, 2$ were discussed before; and for the higher cases it is possible to proceed 'similarly'.

Thus, for semi-simple Lie algebras the cohomology groups with values in $V = K$ are the interesting ones; they correspond to cohomology groups of Lie groups. In the semisimple case nothing is gained by studying cohomology groups over representations. In fact, any representation D decomposes into irreducible representations and this carries with it a direct decomposition of the cohomology groups. Hence $H_D^q(\mathscr{G}, V)$ is isomorphic with the direct sum of several copies $H_{\rho_i}^q(\mathscr{G}, V_i)$.

In Theorem 6.6.1, the non-triviality of ρ was important for $q \neq 1, 2$. This may be illustrated by the following

(6.6.2) THEOREM

If $\mathscr{G}$ is semisimple and $\rho = 0$, $H_0^3(\mathscr{G}, K) \neq 0$.

Proof: To show that $H_0^3 \neq 0$ we look for a non-trivial three-cocycle. Let $\omega \in C^3(\mathscr{G}, K)$ be defined by

$$\omega(X_1, X_2, X_3) = k([X_1, X_2], X_3) \tag{6.6.2}$$

where k is the Killing metric of $\mathscr{G}$ ((1.8.8); see also (1.8.16)). The three-cochain ω is actually a three-cocycle. From (6.1.3) with $\rho = 0$ we obtain

$$\begin{aligned}
s\omega(X_1, X_2, X_3, X_4) = &-k([[X_1, X_2], X_3], X_4) + k([[X_1, X_3], X_2], X_4)\\
&-k([[X_1, X_4], X_2], X_3) - k([[X_2, X_3], X_1], X_4)\\
&+k([[X_2, X_4], X_1], X_3) - k([[X_3, X_4], X_1], X_2)
\end{aligned} \tag{6.6.3}$$

Using now the 'associativity' property of the trace, valid for any three $n \times n$ matrices X, Y, Z and which is the expression of the bi-invariance of the Killing tensor,

$$\mathrm{Tr}([X, Y]Z) = \mathrm{Tr}(X[Y, Z]) \quad , \tag{6.6.4}$$

(6.6.3) is seen to be equal to zero (the bi-invariance of the cochain (6.6.2), $\omega([X, X_1], X_2, X_3) + \omega(X_1, [X, X_2], X_3) + \omega(X_1, X_2, [X, X_3]) = 0$ (cf. (1.8.5) for forms), is in fact sufficient to prove that ω is indeed a cocycle representing a cohomology class in H_0^3, see next footnote). It remains to be checked that ω cannot be identically zero. If the three-cocycle ω is zero, this means that $k([X, Y], Z) = 0$ $\forall X, Y, Z$. The Lie algebra $\mathscr{G}$ being semisimple, this implies $k(X, Z) = 0$ $\forall X, Z$ which contradicts the fact that k is non-singular for a semisimple algebra, q.e.d.

6.7 The Chevalley-Eilenberg formulation of the Lie algebra cohomology and invariant differential forms on a Lie group G

Let V now be R and ρ trivial, and let us discuss the cohomology groups $H^n(\mathcal{G}, R)$ where $\mathcal{G}$ is the Lie algebra of the (connected) group G. The starting point is the definition of the n-dimensional cochain. The skew-symmetric linear mapping in (6.1.1) may be identified with the value $\omega(e)$ at $e \in G$ of an n-covariant tensor field on a Lie group G, and $\mathcal{G}$ with $T_e(G)$ (this passage is sometimes referred to as *localization*). Then the covariant tensor field ω defined by [(1.6.3)],

$$\omega(g) = (L_{g^{-1}})^*(e)\omega(e) \quad , \tag{6.7.1}$$

defines an LI differential n-form on G. It was shown in section 1.8 that, if ω is LI,

$$d\omega(X_1^L, ..., X_{n+1}^L) = \sum_{\substack{i,j \\ i<j}} (-)^{i+j} \omega([X_i^L, X_j^L], X_1^L, ..., \hat{X}_i^L, ..., \hat{X}_j^L, ..., X_{n+1}^L) \quad .$$

$$\tag{6.7.2}$$

It is also clear that the LI n-form ω may be expressed in terms of a linear combination with *constant* coefficients of (exterior) products of n elements of the basis of $\mathcal{X}^{*L}(G)$ defined by the LI one-forms $\omega^{L(i)}$ on G, and that if ω is LI, so is $d\omega$ (the expression of $d\omega$ in terms of the $\omega^{L(i)}$ is most easily obtained from that of ω by using the Maurer-Cartan equations (1.7.4)). Thus, the exterior differential d acting on *left-invariant* forms plays the same rôle as the operator s in (6.1.2) for the trivial representation ρ acting on the cochains on $\mathcal{G}$.

The following definitions translate the Lie algebra cohomology Definitions (6.1.1–3) to the language of differential forms on the Lie group.

(6.7.1) DEFINITIONS (*n-cocycles and n-coboundaries given by differential forms*)

Let G be a Lie group. An n-cochain is a LI differential n-form on G.
   The (vector space) $Z_G^n = \ker d^{(n)}$ is the group of n-cocycles; its elements are the LI n-forms on G which are closed.
   The (vector space) $B_G^n = \mathrm{range}\, d^{(n-1)}$ is the group of the n-coboundaries, or LI n-forms on G which are given by the exterior differential of an LI (n−1)-form on G.

Let us denote the n-th cohomology group Z_G^n/B_G^n by $E^n(G)$. Then the proof of the following theorem is already contained in the previous discussion:

(6.7.1) THEOREM

Let $\mathscr{G}$ be the Lie algebra of a Lie group G. Then $H_0^n(\mathscr{G}, R)$ is isomorphic with the Chevalley-Eilenberg cohomology group $E^n(G)$ obtained by using the LI differential n-forms on G. The ring $H_0(\mathscr{G}, R)$ is isomorphic with the ring $E(G)$ (obtained, respectively, by using the wedge product of antisymmetric tensors and the exterior product of differential forms).

The content of the previous discussion is simply that the Lie algebra cohomology can also be described, via the localization, in terms of differential forms on the group. Let us apply this explicitly to the case of $H_0^2(\mathscr{G}, R)$. We have already seen that C_{ij}^θ and μ_{ij} of (6.4.4) and (6.4.9) define skew-symmetric mappings $\mathscr{G} \times \mathscr{G} \to R$ by (6.4.11) and (6.4.13) respectively. Then the localization process, which takes the Lie algebra elements $X_{(i)} \in T_e(g)$ to LIVFs $X^L(g) \in \mathscr{X}^L(G)$ on the group, transforms the antisymmetric bilinear mappings into LIFs on G for which the cocycle condition just expresses that they are (de Rham) closed forms.

If the group G is in addition compact, the following theorem also holds true.

(6.7.2) THEOREM

Let now G be a compact and connected Lie group, $K = R$ and ρ trivial. Then every de Rham cohomology class of differential forms on G contains precisely one bi-invariant form. The bi-invariant forms span a ring isomorphic with the cohomology ring $H_{DR}(G)$.

As a result, two locally isomorphic compact connected Lie groups have isomorphic cohomology rings.

Example: invariant polynomials
It was shown in Theorems 6.5.2 and 6.6.2 that, for $\mathscr{G}$ semisimple, $H_0^2(\mathscr{G}, R) = 0$, $H_0^3(\mathscr{G}, R) \neq 0$. Let $G = SU(2)$. Because it is compact[*], the cohomology classes defined by the bi-invariant q-forms and the closed q-forms on G are the same. Let ω be a bi-invariant two-form (a two-cocycle). Then eq. (6.7.2) gives

$$\omega([X_1^L, X_2^L], X_3^L) - \omega([X_1^L, X_3^L], X_2^L) + \omega([X_2^L, X_3^L], X_1^L) = 0 \quad . \quad (6.7.3)$$

But the first two terms may be written as $(L_{X_1^L}\omega)(X_2^L, X_3^L)$, which is zero since ω is bi-invariant. Hence $\omega([X_2^L, X_3^L], X_1^L) = 0$ or $\omega(X, Y) = 0 \ \forall X, Y$

---

[*] In fact, as one might expect, semisimplicity is sufficient. Thinking of Weyl's unitary trick it is not difficult to assume that the 'linear properties' of all compact Lie algebras are also shared by the semisimple ones. In particular, as may be also proved directly, every cohomology class in $H_0^n(\mathscr{G}, R)$ for a semisimple algebra contains exactly one bi-invariant multilinear mapping ω_n, and the bi-invariant cocycles generate the cohomology ring $H_0(\mathscr{G}, R)$.

because G is semisimple, as expected from Whitehead's lemma. Then, $H^2_{DR}(SU(2), R) = 0$. Similarly,

$$H^3_0(su(2), R) = H^3_{DR}(SU(2), R) = H^3_{DR}(S^3, R) = R .$$

We thus conclude that $b_2(SU(2)) = 0$ and that $b_3(SU(2)) = 1$, as already well known. But this simple example also exhibits the fact that the bi-invariant differential three-form on the compact G (or the skew-symmetric tensor (6.6.2) on $\mathcal{G}$) was constructed from an invariant symmetric tensor, the Killing tensor (1.8.8). This was done by replacing $k(X_1, X_2) = \mathrm{Tr}(ad\, X_1\, ad\, X_2)$ by $k([X_1, X_2], X_3) \equiv \omega(X_1, X_2, X_3)$, indeed an antisymmetric tensor on account of (6.6.4). The important fact here is that this process may be extended to higher-order invariant tensors: by substituting commutators for all the algebra elements but one as above, it is possible to associate a skew-symmetric tensor of order $2m-1$ (and, after localization, a bi-invariant differential form) with each invariant symmetric tensor of order m. Then a generalization of Theorem 6.7.2 states that the primitive elements of the cohomology ring of a compact simple Lie group G are obtained in this way. Thus, the problem of finding the Betti numbers for the group manifold G is reduced to the problem of finding the invariant irreducible symmetric tensors on $\mathcal{G}$. This, in turn, is associated with the writing of the generalized Casimir invariants of Racah, and it is known that any simple algebra of rank l possesses l of them. The net result is that if the compact simple group has rank l, the de Rham cohomology ring has l independent generators, each one associated with an invariant symmetric tensor. We give in Table 6.7.1 the order $m_1, ..., m_l$ of the generalized Casimir polynomials associated with each simple algebra. The reader may compare the orders $(2m_i - 1)$, $i = 1, ..., l$, of the invariant differential forms associated with the polynomial with the powers of the t's inside the brackets of (1.12.2)–(1.12.5) and with the numbers in (1.12.6), and check that they are the same. In this fascinating way de Rham cohomology, representation theory, homotopy theory (and even characteristic classes) become all inter-related.

Remark

As we have seen, Theorem 6.7.1 relates ideas of the de Rham cohomology on a manifold, here the group G $[H^n_{DR}(G) = $ (closed differential n-forms on G)/(exact differential n-forms on $G)]$ to $E^n(G)$. It is important to realize that, for an arbitrary Lie group G, an n-form ω may be trivial (exact) in the de Rham cohomology and still define a non-trivial element of the Chevalley-Eilenberg (Lie algebra) cohomology $E(G)$; for this, it is sufficient that ω not be the exterior differential of a *left-invariant* $(n-1)$-form on G. Thus, $H^n_{DR}(G)$ may be trivial without $E^n(G)$ being trivial.

Table 6.7.1. Order of the l generalized Casimir invariants

Lie algebra	Dimension	$m_1, m_2, ..., m_l$
A_l $(su(l+1))$	$(l+1)^2 - 1$	$2, 3, ..., l+1$
B_l $(so(2l+1))$	$l(2l+1)$	$2, 4, ..., 2l$
C_l $(sp(2l))$	$l(2l+1)$	$2, 4, ..., 2l$
D_l $(so(2l))$	$l(2l-1)$	$2, 4, ..., 2l-2$ and l
G_2	14	$2, 6$
F_4	52	$2, 6, 8, 12$
E_6	78	$2, 5, 6, 8, 9, 12$
E_7	133	$2, 6, 8, 10, 12, 14, 18$
E_8	248	$2, 8, 12, 14, 18, 20, 24, 30$

Nevertheless, if G is compact, the CE cohomology is equal to the de Rham cohomology of G, as discussed above. These considerations are essential for the definition and classification of Wess-Zumino terms in various theories (Wess-Zumino terms are additional terms in the physical actions with specific transformation properties; they will be discussed in chapters 8 and 10).

We conclude this section by mentioning that the cohomology groups (6.1.5) associated with a non-trivial representation ρ of $\mathcal{G}$ can also be discussed in terms of differential equivariant forms, as the similarity between (2.3.6) and (6.1.2) indicates. This allows us to establish the *Chevalley-Eilenberg approach to the Lie algebra cohomology for $\rho \neq 0$*. In fact the difference between this and the previous case (Definition 6.7.1) is just the non-triviality of the action ρ. Thus, there is again a one-to-one correspondence between the Lie algebra cochains and the equivariant q-forms Ω given by (2.3.5), and the action (2.3.6) of d on these differential forms mimics the action of s (eq. (6.1.2)) on the Lie algebra cochains for a non-trivial ρ. This cohomology may also be defined by using the LI q-forms $\omega^i(g)$ of eq. (2.3.7) if, on account of (2.3.11), d is now replaced by the 'covariant derivative' ∇ defined in (2.3.10).

6.8 The BRST approach to the Lie algebra cohomology

For some physical applications of the Lie algebra cohomology it is convenient to use the so-called BRST operator (for Becchi, Rouet, Stora and Tyutin) acting on BRST cochains which, as we shall see, are easily obtained from the previous cochains ω. To this end, one introduces anticommuting quantities $c^1, ..., c^r$, the *ghosts*,

$$c^i c^j = -c^j c^i \quad , \quad i, j = 1, ..., (\dim \mathcal{G}) \quad , \tag{6.8.1}$$

and defines a new nilpotent operator* $\tilde{s}$ by

$$\tilde{s} = c^i \rho(X_{(i)}) + \frac{1}{2} C_{ij}^k c^j c^i \frac{\partial}{\partial c^k} \quad , \tag{6.8.2}$$

where C_{ij}^k are the structure constants, $[X_{(i)}, X_{(j)}] = C_{ij}^k X_{(k)}$, and ρ is a representation of $\mathscr{G}$ on V. Clearly, this may be generalized to the case of graded Lie groups; then the c's are commuting when they are both associated with odd indices. With this convention, and due to the comment after (6.1.7), the treatment that follows will apply to both the left and right actions by remembering the copy of $\mathscr{G}$ to which the generator X belongs.

(6.8.1) PROPOSITION

The BRST operator is nilpotent,

$$\tilde{s}^2 = 0 \quad . \tag{6.8.3}$$

Proof: First, we write $\tilde{s}$ as

$$\tilde{s} = c^i N_{(i)} \quad , \quad N_{(i)} = \rho(X_{(i)}) + \frac{1}{2} c^j C_{ji}^k \frac{\partial}{\partial c^k} \equiv N_{(i)}^1 + \frac{1}{2} N_{(i)}^2 \quad . \tag{6.8.4}$$

Note that the operator $N_{(i)}$ has two commuting independent pieces, each of them being a representation of the Lie algebra ($N_{(i)}^2 = c^j C_{ji}^k \frac{\partial}{\partial c^k}$ is, in fact, the adjoint representation written in terms of the ghost parameters). This means that

$$[N_{(i)}, N_{(j)}] = C_{ij}^k (N_{(k)}^1 + \frac{1}{4} N_{(k)}^2) \quad . \tag{6.8.5}$$

The nilpotency of $\tilde{s}$ follows now from the anticommuting properties of the c's,

$$\begin{aligned}
\tilde{s}^2 &= c^i N_{(i)} c^j N_{(j)} = \frac{1}{2} c^i c^j [N_{(i)}, N_{(j)}] + \frac{1}{2} c^i (N_{(i)}^2 . c^j) N_{(j)} \\
&= \frac{1}{2} c^i c^j C_{ij}^k \rho_{(k)} + \frac{1}{2} c^i c^j C_{ji}^k \rho_{(k)} = 0 \quad ,
\end{aligned} \tag{6.8.6}$$

since $c^i c^j C_{ji}^k N_{(k)}^2 = 0$ by the Jacobi identity, q.e.d.

---

* The original BRST transformation s was put forward in the context of quantized Yang-Mills theories, where fields $\omega^i(x)$ of unnatural statistics or *ghosts* are introduced to compensate the effect of the quantum propagation of the unphysical degrees of freedom of the gauge fields. The Faddeev-Popov Lagrangian, including the ghost and the gauge fixing terms, is invariant under the BRST transformation ζs where ζ is a spacetime independent Grassmann parameter. On the physical Yang-Mills and spinor fields, the action of the BRST transformation is a gauge transformation of parameters $\zeta \omega^i(x)$, $i = 1,...,$ dim G, so that the BRST invariance includes the ordinary gauge invariance of the 'classical' part of the Lagrangian. The operator $\tilde{s}$ here should not be confused with the complete BRST symmetry operator of Yang-Mills theories although we shall use the same name for it.

To give a description of the Lie algebra cohomology in terms of $\tilde{s}$, the definition of the cochains ω of (6.1.1) is modified. The new BRST n-cochains $\tilde{\omega}$ (we shall use the tilde to refer to BRST cohomology entities) are defined by

$$\tilde{\omega} = \omega(X_{(i_1)}, ..., X_{(i_n)})c^{i_1}...c^{i_n} \equiv c^{i_1}...c^{i_n}\omega_{i_1...i_n} \quad , \tag{6.8.7}$$

where $\omega_{i_1...i_n}$ is antisymmetric. The ω's depend on n ghost parameters and as a result it is said that they have *ghost number* n. The fact that $\tilde{s}$ on $\tilde{\omega}$ reproduces the action of the cohomology operator s on ω is exhibited by

(6.8.1) THEOREM

The action of $\tilde{s}$ on the BRST cochains $\tilde{\omega}$ is consistent with the action of s on the above Lie algebra cochains ω, i.e. the BRST operator satisfies

$$\tilde{s}\tilde{\omega} = \frac{1}{(n+1)}(\widetilde{s\omega}) , \tag{6.8.8}$$

where

$$(\widetilde{s\omega}) = c^{i_1}...c^{i_{n+1}}(s\omega)(X_{(i_1)}, ..., X_{(i_{n+1})}) . \tag{6.8.9}$$

As a result, $\tilde{s}$ and s generate the same Lie algebra cohomology.

Proof : First of all, we compute

$$\tilde{s}\tilde{\omega} \equiv \left(c^i \rho(X_{(i)}) + \frac{1}{2}c^i c^j C_{ji}^k \frac{\partial}{\partial c^k} \right) \omega(X_{(i_1)}, ..., X_{(i_n)})c^{i_1}...c^{i_n}$$

$$= \rho(X_{(i)}).\omega(X_{(i_1)}, ..., X_{(i_n)})c^i c^{i_1}...c^{i_n}$$

$$+ \frac{n}{2}c^i c^j C_{ji}^k \omega(X_{(k)}, X_{(i_2)}, ..., X_{(i_n)})c^{i_2}...c^{i_n} \quad ,$$

where we have used the antisymmetry of ω and of the c's to bring always to the left the ghost variable which is being derived. Thus,

$$\tilde{s}\tilde{\omega} = (-1)^n \rho(X_{(i_{n+1})})\omega(X_{(i_1)}, ..., X_{(i_n)})c^{i_1}...c^{i_{n+1}}$$

$$+ \frac{n}{2}c^i c^j c^{i_2}...c^{i_n}\omega([X_{(j)}, X_{(i)}], X_{(i_2)}, ..., X_{(i_n)}) \quad . \tag{6.8.10}$$

We may now compare this expression with that of $(\widetilde{s\omega})$, which is given by

$$(\widetilde{s\omega}) = c^{i_1}...c^{i_{n+1}}\left\{ \sum_{a=1}^{n+1}(-1)^{a+1}\rho(X_{(i_a)}).\omega(X_{(i_1)}, ..., \hat{X}_{(i_a)}, ..., X_{(i_{n+1})}) \right.$$

$$\left. + \sum_{\substack{a,b \\ a<b}}^{n+1}(-1)^{a+b}\omega([X_{(i_a)}, X_{(i_b)}], X_{(i_1)}, ..., \hat{X}_{(i_a)}, ..., \hat{X}_{(i_b)}, ..., X_{(i_{n+1})}) \right\} .$$

$$\tag{6.8.11}$$

The first term of (6.8.11) can be written as

$$c^{i_1}...c^{i_{n+1}} \sum_{a=1}^{n+1}(-1)^{a+1}(-1)^{n+1-a}\rho(X_{(i_{n+1})}).\omega(X_{(i_1)},...,X_{(i_n)})$$

$$= (n+1)c^{i_1}...c^{i_{n+1}}(-1)^n\rho(X_{(i_{n+1})}).\omega(X_{(i_1)},...,X_{(i_n)})\,,$$

(6.8.12)

where we have renamed the indices i_{n+1}, i_a, which gives a minus sign, and then moved $X_{(i_a)}$ to the left, which produces an additional factor $(-1)^{n-a}$, hence the $(-1)^{n+1-a}$ above. Thus, this term is $(n+1)$ times the first term of (6.8.10). The second term of (6.8.11) gives, interchanging also the indices $i_1 \leftrightarrow i_a$ and $i_{n+1} \leftrightarrow i_b$ and putting again $X_{(i_a)}$, $X_{(i_b)}$ in the natural ordering,

$$c^{i_1}...c^{i_{n+1}} \sum_{\substack{a,b \\ a<b}}^{n+1}(-1)^{a+b}(-1)^{n-b}(-1)^{-a}\omega([X_{(i_1)},X_{(i_{n+1})}],X_{(i_2)},...,X_{(i_n)})$$

$$= (-1)^n\frac{(n+1)n}{2}c^jc^{i_2}...c^{i_n}c^i\omega([X_{(j)},X_{(i)}],X_{(i_2)},...,X_{(i_n)})$$

(6.8.13)

$$= \frac{1}{2}(n+1)nc^ic^jc^{i_2}...c^{i_n}\omega([X_{(j)},X_{(i)}],X_{(i_2)},...,X_{(i_n)})\,.$$

This is $n+1$ times the second term of (6.8.10), and then $\tilde{s}\tilde{\omega} = (n+1)\tilde{s}\tilde{\omega}$ follows, q.e.d.

It is worth emphasizing at this stage the parallel which exists between the BRST and Chevalley-Eilenberg formulations of the Lie algebra cohomology for a non-trivial action ρ on V. First, we notice that from the definition (6.8.2) it follows that

$$\tilde{s}c^k = -\frac{1}{2}C_{ij}^k c^i c^j\,.$$

(6.8.14)

This is, in fact, a rewriting of the Maurer-Cartan equation (eq. (1.7.4)) in terms of the ghosts (notice that, due to the conventions given after (6.8.2), the equation (6.8.14) always has a minus sign on its right hand side since the structure constants would include the additional minus sign for the right algebra if appropriate). Now, if the CE cochains (the invariant n-forms on G, $n \leq r$) are written in terms of the invariant one-forms $\omega^{(i)}(g)$ on G, they have the expression

$$\omega^t(g) = \omega^t_{i_1...i_n}\omega^{(i_1)} \wedge ... \wedge \omega^{(i_n)}\,,$$

(6.8.15)

where t is the V index and $\omega^t_{i_1...i_n}$ are constants. Clearly, eq. (6.8.15) corresponds to (6.8.7), and the action of $\nabla = d + \omega^{(i)}\rho(X_{(i)})$ on $\omega^t(g)$ is the same as that of $\tilde{s}$ on $\tilde{\omega}$, since $d\omega^{(k)}$ in (6.8.15) can be replaced by $-\frac{1}{2}C_{ij}^k\omega^{(i)} \wedge \omega^{(j)}$ (the second term in (6.8.2)) on account of the Maurer-Cartan equations, and $\omega^{(i)}\rho(X_{(i)})$ acts on $\omega^t(g)$ in the same way as $c^i\rho(X_{(i)})$

does on $\tilde{\omega}$, since the anticommuting character of the invariant one-forms and the ghosts plays the same rôle in both cases. Notice, finally, that these considerations may be extended similarly to *graded Lie algebras* since the grading would modify in the same manner the anticommuting properties of the ghosts and those of the invariant one-forms on the graded Lie group.

6.9 Lie algebra cohomology vs. Lie group cohomology

Let G, A be connected Lie groups, and let A be abelian.

(6.9.1) DEFINITION (*Extensions of Lie groups*)

An extension $(\tilde{G}, \pi)$ of G by A is a connected Lie group $\tilde{G}$ such that A is a closed invariant subgroup of $\tilde{G}$, and π is a continuous homomorphism of $\tilde{G}$ into G the kernel of which is A. (Although we take A to be abelian, the definition of extension is of course independent of this fact.)

The extension problem is defined, as usual, by giving a (continuous) homomorphism $\sigma : G \rightarrow \text{Aut } A$. It is then clear that σ induces a mapping

$$\rho : \mathscr{G} \rightarrow \text{End } \mathscr{A} \tag{6.9.1}$$

Let G now be simply connected (see the remark in section 4.3). The discussion in section 6.4 was contained in the following

(6.9.1) THEOREM

Let G, A be connected Lie groups, G simply connected and A abelian and let G act on A (a vector space V) through the representation σ. Then $H^2_\sigma(G, A)$ is isomorphic to $H^2_\rho(\mathscr{G}, \mathscr{A})$.[†]

Since the zero-cohomology V-valued spaces of a Lie group and its algebra are the same (they are the subspace of V on which they operate trivially, eqs. (5.1.15) and (6.1.9)), and if G is simply connected the one-cohomology groups are also the same (the representation ρ of $\mathscr{G}$ may be extended to the representation σ of G), the immediate question is whether the above theorem can be extended for $H^r, r > 2$. It turns out that for the higher cohomology groups a similar correspondence may be established. In fact, Van Est has proved the following

---

[†] An example where the theorem does not apply is the six-parameter, two-dimensional Galilei group, in which the rotation subgroup is $O(2)$. There, $\dim H^2_0$ is 2(3) for the group (algebra).

(6.9.2) THEOREM

Let G be a connected Lie group which is homologically trivial in $1, ..., n$
dimensions. Then its first n cohomology groups on G are canonically isomorphic with the corresponding n cohomology groups on $\mathscr{G}$, whereas the
$(n + 1)$-st cohomology group on G is isomorphic with a subgroup of the
$(n + 1)$-st Lie algebra cohomology group.*[†]

The proof of this theorem is rather long, and the reader may wish to
skip the next few pages till contact is made with the physical application of
section 3.8 at the end. The complete demonstration requires establishing
a mapping b between the A- and $\mathscr{A}$-valued n-cochains ($A \approx \mathscr{A} \approx V$)
$\alpha_n(g_1, ..., g_n)$, $\omega_n(X_1, ..., X_n)$ on G and $\mathscr{G}$ respectively such that

a) b is a homomorphism and $b \circ \delta = s \circ b$, *i.e.* Diagram 6.9.1 is commutative

$$
\begin{array}{ccc}
\alpha_n & \xrightarrow{\ b\ } & \omega_n \\
\delta \downarrow & & \downarrow s \\
\delta\alpha_n & \xrightarrow{\ b\ } & b(\delta\alpha_n)
\end{array}
$$

Diagram 6.9.1

(this property guarantees that cohomologous cocycles are mapped onto
cohomologous cocycles since, if α_{n+1} is a group coboundary, $\alpha_{n+1} = \delta\alpha_n$,
$b(\delta\alpha_n) = s(b\alpha_n)$, which is an algebra coboundary),

b) the mapping b is an isomorphism on the cohomology classes.
We shall not prove here in detail the isomorphism between $H^n(G, A)$ and
$H^n(\mathscr{G}, \mathscr{A})$ and restrict ourselves to deriving the explicit formulae that give
the n-cocycle ω_n on $\mathscr{G}$ (α_n on G) from the n-cocycle α_n on G (ω_n on $\mathscr{G}$).

---

[*] For example, groups for which the exponential mapping $exp : \mathscr{G} \to G$ is a diffeomorphism
between the underlying manifolds of $\mathscr{G}$ and G are called *exponential groups* and as a result they
have trivial homology; any subgroup of an exponential group is also exponential. This is the
case of the connected and simply connected nilpotent groups, for which the Baker-Campbell-
Hausdorff formula has a finite number of terms. All exponential groups are solvable, but not
necessarily nilpotent: the group of transformations $x' = ax+b$, $a > 0$, (algebra $[X_{(a)}, X_{(b)}] = X_{(b)}$)
is exponential and solvable, but not nilpotent. Thus, nilpotent $\Rightarrow$ exponential $\Rightarrow$ solvable (the
oscillator group and the Euclidean group E_2 in two dimensions are solvable but not exponential).
A group may be however homologically trivial without being exponential: this is the case of the
universal covering of E_2, which is contractible ($\sim R^3$) but not exponential.
[†] For instance, since $SU(2)$ is simple $H^{1,2}(su(2), R) = 0$ and $H^3(su(2), R) \neq 0$, but $H^{1,2,3}(SU(2), R) =$
0 (in fact, it may be shown that all cohomology groups H^n, $n \geq 1$, of a compact group with
coefficients in a vector space are trivial).

a) Let $X_{(i)}$ $(i = 1, ..., r)$ be a basis of $\mathcal{G}$, and let $X_k = \zeta_k^{i_k} X_{(i_k)}$ $(k = 1, ..., n;$ $i_k = 1, ..., r)$. Then the mapping $b : \alpha_n \in C^n(G, A) \mapsto b(\alpha_n) = \omega_n \in C^n(\mathcal{G}, \mathcal{A})$ defined* by

$$(b\alpha_n)(X_1, ..., X_n) := \zeta_1^{i_1} ... \zeta_n^{i_n} \partial_{g_1^{[i_1}} ... \partial_{g_n^{i_n]}} \alpha_n(g_1, ..., g_n)|_{g=e} , \tag{6.9.2}$$

where $g = e$ is shorthand for $(g_1 = g_2 = ... = g_n = e)$ and there is an antisymmetrization with respect to $i_1, ..., i_n$ (no weight factorial included), associates an n-dimensional cochain on $\mathcal{G}$ ($b\alpha_n$) to the cochain α_n on G. Clearly b is a homomorphism, $b(\alpha_n + \alpha'_n) = b(\alpha_n) + b(\alpha'_n)$. Let $\sigma(g) \in \text{Aut } A$ be a left action. Then $\rho(X) \in \text{End } \mathcal{A}$ is given by

$$\rho(X_{(k)}) = \partial_{g^k} \sigma(g)|_{g=e} . \tag{6.9.3}$$

The fact that σ is a left group action implies in turn that ρ is an algebra homomorphism:

$$[\rho(X'), \rho(X)] = \zeta'^{[i} \zeta^{j]} \partial_{g'^i} \partial_{g^j} \sigma(g') \sigma(g)|_{g'=g=e}$$

$$= \zeta'^i \zeta^j \partial_{g'^{[i}} \partial_{g^{j]}} \sigma(g'g)|_{g'=g=e} = \zeta'^i \zeta^j \partial_{g''^l} \sigma(g'') \frac{\partial^2 g''^l}{\partial_{g'^{[i}} \partial_{g^{j]}}} \Bigg|_{g'=g=e}$$

$$= \zeta'^i \zeta^j C_{ij}^l \rho(X_{(l)}) = \rho([X', X]) , \tag{6.9.4}$$

where we have written $g'' = g'g$ to use the chain rule and eq. (1.2.16). (A trivial modification of this also shows that if σ defines a right action,

---

* This mapping b can be written in a slightly different way: if $X_1, ..., X_n$ are elements of $\mathcal{G}$ then

$$b\alpha_n(X_1, ..., X_n) = \partial_1(X_{[1}) ... \partial_n(X_{n]}) \alpha_n(g_1, ..., g_n)|_{g=e} , \tag{1}$$

where $\partial_i(X)$ acts on an n-cochain as

$$\partial_i(X) \alpha_n(g_1, ..., g_n) = \frac{d}{dt} [\alpha_n(g_1, ..., g_{i-1}, (\exp tX) g_i, g_{i+1}, ..., g_n)]_{t=0} . \tag{2}$$

Indeed, eq. (2) can be written, with $X = \zeta^k X_{(k)}$, $e^{tX} \equiv e^{h^k X_{(k)}} = h$, $hg_i = g''_i$ and using the chain rule, as

$$\partial_i(X) \alpha_n(g_1, ..., g_n) = \zeta^k \frac{\partial}{\partial h^k} [\alpha_n(g_1, ..., g_{i-1}, hg_i, g_{i+1}, ..., g_n)] \Bigg|_{h=e}$$

$$= \zeta^k \frac{\partial g''^l}{\partial h^k} \Bigg|_{h=e} \frac{\partial}{\partial g''^l} \alpha_n(g_1, ..., g_{i-1}, g''_i, g_{i+1}, ..., g_n) \Bigg|_{g''_i = g_i} . \tag{3}$$

Then, inserting this into (1), we obtain

$$b\alpha_n(X_1, ..., X_n) = \zeta_{[1}^{i_1} ... \zeta_{n]}^{i_n} \frac{\partial}{\partial g''^{l_1}_1} ... \frac{\partial}{\partial g''^{l_n}_n} \alpha_n(g''_1, ..., g''_n) \Bigg|_{g''_1 = ... = g''_n = e} \frac{\partial g''^{l_1}_1}{\partial h_1^{i_1}} \Bigg|_{\substack{h_1 = e \\ g_1 = e}} ... \frac{\partial g''^{l_n}_n}{\partial h_n^{i_n}} \Bigg|_{\substack{h_n = e \\ g_n = e}} , \tag{4}$$

and, by virtue of (1.1.15),

$$b\alpha_n(X_1, ..., X_n) = \zeta_{[1}^{i_1} ... \zeta_{n]}^{i_n} \frac{\partial}{\partial g_1^{i_1}} ... \frac{\partial}{\partial g_n^{i_n}} \alpha_n(g_1, ..., g_n) \Bigg|_{g_1 = ... = g_n = e} \tag{5}$$

which is (6.9.2).

$\sigma(g')\sigma(g) = \sigma(gg')$, then the induced ρ becomes an algebra antiautomorphism, $[\rho(X'), \rho(X)] = -\rho([X', X])$).

Now we have to check the commutativity of Diagram 6.9.1. Using (6.9.2) and (5.1.3), we obtain

$$(b\delta\alpha_n)(X_1, ..., X_{n+1}) = \zeta_1^{i_1}...\zeta_{n+1}^{i_{n+1}} \partial_{g_1^{[i_1}}...\partial_{g_{n+1}^{i_{n+1}]}}(\delta\alpha_n)(g_1, ..., g_{n+1})|_{g=e}$$

$$= \zeta_1^{i_1}...\zeta_{n+1}^{i_{n+1}} \partial_{g_1^{[i_1}}...\partial_{g_{n+1}^{i_{n+1}]}}\{\sigma(g_1)\alpha_n(g_2, ..., g_{n+1})$$

$$+ \sum_{k=1}^{n}(-1)^k\alpha_n(g_1, ..., g_{k-1}, g_kg_{k+1}, g_{k+2}, ..., g_{n+1})\}|_{g=e},$$

$$(6.9.5)$$

where the last term in (5.1.3) has disappeared under the action of $\partial_{g_{n+1}}$ since it does not depend on g_{n+1}. Singling out ∂_{g_1} and removing its index i from the antisymmetrization, the first term in (6.9.5) gives, on account of (6.9.3) and (6.9.2),

$$\zeta_1^{i_1}...\zeta_{n+1}^{i_{n+1}}\{\sum_{k=1}^{n+1}(-1)^{k+1}\partial_{g_1^{i_k}}\partial_{g_2^{[i_1}}...\partial_{g_k^{i_{k-1}}}\partial_{g_{k+1}^{i_{k+1}}}...\partial_{g_{n+1}^{i_{n+1}]}}\sigma(g_1)\alpha_n(g_2, ..., g_{n+1})\}|_{g=e}$$

$$= \zeta_1^{i_1}...\zeta_{n+1}^{i_{n+1}}\sum_{k=1}^{n+1}(-1)^{k+1}\rho(X_{(i_k)})(b\alpha_n)(X_{(i_1)}, ..., \hat{X}_{(i_k)}, ..., X_{(i_{n+1})})$$

$$= \sum_{k=1}^{n+1}(-1)^{k+1}\rho(X_k)(b\alpha_n)(X_1, ..., \hat{X}_k, ..., X_{n+1}),$$

$$(6.9.6)$$

which coincides with the first term of (6.1.2).

For the second term of (6.9.5) we get, writing $g''(g_k, g_{k+1})$ for the group element g_kg_{k+1} and using the chain rule and taking again eq. (1.2.16) into account,

$$\frac{1}{2}\zeta_1^{i_1}...\zeta_{n+1}^{i_{n+1}}\sum_{k=1}^{n}(-1)^k\partial_{g_1^{[i_1}}...\partial_{g_{k-1}^{i_{k-1}}}\left[\frac{\partial g''^l}{\partial g_k^{i_k}\partial g_{k+1}^{i_{k+1}}} - \frac{\partial g''^l}{\partial g_k^{i_{k+1}}\partial g_{k+1}^{i_k}}\right].$$

$$\partial_{g''^l}\partial_{g_{k+2}^{i_{k+2}}}...\partial_{g_{n+1}^{i_{n+1}]}}.\alpha_n(g_1, ..., g_{k-1}, g'', g_{k+2}, ..., g_{n+1})|_{g=e}$$

$$(6.9.7)$$

$$= \frac{1}{2}\zeta_1^{i_1}...\zeta_{n+1}^{i_{n+1}}\sum_{k=1}^{n}(-1)^k\partial_{g_1^{[i_1}}...\partial_{g_{k-1}^{i_{k-1}}}C_{i_ki_{k+1}}^l\partial_{g''^l}\partial_{g_{k+2}^{i_{k+2}}}...\partial_{g_{n+1}^{i_{n+1}]}}.$$

$$\alpha_n(g_1, ..., g_{k-1}, g'', g_{k+2}, ..., g_{n+1})|_{g=e},$$

where the antisymmetrization applies to the indices $i_1, ..., i_k, i_{k+1}, ..., i_{n+1}$ and not to the dummy index l (the term $\frac{\partial^2\alpha}{\partial g''^l\partial g''^m}\frac{\partial g''^l}{\partial g_k^{i_k}}\frac{\partial g''^m}{\partial g_{k+1}^{i_{k+1}}}$, which is also obtained from (6.9.5), has been dropped because at $g_k = e = g_{k+1}$ the derivatives of g'' give $\delta_{i_k}^l\delta_{i_{k+1}}^m$ [(1.1.15)] and so is zero on account of the antisymmetry in $i_k i_{k+1}$). Using the antisymmetrization in $i_1, ..., i_{n+1}$ (l not included) and shifting the C's to the left, eq. (6.9.7) gives

$$\frac{1}{(n-1)!}\frac{1}{2}\zeta_1^{i_1}...\zeta_{n+1}^{i_{n+1}}\sum_{k=1}^{n}(-1)^k C_{[i_1 i_2}^l \partial_{g_1^{[i_3}}...\partial_{g_{k-1}^{i_{k+1}}}\partial_{g''}^l \partial_{g_{k+2}^{i_{k+2}}}...\partial_{g_{n+1}^{i_{n+1}]}}\cdot$$
$$\alpha_n(g_1, ..., g_{k-1}, g'', g_{k+2}, ..., g_{n+1})|_{g=e} \,, \tag{6.9.8}$$

where the additional $1/(n-1)!$ compensates for the additional antisymmetrization introduced in the indices $i_3, ..., i_{k+1}, i_{k+2}, ..., i_{n+1}$. To compare (6.9.8) with the second term in (6.1.2) we first notice that

$$\partial_{g''[l}\partial_{g_3^{i_3}}...\partial_{g_{n+1}^{i_{n+1}]}}\alpha_n(g'', g_3, ..., g_{n+1})|_{g=e}$$
$$= \partial_{g''l}\partial_{g_3^{[i_3}}...\partial_{g_{n+1}^{i_{n+1}]}}\alpha_n(g'', g_3, ..., g_{n+1})|_{g=e}$$
$$+ \sum_{k=2}^{n}(-1)^{k-1}\partial_{g''[i_3}\partial_{g_3^{i_4}}...\partial_{g_k^{i_{k+1}}}\partial_{g''}^l \partial_{g_{k+2}^{i_{k+2}}}...\partial_{g_{n+1}^{i_{n+1}]}}\alpha_n(g'', g_3, ..., g_{n+1})|_{g=e} \,,$$
$$\tag{6.9.9}$$

where in the l.h.s. (r.h.s.) the l is (is not) included in the antisymmetrization. Since the derivatives are taken at $g = e$, the group element labels are unimportant; only the order of the arguments matters. Replacing $(g'', g_3, ..., g_k, g_{k+1})$ by $(g_1, g_2, ..., g_{k-1}, g'')$ in the second term, (6.9.9) gives

$$\partial_{g''l}\partial_{g_3^{[i_3}}...\partial_{g_{n+1}^{i_{n+1}]}}\alpha_n(g'', g_3, ..., g_{n+1})|_{g=e} + \sum_{k=2}^{n}(-1)^{k-1}$$
$$\partial_{g_1^{[i_3}}...\partial_{g_{k-1}^{i_{k+1}}}\partial_{g''}^l\partial_{g_{k+2}^{i_{k+2}}}...\partial_{g_{n+1}^{i_{n+1}]}}\alpha_n(g_1, g_2, ..., g_{k-1}, g'', g_{k+2}, ..., g_{n+1})|_{g=e}$$
$$= -\sum_{k=1}^{n}(-1)^{k}$$
$$\partial_{g_1^{[i_3}}...\partial_{g_{k-1}^{i_{k+1}}}\partial_{g''}^l\partial_{g_{k+2}^{i_{k+2}}}...\partial_{g_{n+1}^{i_{n+1}]}}\alpha_n(g_1, g_2, ..., g_{k-1}, g'', g_{k+2}, ..., g_{n+1})|_{g=e}\cdot$$
$$\tag{6.9.10}$$

Consequently, (6.9.8) is

$$-\frac{1}{(n-1)!}\frac{1}{2}\zeta_1^{i_1}...\zeta_{n+1}^{i_{n+1}}C_{[i_1 i_2}^l\partial_{g''[l}\partial_{g_3^{i_3}}...\partial_{g_{n+1}^{i_{n+1}]}}\alpha_n(g'', g_3, ..., g_{n+1})|_{g=e} \,, \tag{6.9.11}$$

which on using (6.9.2) becomes

$$-\frac{1}{(n-1)!}\frac{1}{2}\zeta_1^{i_1}...\zeta_{n+1}^{i_{n+1}}C_{[i_1 i_2}^l(b\alpha_n)(X_{(l)}, X_{(i_3)}, ..., X_{(i_{n+1})]})$$
$$= -\frac{1}{(n-1)!}\frac{1}{2}\zeta_1^{i_1}...\zeta_{n+1}^{i_{n+1}}(b\alpha_n)([X_{[(i_1)}, X_{(i_2)}], X_{(i_3)}, ..., X_{(i_{n+1})]})\cdot$$
$$\tag{6.9.12}$$

The above expression includes an antisymmetrization in all indices $i_1, ..., i_{n+1}$. The indices in the Lie bracket can be removed from the antisymmetrization by writing (6.9.12) as

$$-\frac{1}{(n-1)!}\frac{1}{2}\zeta_1^{i_1}...\zeta_{n+1}^{i_{n+1}}$$

$$.2\sum_{r<s}(-1)^{(r-1)+(s-2)}(b\alpha_n)([X_{(i_r)}, X_{(i_s)}], X_{[(i_1)}, ..., \hat{X}_{(i_r)}, ..., \hat{X}_{(i_s)}, ..., X_{(i_{n+1})]}) \; ;$$

(6.9.13)

in this form, the antisymmetrization does not affect i_r, i_s. Removing the remaining antisymmetrization in $(b\alpha_n)$ implies adding a factor $(n-1)!$; thus, (6.9.13) is equal to the second term in (6.1.2),

$$\sum_{r<s}(-1)^{r+s}(b\alpha_n)([X_r, X_s], X_1, ..., \hat{X}_r, ..., \hat{X}_s, ..., X_{n+1}) \; .$$

(6.9.14)

This shows that $b\delta\alpha_n = \delta b\alpha_n$.

b) Let us now see, given an n-cocycle ω_n of the Lie algebra, how to construct an n-cocycle (which we will write formally as $\alpha_n = \tilde{b}\omega_n$) of the group, such that, if ω_n is a coboundary, $\tilde{b}\omega_n$ is also a coboundary, and fulfilling the condition $b\tilde{b}\omega_n = \omega_n$. This provides a way of obtaining the finite (group) cocycles from the infinitesimal (Lie algebra) ones.

The formula for $\tilde{b}$ can be obtained as follows. Consider first, given a group element g, the one-parameter group $g(t) = e^{tX}$ (for $g = e^X$) that joins the identity element e with g, and write $h(t, g) \equiv g(t)$. For n group elements $g_1, ..., g_n$, take the 'simplex' $S_n[g_1, ..., g_n]$ with vertices e, g_1, $g_1g_2, ..., g_1...g_n$ that can be described recursively in terms of the parameters $t_k \in [0, 1]$, $k = 1, ..., n$, as the set of elements of G defined by

$$S_1(t_1; g_1) = h(t_1, g_1) = g(t_1) \quad , \quad h(0, g) = e \quad ,$$

$$............$$

(6.9.15)

$$S_n(t_1, ..., t_n; g_1, ..., g_n) = h(t_1, g_1 S_{n-1}(t_2, ..., t_n; g_2, ..., g_n)) \quad .$$

For instance,

$$S_2(t_1, t_2; g_1, g_2) = h(t_1, g_1 h(t_2, g_2)) \quad ,$$

$$S_3(t_1, t_2, t_3; g_1, g_2, g_3) = h(t_1, g_1 h(t_2, g_2 h(t_3, g_3))) \quad ,$$

(6.9.16)

etc. These parameters do not describe univocally the points of $S_n[g_1, ..., g_n]$. For some purposes, a more appropriate set is $\lambda_1 = t_1$, $\lambda_2 = t_1 t_2, ...,$ $\lambda_k = t_1...t_k$, $k = 1, ..., n$.

Let us see the figure described by a simplex $S_n[g_1, ..., g_n]$. For instance, for $S_3[g_1, g_2, g_3]$, this is a 'tetrahedron' with vertices e, g_1, g_1g_2, $g_1g_2g_3$ (which are obtained by setting $t_1 = 0$; $t_1 = 1, t_2 = 0$; $t_1 = t_2 = 1, t_3 = 0$; $t_1 = t_2 = t_3 = 1$ respectively), and 'faces' $g_1 S_2[g_2, g_3]$, $S_2[g_1g_2, g_3]$, $S_2[g_1, g_2g_3]$,

$S_2[g_1, g_2]$ (corresponding to the values $t_1 = 1$, $t_2 = 1$, $t_3 = 1$, $t_3 = 0$ in $S_3(t_1, t_2, t_3; g_1, g_2, g_3)$). This (or direct inspection of $S_n[g_1, ..., g_n]$) shows that in the general case the vertex e then corresponds to the value $t_1 = 0$; the vertex $g_1...g_k$ ($k < n$) is obtained when $t_1 = ... = t_k = 1$, $t_{k+1} = 0$; and, when $t_1 = ... = t_n = 1$, $g_1, ..., g_n$ is obtained. The boundary of $S_n[g_1, ..., g_{n+1}]$ is made of the pieces obtained by any one of the t's equal to 1, plus the piece corresponding to $t_n = 0$. When $t_1 = 1$, one gets $g_1 S_n(t_2, ..., t_{n+1}; g_2, ..., g_{n+1})$ which parametrizes $g_1 S_n[g_2, ..., g_{n+1}]$. For $t_k = 1$ ($k > 1$), the resulting piece is $S_n[g_1, ..., g_{k-2}, g_{k-1}g_k, g_{k+1}, ..., g_n]$, and $t_n = 0$ gives $S_n[g_1, ..., g_n]$. Taking into account the relative orientations, it is possible to write formally for the boundary $\partial S_{n+1}[g_1, ..., g_{n+1}]$ of the n-simplex

$$\partial S_{n+1}[g_1, ..., g_{n+1}] = g_1 S_n[g_2, ..., g_{n+1}]$$
$$+ \sum_{k=1}^{n} (-1)^k S_n[g_1, ..., g_{k-1}, g_k g_{k+1}, g_{k+2}, ..., g_{n+1}] \quad (6.9.17)$$
$$+ (-1)^{n+1} S_n[g_1, ..., g_n] \quad ,$$

the positive (negative) sign meaning that the corresponding piece is oriented outwards (inwards).

We are now in a position to give $\tilde{b}\omega_n$. Let us call $\hat{\omega}^n(g)$ the left-equivariant form on G associated with the Lie algebra n-cochain ω_n as described in section 2.3. Then we define

$$(\tilde{b}\omega_n)(g_1, ..., g_n) \equiv \int_{S_n[g_1,...,g_n]} \hat{\omega}^n(g) \quad . \quad (6.9.18)$$

Let us now prove that, with this definition, the property $\delta\tilde{b}\omega_n = \tilde{b}s\omega_n$ is fulfilled. From eq. (5.1.3) we obtain

$$(\delta\tilde{b}\omega_n)(g_1, ..., g_{n+1}) = \sigma(g_1)(\tilde{b}\omega_n)(g_2, ..., g_{n+1})$$
$$+ \sum_{k=1}^{n} (-1)^k (\tilde{b}\omega_n)(g_1, ..., g_{k-1}, g_k g_{k+1}, g_{k+2}, ..., g_{n+1})$$
$$+ (-1)^{n+1} (\tilde{b}\omega_n)(g_1, ..., g_n)$$
$$= \sigma(g_1) \int_{S_n[g_2,...,g_{n+1}]} \hat{\omega}^n(g)$$
$$+ \sum_{k=1}^{n} (-1)^k \int_{S_n[g_1,...,g_{k-1},g_k g_{k+1},g_{k+2},...,g_{n+1}]} \hat{\omega}^n(g)$$
$$+ (-1)^{n+1} \int_{S_n[g_1,...,g_n]} \hat{\omega}^n(g) \quad .$$

$$(6.9.19)$$

The first term in this expression is (see (2.3.1))

$$\int_{S_n[g_2,...,g_{n+1}]} \sigma(g_1)\hat{\omega}^n(g) = \int_{S_n[g_2,...,g_{n+1}]} (L_{g_1})^* \hat{\omega}^n(g_1 g) \quad . \tag{6.9.20}$$

Now, if we perform the change of variables $g \to g_1 g$ in the integration, the integral is converted into (eq. (1.4.56))

$$\int_{g_1 S_n[g_2,...,g_{n+1}]} \hat{\omega}^n(g) \quad . \tag{6.9.21}$$

Introducing (6.9.21) in (6.9.19), we get, using (6.9.17) and Stokes' theorem,

$$(\delta \tilde{b}\omega_n)(g_1,...,g_{n+1}) = \int_{\partial S_{n+1}[g_1,...,g_{n+1}]} \hat{\omega}^n(g) = \int_{S_{n+1}[g_1,...,g_{n+1}]} d\hat{\omega}^n(g)$$

$$= \int_{S_{n+1}[g_1,...,g_{n+1}]} \widehat{s\omega}^n(g) \equiv (\tilde{b}s\omega_n)(g_1,...,g_{n+1}) \quad , \tag{6.9.22}$$

where we have used that $d\hat{\omega} = \widehat{s\omega}$ (see section 6.7 and Theorem 2.3.2). From this it follows easily that, if ω_n is a Lie algebra cocycle (coboundary), $\tilde{b}\omega_n$ is also a Lie group cocycle (coboundary).

It only remains to show that $b\tilde{b}\omega_n = \omega_n$. Using (6.9.2) we obtain

$$b\tilde{b}\omega_n(X_1,...,X_n) = \zeta_1^{i_1}...\zeta_n^{i_n} \left(\partial_{g_1^{[i_1}}...\partial_{g_n^{i_n]}} \int_{\Lambda_n} \hat{\omega}^n \right) \bigg|_{g_1 = ... = g_n = e} \quad , \tag{6.9.23}$$

where Λ_n is the polyhedron in R^n determined by the n parameters. Since (6.9.23) is taken at the unity, it is seen that, expressing $\hat{\omega}(g)$ in local coordinates,

$$\hat{\omega}^n(g) = \frac{1}{n!}\hat{\omega}^n_{i_1,...,i_n}(g)dg^{i_1} \wedge ... \wedge dg^{i_n} \quad , \tag{6.9.24}$$

the only term that contributes is that linear in all group parameters. Since the coordinates of $g = h(t_1, g_1)$ are given by $g^i = t_1 g_1^i$ and to first order the group law is additive, the lowest order contribution to $g^i(t_1,...,t_n) = S_n^i(t_1,...,t_n; g_1,...,g_n)$ is

$$g^i(t) = t_1 g_1^i + t_1 t_2 g_2^i + ... + t_1...t_n g_n^i \equiv \lambda_1 g_1^i + ... + \lambda_n g_n^i = g^i(\lambda) \quad , \tag{6.9.25}$$

where we are using now $\lambda_k = t_1...t_k$ to characterize the simplex. Since $dg^{i_k} = \frac{\partial g^{i_k}}{\partial \lambda_k} d\lambda_k = g_k^{i_k} d\lambda_k$, the term we are looking for is

$$\int_{\Lambda_n} d\lambda_1 \wedge ... \wedge d\lambda_k \, \hat{\omega}^n_{i_1...i_n}(e)g_1^{i_1}...g_n^{i_n} \quad . \tag{6.9.26}$$

The derivatives $\partial_{g_k^{i_k}}$ of this term reduce the computation of the algebra cocycle to an integral over the λ's. Since $\int_{\Lambda_n} d^n\lambda = \int_0^1 d\lambda_1 \int_0^{\lambda_1} d\lambda_2 \dots \int_0^{\lambda_{n-1}} d\lambda_n = \frac{1}{n!}$, eq. (6.9.23) gives

$$(b\tilde{b}\omega_n)(X_1, ..., X_n) = \zeta_1^{i_1} ... \zeta_n^{i_n} \omega_{i_1...i_n}(e) = \omega(X_1, ..., X_n) \quad , \tag{6.9.27}$$

because the equivariant form at the identity element coincides with the cocycle (there is a factor $n!$ due to the antisymmetrization in (6.9.23), which cancels the factor $\frac{1}{n!}$ in (6.9.24)).

On the cohomologous classes b and $\tilde{b}$ are inverse of each other, q.e.d.

Clearly, had we defined σ to be a right action, the additional minus sign in (6.1.6) would have come from the fact that while the second term in $(b\delta\alpha_n)$ gives $(-1)^{n+1}$ times (6.9.14) (cf. the second term of (5.1.4)), in the first term one has $\sigma(g_{n+1})\alpha_n(g_1, ..., g_n)$ which gives $(-1)^n$ times (6.9.6) due to the fact that it is the group element g_{n+1} which appears now as the argument of σ. For the right action, the formula for $\tilde{b}$ is analogous, with the left-equivariant form replaced by a right equivariant one, and with the integral defined over the 'simplex' of vertices $g_1...g_n$, $g_2...g_n$,..., $g_{n-1}g_n$, g_n, e.

After this long mathematical digression, it may be worthwhile to consider a physical example. A first possibility is to find the expression for the two-cocycle of the Galilei group from the CE algebra two-cocycle which, as will be seen in chapter 8, is proportional to $\omega^{L(V)} \wedge \omega^{L(A)} = d\mathbf{V} \wedge (d\mathbf{A} - \mathbf{V}d\mathbf{B})$ [(3.4.5)]; the result is yet another expression for the Galilei two-cocycle[*]. Another one is the case of the magnetic monopole discussed in section 3.8, where G was the translation group Tr_3 and the infinitesimal three-cocycle was given by

$$\omega_{ijk}(\mathbf{x}) = q\epsilon_{ijk}\nabla \cdot \mathbf{B}(\mathbf{x}) \quad . \tag{6.9.28}$$

Let us assume that we can extend the previous reasonings to the case of $\mathscr{F}(M)$-valued cochains (see next section). Since (6.9.28) is an $\mathscr{F}(R^3)$-valued cochain, the action of the translation $\mathbf{a}$ is given by $(\sigma(\mathbf{a})\mathbf{B})(\mathbf{x}) = \mathbf{B}(\mathbf{x} + \mathbf{a})$, and this means identifying $\omega_{ijk}(\mathbf{x})$ with $\omega_{ijk}(\mathbf{b}, \mathbf{x})$ at $\mathbf{b} = 0$ that

$$\hat{\omega}^3(\mathbf{b}, \mathbf{x}) = \frac{1}{3!}q\epsilon_{ijk}\mathbf{B}(\mathbf{x} + \mathbf{b})db^i \wedge db^j \wedge db^k = q\mathbf{B}(\mathbf{x} + \mathbf{b})db^1 \wedge db^2 \wedge db^3 \quad , \tag{6.9.29}$$

where db^i are the invariant forms on the translation group. Then (6.9.18) gives

$$(\tilde{b}\omega_3)(\mathbf{a}'', \mathbf{a}', \mathbf{a}) = q \int_T d^3b\, \nabla \cdot \mathbf{B}(\mathbf{x} + \mathbf{b}) \quad , \tag{6.9.30}$$

---

[*] Answer: $\frac{1}{2}(\mathbf{V}'\mathbf{A} + \mathbf{V}'^2\mathbf{B} - \mathbf{V}\mathbf{A}') + \frac{1}{3}(\mathbf{V}'\mathbf{V}\mathbf{B}' - \mathbf{V}'^2\mathbf{B}) + \frac{1}{6}(\mathbf{V}^2\mathbf{B}' - \mathbf{V}'\mathbf{V}\mathbf{B})$.

where T is the tetrahedron in the group space of vertices $\mathbf{a}''$, $\mathbf{a}'' + \mathbf{a}'$, $\mathbf{a}'' + \mathbf{a}' + \mathbf{a}$ as we used in section 3.8 (the difference in tetrahedron vertices with respect to section 3.8 is due to the fact that here we are considering the left action on the cochains).

6.10 $\mathscr{F}(M)$-valued Lie algebra cohomology

We conclude this chapter by extending for completeness the Lie algebra cohomology to the case where $V = \mathscr{F}(M^m)$ (M being m-dimensional) in much the same way that it was done in section 5.4 for the Lie group cohomology. We shall restrict ourselves to the appropriate definitions. The n-cochains are mappings

$$\omega_n : \overset{n}{\mathscr{G} \times \cdots \times \mathscr{G}} \to \mathscr{F}(M) \quad , \quad \omega_n : (X_1,...,X_n) \mapsto \omega_n(\,\cdot\,;X_1,...,X_n) \quad (6.10.1)$$

and

$$\omega_n(\,\cdot\,;X_1,...,X_n) : M \to R \quad , \quad \omega_n(\,\cdot\,;X_1,...,X_n)(x) = \omega_n(x;X_1,...,X_n) \quad . \tag{6.10.2}$$

With the natural addition of functions, $C^n(\mathscr{G},\mathscr{F}(M))$ has an abelian group structure.

Let $\rho(X) = Y$ be the vector field associated with the action of G on the manifold M. This induces in a natural way an action on the elements of $C^n(\mathscr{G},\mathscr{F}(M))$ where the vector fields act as derivations:

$$\rho(X)\omega_n(\,\cdot\,;X_1,...,X_n)(x) = Y.\omega_n(x;X_1,...,X_n) \,. \tag{6.10.3}$$

Using (6.10.3), the coboundary operator is defined by

$$(s\omega_n)(x;X_1,...,X_{n+1}) = \sum_{a=1}^{n+1}(-1)^{a+1}Y_a.\omega_n(x;X_1,...,\hat{X}_a,...,X_{n+1})$$

$$+ \sum_{\substack{a,b \\ a<b}}^{n+1}(-1)^{a+b}\omega_n(x;[X_a,X_b],X_1,...,\hat{X}_a,...,\hat{X}_b,...,X_{n+1}) \,. \tag{6.10.4}$$

Notice that Y is a vector field on M, but that $X_1,...,X_n \in \mathscr{G}$.

As in the group case, the $\mathscr{F}(M)$-valued n-cochains ω_n may be generalized to $\Lambda^p(M)$-valued n-cochains. In this case, they will be denoted ω_n^p, and in the above expressions $\rho(X)$ will then be the Lie derivative L_Y (which of course reduces to the ordinary directional derivative when $p = 0$).

Bibliographical notes for chapter 6

General information on the cohomology of Lie algebras is given in the books of Jacobson (1962), Hilton and Stammbach (1970), Greub *et al.* (1972, 1973, 1976) and Knapp (1988). The Lie algebra cohomology in terms of invariant differential forms on the group is given in the classic paper of Chevalley and Eilenberg (1948), where a proof of theorem 6.6.1 may be found; see also Eilenberg (1948). For the generalized Casimir invariants mentioned in section 6.7 see Racah (1950, 1951); see also Gruber and O'Raifeartaigh (1964), and Okubo and Patera (1985). More specific topics on Lie algebra cohomology are discussed in the mathematical papers of Koszul (1950), van Est (1953, 1955), Hochschild and Serre (1953b), Sullivan (1977), and in the book of Guichardet (1980). Advanced general texts are Cartan and Eilenberg (1956) and Mac Lane (1963); this last reference contains a detailed bibliography on the cohomology of groups and algebras. For the cohomology of vector fields on a manifold see Guillemin (1973). The cohomology of finite- and infinite-dimensional Lie algebras in general is discussed in the book by Fuks (1986).

The BRST cohomology was introduced by Becchi *et al.* (1976) and Tyutin (1975); see also Henneaux (1985). For a general description see, *e.g.*, Henneaux and Teitelboim (1992). For the mathematical background of the BRS cohomology see also Koszul (1950) (or Greub *et al.* (1976)). For further developments see Brandt *et al.* (1990b), van Holten (1990) and Stasheff (1992).

The relation between the cohomology of Lie groups and algebras is discussed by Hochschild (1951) and specially in van Est (1953, 1955), where a complete proof of Theorem 6.9.2 and its refinements may be found; see also Guichardet (1980) and Houard (1980). The relative Lie algebra cohomology (not treated in the text), which is useful in these discussions, was introduced in Chevalley-Eilenberg (1948); see also Hochschild and Serre (1953b) and Guichardet (1980).

In connection with the properties of solvable and exponential algebras see Dixmier (1957) and Bernat *et al.* (1972).

The cohomology of superalgebras was introduced by Leĭtes (1975); see also chapter two of Fuks (1986) and the review by Voronov (1992). See also D'Auria *et al.* (1980) in connection with supergravity.

7

Group extensions by non-abelian kernels

This short chapter contains some results of the theory of extensions by non-abelian groups. It is shown that the study of such extensions can be reduced to an abelian cohomology problem: a G-kernel is extendible iff some three-cocycle with values in the centre of K, C_K, is trivial, and the possible extensions of G by K are in one-to-one correspondence with the extensions of G by the abelian group C_K. The obstruction to the extension is given by a three-cocycle.

As a physical example the eight 'covering' groups of the complete Lorentz group are found (although this problem does not necessarily require the use of the techniques explained here).

The chapter is intended to provide a more complete description of the problem of group extensions, but may be omitted in a first reading since it is not necessary for the rest of the book.

7.1 The information contained in the G-kernel (K,σ)

Let us go back to the general problem of extending an *abstract* group G (chapter 4) and consider the case where K is not abelian. The main difference from the abelian case of chapter 5 is that not every G-kernel (K,σ) is associated with one or more extensions $\tilde{G}$ of G by K: not every G-kernel (K,σ) is extendible. In fact, one of the aims of this chapter is to show that the G-kernel (K,σ) determines an *obstruction* to the extension in the form of a certain three-cocycle; the G-kernel (K,σ) is extendible iff this cocycle is trivial.

Before attacking the problem of giving a necessary and sufficient condition for $\tilde{G}$ to exist, let us further analyse the information contained in the data, the G-kernel (K,σ). Consider first the exact sequence (4.2.4),

$$1 \to \text{Int } K \to \text{Aut } K \underset{\to}{\overset{t}{\leftarrow}} \text{Out } K \to 1 \quad , \tag{7.1.1}$$

264

which expresses Aut K as the extension of Out K by Int K. Let t be a trivializing section and let $\alpha = t \circ \sigma$ be defined by

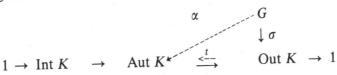

Diagram 7.1.1

It is clear that there exists an element $h(g', g) \in K$ such that

$$\alpha(g')\alpha(g) = [h(g', g)]\alpha(g'g) \tag{7.1.2}$$

since $\alpha(g')\alpha(g)$ and $\alpha(g', g)$ differ by an internal automorphism. Consequently, (7.1.2) defines a mapping

$$[h] : G \times G \to \text{Int } K \quad , \quad [h] : (g', g) \mapsto [h(g', g)] \quad , \tag{7.1.3}$$

$$[h(g', g)]k := h(g', g)\, k\, h^{-1}(g', g) \quad . \tag{7.1.4}$$

The associative property in Aut K, $\alpha(g'')(\alpha(g')\alpha(g)) = (\alpha(g'')\alpha(g'))\alpha(g)$, leads to the two-cocycle property for $[h(g', g)] \in Z^2_\alpha(G, \text{Int } K)$,

$$[\alpha(g'')h(g', g)]\, [h(g'', g'g)] = [h(g'', g')]\, [h(g''g', g)] \quad , \tag{7.1.5}$$

where

$$[\alpha(g'')h(g', g)] = \alpha(g'')\, [h(g', g)]\alpha(g'')^{-1} \quad , \tag{7.1.6}$$

or

$$[(\alpha(g'')h(g', g))h(g'', g'g)] = [h(g'', g')h(g''g', g)] \quad . \tag{7.1.7}$$

Eq. (7.1.7) implies that the elements

$$(\alpha(g'')h(g', g))h(g'', g'g) \quad , \quad h(g'', g')h(g''g', g)$$

of K determine the same element of Int K. Thus, they differ by an element of K which determines the identity of Int K, i.e. they differ by an element of the centre C_K. Thus the equality (7.1.7) in Int K leads to an equality in K,

$$(\alpha(g'')h(g', g))h(g'', g'g) = f_3(g'', g', g)h(g'', g')h(g''g', g) \quad ; \tag{7.1.8}$$

notice that $h(g', g)$ would itself be a two-cocycle for $f_3 = 1$. Eq. (7.1.8) determines a mapping $f_3 : G \times G \times G \to C_K$, i.e. a three-cochain on G with values in the *abelian* group C_K.

(7.1.1) THEOREM

$f_3 \in Z^3_{\sigma_0}(G, C_K)$, *where* $\sigma_0(g) = \sigma(g)$ *acting on* C_K *(Diagram 4.2.3) where it coincides with* $\alpha(g)$.

Proof: To prove the theorem it is sufficient to write f_3 as

$$f_3(g'', g', g) = (\alpha(g'')h(g', g))h(g'', g'g)h^{-1}(g''g', g)h^{-1}(g'', g') \qquad (7.1.9)$$

and use this expression to check that f_3 satisfies the cocycle condition,

$$(\delta f_3)\,(g'', g', g, g_1) = (\alpha(g'')f_3(g', g, g_1))f_3^{-1}(g''g', g, g_1).$$

$$f_3(g'', g'g, g_1) \cdot f_3^{-1}(g'', g', gg_1) \cdot f_3(g'', g', g) = 0 \quad ,$$
$$(7.1.10)$$

which is obtained from (5.1.3) written in multiplicative notation, *q.e.d.*

Having done this, the next step is to prove

(7.1.2) THEOREM

The G-kernel (K, σ) characterizes an element of the third cohomology group $H^3_{\sigma_0}(G, C_K)$.

Proof: We have to show that, if the choice of t (which modifies α) or that of h in (7.1.8) ([$h(g', g)$] is uniquely determined by (7.1.2), but this fixes h only up to an element of C_K) is modified, the new f_3 differs from the previous one by a three-coboundary. The proof is done in two steps.

a) Let h' be another element of the class $h * C_K$. Then $[h'(g', g)] = [h(g', g)]$, and it is related to h by

$$h'(g', g) = c(g', g)h(g', g) \quad , \quad c(g', g) \in C_K \quad . \qquad (7.1.11)$$

Let us now consider (7.1.9) written for f_3' and h' and replace h' in it by its expression (7.1.11),

$$f_3'(g'', g', g) = (\alpha(g'')(c(g', g)h(g', g)))c(g'', g'g)h(g'', g'g).$$
$$h^{-1}(g''g', g)c^{-1}(g''g', g)h^{-1}(g'', g')c^{-1}(g'', g') \quad . \qquad (7.1.12)$$

Since h takes values in K and c in C_K, the location of the c's is irrelevant, and (7.1.12) may be written as

$$f_3'(g'', g', g) = (\alpha(g'')c(g', g))c(g'', g'g)c^{-1}(g''g', g)c^{-1}(g'', g').$$
$$(\alpha(g'')h(g', g))h(g'', g'g)h^{-1}(g''g', g)h^{-1}(g'', g') \qquad (7.1.13)$$
$$\equiv (\delta c)(g'', g', g)f_3(g'', g', g) \quad .$$

b) Now let α' be the mapping associated to a different trivializing section t', and let $[k(g)] \in \text{Int } K$ be the element which satisfies

$$\alpha'(g) = [k(g)]\alpha(g) \quad g \in G \quad . \qquad (7.1.14)$$

A way of proving that the change of α is unimportant is to take advantage of the freedom in h to choose an h' in such a way that f_3 remains unaltered under the combined changes $\alpha \to \alpha'$, $h \to h'$.

First we find, as usual, that

$$
\begin{aligned}
\alpha'(g')\alpha'(g) &= [k(g')]\alpha(g')[k(g)]\alpha(g) \\
&= [k(g')]\ [\alpha(g')k(g)]\alpha(g')\alpha(g) \\
&= [k(g')]\ [\alpha(g')k(g)]\ [h(g',g)]\alpha(g'g) \\
&= [k(g')]\ [\alpha(g')k(g)]\ [h(g',g)]\ [k^{-1}(g'g)]\alpha'(g'g) \\
&\equiv [k(g')\ (\alpha(g')k(g))h(g',g)k^{-1}(g'g)]\alpha'(g'g) \quad .
\end{aligned}
$$

(7.1.15)

Now, looking at (7.1.15), we select an $h'(g',g)$ such that

$$
h'(g',g) = k(g')\ (\alpha(g')k(g))h(g',g)k^{-1}(g'g) \tag{7.1.16}
$$

(one may notice, at this stage, that h and h' differ in a coboundary generated by the cochain $k(g)$; see (4.4.11)).

Let us now write (7.1.9) for α' and h' above,

$$
\begin{aligned}
f'_3(g'',g',g) = &\ \alpha'(g'')\ (k(g')(\alpha(g')k(g))h(g',g)k^{-1}(g'g)). \\
&\ k(g'')\ (\alpha(g'')k(g'g))h(g'',g'g)k^{-1}(g''g'g). \\
&\ k(g''g'g)h^{-1}(g''g',g)\ (\alpha(g''g')k^{-1}(g))k^{-1}(g''g'). \\
&\ k(g''g')h^{-1}(g'',g')\ (\alpha(g'')k^{-1}(g'))k^{-1}(g'') \quad .
\end{aligned}
$$

(7.1.17)

Using the relation implied by (7.1.14),

$$
k(g'') \cdot \alpha(g'')\ (\ \ ) = \alpha'(g'')\ (\ \ ) \cdot k(g'') \quad , \tag{7.1.18}
$$

(7.1.17) becomes equal to

$$
\alpha'(g'')\ (k(g')\ (\alpha(g')k(g))h(g',g)k^{-1}(g'g)) \cdot
$$

$$
(\alpha'(g'')k(g'g))k(g'')h(g'',g'g) \cdot
$$

$$
h^{-1}(g''g',g)\ (\alpha(g''g')k^{-1}(g)) \cdot
$$

$$
h^{-1}(g'',g')\ (\alpha(g'')k^{-1}(g'))k^{-1}(g'') \quad . \tag{7.1.19}
$$

Using (7.1.18) again, (7.1.19) is seen to be equal to

$$
\begin{aligned}
&\alpha'(g'')\ (k(g')\alpha(g')k(g))k(g'') \\
&\left\{(\alpha(g'')h(g',g))h(g'',g'g)h^{-1}(g''g',g)h^{-1}(g'',g')\right\} \cdot \\
&(\alpha(g'')\alpha(g')k^{-1}(g))\ (\alpha(g'')k^{-1}(g'))k^{-1}(g'') \\
&= f_3(g'',g',g)\ k(g'')\ \alpha(g'')(k(g')\alpha(g')k(g)) \cdot \\
&\alpha(g'')\ ((\alpha(g')k^{-1}(g))k^{-1}(g'))\ k^{-1}(g'') = f_3(g'',g',g) \quad ,
\end{aligned}
$$

(7.1.20)

q.e.d.

7.2 The necessary and sufficient condition for a *G*-kernel (K, σ) to be extendible

If the *G*-kernel is extendible (Definition 4.2.2) then we have the diagram

$$
\begin{array}{ccccccc}
1 & \to K & \overset{i}{\to} \tilde{G} & \overset{\pi}{\to} & G & \to 1 \\
& & \downarrow f & & \downarrow \sigma & \\
1 & \to \operatorname{Int} K & \to \operatorname{Aut} K & \to & \operatorname{Out} K & \to 1
\end{array}
$$

Diagram 7.2.1

where $f : \tilde{G} \to \operatorname{Aut} K$ is the natural homomorphism defined by the action of the elements $\tilde{g} \in \tilde{G}$ on its subgroup K,

$$
f : \tilde{g} \mapsto [\tilde{g}] \quad , \quad [\tilde{g}]k = \tilde{g}k\tilde{g}^{-1} \quad . \tag{7.2.1}
$$

If $\tilde{G}$ is an extension, it is clear that the homomorphisms f and σ are compatible. The necessary and sufficient condition to construct the Diagrams 4.2.2, 7.2.1 turns out to be a condition on the 3-cocycle f_3 determined by the g-kernel (K, σ).

(7.2.1) THEOREM (*Eilenberg-Mac Lane*)

The G-kernel (K, σ) *is extendible iff the cocycle* f_3 *which it determines is a three-coboundary.*

Proof:

a) *if* $\tilde{G}$ *exists,* f_3 *is a coboundary:*
Let s be a trivializing section of the principal bundle $\tilde{G}(K, G)$ as in section 4.3, and let $\omega(g', g)$ be as in (4.4.1). Now define $\alpha : G \to \operatorname{Aut} K$,

$$
\begin{array}{ccccccc}
1 & \to K & \to \tilde{G} & \underset{s}{\overset{\pi}{\underset{\longleftarrow}{\rightrightarrows}}} & G & \to 1 \\
& & f\downarrow & \overset{\alpha}{\nearrow} & \downarrow \sigma & \\
1 & \to \operatorname{Int} K & \to \operatorname{Aut} K & \underset{t}{\rightleftarrows} & \operatorname{Out} K & \to 1
\end{array}
$$

Diagram 7.2.2

by (cf. Diagram 7.1.1))

$$
\alpha = f \circ s \quad , \quad \alpha(g) = f(s(g)) \equiv [s(g)] \quad . \tag{7.2.2}
$$

($\tilde{G}$ is supposed here to be known, and so is f through (7.2.1)). Then

$$\alpha(g')\alpha(g) \equiv (f \circ s)(g')(f \circ s)(g) \equiv [s(g')][s(g)] = [\omega(g',g)][s(g'g)]$$

$$\equiv [\omega(g',g)](f \circ s)(g'g) = [\omega(g',g)]\alpha(g'g) \quad . \tag{7.2.3}$$

Then, from (7.1.2), we obtain

$$[\omega(g',g)] = [h(g',g)] \quad , \tag{7.2.4}$$

and an $h(g',g)$ in its class by C_K may be chosen such that

$$\omega(g',g) = h(g',g) \quad . \tag{7.2.5}$$

It is now easy to check that $f_3 = 1$: from its definition (7.1.9) we read

$$f_3(g'',g',g) = ([s(g'')]\omega(g',g))\omega(g'',g'g)\omega^{-1}(g''g',g)\omega^{-1}(g'',g') = 1 \quad , \tag{7.2.6}$$

where the last equality follows from the two-cocycle condition (4.4.5) for ω. (Had we selected another $h(g',g)$ instead of the one which satisfies (7.2.5), we would have obtained a cohomologous cocycle, see (7.1.13).)

b) *if the f_3 determined by the G-kernel (K,σ) is trivial, then the G-kernel is extendible.*
We now follow the lower part of the diagram to define

$$\alpha = t \circ \sigma \quad , \tag{7.2.7}$$

and choose h (Theorem 7.1.2) so that $f_3 = 1$. Let us now label the elements of $\tilde{G}$ by $\tilde{g} = (k,g)$ and define the group law by

$$\tilde{g}'' = (k',g') (k,g) = (k'(\alpha(g)k)h(g',g), g'g) \quad . \tag{7.2.8}$$

Since $f_3 = 1$, $h(g',g)$ is a two-cocycle (see eq. (7.1.8)). Then the two equations

$$\alpha(g')\alpha(g) = [h(g',g)]\alpha(g'g) \quad , \tag{7.2.9}$$

$$(\alpha(g'')h(g',g))h(g'',g'g) = h(g'',g')h(g''g',g) \quad , \tag{7.2.10}$$

show immediately that the group law (7.2.8) is associative, that $\tilde{e} = (1,e)$ is the identity element and that [(4.5.16)]

$$\tilde{g}^{-1} \equiv (k,g)^{-1} = (\alpha(g)^{-1}(k^{-1}h^{-1}(g,g^{-1})) , g^{-1}) \quad . \tag{7.2.11}$$

Since we did not check the associativity property in section 4.5 let us do it here:

$$(k'', g'')((k', g')(k, g))$$
$$= (k''\alpha(g'')\{k'(\alpha(g')k)h(g', g)\}h(g'', g'g) , g''g'g)$$
$$= (k''\alpha(g'')(k'(\alpha(g')k)) (\alpha(g'')h(g', g))h(g'', g'g) , g''g'g) ;$$
$$((k'', g'')(k', g'))(k, g) \hspace{3cm} (7.2.12)$$
$$= (k''(\alpha(g'')k')h(g'', g') , g''g')(k, g)$$
$$= (k''(\alpha(g'')k')h(g'', g')(\alpha(g''g')k)h(g''g', g) , g''g'g) .$$

The terms in K in (7.2.12) have to be equal. This implies

$$(\alpha(g'')\alpha(g')k) . (\alpha(g'')h(g', g)) . h(g'', g'g) = h(g'', g') . (\alpha(g''g')k) . h(g''g', g) .$$
$$\hspace{10cm} (7.2.13)$$

On using (7.2.9) and (7.2.10), this is seen to be

$$([h(g'', g')]\alpha(g''g')k) (\alpha(g'')h(g', g))h(g'', g'g)$$

$$= h(g'', g') (\alpha(g''g')k)h^{-1}(g'', g') (\alpha(g'')h(g', g))h(g'', g'g) ,$$

an equality which is obviously fulfilled.

Thus, $h = \xi$, $t \circ \sigma = f \circ s$, and Diagram 7.2.2 is well defined. The proof is complete, since (7.2.8) obviously implies $K \lhd \tilde{G}$, $\tilde{G}/K = G$, q.e.d.

7.3 Construction of extensions

Once a G-kernel (K, σ) has been shown to be extendible, the next problem is to obtain the extensions $\tilde{G}$. In this section we describe in three steps the Eilenberg-Mac Lane theory, so that the contents of Theorem 7.3.1 below can be fully appreciated.

(a) *Product of kernels of equal centre C_K*

Let (K_1, σ_1), (K_2, σ_2) be two G-kernels *with the same centre* $C_{K_1} = C_{K_2} \equiv C_K$, and let $T = K_1 \otimes K_2$ be the group obtained by taking the direct product of K_1 and K_2. The subgroup $S = \{(z, z^{-1}) | z \in C_K\}$ is obviously an invariant subgroup of $K_1 \otimes K_2$; the quotient group

$$K = (K_1 \otimes K_2)/S \equiv T/S \hspace{3cm} (7.3.1)$$

also has C_K as its centre.

The homomorphisms $\sigma_i : G \to$ Out K_i, $i = 1, 2$, induce automorphisms of $K_1 \otimes K_2$ in the natural way:

$$\sigma_1 \times \sigma_2 : G \to \text{Out } K_1 \otimes \text{Out } K_2 , \quad (\sigma_1 \times \sigma_2)(g) : (k_1, k_2) \mapsto (\sigma_1(g)k_1, \sigma_2(g)k_2) .$$
$$\hspace{10cm} (7.3.2)$$

Since $(\sigma_1(g)z, \sigma_2(g)z^{-1}) = (\sigma_1(g)z, (\sigma_2(g)z)^{-1})$, the class S transforms into itself, and thus $(\sigma_1 \times \sigma_2)(g) \equiv \sigma(g)$ defines an element of Out K. This implies that (K, σ) is a G-kernel:

(7.3.1) DEFINITION ($\wedge$-*product of G-kernels of equal centre*)

Let (K_i, σ_i), $i = 1, 2$, *be G-kernels with the same centre* C_K, *and denote*

$$(K, \sigma) = (K_1, \ \sigma_1) \wedge (K_2, \ \sigma_2) \tag{7.3.3}$$

with K and σ defined as above. Then (K, σ) is called the C_K-product ($\wedge$) of the G-kernels (K_1, σ_1) and (K_2, σ_2) of equal centre.

The C_K-product $\wedge$ of G-kernels is associative and commutative, and the G-kernel (C_K, σ_0) (Definition 4.2.3) is the unit element for the C_K-product $\wedge$.

(b) C_K-*product of extensions*

Let $(\tilde{G}_i, \pi_i)$, $i = 1, 2$, be two extensions of the G-kernels (K_i, σ_i) of equal centre C_K. Let R be the subgroup of the direct product $\tilde{G}_1 \otimes \tilde{G}_2$ defined by

$$R = \{(\tilde{g}_1, \tilde{g}_2) | \ \pi_1(\tilde{g}_1) = \pi_2(\tilde{g}_2) = g \in G\} \quad . \tag{7.3.4}$$

The group S (defined in (a)) is invariant in R. Let $\tilde{G}$ be the quotient group

$$\tilde{G} = R/S \quad . \tag{7.3.5}$$

It is not difficult to see that $K = T/S$ (7.3.1) is invariant in $\tilde{G}$; clearly, T is also invariant in R and $R/T = G$. Then, since $R > T > S$ and $R > S$, the so-called second group isomorphism theorem tells us that it is possible to 'divide' by S so that

$$\tilde{G}/K \equiv (R/S)/(T/S) = R/T = G \quad . \tag{7.3.6}$$

Let π be the projection of $\tilde{G}$ onto G of kernel K. We may now give

(7.3.2) DEFINITION ($\wedge$-*product of extensions by G-kernels of equal centre*)

Let $(\tilde{G}_i, \pi_i)$ *be two extensions of the G-kernels (K_i, σ_i) of equal centre C_K. The group $\tilde{G}$ defined above is called the C_K-product of the extensions, and we write*

$$(\tilde{G}, \pi) \equiv (\tilde{G}_1, \pi_1) \wedge (\tilde{G}_2, \pi_2) \quad . \tag{7.3.7}$$

From the above discussion it follows that the C_K-product $(\tilde{G}, \pi)$ is an extension of G by the G-kernel (K, σ), the latter being the product of the kernels (K_1, σ_1), (K_2, σ_2) of the extensions $\tilde{G}_1, \tilde{G}_2$.

(c) Construction of the extensions

Let (K, σ) be an extendible G-kernel, and assume that the (much easier) abelian problem of finding all the extensions $\tilde{G}'_i$ of G by the G-kernel (C_K, σ_0) has been solved. Suppose also that an extension $\tilde{G}$ of the G-kernel (K, σ) is known. Then $\tilde{G} \wedge \tilde{G}'$ is another extension of G by (K, σ) because the G-kernel of the $\wedge$-product is the $\wedge$-product of the G-kernels and $(C_K, \sigma_0) \wedge (K, \sigma) = (K, \sigma)$. The fact that all the extensions obtained in this way are inequivalent and that this construction gives all of them is guaranteed by the Eilenberg-MacLane theorem:

(7.3.1) THEOREM

Let (K, σ) be an extendible G-kernel. The extensions $(\tilde{G}, \pi)$ of G by (K, σ) may be put in one-to-one correspondence with the elements of the second co-homology group $H^2_{\sigma_0}(G, C_K)$ which characterizes the extensions of the associated abelian problem, namely the extensions of G by the G-kernel (C_K, σ_0).

Proof: Let $\tilde{G}'_j$ be the extensions of G by the abelian G-kernel (C_K, σ_0). The proof of the theorem requires showing that

a) equivalent extensions $\tilde{G}'_j$ of the G-kernel (C_K, σ_o) give rise to equivalent extensions $\tilde{G} \wedge \tilde{G}'_j$ of the G-kernel (K, σ),

b) all extensions of the G-kernel (K, σ) are of the form $\tilde{G} \wedge \tilde{G}'_j$, and

c) inequivalent extensions $\tilde{G}'_j$ of the G-kernel (C_K, σ_o) lead to inequivalent extensions $\tilde{G} \wedge \tilde{G}'_j$ of the G-kernel (K, σ).

Since we are going to prove an equivalence of extensions, we shall not omit i in the image $i(K)$ of K, etc.

a) Let $(\tilde{G}'_1, \pi'_1)$, $(\tilde{G}'_2, \pi'_2)$ be two equivalent extensions of the G-kernel (C_K, σ_0). Then, by Definition 4.2.4, there is an isomorphism $\tilde{f} : \tilde{G}'_1 \to \tilde{G}'_2$ such that, if i'_j $(j = 1, 2)$ are the injections that map C_K into $\tilde{G}'_j$, $\tilde{f} \circ i'_1 = i'_2$ and $\pi'_1 = \pi'_2 \circ \tilde{f}$. Now, consider the mapping $id \times \tilde{f} : \tilde{G} \otimes \tilde{G}'_1 \to \tilde{G} \otimes \tilde{G}'_2$. Let R_1 and R_2 be the subgroups of $\tilde{G} \otimes \tilde{G}'_1$, $\tilde{G} \otimes \tilde{G}'_2$ defined by (7.3.4). After taking the quotients (7.3.5) by the subgroup S, a new isomorphism $\tilde{F} : \tilde{G} \wedge \tilde{G}'_1 \to \tilde{G} \wedge \tilde{G}'_2$ is induced,

$$\tilde{F} : [\tilde{g}, \tilde{h}] \in \tilde{G} \wedge \tilde{G}'_1 \mapsto [\tilde{g}, \tilde{f}(\tilde{h})] \in \tilde{G} \wedge \tilde{G}'_2, \qquad (7.3.8)$$

where $\tilde{g} \in \tilde{G}$, $\tilde{h} \in \tilde{G}'_1$, $\pi(\tilde{g}) = \pi'_1(\tilde{h}) = g \in G$, and $[\tilde{g}, \tilde{h}]$, $[\tilde{g}, \tilde{f}(\tilde{h})]$ characterize, respectively, the *classes* defined by the cosets $(\tilde{g}, \tilde{h})S$, $(\tilde{g}, \tilde{f}(\tilde{h}))S$, eq. (7.3.5). The mapping $\tilde{F}$ is well defined because $\pi'_2(\tilde{f}(\tilde{h})) = \pi'_1(\tilde{h}) = \pi(\tilde{g})$ and, if we take other representatives $\tilde{g}'$, $\tilde{h}'$, and i is the injection of K into the extension $\tilde{G}$, we have

$$(\tilde{g}', \tilde{h}') = (\tilde{g}i(z^{-1}), \tilde{h}i'_1(z))$$

for some $z \in C_K$. As a result,

$$\tilde{F}[(\tilde{g}',\tilde{h}')] = [\tilde{g}', \tilde{f}(\tilde{h}')] = [\tilde{g}i(z^{-1}), \tilde{f}(\tilde{h}i'_1(z))]$$
$$= [\tilde{g}i(z^{-1}), \tilde{f}(\tilde{h})(\tilde{f} \circ i'_1)(z)] = [\tilde{g}i(z^{-1}), \tilde{f}(\tilde{h})i'_2(z)] = [\tilde{g}, \tilde{f}(\tilde{h})] \,, \tag{7.3.9}$$

and thus $\tilde{F}$ is well defined on R_1/S, R_2/S, *i.e.* on $\tilde{G} \wedge \tilde{G}'_1$, $\tilde{G} \wedge \tilde{G}'_2$. Moreover, $\tilde{F}$ is an isomorphism since:

- $\tilde{F}$ is injective: $\tilde{F}[\tilde{g}, \tilde{h}] = \tilde{F}[\tilde{g}', \tilde{h}']$ or $[\tilde{g}, \tilde{f}(\tilde{h})] = [\tilde{g}', \tilde{f}(\tilde{h}')]$ means that there is an element $z \in C_K$ such that

$$\tilde{g} = \tilde{g}'i(z^{-1}) \,, \quad \tilde{f}(\tilde{h}) = \tilde{f}(\tilde{h}')i'_2(z) = \tilde{f}(\tilde{h}')\tilde{f}(i'_1(z)) = \tilde{f}(\tilde{h}'i'_1(z)) \tag{7.3.10}$$

and, since $\tilde{f}$ is an isomorphism, $\tilde{h} = \tilde{h}'i'_1(z)$, $[\tilde{g}, \tilde{h}] = [\tilde{g}', \tilde{h}']$.
- $\tilde{F}$ is surjective: let $[\tilde{g}', \tilde{h}']$ now be an element of $\tilde{G} \wedge \tilde{G}'_2$. Since $\tilde{f}$ is an isomorphism, there is an $\tilde{h} \in \tilde{G}'_1$ such that $\tilde{f}(\tilde{h}) = \tilde{h}'$. Then the element $[\tilde{g}', \tilde{h}] \in \tilde{G} \wedge \tilde{G}'_1$ has $[\tilde{g}', \tilde{h}']$ as its image under $\tilde{F}$ in $\tilde{G} \wedge \tilde{G}'_2$.
- $\tilde{F}$ is a homomorphism: both $\tilde{f}$ and *id* are homomorphisms.

The isomorphism $\tilde{F}$ also defines an equivalence of extensions because, if i_j, π_j denote the injections and projections for the extensions $\tilde{G}_j = \tilde{G} \wedge \tilde{G}'_j$, $j = 1, 2$, we have

- $\tilde{F} \circ i_1 = i_2$: $\forall k \in K$,

$$(\tilde{F} \circ i_1)(k) = \tilde{F}[i(k), e] = [i(k), e] = i_2(k) \in \tilde{G}_2(k) \,, \tag{7.3.11}$$

- $\pi_1 = \pi_2 \circ \tilde{F}$: $\forall [\tilde{g}, \tilde{h}] \in \tilde{G}_1$,

$$\pi_1[\tilde{g}, \tilde{h}] = \pi(\tilde{g}) = \pi_2[\tilde{g}, \tilde{f}(\tilde{h})] = \pi_2 \circ \tilde{F}[\tilde{g}, \tilde{h}] \,. \tag{7.3.12}$$

b) We now have to prove that every extension $\tilde{G}_j$ of the G-kernel (K, σ) is of the form $\tilde{G} \wedge \tilde{G}'_j$. Let s_j and s be trivializing sections in $\tilde{G}_j$ and $\tilde{G}$ such that $[s_j(g)]$ and $[s(g)]$ define the same automorphism $\alpha_j(g) = \alpha(g)$ $\forall g \in G$ (this can be done because $\tilde{G}_j$ and $\tilde{G}$ are extensions of the same G-kernel and, consequently, $\alpha_j(g)$ and $\alpha(g)$ differ by an inner automorphism which can be eliminated changing the definition of one section).[*] Then, calling ω_j and ω the corresponding factor systems and using (7.2.3) repeatedly, we have

$$\alpha_j(g')\alpha_j(g)k = [\omega_j(g', g)]\alpha_j(g'g)k = [\omega_j(g', g)]\alpha(g'g)k$$
$$= [\omega_j(g', g)\omega^{-1}(g', g)]\alpha(g')\alpha(g)k \tag{7.3.13}$$
$$= [\omega_j(g', g)\omega^{-1}(g', g)]\alpha_j(g')\alpha_j(g)k \,,$$

---

[*] The section s defines an element $\alpha \in Aut\,K$ by $i(\alpha(g)k) = [s(g)]i(k)$ $\forall k \in K$. If the image of k in $\tilde{G}$ is identified with k itself, this is just $(\alpha(g))(k) = [s(g)](k)$, eq. (7.2.2).

so that $[\omega_j(g',g)\omega^{-1}(g',g)]k = k \ \forall k \in K$. This means that

$$n_j(g',g) \equiv \omega_j(g',g)\omega^{-1}(g',g) \tag{7.3.14}$$

is an element of C_K. It is a factor system because (see eq. (4.4.5))

$$n_j(g'',g')n_j(g''g',g)$$
$$= \omega_j(g'',g')\omega^{-1}(g'',g')n_j(g''g',g)$$
$$= \omega_j(g'',g')n_j(g''g',g)\omega^{-1}(g'',g')$$
$$= \omega_j(g'',g')\omega_j(g''g',g)(\omega(g'',g')\omega(g''g',g))^{-1}$$
$$= (\alpha_j(g'')\omega_j(g',g))\omega_j(g'',g'g)\omega^{-1}(g'',g'g)(\alpha(g'')\omega(g',g))^{-1}$$
$$= (\alpha(g'')n_j(g',g))n_j(g'',g'g) \ . $$
$$\tag{7.3.15}$$

Thus $n_j(g',g) \in C_K$, being a factor system, defines an extension $\tilde{G}'_j$ of the G-kernel (C_K,σ_0). To prove that $\tilde{G} \wedge \tilde{G}'_j$ is equivalent to $\tilde{G}_j$, we compute the factor system of the $\wedge$-product for a certain trivializing section. Choosing the section s in $\tilde{G}$ and that giving the factor system $n_j(g',g)$ (denoted by v) in $\tilde{G}'_j$, we obtain a trivializing section $w = s \times v$ in $\tilde{G} \wedge \tilde{G}'_j$,

$$w : G \to \tilde{G} \wedge \tilde{G}'_j, \quad g \in G \mapsto w(g) = [s(g),v(g)] \ . \tag{7.3.16}$$

The element of $\mathrm{Aut}\,K$ given by $w(g)$ is identified by its action on $K \wedge C_K \approx K$,

$$[w(g)]([i(k),i'_j(z)]) = [[s(g)]i(k),[v(g)]i'_j(z)] \quad , \tag{7.3.17}$$

and, since the elements k can be identified with $[i(k),e]$, it is clear that the automorphism defined by w is again $\alpha(g)$. On the other hand,

$$w(g')w(g) = [s(g'),v(g')]\,[s(g),v(g)]$$
$$= [i(\omega(g',g))s(g'g),i'_j(n_j(g',g))v(g'g)] \tag{7.3.18}$$
$$= [i(\omega(g',g)),i'_j(n_j(g',g))]\,w(g'g) \ ,$$

and, since $[i(\omega(g',g)),i'_j(n_j(g',g))] = [i(\omega(g',g)n_j(g',g)),e]$, it turns out that the factor system for $\tilde{G} \wedge \tilde{G}'_j$ defined by w is given by

$$\omega(g',g)n_j(g',g) = \omega_j(g',g)\omega^{-1}(g',g)\omega(g',g) = \omega_j(g',g) \ , \tag{7.3.19}$$

which is the factor system of $\tilde{G}_j$. Thus, $\tilde{G}_j$ and $\tilde{G} \wedge \tilde{G}'_j$ are equivalent.

c) It is now sufficient to check that, if $\tilde{G}_1$ and $\tilde{G}_2$ are equivalent, $\tilde{G}'_1$ and $\tilde{G}'_2$ are also equivalent. Let $\tilde{F}$ be an isomorphism between $\tilde{G}_1$, $\tilde{G}_2$ such that $\tilde{F} \circ i_1 = i_2$ and $\pi_1 = \pi_2 \circ \tilde{F}$. Let s_1, s_2 and s be trivializing sections in $\tilde{G}_1$, $\tilde{G}_2$, $\tilde{G}$ respectively such that, as before, $\alpha_1(g) = \alpha_2(g) = \alpha(g)$. Since $\tilde{F}$ is an equivalence, we have $\tilde{F}(s_1(g)) = s_2(g)i_2(\gamma(g))$ for a certain one-cochain

γ, $\gamma(g) \in K$. Now, applying $\tilde{F}$ to the identity $[s_1(g)]i_1(k) = i_1(\alpha_1(g)k)$ and comparing the result with the corresponding identity for $\tilde{G}_2$, we get $[s_2(g)i_2(\gamma(g))] = [s_2(g)]$, *i.e.* $\gamma(g) \in C_K$. The proof is now completed by showing that $\gamma(g)$ generates the coboundary that relates the factor systems of $\tilde{G}'_1$ and $\tilde{G}'_2$. To show this, let $\tilde{F}$ act on

$$s_1(g')s_1(g) = i_1(n_1(g',g)\omega(g',g))s_1(g'g), \qquad (7.3.20)$$

where $n_1(g',g)\omega(g',g) = \omega_1(g',g)$ is the factor system of $\tilde{G}_1$, eq. (7.3.19). This gives

$$i_2(\gamma(g')\gamma(g))s_2(g')s_2(g) = i_2(n_1(g',g)\omega(g',g))s_2(g'g)i_2(\gamma(g'g)) \ . \qquad (7.3.21)$$

Using in (7.3.21) the equivalent of (7.3.20) for $\tilde{G}_2$, we get

$$n_2(g',g) = n_1(g',g)\gamma(g'g)\gamma^{-1}(g')\gamma^{-1}(g), \qquad (7.3.22)$$

which shows that the factor systems of $\tilde{G}'_1$ and $\tilde{G}'_2$ differ by a coboundary, *q.e.d.*

Remark

Since the starting extension $\tilde{G}$ is already one of the possible extensions of the G-kernel (K,σ) and all extensions are of the form $\tilde{G} \wedge \tilde{G}'_j$, there has to be a certain $\tilde{G}'_j$ such that $\tilde{G}$ is equivalent to $\tilde{G} \wedge \tilde{G}'_j$. This extension is precisely the semidirect extension of G by C_K (which always exists since K is abelian) for the mapping $\sigma_0 : G \rightarrow \text{Out } C_K$. This is immediate because the factor system of the extension $\tilde{G} \wedge \tilde{G}'_j$ coincides with that of $\tilde{G}$, once appropriate trivializing sections in $\tilde{G}'_j$ and $\tilde{G}$ are considered, as in the proof of the previous theorem.

7.4 Example: the covering groups of the complete Lorentz group *L*

As is well known, the restricted Lorentz group $L^\uparrow_+$, being a connected group, has a universal (unique) covering group, $SL(2,C)$. As such, $SL(2,C)$ is locally isomorphic to $L^\uparrow_+$, $SL(2,C)/Z_2 \approx L^\uparrow_+$, and simply connected. A question that arises is what happens if, instead of $L^\uparrow_+$, the complete group L including the spacetime reflections is considered.

The complete Lorentz group L is, as a topological group, the union of four disconnected pieces,

$$L = L^\uparrow_+ \cup L^\downarrow_+ \cup L^\uparrow_- \cup L^\downarrow_-, \qquad (7.4.1)$$

where the piece $L^{\uparrow(\downarrow)}_{+(-)}$ consists of 4×4 matrices $\Lambda^\mu_\nu \in L$ with determinant $+1(-1)$ (represented by the $\pm$ subscript) and with component $\Lambda^0_0 \geq +1 (\leq -1)$ (indicated by the up and down arrows which refer to the reversal of time). It is a semidirect extension of the group $V = Z_2 \otimes Z_2$ of spacetime reflections $\{I, P, T, PT\}$ by $L_+^\uparrow$, where

$$I = \begin{pmatrix} 1 & & & \\ & 1 & & \\ & & 1 & \\ & & & 1 \end{pmatrix}, \qquad P = \begin{pmatrix} 1 & & & \\ & -1 & & \\ & & -1 & \\ & & & -1 \end{pmatrix},$$

$$T = \begin{pmatrix} -1 & & & \\ & 1 & & \\ & & 1 & \\ & & & 1 \end{pmatrix}, \qquad PT = \begin{pmatrix} -1 & & & \\ & -1 & & \\ & & -1 & \\ & & & -1 \end{pmatrix}. \tag{7.4.2}$$

Thus, $L = L_+^\uparrow \textcircled{s} V$, $L/L_+^\uparrow = V$, where the action σ of V on $L_+^\uparrow$ is defined by

$$\sigma(R) : \Lambda \in L_+^\uparrow \mapsto R\Lambda R^{-1} \in L_+^\uparrow, \quad R \in V. \tag{7.4.3}$$

Since L is not connected, there is no 'universal' covering group as there is for connected groups. But it is reasonable to take as 'covering groups' the groups $\tilde{L}_j$ defined by being extensions of V by $SL(2, C)$ such that $\tilde{L}/Z_2 \approx L$, because L is itself an extension of V by $L_+^\uparrow$.

Let us calculate the possible extensions of a V-kernel $(SL(2, C), \sigma)$ for a certain action σ. The group V can be parametrized by (μ_1, μ_2), with μ_i $(i = 1, 2)$ equal to 0 or 1. Then $(0, 0) = I$, $(0, 1) = P$, $(1, 1) = T$ and $(1, 0) = PT$ and the group law is expressed by

$$(\mu_1', \mu_2')(\mu_1, \mu_2) = (\mu_1' + \mu_1, \mu_2' + \mu_2). \tag{7.4.4}$$

The elements $\tilde{g}$ of the groups $\tilde{L}_j$ we are looking for can be written as $\tilde{g} = (A, \mu_i)_j$ where $A \in SL(2, C)$. Once a trivializing section t in (7.1.1) is given, the group laws of $\tilde{L}_j$ can be expressed as

$$(A', \mu_i')_j (A, \mu_i)_j = (A'\alpha(\mu_i')A \,\omega_j(\mu_i', \mu_i), \mu_i' + \mu_i)_j \quad, \tag{7.4.5}$$

where $\alpha = t \circ \sigma$, $\omega_j(\mu_i', \mu_i) \in SL(2, C)$ (we shall see that, in fact, $\omega \in Z_2$) and $\sigma : V \to \text{Out}\,SL(2, C)$ is given by $\sigma(\mu_i) = K^{\mu_2}$ where K denotes complex conjugation, $K : A \mapsto A^* = KAK$; K is understood to complex-conjugate everything on its right. Explicitly, the problem is to find all factors $\omega_j(\mu_i', \mu_i)$ in

$$(A', \mu_i')_j (A, \mu_i)_j = (A'K^{\mu_2'}AK^{\mu_2'}\omega_j(\mu_i', \mu_i), \mu_i' + \mu_i)_j, \tag{7.4.6}$$

which define the different extensions $\tilde{L}_j$. The 'covering' homomorphism of $\tilde{L}_j$ onto the complete Lorentz group, $\tilde{L}_j/Z_2 = L$, $(A, \mu_i)_j \in \tilde{L}_j \to \Lambda \in L$, given by

$$(\Lambda^\mu_{.\nu} x^\nu)\sigma_\mu = (-1)^{\mu_1} AK^{\mu_2} x^\mu \sigma_\mu K^{\mu_2} A^\dagger \tag{7.4.7}$$

where $\sigma^\mu = (1_2, \vec{\sigma})$, requires $\omega_j(\mu'_i, \mu_i) \in Z_2$. This homomorphism shows that, if $\zeta \in SL(2, C)$ is given by

$$\zeta = \begin{pmatrix} 0 & 1 \\ -1 & 0 \end{pmatrix}, \quad (A^T)^{-1} = \zeta A \zeta^{-1}, \tag{7.4.8}$$

the elements $(\pm\zeta, 0, 1)_j$, $(\pm\zeta, 1, 1)_j$ and $(\pm 1_2, 1, 0)_j$ correspond in $\check{L}_j$ to P, T and PT respectively; P and T require the complex conjugation. Eq. (7.4.7) does define a two-to-one homomorphism since $(\pm A, \mu_1, \mu_2)_j$ give the same element of L. Of course, for $\mu_1 = \mu_2 = 0$ (7.4.7) reduces to the familiar homomorphism $SL(2, C)/Z_2 = L^\uparrow_+$ between the restricted Lorentz group and its universal covering group.

Since $\omega_j \in Z_2$, the complex conjugation does not act on the factors, and the compatibility condition (4.4.5) reduces to

$$\omega_j(\mu''_1 + \mu'_i, \mu_i)\omega_j(\mu''_i, \mu'_i) = \omega_j(\mu'_i, \mu_i)\omega_j(\mu''_i, \mu'_i + \mu_i). \tag{7.4.9}$$

We see here already that the problem has been reduced to finding the cohomology groups of $H^2_0(V, Z_2)$. But, following the path given by the general theory already discussed, we shall obtain the same result without using that $\omega_j(\mu'_i, \mu_i) \in Z_2$: the extensions of the V-kernel $(SL(2, C), \sigma)$ are in one-to-one correspondence with the abelian problem of extending the V-kernel $(C_{SL(2,C)} = Z_2, \sigma_0 = e)$, the solutions of which are given by $H^2_0(V, Z_2)$.

Following the general theory, let us then consider first whether the V-kernel $(SL(2, C), \sigma)$ is extendible. For it, we construct the three-cochain associated with the particular trivializing section in (7.1.1) that gives (7.4.6). Theorem 7.1.2 states that this defines an element of the third cohomology group of V with values in $C_{SL(2,C)} = Z_2$. This element can be taken to be trivial since, with the trivializing section t that we are considering, we have $\alpha(\mu_i) = K^{\mu_2}$ and

$$\alpha(\mu'_i)\alpha(\mu_i) = [h(\mu'_i, \mu_i)]\alpha(\mu'_i + \mu_i) \tag{7.4.10}$$

shows that $[h(\mu'_i, \mu_i)]$ is the identity automorphism and $h(\mu'_i, \mu_i)$ can be taken as the unit element of $SL(2, C)$. Thus, the three-cocycle constructed out of it is trivial and the V-kernel is extendible by Theorem 7.2.1.

Following Theorem 7.3.1, we first find one extension $\check{L}$, and then obtain the rest by $\wedge$-product with all in equivalent extensions $\tilde{V}'_j{}^*$ of the V-kernel $(Z_2, \sigma_0 = e)$ (since $\sigma_0 = e$, these are central extensions). Let the

---

* We write $\tilde{V}'_j$ following the notation of Theorem 7.3.1, although the prime is not necessary since here we are using $\check{L}_j$ (rather than $\tilde{V}_j$) to denote the extensions of V by $SL(2, C)$.

Table 7.4.1.

Group	$\omega_j(\mu'_i, \mu_i)$	Property of the Extension
Q	$(-1)^{\mu'_1\mu_1+\mu'_1\mu_2+\mu'_2\mu_2}$	
$Z_2 \otimes Z_2 \otimes Z_2$	1	
$Z_2 \otimes Z_4$	$\begin{cases} (-1)^{\mu'_1\mu_1} \\ (-1)^{\mu'_1\mu_1+\mu'_2\mu_2} \\ (-1)^{\mu'_1\mu_2+\mu'_2\mu_1+\mu'_2\mu_2} \end{cases}$	$(z,0,1)$ involutive $(z,1,1)$ involutive $(z,1,0)$ involutive
D_4	$\begin{cases} (-1)^{\mu'_1\mu_1+\mu'_1\mu_2} \\ (-1)^{\mu'_2\mu_1} \\ (-1)^{\mu'_2\mu_1+\mu'_2\mu_2} \end{cases}$	$(z,1,0)$ tetracyclic $(z,1,1)$ tetracyclic $(z,0,1)$ tetracyclic

starting point $\tilde{L}$ be the semidirect extension $\tilde{L} = L_+^\uparrow \otimes V$ given by

$$(A', \mu_i')(A, \mu_i) = (A'K^{\mu_2}AK^{\mu'_2}, \mu'_i + \mu_i), \qquad (7.4.11)$$

which happens to exist in this case, although for non-abelian kernels this is not always guaranteed.

The possible extensions $\tilde{V}'_j$ of V by Z_2 are most easily found by looking for all groups of eight elements that have an invariant Z_2 subgroup in their centre and such that $\tilde{V}'_j/Z_2 \approx V$. These are

$$\tilde{V}'_j = \{ Z_4 \oplus Z_2, \quad Q, \quad D_4, \quad Z_2 \otimes Z_2 \otimes Z_2 \}, \qquad (7.4.12)$$

where Q denotes the quaternionic group and D_4 is the dihedral group of eight elements[*]. There are four groups, but the same group can define different extensions depending on the way V is obtained when taking the quotient $\tilde{V}'_j/Z_2$. In this way, we obtain three different extensions from D_4 and also from $Z_4 \otimes Z_2$, due to the three projections which can be used in each case to obtain V.

Writing $(z, \mu_i)_j$, $z \in Z_2$, $j = 1, ..., 8$, for the elements of the central, non-equivalent extensions $\tilde{V}'_j$ of V by Z_2, their group laws are given by

$$(z', \mu'_i)_j(z, \mu_i)_j = (z'z\omega_j(\mu'_i, \mu_i), \mu'_i + \mu_i)_j, \qquad (7.4.13)$$

where $\omega_j(\mu'_i, \mu_i) \in Z_2$. A set of solutions $\omega_j(\mu'_i, \mu_i)$ satisfying (7.4.9) is given in Table 7.4.1. There we see that $\omega_j(0, 0) = 1$. Once the extensions $\tilde{V}'_j$ are known, all the extensions $\tilde{L}_j$ are given by $\tilde{L} \wedge \tilde{V}'_j$. More precisely, the elements of the extensions $\tilde{L}_j$ of the V-kernel $(SL(2, C), \sigma)$ are given by the elements

$$[(A, \mu_i), (z, \mu_i)_j] \qquad (7.4.14)$$

---

[*] There are five groups of eight elements. The fifth, the cyclic group of eight elements Z_8, cannot be one of the extensions sought since $Z_8/Z_2 = Z_4 \neq V$.

of the quotient R/S, where S is the invariant subgroup $S = \{[(z,0),(z,0)_j]|$ $z \in Z_2\}$ ($z^{-1} = z$ here). These extensions are equivalent to (7.4.6) with ω_j given as in Table 7.4.1. This is because from the proof of Theorem 7.3.1 we know that the factor system associated to the extensions (7.4.14) is, for a certain trivializing section, the composition of the systems of the factor extensions in the $\wedge$-product. In our case this reduces to $\omega_j(\mu'_i, \mu_i)$ and gives (7.4.6). Thus, having the same factor systems, both extensions are equivalent.

These considerations may be generalized, with not too many changes, to define the covering groups of the complete Poincaré group. This, taking into account the antiunitary nature of time reversal in quantum mechanics, leads to the study of the 'full quantum-mechanical Poincaré groups'.

7.5 A brief comment on the meaning of the higher cohomology groups $H^n(G,A)$

The interpretation of $H^n(G,A)$ for the cases $n = 0,1,2$ was given in section 5.1. We saw there, in particular, that H^2 characterizes the extensions of G by the G-module A. It was also shown in chapter 3 that H^2 equivalently appears when the group law is 'relaxed' to allow for a phase (see (3.2.2)). H^3 has been interpreted in this chapter through the Eilenberg-MacLane theorem, where it appears as a possible obstruction to the extension of G by a non-abelian kernel K. Also, by extending the reasonings which led to the above interpretation of H^2, it was shown in section 3.8 that H^3 appears in the description of non-associative systems, as Eilenberg and MacLane themselves pointed out. (It is curious to note that later MacLane remarked that such a description, which does not have an obvious meaning for higher values of n, 'did not seem very satisfactory'.)

The discussion leading to (3.8.3) may be formulated in purely abstract terms as follows. Let L be a 'non-associative group' or 'loop' (this may be used to replace the notion of group extension by the more general of 'loop prolongation'). In Eilenberg and Mac Lane's terminology, a *loop* (there will be no confusion with the loops in chapters 8 and 10) is a multiplicative system of elements such that (a) $\forall a,b \in L$, the element $ab \in L$ is uniquely defined; (b) $\exists 1 \in L$ such that $1a = a = a1$ $\forall a \in L$ and c) $\forall a,b \in L$ the equations $ax = b$ and $ya = b$ have unique solutions $x, y \in L$. It follows that the unit element is unique. The composition law defined in L is non-associative; thus, for $a,b,c \in L$, $a(bc) \neq (ab)c$. The *associator* $A(a,b,c)$ is defined by

$$a(bc) = A(a,b,c)[(ab)c] \quad ; \tag{7.5.1}$$

clearly, $A(a,b,c)$ is a measure of the deviation from associativity. Assume for simplicity that L is commutative and that the associativity law holds whenever one of the three elements involved is an associator. Then

$$a[b(cd)] = A(a, b, cd)(ab)(cd) = A(a, b, cd)A(ab, c, d)[(ab)c]d \quad ;$$
$$a[b(cd)] = A(b, c, d)a[(bc)d] = A(b, c, d)A(a, bc, d)[a(bc)]d \quad (7.5.2)$$
$$= A(b, c, d)A(a, bc, d)A(a, b, c)[(ab)c]d \quad .$$

Thus, using an additive notation we find (cf. (3.8.3) or (5.1.5), (5.1.6) for the trivial action)

$$A(b, c, d) - A(ab, c, d) + A(a, bc, d) - A(a, b, cd) + A(a, b, c) = 0 \quad . \quad (7.5.3)$$

Thus, an A satisfying (7.5.3) is a suitable generalization of a three-cocycle. Higher order associators may be defined inductively,

$$A(a_1, ..., a_{2n+1}) = A(a_1, ..., a_{2n-2}, A(a_{2n-1}, a_{2n}, a_{2n+1})) \quad . \quad (7.5.4)$$

In Eilenberg's words, associators 'have a tendency to behave like cocycles'.
There is another interpretation for $H^3(G, A)$ which extends easily to higher n. It may be shown that $H^3(G, A)$ is in one-to-one correspondence with the equivalence classes of sequences of the form

$$0 \to A \xrightarrow{i} N \xrightarrow{\alpha} E \xrightarrow{\pi} G \to 1$$

where N is a 'crossed module' over E and A is a G-module. The generalization is now (conceptually) simple: the elements of $H^n(G, A)$ can be interpreted as the equivalence classes of long exact sequences starting with ordinary G-modules and ending with a crossed sequence.

Bibliographical notes for chapter 7

The cup product of two G-kernels of equal centre is given in Eilenberg and Mac Lane (1947aII); this paper also contains the interpretation of the three-cocycle as an obstruction to the extensions by a non-abelian kernel (see also Kurosh (1960)). More details on cohomology theory of groups are given in Eilenberg and Mac Lane (1947b), in Eilenberg (1949) and in the references given in the bibliographical notes for chapters 4 and 5.
   For the extensions of the Lorentz and Poincaré groups in section 7.4 see Shirokov (1960), Wightman (1960), Wigner (1964), Michel (1964) (see also Galindo (1967) for the Poincaré algebra), Parthasarathy (1969a,b) and the book by Cornwell (1984).
   The mathematical interpretation of associators, H^3, H^n and of higher order cocycles is discussed in Eilenberg and Mac Lane (1947b), Eilenberg (1949), Holt (1979), Mac Lane (1979); see also Brown (1982). For a physical discussion of three-cocycles see the references given in the bibliographical notes for chapter 3.

8

Cohomology and Wess-Zumino
terms: an introduction

The relation between mechanics and cohomology already sketched in chapter 3 is re-analysed here from the Lagrangian point of view, itself initially introduced in the 1-jet bundle framework. It is seen how quasi-invariance for Lagrangians is related to group (or Lie algebra) central extensions. This point of view is extended by considering $\mathscr{F}(J^1(E))$-valued cocycles in such a way that the problem of quasi-invariant Lagrangians is described in terms of one-cocycles as well as two-cocycles, and they are seen to be related by the cohomological descent procedure. This general scheme will appear again in chapter 10 in the context of non-abelian, consistent, chiral gauge anomalies.

The complementary aspect of obtaining physical actions (or terms in them) from non-trivial cohomology is also studied. This leads to the concept of Wess-Zumino term on a group manifold. The cohomological descent is then applied to this picture, which exhibits the different rôle of the left and right versions of the symmetry Lie algebra involved.

Examples of these two aspects are given by using the Galilei group (also studied in chapter 3) and the supersymmetric extended objects (which may be omitted by readers not interested in supersymmetry). Finally, a Lagrangian action for the monopole (chapter 2) is studied as a different kind of Wess-Zumino term associated with quasi-invariance under gauge (rather than rigid) transformations.

8.1 A short review of the variational principle and of the Noether theorem in Newtonian mechanics

In the variational formulation of dynamical systems the starting point is the definition of the Lagrangian. Since Newton's equations are of second order in time, the Lagrangian depends on the generalized coordinates and velocities (we shall not consider here variational principles of higher order which are based on Lagrangians that depend on the second or higher order derivatives of the coordinates). Let $E = R \times N$ be the manifold parametrized by the time t (the R coordinate) and the degrees of freedom q^i (the coordinates of N). To incorporate velocities into the picture it is convenient to introduce the tangent manifold $T(N)$ to N and consider the manifold $R \times T(N) \equiv J^1(E)$, labelled by $(t, q^i, \dot{q}^i)$. Trajectories in this general *evolution* space are sections $s^1 : t \mapsto s^1(t) = (q^i(t), \dot{q}^i(t))$ of the vector bundle $J^1(E) \xrightarrow{\pi} R$ or *jet bundle*; at this stage, $q^i(t)$ and $\dot{q}^i(t)$ are unrelated.

An appropriate mathematical framework for discussing variational principles is the jet-bundle theory (see example (e) after Definition 1.3.3), and we shall use some of its techniques in what follows. The main advantage of this formalism is that it allows us to work directly on the $(2n + 1)$-dimensional manifold $R \times T(N) \equiv J^1(E)$ rather than on the space of functions $q^i(t)$, $dq^i(t)/dt$. In fact, the above vector bundle $J^1(E) \xrightarrow{\pi} R$ is called the bundle of the 1-jets of the bundle $E \xrightarrow{\pi} R$. Sections s of the bundle $E \xrightarrow{\pi} R$ are given by $s(t) = q^i(t)$; these can be prolonged to sections $\bar{s}^1$ of $J^1(E) \xrightarrow{\pi} R$, $\bar{s}^1 \equiv j^1(s)$ (where j^1 or the bar over a section s indicates 1-*jet prolongation*), which are given by $\bar{s}^1(t) = (q^i(t), \dot{q}^i(t) = dq^i(t)/dt)$. Thus, a section s^1 of $J^1(E) \xrightarrow{\pi} R$ is *the* jet prolongation $\bar{s}^1$ of a section $s(t)$ of E if its second component is the derivative of the first. Physical trajectories or 'histories' are sections $s(t)$; the velocities $dq^i(t)/dt$ are included in the '1-jet' sections $\bar{s}^1(t)$. For $\bar{s}^1(t)$, the dot in $\dot{q}^i(t)$ does indicate time derivative; for $s^1(t)$, it does not.

Similar considerations can be made for vector fields. An arbitrary vector field X^1 on $J^1(E)$ is a section of the tangent bundle $\tau(J^1(E))$. In local coordinates $(t, q^i, \dot{q}^i)$, a general vector field X^1 is written

$$X^1 = X^t \frac{\partial}{\partial t} + X^i \frac{\partial}{\partial q^i} + \dot{X}^i \frac{\partial}{\partial \dot{q}^i} \quad , \tag{8.1.1}$$

where X^t, X^i and $\dot{X}^i$ are, in general, functions of $(t, q^i, \dot{q}^i)$. If X^t and X^i do not depend on $\dot{q}^i$, the vector field X^1 on $J^1(E)$ is of the form

$$X^1 = X + \dot{X}^i \frac{\partial}{\partial \dot{q}^i} \quad , \tag{8.1.2}$$

where $X \in \mathscr{X}(E)$, *i.e.* X is a vector field on E, and $\dot{X}^i$ is not related to the components of X. When the components $\dot{X}^i$ of the vector field X^1 are determined by the components of X through the expression

$$\dot{X}^i = (\frac{\partial}{\partial t} + \dot{q}^j \frac{\partial}{\partial q^j})(X^i - \dot{q}^i X^t) = \frac{\partial X^i}{\partial t} - \dot{q}^i \frac{\partial X^t}{\partial t} + \dot{q}^j \frac{\partial X^i}{\partial q^j} - \dot{q}^j \frac{\partial X^t}{\partial q^j} \dot{q}^i \quad , \tag{8.1.3}$$

it is said that X^1 is the 1-jet prolongation of the vector field X on E, and one writes $X^1 = j^1(X) \equiv \bar{X}^1$. In other words, the components $\dot{X}^i$ of $\bar{X}^1$ are defined (from the components X^t and X^i of X) in such a way that its action on the velocities $\dot{q}^i$ is the one induced from that of X on the coordinates q^i, and not an independent one as in the general case (8.1.1) above (an example will be given below). As a result, and as a little check using (8.1.2) with $\dot{X}^i$ given by (8.1.3) shows, *the 1-jet prolongation of vector fields preserves their Lie algebra structure*:

$$j^1([X,Y]) = [j^1(X), j^1(Y)], \quad \forall X, Y \in \mathscr{X}(E) \quad . \quad (8.1.4)$$

This important property simply states that the action of a Lie group G on E can be lifted to $J^1(E)$ in a canonical way, and that the prolonged vector fields $\bar{X}^1, \bar{Y}^1 \in \mathscr{X}(J^1(E))$ generate the same algebra $\mathscr{G}$.

Let us now review the essential ingredients of the jet-bundle approach to mechanics starting with

(8.1.1) DEFINITION (*Lagrangian on $J^1(E)$*)

A (first order) Lagrangian L is a mapping $L : J^1(E) \rightarrow R$, i.e., an element of $\mathscr{F}(T(N) \times R)$.

Thus, in the jet-bundle framework, Lagrangians in mechanics are real functions $L(t, q^i, \dot{q}^i)$ on $J^1(E)$. One may also interpret the Lagrangian as a *Lagrangian form** by writing $L(t, q^i, \dot{q}^i)dt$. To consider the rôle of symmetries and derive the Noether theorem, it is convenient to use the *Poincaré-Cartan form* Θ_{PC} on $J^1(E)$ instead of the Lagrangian form. It is defined by

$$\Theta_{PC} := \frac{\partial L}{\partial \dot{q}^i}(dq^i - \dot{q}^i dt) + Ldt \equiv \frac{\partial L}{\partial \dot{q}^i}\theta^i + Ldt \quad , \quad (8.1.5)$$

where $\theta^i := (dq^i - \dot{q}^i dt)$ are the 'contact forms'. It is worth noticing that the pull-back $(\bar{s}^1)^*(\Theta_{PC})$ of Θ_{PC} to R by a 1-jet section $\bar{s}^1$ just gives the ordinary Lagrangian because then dq^i becomes $(\bar{s}^1)^*(dq^i) = (dq(t)/dt)dt = \dot{q}(t)dt$, the first term in (8.1.5) is zero since the pull-back of the contact forms by a 1-jet cross section is zero $((\bar{s}^1)^*(dq^i - \dot{q}^i dt) = \frac{dq^i}{dt}dt - \frac{dq^i}{dt}dt = 0)$ and $L(t, q^i, \dot{q}^i)$ becomes $L(t, q^i(t), dq^i(t)/dt)$.[†]

---

* This can be extended without difficulty to field theory by replacing E by a vector bundle over spacetime with fibre R^n (for real fields) or C^n (for complex fields). In this case the Lagrangian *density* is a D-form, where D is the dimension of spacetime. Also, we note in passing, higher order Lagrangians (i.e., depending on q^i, $\dot{q}^i$, $\ddot{q}^i$,...) may be defined by replacing J^1 by the appropriate higher order jet bundle J^r.

† One may say that the 1-jets are integral submanifolds of the Pfaffian system defined by the contact one-forms $\theta^i = dq^i - \dot{q}^i dt$, a fact which may also be used to derive the expression (8.1.3) which determines $\dot{X}$ in (8.1.2). This is easily seen by using the generic expression (8.1.1) to compute $L_{\bar{X}^1}(dq^i - \dot{q}^i dt) \equiv L_{\bar{X}^1}\theta^i$; notice that $\frac{\partial}{\partial t} + \dot{q}^j \frac{\partial}{\partial q^j}$ may be considered as a formal total derivative (it is $\frac{d}{dt}$ when $\dot{q}^j = \frac{dq^j}{dt}$). In fact, the assumption that $\bar{X}^1$ is a 1-jet prolongation implies that $L_{\bar{X}^1}\theta^i = A^i_j \theta^j$ (so that the pull-back by $\bar{s}^1$ is zero before and after the action of the Lie derivative on θ^i), and this condition suffices to determine both A^i_j and (8.1.3).

Using (8.1.5) the action functional for the *ordinary variational principle* is written

$$I[s] = \int_{\bar{s}^1(M)} \Theta_{PC} \tag{8.1.6}$$

where M here is an interval of the real line. Then we can introduce in this language the

ORDINARY VARIATIONAL PRINCIPLE
The critical trajectories are those for which $\delta I = 0$, *i.e.*,

$$\int_M (\bar{s}^1)^*(L_{\bar{X}^1}\Theta_{PC}) = 0 \quad , \quad \forall \bar{X}^1 \quad , \tag{8.1.7}$$

where the arbitrary variations δ *are produced by the Lie derivative with respect to an arbitrary (1-jet) vector field* $\bar{X}^1$ *and* $(\bar{s}^1)^*$ *is the pull-back by the 1-jet section* $\bar{s}^1$.

The word 'ordinary' just indicates that $\bar{X}^1$ is a 1-jet vector field, since it is possible to define a more general principle by removing this restriction (which turns out to be equivalent to the ordinary variational principle (8.1.7) for *non-singular* Lagrangians for which the Hessian det $\left(\frac{\partial^2 L}{\partial \dot{q}^i \partial \dot{q}^j}\right) \neq 0$).

(8.1.1) COROLLARY (*Lagrange equations*)

The critical sections (or 'physical' trajectories) are the solutions of

$$(\bar{s}^1)^*(i_{\bar{X}^1}d\Theta_{PC}) = 0 \quad . \tag{8.1.8}$$

Proof: Using that $L_{\bar{X}^1} = i_{\bar{X}^1}d + di_{\bar{X}^1}$, we see that the second term in the integrand gives zero contribution on account of the boundary conditions so that $\delta I = 0$ implies (8.1.8), *q.e.d*

It is not difficult to show that (8.1.8) is equivalent to the Lagrange equations. From (8.1.5) we obtain

$$d\Theta_{PC} = d\left(\frac{\partial L}{\partial \dot{q}^i}\right) \wedge (dq^i - \dot{q}^i dt) + \frac{\partial L}{\partial q^i}dq^i \wedge dt \quad ;$$

$$i_{\bar{X}^1}d\Theta_{PC} = \left(i_{\bar{X}^1}\left[d\left(\frac{\partial L}{\partial \dot{q}^i}\right)\right]\right)(dq^i - \dot{q}^i dt) - d\left(\frac{\partial L}{\partial \dot{q}^i}\right)(X^i - \dot{q}^i X^t) \tag{8.1.9}$$

$$+ \frac{\partial L}{\partial q^i}X^i dt - \frac{\partial L}{\partial q^i}dq^i X^t \quad .$$

There is no need of calculating explicitly the first term in $i_{\bar{X}^1}d\Theta_{PC}$ above since it does not contribute to $(\bar{s}^1)^*(i_{\bar{X}^1}d\Theta_{PC})$. Thus, eq. (8.1.8) implies

$$X^i\left(\frac{\partial L}{\partial q^i} - \frac{d}{dt}\left(\frac{\partial L}{\partial \dot{q}^i}\right)\right) - \dot{q}^i X^t\left(\frac{\partial L}{\partial q^i} - \frac{d}{dt}\left(\frac{\partial L}{\partial \dot{q}^i}\right)\right) = 0 \tag{8.1.10}$$

and, since the components X^t and X^i are arbitrary, we see that (8.1.8) is another form of expressing the familiar Lagrange equations,

$$\frac{\partial L}{\partial q^i} - \frac{d}{dt}\left(\frac{\partial L}{\partial \dot{q}^i}\right) = 0 \quad . \tag{8.1.11}$$

Notice that, *in* (8.1.10) and (8.1.11), $L = L(t, q^i(t), \dot{q}^i(t))$ and $\dot{q}^i(t)$ means $dq^i(t)/dt$. Thus, (8.1.7) is equivalent to the more familiar formulation of the ordinary variational principle in terms of the Lagrangian,

$$\delta I = 0 \quad , \quad I := \int_1^2 dt \, L(t, q^i(t), dq^i(t)/dt) \quad . \tag{8.1.12}$$

Let us nevertheless still refer to the Poincaré-Cartan form Θ_{PC} on $J^1(E)$ to introduce the symmetries of dynamical systems.

(8.1.2) DEFINITION

A vector field X on E generates a symmetry of the dynamical system described by (8.1.5) if its prolongation $\bar{X}^1$ to $J^1(E) = R \times T(N)$ satisfies the (quasi-)invariance condition

$$L_{\bar{X}^1}\Theta_{PC} = d\Delta \tag{8.1.13}$$

on any section $\bar{s}^1$, where $\Delta \neq \Delta(\dot{q}^i)$.

Note that setting the r.h.s. of (8.1.13) equal to zero would be too restrictive in general, since the addition of $d\Delta$ to L does not modify the Lagrange equations.

(8.1.1) THEOREM (*Noether theorem for mechanics*)

The quantity

$$\mathcal{N} = i_{\bar{X}^1}\Theta_{PC} - \Delta \tag{8.1.14}$$

is conserved on the critical trajectories (solutions of the Lagrange equation (8.1.8)). In other words, $\mathcal{N}$ is a constant of the motion.

Proof: It is sufficient to write (8.1.13) in the form

$$d(i_{\bar{X}^1}\Theta_{PC} - \Delta) + i_{\bar{X}^1}d\Theta_{PC} = 0 \tag{8.1.15}$$

and then use (8.1.8), *q.e.d.*

Explicitly, (8.1.14) reads

$$\mathcal{N} = X^i\frac{\partial L}{\partial \dot{q}^i} - (\dot{q}^i\frac{\partial L}{\partial \dot{q}^i} - L)X^t - \Delta \quad . \tag{8.1.16}$$

A simple example (free Newtonian particle)
Let G be the Galilei group, $L = \frac{1}{2}m\dot{\mathbf{x}}^2$ and the action of G on $E = R \times T(R^3)$ given by

$$X_{(B)} = \frac{\partial}{\partial t} \quad , \quad X_{(A)} = \frac{\partial}{\partial \mathbf{x}} \quad , \quad X_{(V)} = \frac{\partial}{\partial \dot{\mathbf{x}}} + t\frac{\partial}{\partial \mathbf{x}} \quad , \tag{8.1.17}$$

where the last vector field is the only one for which (8.1.3) produces an additional term when moving from $E = R \times R^3$ (where $X_{(V)} = t\frac{\partial}{\partial \mathbf{x}}$) to $J^1(E)$. Then it is clear that $\Delta_{(V)} = m\mathbf{x}$, $\Delta_{(B)} = 0$, $\Delta_{(A)} = \mathbf{0}$, and (8.1.16) gives (adding a minus sign for convenience)

$$\mathcal{N}_{(B)} = \dot{\mathbf{q}} \cdot \mathbf{p} - L = H \quad , \quad \mathcal{N}_{(A)} = -m\dot{\mathbf{x}} \equiv -\mathbf{p} \quad , \quad \mathcal{N}_{(V)} = m\mathbf{x} - \mathbf{p}t \quad , \tag{8.1.18}$$

where we have put $\partial L/\partial \dot{\mathbf{x}} = \mathbf{p}$.

We might think for a moment that the Poisson brackets of (8.1.18), taken as the generators of contact transformations, should reproduce the Galilei algebra. But from (8.1.18) it follows that, as is well known,

$$\{\mathcal{N}_{(A^i)}, \mathcal{N}_{(V^j)}\} = m\delta_{ij} \tag{8.1.19}$$

instead of being zero. This is a consequence of the additional term $\Delta_{(V)}$ in the conserved 'charge' $\mathcal{N}_{(V)}$ and motivates the subsequent discussion of 'classical anomalies' in the Lagrangian framework. As already mentioned, however, these terms, rather than being anomalous, just indicate that the group relevant for the physical problem is an extension of the group originally considered.

To conclude this section, let us mention that we could also have formulated the variational principle in a Hamiltonian form. Notice that by introducing the momenta $p_i = \partial L/\partial \dot{q}^i$ in eq. (8.1.5) the Poincaré-Cartan form becomes

$$\Theta_{PC} = p_i dq^i - H dt \equiv \Lambda - H dt \tag{8.1.20}$$

where Λ is the Liouville form already mentioned in section 3.6. Using (8.1.20) means that the space $J^1(E) = R \times T(N)$ has been replaced by $J^{1*} = R \times T^*(N)$ of coordinates (t, q^i, p_i). The variational principle leads now to the Hamilton equations, but we shall not discuss this or the Noether theorem in this formulation.

8.2 Invariant forms on a manifold and cohomology; WZ terms

It was proven in Theorem 1.8.1 that, if ω is a LI form which is also RI, then it is closed. It was also shown in Theorem 6.7.1 that there is a one-to-one correspondence between the Lie algebra cohomology groups $H_0^n(\mathcal{G}, R)$ and the *classes* of closed LI forms on the group G (closed LI forms modulo those given by the exterior differential of a LI form), *i.e.* between the Lie algebra cohomology groups and the elements of the corresponding Chevalley-Eilenberg (CE) cohomology group $E^n(G)$.

This may be easily applied to the case of certain group manifolds of physical interest. In physics, manifolds appear quite often as homogeneous spaces which themselves have a group structure (they are homogeneous spaces of groups with a semidirect structure). This is the case of spacetime in both Newtonian and Einsteinian mechanics, which are governed by the Galilei and Poincaré groups respectively*, as well as in supersymmetry, where the relevant group is the super-Poincaré group. For instance, both Minkowski space $\mathcal{M}$ ($\sim R^4$) and superspace Σ can be viewed as the homogeneous spaces $\mathcal{P}/L$ and $\mathcal{S}\mathcal{P}/L$ (Poincaré and super-Poincaré, respectively, modulo the Lorentz group). These kinematical groups have the important property of being semidirect or splitting extensions (see section 5.2) $G = H \circ S$ of a group S by the homogeneous space H, which happens to be an invariant subgroup of the kinematical group (or supergroup) which constitutes the extension. For our purposes it will be sufficient to observe that the homogeneous space $H = G/S$ is a subgroup of G. We shall also assume that H has trivial de Rham cohomology; this is certainly the case for ordinary spacetime (R^4) and, inasmuch as these notions can be extended to supermanifolds†, also for flat superspace, which means that every closed differential form on Σ is exact.

Let us consider, for the sake of definiteness, a *closed* two-form $\omega^{(2)}$ made out of LI one-forms on H. We may also require it to be invariant under S, if the physical content of $\omega^{(2)}$ has to be invariant under the whole kinematical group. (This may imply, for instance, that $\omega^{(2)}$ is a scalar under the rotations or the Lorentz group.) We shall also assume that $\omega^{(2)} \neq d\omega^{(1)}$, where $\omega^{(1)}$ is an LI one-form on H. Under these conditions, $\omega^{(2)}$ defines a non-trivial element of the second CE cohomology: it is a closed form which cannot be expressed as the exterior differential *of an LI form*. Assuming that $\omega^{(2)}$ takes values in R, then $\omega^{(2)}$ defines a central extension through (6.3.8). If we denote, as before, the generators of the extended group by $\bar{X}^L$, the modified commutators of the algebra are those given by

$$[\bar{X}^L_{(i)}, \bar{X}^L_{(j)}] = C^k_{ij}\bar{X}^L_{(k)} + \omega^{(2)}(X^L_{(i)}, X^L_{(j)})\bar{X}_{(\theta)} \equiv C^k_{ij}\bar{X}^L_{(k)} + C^\theta_{ij}\bar{X}_{(\theta)} \quad \forall X_i \in H \tag{8.2.1}$$

where $\bar{X}_{(\theta)} = \partial/\partial\theta$ is the central generator associated with the additional central parameter θ. The generator $\bar{X}_{(\theta)}$ is sometimes written as 1 because

---

* We are not considering general relativity here.
† Questions on global properties only arise for the bosonic variables; everything seems to be purely algebraic in the Grassmann variables.

of its commuting properties, a notation which is not quite correct because 1 is the coordinate of the central vector field, but not the vector field itself.

There is another way of expressing the same result. Since we are assuming trivial de Rham cohomology and $\omega^{(2)}$ is closed we may write

$$\omega^{(2)} = d\beta^{(1)} \quad , \tag{8.2.2}$$

where $\beta^{(1)}$ is not LI, $\beta^{(1)} \notin \mathcal{X}^{*L}(H)$, since by hypothesis $\omega^{(2)}$ is a nontrivial CE two-cocycle. Using (8.2.1) and (1.7.2) we find that the structure constants which differ from those of the unextended algebra are given by the non-zero values of

$$C_{ij}^{\theta} = X_{(i)}^{L}.\beta^{(1)}(X_{(j)}^{L}) - X_{(j)}^{L}.\beta^{(1)}(X_{(i)}^{L}) - \beta^{(1)}([X_{(i)}^{L}, X_{(j)}^{L}]) \quad . \tag{8.2.3}$$

It is readily seen in (8.2.3) that the ambiguity in the definition of $\beta^{(1)}$, which arises because it is always possible to replace $\beta^{(1)}$ by $\beta^{(1)} + df$, where f is a function on H, does not affect the C_{ij}^{θ}. Further, since by hypothesis $\omega^{(2)}$ is LI, $L_{X^R}\omega^{(2)} = 0$. Thus,

$$L_{X_{(i)}^R}\omega^{(2)} = dL_{X_{(i)}^R}\beta^{(1)} = 0 \tag{8.2.4}$$

and, since we assumed that $L_{X_{(i)}^R}\beta^{(1)}$ is non-zero, it follows that

$$L_{X_{(i)}^R}\beta^{(1)} = d\Delta_{(i)} \quad . \tag{8.2.5}$$

Thus, and in contrast with $\omega^{(2)}$ which was invariant under left translations, $\beta^{(1)}$ is only quasi-invariant and associates a function $\Delta_{(i)}$ on H to each $X_{(i)}^R$. Moreover the identity $[L_X, L_Y] = L_{[X,Y]}$ implies the obvious consistency condition

$$L_{X_{(i)}^R}d\Delta_{(j)} - L_{X_{(j)}^R}d\Delta_{(i)} - d\Delta_{[i,j]} = 0 \quad , \tag{8.2.6}$$

which in turn implies

$$d(L_{X_{(j)}^R}\Delta_{(i)} - L_{X_{(i)}^R}\Delta_{(j)} + \Delta_{[i,j]}) = 0 \quad ; \tag{8.2.7}$$

this will be called below the Wess-Zumino condition (in analogy with a condition of similar structure, to be discussed in chapter 10, that appears in anomalous gauge theories and that originally received this name). Eq. (8.2.7) also tells us that the term inside the bracket is a constant. Thus, the bracket may be understood as a mapping $\omega_2 : \mathcal{G} \times \mathcal{G} \to R$ which, in fact, defines the two-cocycle of a central extension (it satisfies $s\omega_2 = 0$ for the trivial action $\rho = 0$). Thus, one can also use

$$X_{(j)}^R.\Delta_{(i)} - X_{(i)}^R.\Delta_{(j)} + \Delta_{[i,j]} \tag{8.2.8}$$

to define the structure constants of the centrally extended algebra involving the central generator. We shall show that these correspond to the structure constants $C_{ij}^{R\theta}$ of the 'right' algebra, and that hence $C_{ij}^{R\theta} = -C_{ij}^{\theta}$.

In view of the above discussion, we may now introduce the 'Wess-Zumino' forms by means of the following

(8.2.1) DEFINITION (*Wess-Zumino forms on a Lie group G*)

A Wess-Zumino form is the potential $(p+1)$-form β on H of a $(p+2)$-form ω that is a non-trivial element of the $(p+2)$-nd CE cohomology group.

Wess-Zumino terms are thus given by differential forms, and do not depend on the metric. They may be called *topological* in the sense that they depend only on the properties of the manifold on which they are defined. The case $p = 0$ is exemplified by the Wess-Zumino forms in mechanics, where the pull-back by sections produces one-forms on R depending on the time coordinate t. The case $p \neq 0$ determines higher order forms; their pull-back produces $(p + 1)$-forms on a manifold parametrized by the time coordinate plus p other coordinates. This permits, for instance, the description of p-extended supersymmetric objects, as we shall briefly discuss in sections 8.7, 8.8 (the value of p corresponds to the extension of the object; $p = 0$ indicates point-like objects).

The potential Wess-Zumino $(p + 1)$-form β is quasi-invariant, which is the content of the following

(8.2.1) LEMMA (*Wess-Zumino consistency condition*)

The potential form β is quasi-invariant, i.e. its Lie derivative is an exact form $d\alpha$,

$$L_{\tilde{X}^1}\beta = d\alpha \quad , \tag{8.2.9}$$

and α satisfies the consistency condition

$$d(L_{\tilde{X}^1_{(i)}}\alpha_{(j)} - L_{\tilde{X}^1_{(j)}}\alpha_{(i)} - \alpha_{[i,j]}) = 0 \quad , \quad \forall X_{(i)} \in \mathcal{G} \quad . \tag{8.2.10}$$

Proof: Eq. (8.2.9) follows from the commutativity of the exterior and Lie derivatives, and (8.2.10) from the fact that $[L_X, L_Y] - L_{[X,Y]} = 0$, q.e.d.

In physical terms, a Wess-Zumino term is a quasi-invariant [(8.2.5)] term in an action integral which comes from a closed form which is invariant under certain transformations. Because its variation is given by an exact form, its presence will not spoil the invariance properties of the other part of the action (in the case where there was already another one, see next section). The case discussed above, which corresponds to $p = 0$, is the simplest in two respects. First, it considers a Wess-Zumino term which comes from a two-form; thus β is a one-form and $\alpha = \Delta$ is a function on the coset H, so that in this case (8.2.8) are real numbers defining the structure constants of the extended algebra. Secondly, we assumed trivial *de Rham* cohomology for H; in general this may not be trivial, and (8.2.2) will only be a *local* expression. Thus, in the more general cases (higher order Wess-Zumino forms and/or non-trivial cohomology) one should expect new features to appear and this is indeed what happens.

A specially important consequence of the appearance of a de Rham non-trivial Wess-Zumino term in a physical action is the quantization of the coefficient accompanying it, a point specially emphasized by Witten. In this case ω is not exact (this happens, for instance, in models where ω is related to the volume form of *compact* groups, since then the Chevalley-Eilenberg and de Rham cohomologies coincide (section 6.7)). In that case, the path integral formulation of the quantum theory or the wavefunction suffer from a potential lack of univaluedness, since they depend on the exponential of the action. The quantum theory is then well defined only for certain quantized values of the proportionality constant in the Wess-Zumino-Witten term, as we shall discuss further in section 10.11.

We shall see that when a Wess-Zumino term is present, the structure of the Lagrangian is determined by the splitting $\mathscr{L} \propto \mathscr{L}_0 + \mathscr{L}_{WZ}$, where $\mathscr{L}_0$ is globally defined and $\mathscr{L}_{WZ}$ is the Wess-Zumino term. The properties of $\mathscr{L}_{WZ}$ may be summarized by saying that (a) it is obtained from a differential form and does not depend on the metric, (b) it is quasi-invariant and its variation satisfies the Wess-Zumino consistency condition and (c) its expression may require local charts, in which case it will lead to (topological) quantization conditions for the physical parameter accompanying it.

8.3 Newtonian mechanics and Wess-Zumino terms: two simple examples

In contrast with the Poincaré group, $\mathscr{P} = Tr_4 \circ L$, the Galilei group admits several semidirect structures. The most obvious is given by $G = Tr_4 \circ (B \circ \mathscr{R})$ ((spacetime translations) $\circ$ (boosts $\circ$ rotations)) where, recalling our conventions, the group on the right in a semidirect product acts on the group at its left. Another is given by factorizing the maximal abelian subgroup $Tr_3 \otimes B_3$, $G = (Tr_3 \otimes B_3) \circ (T \otimes \mathscr{R})$ ((space translations $\times$ boosts) $\circ$ (time translations $\times$ rotations)). However, to define Lagrangians as introduced in section 8.1, we require the *evolution* space which includes time, space and velocities. Thus, we shall use here the decomposition

$$G = ((Tr_3 \otimes B_3) \circ T)) \circ \mathscr{R} \equiv H \circ \mathscr{R} \quad , \tag{8.3.1}$$

which allows us to define the evolution space as the homogeneous space parametrized by $(t, \mathbf{x}, \dot{\mathbf{x}})$. (It is interesting to note that (8.3.1) is also singled out by being the Levi-Mal'čev splitting (Theorem 6.5.3) of G.) The fact that the evolution space is not a homogeneous space of the Poincaré group is due to (or leads to) the different cohomological properties of the Galilei and Poincaré groups ($H_0^2(G, U(1)) = R$, $H_0^2(\bar{\mathscr{P}}, U(1)) = 0$) and has far reaching implications. It explains, for instance, why there is not a simple relativistic version of the quantum non-relativistic position operator (which belongs to the extended Galilei algebra where the commutator [boosts, translations] produces the central element) and why as a result the Newton-Wigner one has to be introduced (the triviality of $H_0^2(\bar{\mathscr{P}}, U(1))$ precludes having an analogous commutator in the Poincaré algebra). The different structures of the two groups also explain why the invariant parameter which is used in the variational principle for relativistic mechanics, τ, is not really an independent variable and ends up being related to time by a constraint, something which is absent in the 'non-relativistic' case. In fact, if the possible kinematical groups are classified under rather general assumptions, it is found that those with absolute time – as G – have non-trivial cohomology, and those with relative time – as $\mathscr{P}$ – have trivial cohomology. Finally, another consequence of the different structures of G and $\mathscr{P}$ is the absence of any simple 'covariant' formulation of Newtonian mechanics in sharp contrast with the Einstenian case, where the basic physical laws and quantities are formulated in terms of Lorentz vectors, tensors, etc.

Let us come back to Newtonian mechanics. The group law of H, *i.e.* the group law of the Galilei group without rotations, was already considered in section 3.1, and can be used to define a (left) action on the evolution space of section 8.1. With the simple replacements

$$(B'', \mathbf{A}'', \mathbf{V}'') \mapsto (t', \mathbf{x}', \dot{\mathbf{x}}'), \ (B', \mathbf{A}', \mathbf{V}') \mapsto (B, \mathbf{A}, \mathbf{V}) \text{ and } (B, \mathbf{A}, \mathbf{V}) \mapsto (t, \mathbf{x}, \dot{\mathbf{x}}),$$

eq. (3.1.1) leads to

$$t' = t + B \ , \quad \mathbf{x}' = \mathbf{x} + \mathbf{V}t + \mathbf{A} \ , \tag{8.3.2}$$

$$\dot{\mathbf{x}} = \dot{\mathbf{x}} + \mathbf{V} \ . \tag{8.3.3}$$

The generators of the action defined by (8.3.2), (8.3.3) are given by (8.1.17). They are, of course, the RI vector fields of the (left) action of H on itself; see (3.4.6) with $m = 0$. They were also considered as the 1-jet prolongations of the vector fields of the action of H on E (take (8.3.2) *alone*) to $J^1(E)$ given by (8.1.3). By now it is clear that these vector fields leave invariant (*i.e.*, give zero Lie derivative when acting on) the forms (cf. (3.4.5))

$$\omega^{(\mathbf{A})} = d\mathbf{x} - \dot{\mathbf{x}}dt, \qquad \omega^{(\mathbf{V})} = d\dot{\mathbf{x}} \ . \tag{8.3.4}$$

These are all the ingredients needed to apply the general scheme developed in the previous two sections. The two-form

$$\omega^{(2)} = \omega^{(\mathbf{V})} \wedge \omega^{(\mathbf{A})} \tag{8.3.5}$$

is both invariant under the group action (it is the product of invariant forms), and closed. We may write it as

$$\omega^{(V)} \wedge \omega^{(A)} = d\beta^{(1)}, \qquad \beta^{(1)} = \dot{x}(dx - \frac{1}{2}\dot{x}dt) \quad . \tag{8.3.6}$$

Clearly $\delta_{(V)}\beta^{(1)} \neq 0$. Thus, $\beta^{(1)}$ is not an LI form and accordingly $\omega^{(2)}$ defines a non-trivial (second) CE cohomology class. We observe now that the Wess-Zumino form $\beta^{(1)}$ or, more precisely, $m\beta^{(1)}$ given by

$$m\beta^{(1)} = \frac{\partial L}{\partial \dot{x}}(dx - \dot{x}dt) + Ldt \equiv \Theta_{PC} \tag{8.3.7}$$

is precisely the Poincaré-Cartan form (8.1.5) for $L = \frac{1}{2}m\dot{x}^2$. Thus, *the Poincaré-Cartan form for the Newtonian particle is itself a Wess-Zumino term.* Moreover, in this simple example, *the action is just reduced to the Wess-Zumino piece.*[*] We shall see in Proposition 8.5.1 that this corresponds to the fact that, in the usual formulation of the free particle action principle where $L = \frac{1}{2}m(\frac{dx(t)}{dt})^2$, $\delta_{(v)}L = \frac{d}{dt}m(x(t)V + \frac{1}{2}V^2 t)$, i.e. that L is quasi-invariant. Thus, if the Poincaré-Cartan form on $J^1(E)$ is quasi-invariant, the variation of the associated Lagrangian Ldt on R is given by dt times a time derivative.

The quasi-invariance of $\beta^{(1)}$ is exhibited by the fact that, for a boost V,

$$\delta_{(v)}(m\beta^{(1)}) \equiv m(\dot{x} + V)((dx + Vdt) - \frac{1}{2}(\dot{x} + V)^2 dt) - \beta^{(1)}$$
$$= d\{m(V.x + \frac{1}{2}V^2 t)\} \tag{8.3.8}$$

(the Lie derivative, $L_{X_{(v)}}\beta^{(1)}$, gives of course the same result to first order, $d(mV.x)$). The (left) Lie algebra of the extended group is now read from (8.3.6) and (8.2.1) or (8.2.3). The only new commutator is the one given by

$$C^{\theta}_{A^i, V^j} = -m\delta_{ij} \quad . \tag{8.3.9}$$

The generators of the action on the extended evolution space are given by (cf. (3.7.3), (8.1.17))

$$\bar{X}_{(B)} = X_{(B)} \ , \ \bar{X}_{(A)} = X_{(A)} \ , \ \bar{X}_{(V)} = X_{(V)} + mx\frac{\partial}{\partial \chi} \ , \ \bar{X}_{(\theta)} = \frac{\partial}{\partial \chi} \ ; \tag{8.3.10}$$

---

[*] In fact, this was to be expected on account of the analysis of section 3.7 of the coadjoint orbits for the Galilei group and the derivation of the symplectic form, and constitutes a particular case of a more general situation, which relates the Wess-Zumino terms to the symplectic forms on an orbit. The form (8.3.5) is presymplectic, but it leads to a symplectic form on the manifold which is obtained by removing the time evolution $X = \frac{\partial}{\partial t} + \dot{x}\frac{\partial}{\partial x}$ (in fact, this vector field generates the characteristic module $\{X | i_X \omega^{(2)} = 0\}$). The manifold of solutions S is then characterized by the initial constants x_0, $\dot{x}_0$ that determine a solution $x = \dot{x}t + x_0$, $\dot{x} = \dot{x}_0$ (an integral of X). On this manifold, $\omega^{(2)}|_S = d\dot{x}_0 \wedge dx_0$.

both the new variable χ and the new group parameter θ have the dimensions of an action (and now $[\bar{X}_{(A^i)}, \bar{X}_{(V^j)}] = m\delta_{ij}\bar{X}_{(\theta)}$).

Having reached this point, we might ask ourselves whether it would be possible to modify the Lagrangian by replacing $\beta^{(1)}$ by another one-form $\bar{\beta}^{(1)}$ which would be LI. This may be achieved by defining $\bar{\beta}^{(1)}$ not on the evolution space, but on the homogeneous space obtained by adding to $(t, \mathbf{x}, \dot{\mathbf{x}})$ the new variable χ, *i.e.* on $\bar{G}_{(m)}/\mathcal{R}$ where $\bar{G}_{(m)}$ is now the extended Galilei group (section 3.4). In other words, and proceeding as before, we now make the replacements $(\theta, B, \mathbf{A}, \mathbf{V}) \to (\chi, t, \mathbf{x}, \dot{\mathbf{x}})$, etc. Ignoring the rotations, the action of $\bar{G}_{(m)}$ on χ is given by

$$\chi' = \chi + \theta + m(\mathbf{V}.\mathbf{x} + \frac{1}{2}\mathbf{V}^2 t) \quad . \tag{8.3.11}$$

Eq. (8.3.11) and eqs. (8.3.2), (8.3.3) define the group action on the *enlarged evolution space* $(\chi, t, \mathbf{x}, \dot{\mathbf{x}})$. Using (8.3.11) it is clear that the new *invariant* form is the one associated with the new parameter,

$$\bar{\beta}^{(1)} = \Theta_{PC} - d\chi = m\dot{\mathbf{x}}.(d\mathbf{x} - \frac{1}{2}\dot{\mathbf{x}}dt) - d\chi \equiv \bar{\Theta}_{PC} \tag{8.3.12}$$

(compare with $\bar{\omega}^{L(\theta)}$ in (3.4.5) adding a global sign) since the variations of χ and Θ_{PC} compensate each other. This of course implies that $L_{\bar{X}_{(V)}}\bar{\beta}^{(1)} = 0$; we can no longer speak of a Wess-Zumino term, and we cannot further extend the already extended evolution space because the available cohomology has been 'exhausted' in moving from G to $\bar{G}$. In this enlarged evolution space* there is no equation of motion for χ ($i_{\bar{X}^1}d\bar{\Theta}_{PC}$ does not contain χ), and the Lagrange equations for $\mathbf{x}$ are still the same. However, since $\bar{\beta}^{(1)}$ is now an LI form, there is no need of the additional piece Δ in the Noether charge (8.1.16). It is, of course, still there, but now it comes from the extra term of $\bar{X}_{(V)}$ in (8.3.10). Thus, we now have

$$L_{\bar{X}_{(V)}}\bar{\beta}^{(1)} = 0 \tag{8.3.13}$$

so that on any 1-jet prolongation of a critical section,

$$\mathcal{N}_{(i)} = i_{\bar{X}_{(i)}}\bar{\beta}^{(1)} = i_{\bar{X}^1_{(i)}}\bar{\Theta}_{PC} - X^\chi_{(i)} \tag{8.3.14}$$

is conserved. Of course, (8.3.14) and (8.1.18) are identical because only $\bar{X}^\chi_{(V)} = m\mathbf{x}$ is non-zero[†].

---

* It is also possible to define the action of $(\theta, B, \mathbf{A}, \mathbf{V})$ on $\dot{\chi}$. In this case, the generators of θ, B and $\mathbf{A}$ remain unaltered, and $\bar{X}_{(V)}$ picks up another extra term, $\dot{\mathbf{x}}\frac{\partial}{\partial\dot{\chi}}$ (see (8.1.3)).

† We would like to mention that all the discussion was based on $J^1(E) = R \times T(N)$ (or on the associated extended evolution space). This is the appropriate place to discuss systems which depend on time (contact structures). Similarly, we could have worked on $J^{1*}(E) = R \times T^*(N)$, *i.e.* on the phase space plus time (coordinates (t, q^i, p_i)). The prolongation of the action on E to $J^{1*}(E)$ requires more care, but can still be performed, and we would have obtained a translation of the above results to the new variables.

As a second example, consider the non-relativistic motion of a particle of unit charge in a constant magnetic field B. The two-dimensional dynamics is governed by the Lagrangian

$$L = \frac{1}{2}m\dot{\mathbf{x}}^2 - \frac{B}{2}\epsilon_{ij}\dot{x}^i x^j = \frac{1}{2}m\dot{\mathbf{x}}^2 - \dot{x}^i A_i \quad , \quad A_i = \frac{B}{2}\epsilon_{ij}x^j \ (i,j = 1, 2) \ (8.3.15)$$

or, in terms of the canonical momentum $p_i = m\dot{x}_i - A_i$, by the Hamiltonian

$$H = \frac{1}{2m}(\mathbf{p} + \mathbf{A})^2 . \tag{8.3.16}$$

To proceed as in the free case, it is convenient to extract from (8.3.15) the symmetry of the problem. Clearly, (8.3.15) is invariant under time translations b and quasi-invariant under space translations a_i (to avoid confusion with the magnetic field B and its potential A, the time and space translations will be denoted with small letters in the remainder of this section). To find the transformations corresponding to the boosts, we assume that they are of the form $x'^i = x^i + f^i$ where f^i is a function to be found, and obtain

$$\delta L \simeq \dot{x}^i(m\dot{f}_i - B\epsilon_{ij}f^j) - \frac{d}{dt}(\frac{1}{2}B\epsilon_{ij}x^i f^j) \quad . \tag{8.3.17}$$

Thus, L will be quasi-invariant if f is a function of t such that

$$m\dot{f}_i - B\epsilon_{ij}f^j = C^i \quad , \tag{8.3.18}$$

the C^i being constants. This leads to

$$f_i = D_i \sin\frac{B}{m}t - \epsilon_{ij}D^j \cos\frac{B}{m}t + \frac{1}{B}\epsilon_{ij}C^j. \tag{8.3.19}$$

The constants D_i, C_i are determined by the condition that f^i becomes $V^i t$ in the free case ($B = 0$); this gives $D_i = \frac{m}{B}V_i$ and $C_i = mV_i$. The Lagrangian (8.3.15) is thus quasi-invariant under the symmetry transformations

$$t' = t + b \quad ,$$

$$x'_i = x_i + a_i + \frac{m}{B}V_i \sin\frac{B}{m}t + \frac{m}{B}\epsilon_{ij}V^j\left(1 - \cos\frac{B}{m}t\right) \quad , \tag{8.3.20}$$

$$\dot{x}'_i = \dot{x}_i + V_i \cos\frac{B}{m}t + \epsilon_{ij}V^j \sin\frac{B}{m}t \quad .$$

These transformations determine a group $G_{(B)}$ of parameters $g = (b, a_i, V_i)$ (i=1,2) and group law given by

$$b'' = b' + b \quad ,$$

$$a''_i = a'_i + a_i + \frac{m}{B}V'_i \sin\frac{B}{m}b + \frac{m}{B}\epsilon_{ij}V'^j\left(1 - \cos\frac{B}{m}b\right) \quad , \tag{8.3.21}$$

$$V''_i = V'_i \cos\frac{B}{m}b + V_i + \epsilon_{ij}V'^j \sin\frac{B}{m}b \quad ,$$

the inverse element $(b, a_i, V_i)^{-1} \equiv (\hat{b}, \hat{a}_i, \hat{v}_i)$ being

$$\hat{b} = -b \quad ,$$

$$\hat{a}_i = -a_i + \frac{m}{2B} V_i \sin \frac{B}{m} b - \frac{m}{B} \epsilon_{ij} V^j \left(1 - \cos \frac{B}{m} b\right) \quad , \qquad (8.3.22)$$

$$\hat{V}_i = -V_i \cos \frac{B}{m} b + \epsilon_{ij} V^j \sin \frac{B}{m} b \quad .$$

The corresponding Lie algebra is found easily. For instance, the generators of the group action (8.3.20) on evolution space $(t, \mathbf{x}, \dot{\mathbf{x}})$ are

$$X_{(b)} = \frac{\partial}{\partial t} \quad , \quad X_{(a^i)} = \frac{\partial}{\partial x^i} \quad ,$$

$$X_{(V^i)} = \frac{m}{B} \left[\sin \left(\frac{b}{m} t\right) \delta_{ij} + \epsilon_{ji} \left(1 - \cos \frac{B}{m} t\right) \right] \frac{\partial}{\partial x^j} \qquad (8.3.23)$$

$$+ \cos \left(\frac{B}{m} t\right) \frac{\partial}{\partial \dot{x}^i} - \epsilon_{ij} \sin \left(\frac{B}{m} t\right) \frac{\partial}{\partial \dot{x}^j} \quad ,$$

and determine the algebra $\mathscr{G}_{(B)}$ of $G_{(B)}$

$$[X_{(b)}, X_{(a^i)}] = 0 \quad , \qquad [X_{(b)}, X_{(V^i)}] = X_{(a^i)} - \frac{B}{m} \epsilon_{ij} X_{(V^j)} \quad , \qquad (8.3.24)$$

$$[X_{(a^i)}, X_{(a^j)}] = 0 \quad , \qquad [X_{(V^i)}, X_{(V^j)}] = 0 \quad .$$

Let us now compute the conserved quantities associated with the symmetry (8.3.20). Using (8.1.16) (which of course may be derived from L without using Θ_{PC}) and

$$\Delta_{(b)} = 0 \quad , \qquad \Delta_{(a_i)} = \frac{B}{2} \epsilon_{ij} x^j \quad ,$$

$$\Delta_{(V_i)} = \frac{m}{2} \left[x_i \left(1 + \cos \frac{B}{m} t\right) - \epsilon_{ij} x^j \sin \frac{B}{m} t \right] \quad , \qquad (8.3.25)$$

we obtain

$$\mathscr{N}_{(a_i)} \equiv P_i = -p_i + \frac{B}{2} \epsilon_{ij} x^j = -p_i + A_i \; (= -m\dot{x}_i + 2A_i) \quad ,$$

$$\mathscr{N}_{(b)} \equiv H = \frac{1}{2m} \mathbf{p}^2 + \frac{B^2}{8m} \mathbf{x}^2 + \frac{B}{2m} \epsilon_{ij} p^i x^j = \frac{(\mathbf{p} + \mathbf{A})^2}{2m} \quad , \qquad (8.3.26)$$

$$\mathscr{N}_{(V_i)} = -\frac{m}{B} \left[\delta_{ij} \sin \frac{B}{m} t - \epsilon_{ij} \left(1 - \cos \frac{B}{m} t\right) \right] p^j$$

$$+ \frac{m}{2} \left[x_i \left(1 + \cos \frac{B}{m} t\right) - \epsilon_{ij} x^j \sin \frac{B}{m} t \right] \quad .$$

For $B = 0$, all the previous formulae reduce to the Galilei case expressions.

If we now compute the Poisson brackets $\{\mathscr{N}_{(g^i)}, \mathscr{N}_{(g^j)}\}$ of the conserved quantities by using $\{x^i, p_j\} = \delta^i_j$, we again find that they realize the algebra $\bar{\mathcal{G}}_{(B)}$ of a central extension $\bar{G}_{(B)}$ of the original group rather than the original Lie algebra $G_{(B)}$. The generators $\bar{X}_{(g^i)}$ of $\bar{\mathcal{G}}_{(B)}$ are easily read from the conserved quantities, and are given by

$$\bar{X}_{(b)} = \frac{\partial}{\partial b} \quad , \quad \bar{X}_{(a^i)} = \frac{\partial}{\partial x^i} + \frac{B}{2}\epsilon_{ij}x^j\frac{\partial}{\partial \chi} \quad ,$$

$$\bar{X}_{(V_i)} = \frac{m}{B}[\delta_{ij}\sin\frac{B}{m}t - \epsilon_{ij}(1 - \cos\frac{B}{m}t)]\frac{\partial}{\partial x^j} + \Delta_{(V_i)}\frac{\partial}{\partial \chi} \quad , \quad \bar{X}_{(\theta)} = \frac{\partial}{\partial \chi}$$

$$(8.3.27)$$

(the term in $\partial/\partial \dot{x}^i$ in $\bar{X}_{(V^i)}$ has been dropped but it could be retained if desired). The Lie algebra $\bar{\mathcal{G}}_{(B)}$ of $\bar{G}_{(B)}$ is given by

$$[\bar{X}_{(b)}, \bar{X}_{(a^i)}] = 0 \quad , \quad [\bar{X}_{(a^i)}, \bar{X}_{(V^j)}] = m\delta_{ij}\bar{X}_{(\theta)} \quad ,$$

$$[\bar{X}_{(b)}, \bar{X}_{(V^i)}] = \bar{X}_{(a^i)} - \frac{B}{m}\epsilon_{ij}\bar{X}_{(V^j)} \quad , \quad [\bar{X}_{(V^i)}, \bar{X}_{(V^j)}] = 0 \quad , \qquad (8.3.28)$$

$$[\bar{X}_{(a^i)}, \bar{X}_{(a^j)}] = -B\epsilon_{ij}\bar{X}_{(\theta)} \quad .$$

This centrally extended algebra $\bar{\mathcal{G}}_{(B)}$ includes two commutators that differ from (8.3.24). Besides the familiar (translations, boosts) commutator, which contains the mass, the translations commutator determines another Weyl-Heisenberg subalgebra in $\bar{\mathcal{G}}_{(B)}$, its central charge being the magnetic field B which characterizes the central extension of the original abelian translation algebra. Again, this is already present at the Poisson bracket level, as exhibited by the isomorphism $\bar{X}_{(g^i)} \leftrightarrow \mathcal{N}_{(g^i)}$, $[\bar{X}_{(g^i)}, \bar{X}_{(g^j)}] \leftrightarrow \{\mathcal{N}_{(g^i)}, \mathcal{N}_{(g^j)}\}$; thus, this commutator does not reflect any departure from the classical dynamics. Indeed, the fact that the translations are a symmetry of the classical and quantum dynamics ($\{H, P_{(a^i)}\} = 0$, $[\bar{X}_{(b)}, \bar{X}_{(a^i)}] = 0$) is compatible with the non-trivial Poisson bracket $\{P_i, P_j\} = -B\epsilon_{ij}$ or with the quantum operators commutator $[\hat{P}_i, \hat{P}_j] = -iB\epsilon_{ij}$ where $\hat{P}_i = i\bar{X}_{(a^i)}$ (quantum operators are obtained from (8.3.27) by including a factor $i\hbar$ or i if $\hbar = 1$).

Let us now show that the Lagrangian (8.3.15) may be also obtained from a WZ term on $G_{(B)}$. To this end we look for a closed invariant two-form $\omega^{(2)}$ on $G_{(B)}$ the potential form $\beta^{(1)}$ of which is not invariant. Since the LI forms on $G_{(B)}$ are easily found to be (cf. (3.4.5))

$$\omega^{L(b)} = db \quad , \quad \omega^{L(a^i)} = da^i - V^i db \quad , \quad \omega^{L(V^i)} = dV^i - \frac{B}{m}\epsilon^{ij}V_j db \quad ,$$

$$(8.3.29)$$

we conclude by looking at the commutators (8.3.28) that

$$\omega^{(2)}_{(B)} = m\omega^{L(V^i)} \wedge \omega^{L(a_i)} + \frac{B}{2}\epsilon_{ij}\omega^{L(a_i)} \wedge \omega^{L(a_j)} \qquad (8.3.30)$$

is a non-trivial CE cocycle, and that a potential one-form may be given by

$$\beta_{(B)}^{(1)} = m\dot{x}^i dx_i - \frac{1}{2}m\dot{x}^2 dt + \frac{B}{2}\epsilon_{ij}dx^i \wedge dx^j \qquad (8.3.31)$$

once (b, a^i, V^i) are identified with $(t, x^i, \dot{x}^i)$. This WZ form may be rewritten as the Poincaré-Cartan one-form Θ_{PC} for the dynamics of a particle in a constant magnetic field,

$$\beta_{(B)}^{(1)} = \Theta_{PC} = \frac{\partial L}{\partial \dot{x}^i}(dx^i - \dot{x}^i dt) + L dt \quad , \qquad (8.3.32)$$

where L is the Lagrangian in (8.3.15). The physical trajectories are thus the solution of (8.3.22), and are of course equivalent to the Lagrange equation for (8.3.15), $m d^2 x^i(t)/dt^2 = B\epsilon_{ij}\dot{x}^j(t)$. In fact, its solutions are (once $(B, a^i, V^i) \rightarrow (t, x^i, \dot{x}^i)$) the integral curves of the LI vector field on $G_{(B)}$ (cf. (3.4.5))

$$X_{(b)}^L = \frac{\partial}{\partial b} + V_i\frac{\partial}{\partial a^i} + \frac{B}{m}\epsilon_{ij}V^j\frac{\partial}{\partial V^i} \qquad (8.3.33)$$

generating the time translations . This is as it should be: since $i_X(\alpha \wedge \gamma) = (i_X\alpha)\gamma - \alpha(i_X\gamma)$ for two one-forms α, γ, it follows that $i_{X_{(b)}^L}d\Theta_{PC} = 0$ since, by LIF-LIVF duality, $\theta^{L(a^i)}(X_{(b)}^L) = 0 = \theta^{L(V^i)}(X_{(b)}^L)$. The new commutators just imply that the Poincaré-Cartan one-form is a Wess-Zumino form with two terms, each one multiplied by the central charge of the relevant commutator in the extension.

8.4 Cohomology and classical mechanics: preliminaries

Let us consider the problem of cohomology in the more familiar formulation where the Lagrangian is given by $L = L(q(t), \dot{q}(t), t)$ (we shall omit the i superscripts). In this section the dot will always mean time derivative, $\dot{q}(t) = dq(t)/dt$ (we could also take q, t depending on an evolution parameter τ parametrizing the worldline of the particle); the general discussion on $J^1(E)$ will be given in section 8.5. Consider the action of the ordinary formulation of the variational principle

$$I[s] = \int_1^2 L(q(t), \ \dot{q}(t), \ t)dt \quad , \quad \delta I = 0 \quad , \qquad (8.4.1)$$

which leads to (8.1.11). We can look at the action as a mapping

$$I : (q_1, q_2) \mapsto I(q_1, q_2) \in R \quad , \qquad (8.4.2)$$

where $q_1 = q(t_1)$ and $q_2 = q(t_2)$ are the end points. Let L now be quasi-invariant under the action of a symmetry group G,

$$L(gt, gq, g\dot{q}) = L(t, q, \dot{q}) + \frac{d}{dt}\Delta(t, q; g) \quad . \tag{8.4.3}$$

The function Δ does not depend on $\dot{q}$ because we assume that the generators which define the action of G on the evolution space do not depend on $\ddot{q}$; it depends on the group parameters since the action of g takes L to L'. As a result, $\delta L(t, q, \dot{q})$ does not contain terms in $\ddot{q}$, and so δL and Δ cannot depend on $\ddot{q}$ and $\dot{q}$ respectively. Lagrangians differing in a total derivative lead to the same Lagrange equations. Thus, the G-invariance of the equations of motion means that

$$\exists \Delta \ , \quad \Delta : (t, q; g) \mapsto \Delta(t, q; g) \quad , \tag{8.4.4}$$

$$I(gq_1, \ gq_2) = I(q_1, q_2) + \Delta(t_2, q_2; g) - \Delta(t_1, q_1; g) \quad . \tag{8.4.5}$$

Since the variation (8.4.5) does not affect the equations of motion, the function I may be called a *gauge-variant function* and the real functions $\Delta(t, q; g)$ on (spacetime)$\times G$, *gauge functions*. Then we have the following

(8.4.1) LEMMA (*Lévy-Leblond*)

The mapping $\xi : G \times G \to R$, defined by

$$\xi(g', g) := \Delta(gt, gq; g') - \Delta(t, q; \ g'g) + \Delta(t, q; g) \quad , \tag{8.4.6}$$

is an R-valued function on $G \times G$, independent of the spacetime coordinates, which defines a two-cocycle of G.

Proof: It is sufficient to compare

$$I((g'g)q_1, \ (g'g)q_2) = I(q_1, \ q_2) + \Delta(t_2, q_2; \ g'g) - \Delta(t_1, q_1; g'g) \tag{8.4.7}$$

with

$$\begin{aligned} I(g'(gq_1), g'(gq_2)) &= I(gq_1, gq_2) + \Delta(gt_2, gq_2; g') - \Delta(gt_1, gq_1; g') \\ &= I(q_1, q_2) + \Delta(t_2, q_2; g) - \Delta(t_1, q_1; g) \\ &\quad + \Delta(gt_2, gq_2; g') - \Delta(gt_1, gq_1; g') \quad , \end{aligned} \tag{8.4.8}$$

which have to be equal. This implies

$$[\Delta(gt_2, gq_2, g') - \Delta(gt_1, gq_1, g')] - [\Delta(t_2, q_2, gg') - \Delta(t_1, q_1, gg')] \\ + [\Delta(t_2, q_2, g) - \Delta(t_1, q_1, g)] = 0 ,$$

$$\begin{aligned} \Delta(gt_2, gq_2; g') &- \Delta(t_2, q_2; g'g) + \Delta(t_2, q_2; g) = \\ &\Delta(gt_1, gq_1; g') - \Delta(t_1, q_1; g'g) + \Delta(t_1, q_1; g) , \end{aligned} \tag{8.4.9}$$

i.e., that $\xi \neq \xi(t, q)$. By substituting (8.4.6) in the cocycle condition (5.1.20) with σ trivial, we see that ξ is a two-cocycle on G. Also, if $g = e$, (8.4.5) implies $\Delta(t, q; e)$ constant; by selecting Δ in such a way that $\Delta(t, q; e) = 0$ we get $\xi(e, g) = 0$, etc.; ξ is then *normalized*, *q.e.d.*

We may now recover the results obtained in section 3.5 for a specific example. If a term that is a derivative is added to the Lagrangian L in (8.4.1), the new Lagrangian is $\tilde{L} = L + \frac{d}{dt}\phi(t, q(t))$. The equations of the motion remain unchanged, and for this reason $\tilde{L}$ is equivalent to L. Nevertheless, the variation of $\tilde{L}$ gives an extra piece that has to be added to the original gauge function $\Delta(t, q; g)$. Denoting this piece by $\Delta_{cob}(t, q; g)$ it is found that

$$\Delta_{cob}(t, q; g) = \phi(gt, gq) - \phi(t, q) + \eta(g), \qquad (8.4.10)$$

since a time-independent $\eta(g)$ can always be added under the action of $\frac{d}{dt}$ in (8.4.3). In this way, all the freedom in Δ is contained in (8.4.10). We then have

(8.4.1) COROLLARY

The mapping $\xi_{cob} : G \times G \to R$ *defined by*

$$\xi_{cob}(g', g) \equiv \Delta_{cob}(gt, gq; g') - \Delta_{cob}(t, q; g'g) + \Delta_{cob}(t, q; g) \qquad (8.4.11)$$

is a two-coboundary generated by $\eta(g)$.

Proof: Substituting (8.4.10) in (8.4.11) we get

$$\begin{aligned}\xi_{cob}(g', g) &= \phi(g'gt, g'gq) - \phi(gt, gq) + \eta(g') - \phi(g'gt, g'gq) + \phi(t, q) \\ &\quad - \eta(g'g) + \phi(gt, gq) - \phi(t, q) + \eta(g) \\ &= \eta(g) + \eta(g') - \eta(g'g) \equiv (\delta\eta)(g', g),\end{aligned} \qquad (8.4.12)$$

as is seen from eq. (5.1.3) with $n = 1$ and σ trivial, *q.e.d.*

Thus, ϕ does not intervene in the expression of ξ_{cob}, so that a modification of the Lagrangian by a total derivative $\frac{d}{dt}\phi$ does not change the cocycle that it defines. Rather, the origin of the coboundaries is tied to the freedom in the choice of the gauge functions, which can be modified by the addition of an $\eta(g)$ (section 3.5). Lemma 8.4.1 and its corollary may be strengthened in the form of a theorem which will be given without proof:

(8.4.1) THEOREM (*Lévy-Leblond*)

Let G *act* (*transitively*) *on the homogeneous space* $M = G/S$ (*M may be the spacetime manifold*), *and let* $\Delta(x, g)$ ($x \in M, g \in G$) *be the gauge* functions of G for this homogeneous space. Then, a) the restriction of the two-cocycle (8.4.6) to the subgroup S is equivalent to zero (the condition is in fact stronger: a two-cocycle ξ equivalent to zero on S is also equivalent to zero on $G \times S$, i.e. $\exists \eta$ on G such that $\xi(g,s) = \eta(g) + \eta(s) - \eta(gs)$ $\forall g \in G$, $\forall s \in S$) and b) the classes of gauge functions Δ on $M \times G$ are in one-to one correspondence with the classes of two-cocycles equivalent to zero on S, the elements of the same class differing by two-coboundaries generated by functions η which are equal to zero on S.

The reader will notice that, for Galilean mechanics, eqs. (8.4.3), (8.4.8) are just the classical counterparts of (3.1.22), (3.1.24). This had to be so, since the wave equation of a quasi-classical physical system has the form

$$\psi \propto \exp \frac{i}{\hbar} I \quad . \tag{8.4.13}$$

The classical limit, which corresponds to a large phase, can be described as the ($h \to 0$) passage which makes the $U(1) \sim S^1$ radius infinite. In the language of chapter 3, this means replacing $\tilde{G}$ by $\bar{G}$, i.e. the Galilei group extended by $U(1)$ by the Galilei group extended by R. This allows us to establish a complete parallel between the classical and the quantum case. The addition of a new variable χ, with the dimensions of an action, allowed us to obtain a classical Lagrangian on the enlarged evolution space invariant under $\bar{G}$; the quantum version of the theory is invariant under the quantum group $\tilde{G}$. The classical variable χ does not appear in the equations of motion, since from (8.3.12) we get that the new Lagrangian is given by

$$(\bar{s}^1)^* \bar{\Theta}_{PC} = \frac{1}{2} m \left(\frac{dx}{dt} \right)^2 - \frac{d\chi}{dt} \tag{8.4.14}$$

and $\partial L / \partial \dot{\chi} = -1$. Thus, this degree of freedom is decoupled classically; the time dependence of χ is completely arbitrary. We have already mentioned in chapter 3 how essential the *extended* Galilei group is for a consistent quantization process. We see now why there is no anomaly in the appearance of the extension $\tilde{G}$ of G in the quantum formulation, since the extension $\bar{G}$ (which is locally isomorphic to $\tilde{G}$) was already present in the classical case: the extended Lie algebra and the classical Poisson algebra are isomorphic. The classical degree of freedom represented by χ corresponds to the quantum phase degree of freedom of ψ; only by ignoring χ does the quantum version of the theory look 'anomalous'.

Having constructed the group cohomology starting from a quasi-invariant Lagrangian, we may now do the same for the Lie algebra cohomology associated with infinitesimal variations of the group. We start by introducing the infinitesimal variations $\delta_{(i)}$, $i = 1, ..., \dim(G)$, that determine the action of the group G on the functions $q(t)$, $\dot{q}(t)$ and t which the Lagrangian depends on. For instance, if the mechanical problem is initially defined on $J^1(E)$, these variations are induced by the vector fields $Y_{(i)}$ that represent $X_{(i)}$ on $J^1(E)$, *i.e.* by their 1-jet prolongation to $J^1(E)$ (denoted here $Y_{(i)}$). Nevertheless, in this section we shall not worry about this and will represent the variation of the Lagrangian generically by $\delta_{(i)}L$[†].

In general, the variation of a function $F(q, \dot{q})$ under an infinitesimal transformation g of parameters ζ^i is given by (see section 1.1)

$$\delta F \simeq \zeta^i \delta_{(i)} F(q, \dot{q}) = \zeta^i \frac{\partial}{\partial g^i} F(gq, g\dot{q})\big|_{g=e} \quad , \qquad (8.4.15)$$

where gq, $g\dot{q}$ are shorthand for the action of g. As a result, the infinitesimal variation of the Lagrangian L is given by

$$L(gq, g\dot{q}) - L(q, \dot{q}) \simeq \zeta^i \frac{\partial L(gq, g\dot{q})}{\partial g^i}\bigg|_{g=e} . \qquad (8.4.16)$$

Since by hypothesis the Lagrangian considered is quasi-invariant, its infinitesimal variation $\delta_{(i)}L$ may be written in the form

$$\delta_{(i)}L = \frac{\partial}{\partial g^i}(L(gq, g\dot{q}) - L(q, \dot{q}))\big|_{g=e} = \frac{d}{dt}\left(\frac{\partial}{\partial g^i}\Delta(q; g)\big|_{g=e}\right) = \frac{d}{dt}\Delta_{(i)}(q) ,$$
$$(8.4.17)$$

where

$$\Delta_{(i)}(q) := \frac{\partial \Delta(q; g)}{\partial g^i}\bigg|_{g=e} . \qquad (8.4.18)$$

Thus, by the same reasoning that led to the Wess-Zumino consistency condition (8.2.7) it follows that the combination

$$\delta_{(i)}.\Delta_{(j)}(q) - \delta_{(j)}.\Delta_{(i)}(q) - C_{ij}^{Rk}\Delta_{(k)}(q) \equiv C_{ij}^{R\theta} , \quad \frac{d}{dt}C_{ij}^{R\theta} = 0 \quad , \qquad (8.4.19)$$

is a constant, real valued cocycle of the Lie algebra $\mathcal{G}$ of G that can be used to define the structure constants $C_{ij}^{R\theta}$ that give the corresponding extension. On the other hand, if we modify the Lagrangian to $L + \frac{d}{dt}\phi$, $C_{ij}^{R\theta}$ remains unchanged; and if a new $\Delta'_{(i)}(q) = \Delta_{(i)}(q) + \eta_{(i)}$ is taken, where η_i is constant, then

$$C'^{R\theta}_{ij} = C^{R\theta}_{ij} - C^{Rk}_{ij}\eta_k , \qquad (8.4.20)$$

which means that the coboundaries are generated, as in the Lie group case, by the constant terms that can be added to the right of the derivative in

---

† Moreover, from now on we shall not write the t dependence of L, Δ, etc. explicitly; it will be assumed included in the q coordinates with the exception of Proposition 8.5.1.

(8.4.17). The η_i themselves can be related to $\eta(g)$ in a simple way:

$$\delta_{(i)}.L = \frac{\partial L(gq, g\dot q)}{\partial g^i}\bigg|_{g=e} = \frac{d}{dt}\left[\left(\frac{\partial \Delta(q;g)}{\partial g^i}\bigg|_{g=e}\right) + \frac{\partial \eta(g)}{\partial g^i}\bigg|_{g=e}\right]$$

$$\equiv \frac{d}{dt}(\Delta_i(q) + \eta_i) .$$

(8.4.21)

Eq. (8.4.20) should be compared with (6.4.7).

We know that a quasi-invariant Lagrangian determines elements of both the Lie group cohomology ($\xi(g',g)$) and the Lie algebra cohomology ($C_{ij}^{R\theta}$). The following theorem (cf. Proposition 6.4.1 and eq. (6.4.4)) shows how they are related:

(8.4.2) THEOREM

The structure constants $C_{ij}^{R\theta} = \left[\dfrac{\partial^2 \xi(g',g)}{\partial g'^j \partial g^i} - \dfrac{\partial^2 \xi(g',g)}{\partial g'^i \partial g^j}\right]_{g'=g=e}$ that are obtained from the central extension two-cocycle $\xi(g',g) = \Delta(gq, g') - \Delta(q, g'g) + \Delta(q,g)$ are precisely the structure constants $C_{ij}^{R\theta}$ in the extended algebra of the generators of the left action (cf. (1.2.11) and (1.2.16)),

$$[X_{(i)}^R, X_{(j)}^R] = C_{ij}^{Rk} X_{(k)}^R + C_{ij}^{R\theta} \frac{\partial}{\partial\theta} \quad , \qquad (8.4.22)$$

resulting from the extended group law (3.2.15).

Proof: It is sufficient to substitute (8.4.6) into the expression for $C_{ij}^{R\theta}$ above and to take into account (8.4.18) to get

$$\left[\frac{\partial^2 \xi(g',g)}{\partial g'^j \partial g^i} - \frac{\partial^2 \xi(g',g)}{\partial g'^i \partial g^j}\right]_{g'=g=e}$$

$$= \frac{\partial^2 \Delta(gq, g')}{\partial g'^j \partial g^i}\bigg|_{g'=g=e} - \frac{\partial^2 \Delta(q, g'g)}{\partial g'^j \partial g^i}\bigg|_{g'=g=e} + \frac{\partial^2 \Delta(q, g)}{\partial g'^j \partial g^i}\bigg|_{g'=g=e} - i \leftrightarrow j$$

$$= Y_{(i)}.\Delta_{(j)} - \frac{\partial^2 \Delta(q, g'')}{\partial g''^l \partial g''^k}\frac{\partial g''^l}{\partial g'^j}\frac{\partial g''^k}{\partial g^i}\bigg|_{g=g'=e} - \frac{\partial \Delta(q;g'')}{\partial g''^k}\frac{\partial g''^k}{\partial g'^j \partial g^i}\bigg|_{g=g'=e} - i \leftrightarrow j$$

$$= Y_{(i)}.\Delta_{(j)} - Y_{(j)}.\Delta_{(i)} - \Delta_{(k)}\left(\frac{\partial g''^k}{\partial g'^j \partial g^i} - \frac{\partial g''^k}{\partial g'^i \partial g^j}\right)\bigg|_{g=g'=e}$$

$$= Y_{(i)}.\Delta_{(j)} - Y_{(j)}.\Delta_{(i)} - C^R{}_i{}^k{}_j\Delta_{(k)} = C_{ij}^{R\theta} ,$$

(8.4.23)

since the last (second) term in the second (third) line is zero, and we have used that (1.2.16) gave the structure constants C_{ij}^k of the left algebra ($C^R{}_i{}^k{}_j = -C_i{}^k{}_j$), q.e.d. Of course, if Δ is modified as in (8.4.10), the above expression applied to

$$\Delta'_{(i)}(q) = \Delta_{(i)}(q) + Y_{(i)}\cdot\phi(q) + \left.\frac{\partial\eta(g)}{\partial g^i}\right|_{g=e} \tag{8.4.24}$$

reproduces eq. (8.4.20) since $[Y_{(i)}, Y_{(j)}] = C_{ij}^{Rk} Y_{(k)}$ (again, ϕ does not contribute to $C'^{R\theta}_{ij}$ and only the term $(\partial\eta(g)/\partial g^i)|_{g=e} \equiv \eta_i$ is relevant in this respect).

The reader will recognize that the contents of this section and of section 3.1 are the physical counterpart of the more abstract treatment of section 6.4: *the quasi-invariance of the Lagrangian under a group of transformations G is tied to the existence of a non-trivial group cohomology for G.*

8.5 The cohomological descent approach to classical anomalies

Let us now return to the formulation of mechanics on $R \times T(N) = J^1(E)$. It has been shown how a quasi-invariant Lagrangian defines an element of the cohomology groups $H^2(G, R)$ and $H^2(\mathscr{G}, R)$ given by $\xi(g', g)$ and $C_{ij}^{R\theta}$ respectively. They have been constructed starting from Δ's which can also be formulated as one-cochains $\Delta(q; g)$ and $\Delta_{(i)}(q)$ on $R \times T(N)$. Nevertheless, $\Delta(q; g)$ and $\Delta_{(i)}(q)$ are not cochains for the above R-valued cohomologies. However, since they can be understood as one-cochains for the function-valued cohomologies on G and $\mathscr{G}$, respectively, it is possible to give a formulation of the problem in a more general cohomological setting that includes both $(\xi(g', g), C_{ij}^k)$ and $(\Delta(q; g), \Delta_{(i)}(q))$ as cochains. In this way, the problem of *extensions* or modifications of the original algebra $\mathscr{G}$, i.e., what may be called the problem of classical 'anomalies', can be formulated by using the *cohomological descent procedure*. This procedure, which involves two commuting cohomology operators, will also be used in chapter 10 to discuss the geometry of the non-abelian gauge anomalies, and thus it is convenient to describe it here in a simpler setting.

Let the action of the Lie group G on the space $R \times T(N)$ parametrized by t, q^i and $\dot{q}^i$ be denoted as before by gq, $g\dot{q}$. We shall again assume that the action of G on $\dot{q}$ is the one induced by the action of G on q. The action σ of G on a function $F(q, \dot{q}) \in \mathscr{F}(J^1(E))$ is then defined by

$$\sigma(g)F(q, \dot{q}) = F(gq, g\dot{q}) . \tag{8.5.1}$$

The infinitesimal generators associated with this action will be denoted by $Y_{(i)}$. Thus, the action ρ of the Lie algebra elements $X_{(i)} \in \mathscr{G}$ on $\mathscr{F}(J^1(E))$ is given by

$$\rho(X_{(i)})F(q, \dot{q}) = Y_{(i)}\cdot F(q, \dot{q}) , \tag{8.5.2}$$

where $\rho(X_{(i)}) = j^1(X_{(i)}) = Y_{(i)}$.

Following sections 5.4 and 6.10 we now define the group and algebra cochains with values in $\mathscr{F}(J^1(E))$. A Lie group G n-cochain $\Omega_n \in C_n(G, \mathscr{F}(J^1(E)))$ is a mapping $\Omega_n : G \times \overset{n}{...} \times G \to \mathscr{F}(J^1(E))$, $\Omega_n : (g_1, ..., g_n) \mapsto \Omega_n(\cdot \ ; g_1, ..., g_n)$, which defines a function on $J^1(E)$, $\Omega_n(\cdot \ ; g_1, ..., g_n) : (q, \dot{q}) \mapsto \Omega_n(q, \dot{q} \ ; g_1, ..., g_n)$. A Lie algebra $\mathscr{G}$ n-cochain $\omega_n \in C_n(\mathscr{G}, \mathscr{F}(J^1(E)))$ is a skewsymmetric linear mapping $\omega_n : \mathscr{G} \times \overset{n}{...} \times \mathscr{G} \to \mathscr{F}(J^1(E))$, $\omega_n : (X_1, ..., X_n) \mapsto \omega_n(\cdot \ ; X_1, ..., X_n)$, which defines the element of $\mathscr{F}(J^1(E))$ given by $\omega_n(\cdot \ ; X_1, ..., X_n) : (q, \dot{q}) \mapsto \omega_n(q, \dot{q} \ ; X_1, ..., X_n)$. Using this notation we have

$$\sigma(g)\Omega_n(q, \dot{q}; g_1, ..., g_n) = \Omega_n(gq, g\dot{q}; g_1, ..., g_n) \,,$$
$$\rho(X)\omega_n(q, \dot{q}; X_1, ..., X_n) = Y.\omega_n(q, \dot{q}; X_1, ..., X_n) \,, \tag{8.5.3}$$

and the action of the Lie group and the Lie algebra coboundary operators on these cochains will be given by (5.4.10) and (6.10.4). In this way[*], the cohomology groups $H^2_\sigma(G, \mathscr{F}(J^1(E)))$ and $H^2_\rho(\mathscr{G}, \mathscr{F}(J^1(E)))$ are defined; constant two-cocycles on these groups (*i.e.*, that do not depend on $q, \dot{q}$) define elements of $H^2(G, R)$ and $H^2(\mathscr{G}, R)$.

This provides a convenient framework to discuss with generality the appearance of cohomology in mechanics. To this end we shall work with 'cochains' that take values in the space of differential forms $\Lambda(J^1(E))$ that depend on q and $\dot{q}$ rather than in the space of functions $\mathscr{F}(J^1(E))$. The key point is that the quasi-invariance of the Lagrangian $(\bar{s}^1)^*(L(q, \dot{q})) = L(q(t), dq(t)/dt)$ on R under a symmetry transformation, which is expressed by (8.4.3), implies the quasi-invariance of the Poincaré-Cartan form Θ_{PC} (8.1.5) on $J^1(E)$ associated with $L(q, \dot{q})$. If the variation of the Lagrangian is given by $\delta_{(i)}((\bar{s}^1)^*L) = \frac{d}{dt}\Delta_{(i)}$, then $L_{\bar{X}^1_{(i)}}\Theta_{PC} = d\Delta_{(i)}$, where in the first expression Δ depends on the functions $q(t)$ and in the second Δ is a function on E. This is the content of

(8.5.1) PROPOSITION

Let $\bar{X}^1$ be a jet-prolongation vector field and Ldt a Lagrangian one-form on $J^1(E)$ such that

$$L_{\bar{X}^1}(Ldt) = d\Delta + B_i(dq^i - \dot{q}^i dt) \equiv d\Delta + B_i \theta^i \ , \tag{8.5.4}$$

where $\Delta = \Delta(t, q^i)$ and $B_i = B_i(t, q^i, \dot{q}^i)$. Then the Poincaré-Cartan form $\Theta_{PC} = \frac{\partial L}{\partial \dot{q}^i}\theta^i + Ldt$ satisfies

$$L_{\bar{X}^1}\Theta_{PC} = d\Delta \ . \tag{8.5.5}$$

---

[*] $\mathscr{F}(J^1(E))$ can be endowed with an obvious vector space structure with the composition law given by the sum of functions, so that $\mathscr{F}(J^1(E))$ is a ρ-module.

Proof: First, we notice that if the pull-back of the variation of L has to be a time derivative, the r.h.s. of (8.5.4) has to have the proposed form since $(\bar{s}^1)^*(d\Delta) = \frac{d}{dt}\Delta(t, q(t))$ and $(\bar{s}^1)^*\theta^i = 0$. Let $\bar{X}^1$ be as in (8.1.2), (8.1.3). Then eq. (8.5.4) reads

$$(\bar{X}^1.L)dt + L(\frac{\partial X^t}{\partial t}dt + \frac{\partial X^t}{\partial q^i}dq^i) = \frac{\partial\Delta}{\partial t}dt + \frac{\partial\Delta}{\partial q^i}dq^i + B_i dq^i - B_i\dot{q}^i dt \quad (8.5.6)$$

and implies

$$\bar{X}^1.L + L\frac{\partial X^t}{\partial t} = \frac{\partial\Delta}{\partial t} - B_i\dot{q}^i \quad , \quad B_i = L\frac{\partial X^t}{\partial q^i} - \frac{\partial\Delta}{\partial q^i} \quad (8.5.7)$$

and, eliminating B_i, that

$$\bar{X}^1.L = \frac{\partial\Delta}{\partial t} + \frac{\partial\Delta}{\partial q^i}\dot{q}^i - L(\frac{\partial X^t}{\partial t} + \dot{q}^i\frac{\partial X^t}{\partial q^i}) \quad . \quad (8.5.8)$$

On the other hand,

$$L_{\bar{X}^1}\Theta_{PC} = (\bar{X}^1.\frac{\partial L}{\partial\dot{q}^i})\theta^i + \frac{\partial L}{\partial\dot{q}^i}L_{\bar{X}^1}\theta^i + L_{\bar{X}^1}(Ldt) \quad . \quad (8.5.9)$$

To compute the first term in this expression we use that $\bar{X}^1\frac{\partial}{\partial\dot{q}^i} = \frac{\partial}{\partial\dot{q}^i}\bar{X}^1 + [\bar{X}^1, \frac{\partial}{\partial\dot{q}^i}]$ and that

$$[\bar{X}^1, \frac{\partial}{\partial\dot{q}^i}] = (\delta_i^j\frac{\partial X^t}{\partial t} - \frac{\partial X^j}{\partial q^i} + \frac{\partial X^t}{\partial q^i}\dot{q}^j + \dot{q}^k\frac{\partial X^t}{\partial q^k}\delta_i^j)\frac{\partial}{\partial\dot{q}^j} \quad , \quad (8.5.10)$$

from which we get using (8.5.8)

$$(\bar{X}^1.\frac{\partial L}{\partial\dot{q}^i})\theta^i = (\frac{\partial\Delta}{\partial q^i} - L\frac{\partial X^t}{\partial q^i} - \frac{\partial L}{\partial\dot{q}^j}[\frac{\partial X^j}{\partial q^i} - \frac{\partial X^t}{\partial q^i}\dot{q}^j])\theta^i \quad . \quad (8.5.11)$$

Since the second term of (8.5.9) contains

$$L_{\bar{X}^1}\theta^i = dX^i - \dot{X}^i dt - \dot{q}^i dX^t = (\frac{\partial X^i}{\partial q^j} - \dot{q}^i\frac{\partial X^t}{\partial q^j})\theta^j \quad (8.5.12)$$

and the third is given by (8.5.4) which using (8.5.7) for B_i reads

$$L_{\bar{X}^1}(Ldt) = d\Delta + (L\frac{\partial X^t}{\partial q^i} - \frac{\partial\Delta}{\partial q^i})\theta^i \quad , \quad (8.5.13)$$

we find from (8.5.11), (8.5.12) and (8.5.13) that all terms in the r.h.s. of (8.5.9) cancel with the exception of $d\Delta$, *q.e.d.*

The definition of the cohomology operators s and δ on cochains that give elements of $\Lambda^m(J^1(E))$ (and that accordingly have an additional index m indicating the degree of the form) is the same as in the usual cases, except for the fact that $Y_{(i)}.\omega$ is replaced by $L_{Y_{(i)}}\omega^m$ in the Lie algebra case. These cochains will be denoted by Ω_n^m and ω_n^m, although in the rest of this section m will only take the values $m = 0, 1, 2$. They define mappings $\Omega_n^m : \overset{n}{G \times ... \times G} \to \Lambda^m(J^1(E))$ and $\omega_n^m : \overset{n}{\mathcal{G} \times ... \times \mathcal{G}} \to \Lambda^m(J^1(E))$.

The two cohomology operators δ and s acting on Ω_n^m and ω_n^m, respectively, commute with d:

(8.5.2) PROPOSITION

The action of δ (s) on the forms Ω_n^m (ω_n^m) commutes with the exterior derivative,

$$[\delta, d] = 0 , \quad [s, d] = 0 . \tag{8.5.14}$$

Proof: The proof is trivial, since d acts only on the $J^1(E)$ manifold variables, and thus the commutativity is just the familiar statement that the derivative of the variation is the variation of the derivative, q.e.d.

Consider now the Poincaré-Cartan one-form Θ_{PC}. Since it does not depend on g or on the Lie algebra indices, it can be viewed as a zero (group or algebra) cochain. Let us then write $\Theta_{PC}(q, \dot{q}) \equiv \Omega_0^1(q, \dot{q}) = \omega_0^1(q, \dot{q})$. The zero-forms $\Delta(q; g)$ and $\Delta_{(i)}(q)$ are, respectively, group and algebra one-cochains, so that they can be written as $\Delta(q; g) \equiv \Omega_1^0(q; g)$, $\Delta_{(i)}(q) \equiv \omega_1^0(q; X_{(i)})$. With this notation, the equations which correspond to (8.4.3) and (8.4.17) in this formalism are

$$\Omega_0^1(gq, g\dot{q}) - \Omega_0^1(q, \dot{q}) = d\Omega_1^0(q; g) ,$$
$$L_{\tilde{X}_{(i)}^1}\omega_0^1(q, \dot{q}) = d\omega_1^0(q; X_{(i)}) . \tag{8.5.15}$$

Note that Ω_1^0 and ω_1^0 do not depend on $\dot{q}$ because the gauge functions Δ depend only on q, as we mentioned in section 8.4. Looking at (5.4.10) and (6.10.4), it is clear that equations (8.5.15) read

$$\delta\omega_0^1 = d\Omega_1^0 , \quad s\omega_0^1 = d\omega_1^0 . \tag{8.5.16}$$

This is the starting point of a process that gives elements of the different cohomology groups called the descent method. It arises when there are two nilpotent cohomology operators that commute or anticommute, and a basic equation of the type (8.5.16); this permits the construction of a double cohomology chain.

The descent method proceeds by applying the coboundary operators δ and s to the equations in (8.5.16) and by using $\delta^2 = 0$, $s^2 = 0$ and

$[\delta, d] = 0 = [s, d]$ to obtain

$$d\delta\Omega_1^0 = 0, \quad ds\omega_1^0 = 0. \tag{8.5.17}$$

These equations show that both $\delta\Omega_1^0$ and $s\omega_1^0$ are closed forms. Since they are zero-forms, this means that $\delta\Omega_1^0$ and $s\omega_1^0$ are constant, so that they do not depend on q, $\dot{q}$, as we already know. Explicitly,

$$\xi(g', g) = \delta\Omega_1^0 = \Omega_1^0(gq; g') - \Omega_1^0(q; g'g) + \Omega_1^0(q; g),$$
$$C_{ij}^{R\theta} = s\omega_1^0(X_{(i)}, X_{(j)}) = Y_{(i)}.\omega_1^0(q; X_{(j)}) - Y_{(j)}.\omega_1^0(q; X_{(i)}) \tag{8.5.18}$$
$$- \omega_1^0(q; [X_{(i)}, X_{(j)}])$$

(notice that $\rho = j^1$ is a homomorphism, eq. (8.1.4), that the $X_{(i)}$'s generate the left action on $R \times N$ and satisfy the 'right' commutation relations, and that the comment after (6.1.7) applies here). The fact that they are two-cocycles is now an immediate consequence of the nilpotency of δ and s which implies that $\delta(\delta\Omega_1^0) = 0 = s(s\omega_1^0)$. Thus, $\delta\Omega_1^0$ and $s\omega_1^0$ are two-cocycles in the group and algebra cohomologies respectively, $H_\sigma^2(G, \mathcal{F}(J^1(E)))$, $H_\rho^2(\mathcal{G}, \mathcal{F}(J^1(E)))$. But, since $\delta\Omega_1^0$ and $s\omega_1^0$ are constant, the terms in δ and s due to the action of σ and ρ on the elements of $\mathcal{F}(J^1(E))$ disappear, and what is left are the cocycle conditions for $H^2(G, R)$, $H^2(\mathcal{G}, R)$. Notice that $\delta\Omega_1^0$ and $s\omega_1^0$ are trivial cocycles in the $\mathcal{F}(J^1(E))$-valued cohomologies (obviously they are generated by Ω_1^0 and ω_1^0), but *not* in the R-valued cohomologies, because for that we would need $\delta\Omega_1^0$ ($s\omega_1^0$) to be equal to δ (s) acting on one-cochains *independent* of q and $\dot{q}$.

The above discussion can be summarized in the following commutative diagrams:

$$
\begin{array}{ccc}
\delta\Omega_0^1 & \xleftarrow{d} & \Omega_1^0 \\
\downarrow{\scriptstyle\delta} & & \downarrow{\scriptstyle\delta} \\
0 & \xleftarrow{d} & \delta\Omega_1^0 \\
& & \downarrow{\scriptstyle\delta} \\
& & 0
\end{array}
\qquad\qquad
\begin{array}{ccc}
s\omega_0^1 & \xleftarrow{d} & \omega_1^0 \\
\downarrow{\scriptstyle s} & & \downarrow{\scriptstyle s} \\
0 & \xleftarrow{d} & s\omega_1^0 \\
& & \downarrow{\scriptstyle s} \\
& & 0
\end{array}
$$

Diagrams 8.5.1

in which $\delta\Omega_1^0$ and $s\omega_1^0$ determine the group extension cocycle ξ and the extension structure constants $C_{ij}^{R\theta}$ respectively through the expressions (8.5.18).

For many purposes it is useful formulating the $H_\rho^2(\mathcal{G}, \mathcal{F}(J^1(E)))$ cohomology in such a way that the coboundary operator is constructed directly from the expression of the vector field $Y_{(i)}$. This is achieved by using the BRST technique described in section 6.8, which introduces r anticommuting ghost parameters c^i and which is extended easily to the $\mathscr{F}(J^1(E))$-valued case. To this end, tilded quantities $\tilde{\omega}_n^m$ with ghost number n for every n-cochain ω_n^m are introduced as follows:

$$\tilde{\omega}_n^m(q,\dot{q}) \equiv \omega_n^m(q,\dot{q};X_{(i_1)},...,X_{(i_n)})c^{i_1}...c^{i_n} . \qquad (8.5.19)$$

The BRST operator, which is given by

$$\tilde{s} = c^i Y_{(i)} + \frac{1}{2}c^i c^j C_{ji}^{Rk}\frac{\partial}{\partial c^k} \quad , \qquad (8.5.20)$$

is nilpotent by Proposition 6.8.1 and acts on the n-cochains $\tilde{\omega}_n^m$ in a way that is consistent with its action on the cochains ω_n^m by a simple extension of Theorem 6.8.1. As a result, the reasonings that led to the second of Diagrams 8.5.1 can now be repeated by substituting $\tilde{s}$ for s and $\tilde{\omega}_n^m$ for ω_n^m.

8.6 A simple application: the free Newtonian particle

Let us begin with the description of the free particle Newtonian 'anomaly' by using the cohomological descent method and the operators $\tilde{s}$ and d, $d^2 = 0 = \tilde{s}^2$, $[d,\tilde{s}] = 0$, that generate the double cochain complex. The starting point is, as in section 8.3, the left-invariant closed two-form $m\omega^{(2)}$ given by (8.3.5). Clearly, it has ghost number zero, so that $m\omega^{(2)}$ can also be taken as the form $\tilde{\omega}_0^2$ in the BRST framework. Since $\tilde{\omega}_0^2 = d\dot{\mathbf{x}} \wedge (d\mathbf{x} - \dot{\mathbf{x}}dt)$ is invariant and closed, we have

$$\tilde{s}\tilde{\omega}_0^2 = 0 , \quad d\tilde{\omega}_0^2 = 0 . \qquad (8.6.1)$$

As before, $d\tilde{\omega}_0^2 = 0$ tells us that there exists a 'potential' one-form $\tilde{\omega}_0^1$ such that $\tilde{\omega}_0^2 = d\tilde{\omega}_0^1$; indeed, $\tilde{\omega}_0^1$ is given by $m\beta^{(1)}$, eq. (8.3.7). The action of $\tilde{s}$ on both sides of the equation $d\tilde{\omega}_0^1 = \tilde{\omega}_0^2$ gives $d\tilde{s}\tilde{\omega}_0^1 = 0$. Thus

$$\tilde{s}\tilde{\omega}_0^1 = d\tilde{\omega}_1^0 \qquad (8.6.2)$$

for some zero-form of ghost number one, $\tilde{\omega}_1^0$. This is given by (see (8.3.8))

$$\tilde{\omega}_1^0 = c_{V_i}(mx_i) = c_{V_i}\Delta_{(i)} , \qquad (8.6.3)$$

where c_{V_i} are the ghost parameters corresponding to the boost parameters V_i. Equation (8.6.2) expresses the fact that $m\beta^{(1)}$, which is the Poincaré-Cartan one-form, is quasi-invariant. The next step is to obtain the structure constants of the extension given by this procedure. The application of $\tilde{s}$ to (8.6.2) gives $d\tilde{s}\tilde{\omega}_1^0 = 0$; thus, $\tilde{s}\tilde{\omega}_1^0$ is constant, *i.e.*, it depends on the ghost parameters c only. Explicitly,

$$\tilde{s}\tilde{\omega}_1^0 = \tilde{s}[c_{V_i}(mx_i)]$$

$$= [c_{V_i}X_{V_i}^R + c_{A_i}X_{A_i}^R + c_B X_B^R + \frac{1}{2}c_{V_i}c_B\frac{\partial}{\partial c_{A_i}}](c_{V_i}(mx_i)) \qquad (8.6.4)$$

$$= mc_{A_i}c_{V_i} = \tilde{\omega}_2^0 \, ,$$

which, by virtue of (8.5.18), defines the Lie algebra cocycle given by $C_{A_i,V_j}^{R\theta} = m\delta_{ij}$, as we know from formula (8.3.9). The example that we are considering belongs to the class of Poincaré-Cartan one-forms that are potential forms of an invariant two-form $\tilde{\omega}_0^2$. In all these cases, Diagram 8.5.1 can be completed as follows:

$$
\begin{array}{ccccccl}
 & & [2] & & [1] & & [0] \quad \text{degree of form} \\[4pt]
[0] & 0 & \xleftarrow{d} & \tilde{\omega}_0^2 & \xleftarrow{d} & \tilde{\omega}_0^1 & \\
 & & & \tilde{s}\downarrow & & \tilde{s}\downarrow & \\
[1] & & 0 & \xleftarrow{d} & d\tilde{\omega}_1^0 & \xleftarrow{d} \tilde{\omega}_1^0 & \quad \tilde{\omega}_1^0 = c^i\Delta_{(i)} \\
 & & & & \tilde{s}\downarrow & \tilde{s}\downarrow & \\
[2] & & & 0 & \xleftarrow{d} & \tilde{\omega}_2^0 & \quad \tilde{\omega}_2^0 = c^i c^j C_{ij}^{R\theta} \\
 & & & & & \tilde{s}\downarrow & \\
 & & & & & 0 &
\end{array}
$$

ghost
number

Diagram 8.6.1

The diagram exhibits the fact that the action of $\tilde{s}$ adds a ghost number unit and that of d a unit of degree form. At this point, it is interesting to note that if we calculate the structure constants of the extension directly from the LI two-form $m\omega^{(2)}$, we obtain

$$m\omega^{(2)}(X_{A_i}^L, X_{V_j}^L) = -m\delta_{ij} = C_{A_i,V_j}^\theta \, , \qquad (8.6.5)$$

i.e., the structure constants that characterize the extended algebra of the LI vector fields.

The appearance of the above minus sign can be easily explained by the double cohomology sequences. As the reasoning is general, we shall give the construction for a general symmetry group and the case where the action integrand contains a Wess-Zumino term defined by a left-invariant closed two-form h on the group manifold H, such that $h = db$ and b is only quasi-invariant. If $C_{ij}^{R\theta}$ are the structure constants obtained from the cohomological descent method, then we can define a BRST operator $\tilde{s}$ for the *extended* group of coordinates of $(\theta; g^i)$ by

$$\bar{s} = c^i \bar{X}^R_{(i)} + \frac{1}{2} c^i c^j C^{Rk}_{ji} \frac{\partial}{\partial c^k} + \frac{1}{2} c^i c^j C^{R\theta}_{ji} \frac{\partial}{\partial c^\theta}, \tag{8.6.6}$$

where

$$\bar{X}^R_{(i)} \equiv X^R_{(i)} + \Delta_{(i)} \frac{\partial}{\partial \theta} \tag{8.6.7}$$

are the generators of the left action of the extended group. Clearly, the $C^{R\theta}_{ij}$ may also be obtained from the two-form $d\gamma$, $C^{R\theta}_{ij} = d\gamma(X^R_{(i)}, X^R_{(j)})$ where γ is defined by $\gamma(X^R_{(i)}) = \Delta_{(i)}$, $\gamma = \Delta_{(i)} \omega^{R(i)}$, since then $d\gamma(X^R_{(i)}, X^R_{(j)})$ is just $X^R_{(i)} \Delta_{(j)} - X^R_{(j)} \Delta_{(i)} - \Delta_{[i,j]}$; eq. (8.5.18) expresses the same relation in terms of the s one-cochain ω^0_1, $\omega^0_1(X_j) = \Delta_{(j)}$. Since the new $\bar{s}$ is a BRST operator associated with a group, Proposition 6.8.1 also applies here and $\bar{s}^2 = 0$.

Let us introduce the differential operator $\bar{d}$ on the extended group manifold by

$$\bar{d} = dg^i \frac{\partial}{\partial g^i} + d\theta \frac{\partial}{\partial \theta}. \tag{8.6.8}$$

Obviously, $\bar{d}$ is simply the ordinary exterior derivative acting on the manifold of the extended group; the bar has been added just to distinguish $\bar{d}$ from d, *now* defined as $d := dg^i \frac{\partial}{\partial g^i}$. Using the basis of the LI one-forms $(\bar{\omega}^{L(i)}; \bar{\omega}^{L(\theta)})$, $\bar{d}$ can be written as

$$\begin{aligned}
\bar{d} &= \bar{\omega}^{L(i)} \bar{X}^L_{(i)} + \bar{\omega}^{L(\theta)} \frac{\partial}{\partial \theta} \\
&= \omega^{L(i)} X^L_{(i)} + \omega^{L(i)} A_{(i)} \frac{\partial}{\partial \theta} + \bar{\omega}^{L(\theta)} \frac{\partial}{\partial \theta} \\
&\equiv d + \omega^{L(i)} A_{(i)} \frac{\partial}{\partial \theta} + \bar{\omega}^{L(\theta)} \frac{\partial}{\partial \theta},
\end{aligned} \tag{8.6.9}$$

where we have called $A_{(i)}$ the $\bar{X}^\theta_{(i)}$ component of the generator $\bar{X}^L_{(i)}$ of the extended algebra,

$$\bar{X}^L_{(i)} = X^L_{(i)} + A_{(i)} \frac{\partial}{\partial \theta}, \tag{8.6.10}$$

and used the fact that $\bar{\omega}^{L(i)}$ and $\omega^{L(i)}$ coincide. Now, since $\bar{d}$ and $\bar{s}$ commute as well as d and $\bar{s}$, we find

$$\begin{aligned}
[\bar{d}, \bar{s}] &= [d + \omega^{L(i)} A_{(i)} \frac{\partial}{\partial \theta} + \bar{\omega}^{L(\theta)} \frac{\partial}{\partial \theta}, \bar{s} + c^i \Delta_{(i)} \frac{\partial}{\partial \theta} + \frac{1}{2} c^i c^j C^{R\theta}_{ji} \frac{\partial}{\partial c^\theta}] \\
&= d(c^i \Delta_{(i)}) \frac{\partial}{\partial \theta} - \bar{s}(\omega^{L(i)} A_{(i)}) \frac{\partial}{\partial \theta} = 0,
\end{aligned} \tag{8.6.11}$$

so that we arrive at

$$\mathfrak{z}(\omega^{L(i)} A_{(i)}) = d(c^i \Delta_{(i)}) \,. \tag{8.6.12}$$

Since b does not depend on the extended group variable θ, $\mathfrak{z}b = \mathfrak{z}b = d\tilde{\Delta} \equiv d(c^i \Delta_{(i)})$ and using (8.6.12) we shall make the identification $b = A_{(i)} \omega^{L(i)}$. Thus, the coefficients of the one-form b in the basis of the LI forms are the terms that modify X^L to give the LI vector fields $\bar{X}^L$ of the LI extended Lie algebra. When we calculate $h(X_i^L, X_j^L)$, we are in fact computing

$$db(X_i^L, X_j^L) = X_i^L.b(X_j^L) - X_j^L.b(X_i^L) - b([X_i^L, X_j^L]) \tag{8.6.13}$$
$$= X_i^L.A_{(j)} - X_j^L.A_{(i)} - A_{[i,j]} \,,$$

i.e., the new structure constants of the LI algebra of the extended group. Since the structure constants of the LI algebra are equal but for a minus sign to those of the right-invariant algebra, we conclude that

$$h(X_i^L, X_j^L) = C_{ij}^\theta = -C_{ij}^{R\theta} \,, \tag{8.6.14}$$

as we found in the simple particle example of (8.6.5). The modification of the RI generators is obtained from the quasi-invariance function Δ; that of the LI generators, from the components of the Wess-Zumino term b itself in the basis of LI forms.

8.7 The massive superparticle and Wess-Zumino terms for supersymmetric extended objects

We have discussed in section 5.3 the structure of the $N = 2$ super-Poincaré group extended by a central charge. It is now simple to find a Lagrangian describing a *massive superparticle*. Using units in which $c = 1$, the first term of the Lagrangian is given by

$$L_0 = -m(\dot{\omega}_\mu \dot{\omega}^\mu)^{\frac{1}{2}} \,, \quad \dot{\omega}^\mu = \frac{dx^\mu}{d\tau} + i\left(\frac{d\bar{\theta}}{d\tau}\gamma^\mu\theta - \bar{\theta}\gamma^\mu\frac{d\theta}{d\tau}\right) \tag{8.7.1}$$

where $\dot{\omega}^\mu(\tau)d\tau$ ($\mu = 0, 1, 2, 3$; $D = 4$) is the one-form on R induced from the LI form $\Pi^\mu = dx^\mu + i(d\bar{\theta}\gamma^\mu\theta - \bar{\theta}\gamma^\mu d\theta)$ by the section of the bundle $\Sigma_2 \to R$ which describes the trajectory (parameter τ) of the superparticle in $N = 2$ superspace, Σ_2.* This term is super-Poincaré invariant by construction, and (in the same way that the relativistic particle Lagrangian $-m(\dot{x}_\mu \dot{x}^\mu)^{1/2}$ does) will produce the constraint associated with reparametrization invariance.

---

* The form of Π^μ should be compared with the expression (1.10.10) which corresponds to $N = 1$ superspace, *i.e.* θ Majorana. For non-selfconjugate spinors the first line of the group law (1.10.1) is replaced by $x''^\mu = x'^\mu + x^\mu + i(\bar{\theta}'\gamma^\mu\theta - \bar{\theta}\gamma^\mu\theta')$, which leads to the new LI one-form Π^μ in the text above (see eq. (5.3.3)).

The second term introduces the mass in the Fermi sector. It is a Wess-Zumino term,

$$L_{WZ} = im(\dot{\bar{\theta}}\theta - \bar{\theta}\dot{\theta}) \quad , \tag{8.7.2}$$

to be added to (8.7.1). Eq. (8.7.2) requires that $N = 2$; if θ^α ($\alpha = 1,2,3,4$) were a Majorana spinor, L_{WZ} would be identically zero since, if θ, θ' are Grassmann and Majorana, $\bar{\theta}\theta' = \bar{\theta}'\theta$. The term L_{WZ} is obtained by pulling back to R the quasi-invariant one-form on Σ_2,

$$im(d\bar{\theta}\theta - \bar{\theta}d\theta) \quad , \tag{8.7.3}$$

the differential of which is proportional to $d\bar{\theta} \wedge d\theta$, the (symmetric) two-form defining the extension two-cocycle (section 5.3); the full Lagrangian is then given by $L = L_0 + L_{WZ}$. There is no reason, a priori, why the mass parameters in (8.7.1) and (8.7.2) should be the same. The mass in (8.7.1) is related to the familiar bosonic constraint $p^2 = m^2$, $p_\mu = \frac{\partial L}{\partial \dot{x}^\mu}$, whereas the mass in (8.7.3) is the parameter $m \in H_0^2(\Sigma, R)$ which characterizes the central extension, eq. (5.3.3), and that after quantization describes the mass of a spin $1/2$ Dirac particle. Introducing a (dimensionless) proportionality constant λ, however, would destroy the symmetry between the 'bosonic' and 'fermionic' masses, as well as the appropriate balance of bosonic and fermionic degrees of freedom. Indeed, the action $I_0 + \lambda I_{WZ}$ has a *fermionic* gauge invariance or 'κ-symmetry' only for $\lambda = 1$. This is often referred to as 'Siegel invariance' and will not be discussed here.

As a result of the presence of (8.7.3), the Noether charges acquire additional pieces. Thus, by starting with the $N = 2$ super-Poincaré algebra, one again finds that the algebra realized by the Poisson brackets (or the quantum operators) is the supersymmetry algebra extended by a central charge. Of course, one may again replace (8.7.3) by

$$\Pi^\chi = d\chi + im(d\bar{\theta}\theta - \bar{\theta}d\theta) \tag{8.7.4}$$

and L_{WZ} by

$$L' = \dot{\chi} + im(\dot{\bar{\theta}}\theta - \bar{\theta}\dot{\theta}) \tag{8.7.5}$$

which, in contrast with (8.7.2), is invariant. All that is needed is to replace Σ_2 by the centrally extended superspace; then the new (decoupled) degree of freedom χ corresponds to the additional central parameter ϕ.

The complete Lagrangian of a massive superparticle above can be written, factoring out the mass m, as $L = m(L_0 + L_{WZ})$ where *now* L_0 and L_{WZ} are given by (8.7.1) and (8.7.2) without the m factor. In this form, it constitutes the simplest example of a series of Lagrangians for supersymmetric structureless p-extended objects in general D-dimensional spacetimes, all of which have the common structure

$$\mathscr{L} = T(\mathscr{L}_0 + \mathscr{L}_{WZ}) \quad , \tag{8.7.6}$$

where T is a generalized 'mass' or 'tension', with dimensions $[T] = ML^{-p}$ so that it is indeed a mass (tension) for the particle (string). The term $\mathscr{L}_0$ (as we have seen in the example (8.7.1) and will be discussed further below) depends on the metric; the term $\mathscr{L}_{WZ}$ *does not*. The constant T is, again, a *common* factor; since the object is assumed to be structureless, there should be no *relative*, scale setting dimensionful constant, between $\mathscr{L}_0$ and $\mathscr{L}_{WZ}$. The Lagrangian densities $\mathscr{L}_0$ and $\mathscr{L}_{WZ}$ depend on 'fields' $z^A(\tau, \underline{\sigma})$ $(z^A = (x^\mu, \theta^\alpha))$ that are functions of the 'time' variable τ and of the coordinates $\underline{\sigma}$ that describe the spatial extension p of the supersymmetric object and the vector and spinor indices (μ, α) take the appropriate values for a D-dimensional superspace. Since $[\mathscr{L}_0 d\tau d^p\sigma] = [\mathscr{L}_{WZ} d\tau d^p\sigma] = L^{p+1}$, the action I has the correct dimensions, $[I] = [T\mathscr{L}_0 d\tau d^p\sigma] = [T\mathscr{L}_{WZ} d\tau d^p\sigma] = ML$ $(c = 1)$. The limiting case $p = 0$ corresponds to the massive superparticle $(\mathscr{L} = L)$, which in its evolution describes a worldline parametrized by τ; the case $p = 1$ is the superstring, which in its motion sweeps out a worldsheet; $p = 2$ corresponds to a supermembrane, etc. In general, an extended object sweeps out a $(p + 1)$-dimensional worldvolume W in the course of its evolution.

Let $\xi^i = (\tau, \underline{\sigma})$, $i = 0, 1, ..., p$, be the coordinates of W and let $\phi : W \to \Sigma$ be the mapping that defines the functions $z^A(\tau, \underline{\sigma}) = (x^\mu(\tau, \underline{\sigma}), \theta^\alpha(\tau, \underline{\sigma}))$ on W. Then the pull-back of the LI one-forms Π^μ, $d\theta^\alpha$ on superspace Σ by ϕ^* defines one-forms on W: $\phi^*(dx^\mu) = \partial_i x^\mu(\xi) d\xi^i$, $\phi^*(d\theta^\alpha) = \partial_i \theta^\alpha(\xi) d\xi^i$. How could one guess the form of $\mathscr{L}$ for a general, structureless, supersymmetric extended object? It is not difficult to write the $\mathscr{L}_0$ part by extending the action that can be written for the bosonic case where Σ is reduced to spacetime M, *i.e.* to the ordinary D-dimensional translation group (no θ's). For the case $p = 0$, the Lagrangian of the bosonic particle is proportional to the reparametrization invariant worldline length element, $ds = d\tau |\dot{x}^\mu \dot{x}_\mu|^{1/2}$; to obtain (8.7.1) from it, it suffices to replace $\phi^*(dx^\mu) = \dot{x}^\mu d\tau$ by $\phi^*(\Pi^\mu) = \dot{\omega}^\mu d\tau$. For the bosonic string $(p = 1)$, the first part of the action, known as the Nambu-Goto action, is proportional to the area of the worldsheet swept out by the motion of the string; the $\mathscr{L}_0$ part is thus proportional to $|\det m_{ij}|^{1/2}$, where m_{ij} $(i, j = 0, 1)$ is the metric on W induced by the Minkowski metric $g_{\mu\nu}$ on spacetime, $m_{ij} = \partial_i x^\mu \partial_j x^\nu g_{\mu\nu}$, and $I \sim \int d\tau d\sigma \sqrt{|\dot{x}^2 x'^2 - (\dot{x}x')^2|}$, where the integral is the area element and the prime indicates differentiation with respect to σ. Then, to obtain the corresponding term for the superstring, it suffices to replace $\phi^*(g_{\mu\nu} dx^\mu \otimes dx^\nu) = m_{ij} d\xi^i \otimes d\xi^j$ by $\phi^*(g_{\mu\nu} \Pi^\mu \otimes \Pi^\nu) = M_{ij} d\xi^i \otimes d\xi^j$ so that in the supersymmetric case the integrand becomes $d\tau d\sigma |\det M_{ij}|^{1/2}$ where $M_{ij} = \omega_i^\mu \omega_j^\nu g_{\mu\nu}$ $(i, j = 1, 2)$ and $\omega_i^\mu d\xi^i = \phi^*(\Pi^\mu)$. Clearly the process is analogous for higher p; for $p = 2$ we would obtain $\mathscr{L}_0$ from the Dirac action for the bosonic membrane, which is given by the worldvolume element, etc.

For $p = 1$ the second part of the Lagrangian (8.7.6) determines the celebrated Green-Schwarz superstring action. We see from the previous reasonings that, since T is a generalization of m and the *structure* of the Lagrangian is the same, it is the massive superparticle rather than the massless one that is the point particle ($p = 0$) analogue of the Green-Schwarz superstring. As in the case of the superparticle, the important physical consequence of the additional term $\mathscr{L}_{WZ}$ is that the resulting system has the appropriate degrees of freedom determined by the 'κ-symmetry'. This is again due to the presence of the fermionic gauge invariance already mentioned: the local supersymmetry gauges away the unphysical degrees of freedom. But let us ignore the physical motivation for this term and, following the lead provided by the example of the massive superparticle, let us assume that 1) a second term is needed in the action of a p-extended object and that 2) it is given by the pull-back to W of a Wess-Zumino form (in the sense of Definition 8.2.1) defined on superspace Σ identified with the supertranslation group. This implies, clearly, that only variables corresponding to superspace coordinates (*i.e.*, x, θ) are allowed in the action; higher spin variables are therefore excluded. Under these assumptions, the problem of finding all the possible *classical* actions or of classifying all possible supersymmetric extended objects reduces to finding the non-trivial Chevalley-Eilenberg ($p + 2$)-cocycles h on Σ, *i.e.* the closed, left-invariant ($p + 2$)-forms the potentials b of which are not invariant.

To show how this is so, let us see that the above requirements indeed determine a ($p + 2$)-cocycle h on Σ for certain values of (D, p). Specifically,

a) the ($p + 2$)-form h should be Σ- and Lorentz-invariant, *i.e.* super-Poincaré-invariant, which implies that h should also be a Lorentz scalar constructed out of the LI forms Π^A on Σ;

b) h should have the required physical dimension $[h] = L^{(p+1)}$ so that, since $[h]=[b]$, $[Tb]=[ML]=$[action]; and

c) h should be closed. Since the topology of superspace is assumed trivial, this implies that h should be exact, $h = db$ where b is a ($p + 1$)-form.

By Theorem 1.6.1 and condition a), h has to be the exterior product of LI forms on Σ *i.e.*, of the form $h = c_{A_1...A_{p+2}} \Pi^{A_{p+2}} \wedge ... \wedge \Pi^{A_1}$ where the c's are *constant* coefficients. If h is CE-trivial, then $h = d\alpha$ *and* α is LI. We may then express α as $\alpha = c'_{A_1...A_{p+1}} \Pi^{A_{p+1}} \wedge ... \wedge \Pi^{A_1}$. Assume that α contains q factors Π^α and hence ($p + 1 - q$) factors Π^μ and recall that the usual dimensional assignments (section 5.3) imply that $[\Pi^\mu] = L$, $[\theta] = L^{1/2}$. Then, since the physical dimensions of α are ($q/2 + [p + 1 - q]$), condition b) implies $q=0$. The requirement a) implies $\alpha = c'_{\mu_1...\mu_{p+1}} \Pi^{\mu_{p+1}} \wedge ... \wedge \Pi^{\mu_1}$ and that $c'_{\mu_1...\mu_{p+1}}$

must be the coefficients of a Lorentz-invariant antisymmetric tensor. This is possible only if $D = p + 1$, which corresponds to the degenerate case of a p-dimensional object moving in a $(p + 1)$-dimensional spacetime, and will be excluded. Thus, $h \neq d\alpha$ with α LI: *a form h satisfying* a),b) *and* c) *determines a non-trivial element of the Chevalley-Eilenberg group* $E_{CE}^{(p+2)}(\Sigma)$.

The form of a possible h may now be found easily. Assuming it contains q factors $d\theta$ and hence $p + 2 - q$ factors Π^μ, its dimension is $p + 2 - q/2$, and condition b) gives $q=2$. It then follows that the only super-Poincaré-invariant $(p+2)$-form h of the required dimension is *proportional* to

$$(d\bar\theta \Gamma_{\mu_1 \dots \mu_p} \wedge d\theta) \wedge \Pi^{\mu_p} \wedge \dots \wedge \Pi^{\mu_1} \quad , \tag{8.7.7}$$

where $\Gamma_{\mu_1 \dots \mu_p}$ is the antisymmetrized product of Dirac matrices. To conclude, it is still necessary to impose condition c). The case of $p = 0$ is trivial; for instance, for $D = 4$, it requires that θ be a Dirac (not Majorana) spinor or '$N = 2$ supersymmetry', $\Sigma = \Sigma_2$. For higher values of p (superstrings, supermembranes, etc.), checking that h is closed is much more complicated. Without entering into any details, it will suffice to say that condition c) turns out to be equivalent to the validity of the identity $(d\bar\theta \Gamma_\mu \wedge d\theta) \wedge (d\bar\theta \Gamma^{\mu\nu_1 \dots \nu_{p-1}} \wedge d\theta)=0$, which depends crucially on the properties of Γ matrices and spinors for arbitrary D-dimensional spacetimes and which will not be discussed here.

The important result that emerges from the above analysis is that the above identity is satisfied *only* for certain values of (p, D), which determine the 'extension' p and the spacetime dimension D for which supersymmetric extended object actions exist. For instance, the classical Green-Schwarz superstring $(p = 1)$ action exists in $D = 3, 4, 6$ or 10 dimensions (although the value $D = 10$ is singled out by quantum considerations pertaining to anomaly cancellation). Once a suitable potential form b of h is known (its explicit expression in the case of general p is quite complicated), the Wess-Zumino part of the action is given (apart from the constant T) by $\int d\tau d^p \sigma \mathscr{L}_{WZ} = \int \phi^*(b)$. Clearly, these reasonings still allow for a numerical, dimensionless coefficient in b coming from the corresponding ambiguity in h since both h and λh define non-trivial CE classes. With the proper normalization of b, λ is fixed to be one by requiring, as in the massive superparticle case, the κ-invariance of the total action.

8.8 Supersymmetric extended objects and the supersymmetry algebra

By construction, the Wess-Zumino form b is quasi-invariant (lemma 8.2.1) and, consequently, the presence of $\mathscr{L}_{WZ}$ in the action will modify the supersymmetry algebra. We shall devote this section to the analysis of this effect, which can be done without making explicit the calculations involved, to see how the cohomological descent procedure can be used here. To this end, it is convenient to write the action for the extended supersymmetric object in terms of the Lagrangian L, as in the case of mechanics, rather than in terms of the density $\mathscr{L}$; in this way, selecting the 'time' variable τ, the action reads

$$I = \int d\tau d^p \sigma \mathscr{L}(z^A(\tau,\underline{\sigma}), \partial_i z^A(\tau,\underline{\sigma})) = \int d\tau L[z^A, \dot{z}^A] \quad (i = 0, 1, ..., p) \quad ,$$

where
(8.8.1)

$$L[z^A, \dot{z}^A] = \int d^p \sigma \mathscr{L}(z^A(\tau,\underline{\sigma}), \partial_i z^A(\tau,\underline{\sigma})) . \qquad (8.8.2)$$

Obviously, $\mathscr{L} = L$ for $p = 0$. For $p = 1$, we can look at $L(\tau)$ as a *functional* of the fields $z^A(\tau,\sigma)$ at a given τ. For instance, for a closed string, $\sigma \in [0, 2\pi]$, the fields $z^A(\tau,\sigma)$ at fixed τ determine the space of mappings (loops) $z^A : U(1) \to \Sigma$ or *loop superspace* $L^1\Sigma \equiv L\Sigma$. From this point of view, a closed superstring can be described by its loop superspace coordinates $z^A(\sigma)$ which are functions of the evolution parameter τ, $(z^A(\sigma))(\tau) = z^A(\tau,\sigma)$. By extending this argument *formally*, a generalized p-'loop' superspace $L^p\Sigma$ may be defined as the space of mappings $z^A(\underline{\sigma})$ of the p-dimensional object into Σ; $L^0\Sigma$ may be identified with Σ itself[*]. But the important point in this analysis is the replacement of the density $\mathscr{L}$ by the Lagrangian L given, at a certain τ, by a ($\underline{\sigma}$-space) functional of the fields $z^A(\underline{\sigma})$. Having written everything as in the case of mechanics, the analysis may proceed similarly to the massive superparticle case by replacing superspace Σ by the infinite-dimensional space $L^p\Sigma$, and the mapping $\phi : W \to \Sigma$ by the mapping $\tilde{\phi} : w \to L^p\Sigma$ where w is the 'timeline', defined by $\tilde{\phi} : \tau \mapsto [z^A(\quad)](\tau) \in L^p\Sigma$, $[z^A(\underline{\sigma})](\tau) = z^A(\tau,\underline{\sigma})$.

This requires extending to this functional space some ideas of ordinary superspace. Clearly, the exterior derivative on superspace Σ (section 1.10),

$$d = dz^A \frac{\partial}{\partial z^A} = \Pi^{(A)} D_{(A)} \quad , \quad dz^A \left(\frac{\partial}{\partial z^B} \right) = \Pi^{(A)}(D_{(B)}) = \delta^A{}_B , \qquad (8.8.3)$$

where the Π's and D's are, respectively, the LIFs and LIVFs on Σ (cf. (1.10.10), (1.10.6)), generalizes to the exterior *functional* derivative $\mathbf{d}$ on loop superspace $L^p\Sigma$, say, in the form

$$\mathbf{d} := \oint d^p \sigma dz^A(\underline{\sigma}) \frac{\delta}{\delta z^A(\underline{\sigma})} = \oint d^p \sigma \mathbf{E}^{(A)}(\underline{\sigma}) \mathscr{D}_{(A)}(\underline{\sigma}) , \quad \mathbf{d}^2 = 0 \quad , \qquad (8.8.4)$$

---

[*] In fact, this is a particular example of a more general situation in which a finite-dimensional Lie group G is replaced by the infinite-dimensional group $G(M) = Map(M, G)$, as we shall discuss in more detail in chapter 9.

where

$$\mathbf{E}^{(A)}(\underline{\sigma}) = dz^{B}(\underline{\sigma})E_{B}^{\ (A)}(\underline{\sigma}) \quad , \quad \mathscr{D}_{(A)}(\underline{\sigma}) = (X^{L})_{(A)}^{\ B}(\underline{\sigma})\frac{\delta}{\delta z^{B}(\underline{\sigma})} \qquad (8.8.5)$$

are the basis of one-forms (resp. functional vector fields) on $L^{p}\Sigma$, and $E_{B}^{\ (A)}$ (resp. $(X^{L})_{(A)}^{\ B}$) are the components of the LI forms (resp. vector fields) on the group Σ; clearly, $E_{B}^{\ (A)}(X^{L})_{(A)}^{\ B'} = \delta_{B}^{\ B'}$ and $dz^{A}(\underline{\sigma})(\frac{\delta}{\delta z^{B}(\underline{\sigma}')})$ $= \mathbf{E}^{(A)}(\underline{\sigma})(\mathscr{D}_{(B)}(\underline{\sigma}')) = \delta^{A}_{\ B}\delta(\underline{\sigma} - \underline{\sigma}')$, and $\mathbf{E}^{(A)}$, $\mathscr{D}_{(A)}$ become $\Pi^{(A)}$, $D_{(A)}$ for $p=0$. The functional operators $\mathscr{D}_{(A)}(\underline{\sigma})$ satisfy the graded commutation relations

$$\{\mathscr{D}_{(A)}(\underline{\sigma}), \mathscr{D}_{(B)}(\underline{\sigma})\} = \delta(\underline{\sigma} - \underline{\sigma}')(-C_{AB}^{\ \ C})\mathscr{D}_{(C)}(\underline{\sigma}) \qquad (8.8.6)$$

where the bracket is an anticommutator if both indices are fermionic and the structure constants are labelled $(-C_{AB}^{\ \ C})$ because, following the customary convention, the supersymmetry algebra is defined by the structure constants $C_{AB}^{\ \ C}$ for the (*right* algebra) generators of the left action $Q_{(A)}$. For $D = 4$ the structure constants are minus those of the (right) algebra of supersymmetry generators (1.10.5), but we do not need their explicit expression here. The (graded) Maurer-Cartan equations for $\mathbf{E}^{(A)}(\underline{\sigma})$ can be deduced from (8.8.5) and are given by

$$\mathbf{dE}^{(C)}(\underline{\sigma}) = \frac{1}{2}\mathbf{E}^{(B)}(\underline{\sigma}) \wedge \mathbf{E}^{(A)}(\underline{\sigma})(-C_{AB}^{\ \ C}) \qquad (8.8.7)$$

(or, to mimic the finite-dimensional case notation, by $\mathbf{dE}^{(C)} = \frac{1}{2}\oint d^{p}\sigma' \oint d^{p}\sigma'' \mathbf{E}^{(B)}(\underline{\sigma}') \wedge \mathbf{E}^{(A)}(\underline{\sigma}'')\delta(\underline{\sigma} - \underline{\sigma}')\delta(\underline{\sigma} - \underline{\sigma}'')(-C_{AB}^{\ \ C}))$. Note that, with all indices even, (8.8.7) corresponds to (1.7.4) if $(-C_{AB}^{\ \ C})$ is identified with (the left algebra structure constants) $C_{ab}^{\ \ c}$ in (1.7.4).

The Wess-Zumino term is only quasi-invariant under supersymmetry transformations. To see its relation to the WZ Lagrangian density, let us write the form b on Σ in the basis of LI forms $\Pi^{(A)}$ as

$$b = b_{A_{p+1}...A_{1}}(z^{A})\Pi^{(A_{1})} \wedge ... \wedge \Pi^{(A_{p+1})} ; \qquad (8.8.8)$$

for the reasonings below, we do not need its explicit form. Notice that the coefficients of b depend on the superspace coordinates z^{A} since b is not LI. Then the pull-back of b by $\phi : W \rightarrow \Sigma$ gives

$$\phi^{*}b = b_{A_{p+1}...A_{1}}(z^{A}(\xi))\omega_{i_{1}}^{(A_{1})}...\omega_{i_{p+1}}^{(A_{p+1})}d\xi^{i_{1}} \wedge ... \wedge d\xi^{i_{p+1}} , \qquad (8.8.9)$$

and so the Wess-Zumino Lagrangian *density* is given by

$$\mathscr{L}_{WZ} \propto b_{A_{p+1}...A_{1}}(z^{A}(\xi))\epsilon^{i_{1}...i_{p+1}}\omega_{i_{1}}^{(A_{1})}...\omega_{i_{p+1}}^{(A_{p+1})} . \qquad (8.8.10)$$

Now, the pull-back of $L_{X^R_{(A)}} Tb = d\triangle_{(A)}$, where in general $\triangle_{(A)}$ is non-zero due to the quasi-invariance of b, gives

$$\delta_A T \mathscr{L}_{WZ} d\tau \wedge d\sigma^1 \wedge ... \wedge d\sigma^p \equiv \phi^* L_{X^R_{(A)}} Tb = \phi^* d\triangle_{(A)} = d\phi^* \triangle_{(A)} \quad (8.8.11)$$

and, writing $\phi^* \triangle_{(A)} = (-1)^i \triangle^i_{(A)} d\xi^0 \wedge ... \wedge \widehat{d\xi^i} \wedge ... \wedge d\xi^p$ so that $d\phi^* \triangle_{(A)} = \partial_i \triangle^i_{(A)} d^{p+1}\xi$, we find that

$$\delta_{(A)} T \mathscr{L}_{WZ} = \partial_i \triangle^i_{(A)} \quad , \quad (8.8.12)$$

i.e., the variation of $\mathscr{L}_{WZ}$ is given by a divergence. Since in (8.8.10) all terms include $\omega_0^{(A)} \equiv \dot{\omega}^{(A)}$, we can factor this term out and write

$$I_{WZ} = T \int d\tau d^p\sigma \mathscr{L}_{WZ} = \int d\tau \oint d^p\sigma \dot{\omega}^{(A)} \mathscr{A}_{(A)} \quad (8.8.13)$$

where $\mathscr{A}_{(A)}$ does not contain τ derivatives (recall that the constant T was factored out in (8.7.6) and that accordingly $\triangle$ and $\mathscr{A}$ include T). Eq. (8.8.13) means that L_{WZ} can be expressed in terms of a one-form $\mathbf{A}$ on generalized loop superspace $L^p\Sigma$

$$\mathbf{A} = \oint d^p\sigma E^{(A)}(\underline{\sigma}) \mathscr{A}_{(A)}(\underline{\sigma}) \quad , \quad (8.8.14)$$

because, using the mapping $\tilde{\phi}$ previously introduced, I_{WZ} can be expressed as

$$I_{WZ} = \int d\tau \tilde{\phi}^* \mathbf{A} \equiv T \int d\tau L_{WZ} \quad (8.8.15)$$

since $\oint d^p\sigma \tilde{\phi}^* (E^{(A)}(\underline{\sigma}) \mathscr{A}_{(A)}(\underline{\sigma})) = \left[\oint d^p\sigma \dot{\omega}^{(A)}(\tau, \underline{\sigma}) \mathscr{A}_{(A)}(\tau, \underline{\sigma}) \right] d\tau$.

The fact that $\mathscr{L}_{WZ}$ is quasi-invariant implies that $\mathbf{A}$ is also quasi-invariant. Before showing this, let us consider the BRST operator for the action of supersymmetry on $L^p\Sigma$,

$$\tilde{s} = c^A \oint d^p\sigma \mathscr{Q}_{(A)}(\underline{\sigma}) + c^B c^A C_{AB}{}^C \frac{\partial}{\partial c^C} \equiv c^A Q_{(A)} + c^B c^A C_{AB}{}^C \frac{\partial}{\partial c^C} \quad (8.8.16)$$

where $C_{AB}{}^C$ are the structure constants of the ($Q^{(A)}$-generated) supersymmetry algebra, the graded 'ghosts' c^A are functionally constant, $\mathbf{d}c^A = 0$, and

$$\mathscr{Q}_{(A)}(\underline{\sigma}) = (X^R)_{(A)}{}^B(\underline{\sigma}) \frac{\delta}{\delta z^B(\underline{\sigma})} \quad , \quad (8.8.17)$$

where $(X^R)_{(A)}{}^B$ have the form of the components of the right-invariant vector fields on the group (cf. (1.10.4)). The BRST operator clearly commutes with the functional differential $\mathbf{d}$,

$$[\tilde{s}, \mathbf{d}] = 0 \quad . \quad (8.8.18)$$

The integrated operators $Q_{(A)} = \int d^p\sigma \mathscr{Q}_{(A)}(\underline{\sigma})$ satisfy the supersymmetry algebra graded commutation relations (cf. (1.10.5)),

$$\{Q_{(A)}, Q_{(B)}\} = C_{AB}{}^C Q_{(C)} \quad . \tag{8.8.19}$$

Clearly, since $\mathscr{Q}_{(A)}(\underline{\sigma})$ and $\mathscr{D}_{(A)}(\underline{\sigma})$ correspond to right and left copies of the supersymmetry functional algebra, they satisfy

$$\{\mathscr{Q}_{(A)}(\underline{\sigma}), \mathscr{D}_{(B)}(\underline{\sigma}')\} = 0 \quad . \tag{8.8.20}$$

The above expressions for the left ($\mathscr{D}_{(A)}(\underline{\sigma})$) and right ($\mathscr{Q}_{(A)}(\underline{\sigma})$) generators have been introduced without making reference to the Lagrangian describing the extended supersymmetric object. But, as in the superparticle case, $\mathscr{Q}_{(A)}(\underline{\sigma})$ may be extracted from the first part of Noether charge densities $\mathscr{Q}_{(A)}^{WZ}(\underline{\sigma})$ (when they are realized as functional differential operators), which include a contribution due to the Wess-Zumino term: $\mathscr{Q}_{(A)}(\underline{\sigma}) = \mathscr{Q}_{(A)}^{WZ=0}(\underline{\sigma})$. In the same way, the $\mathscr{D}_{(A)}(\underline{\sigma})$'s are the first part of $\mathscr{D}_{(A)}^{WZ}(\underline{\sigma})$ ($\mathscr{D}_{(A)}(\underline{\sigma}) = \mathscr{D}_{(A)}^{WZ=0}(\underline{\sigma})$) which for fermionic A determine the fermionic constraints of the supersymmetric object. To compute the algebra generated by the Noether charges $Q_{(A)}^{WZ} \equiv \int d^p\sigma\, \mathscr{Q}_{(A)}^{WZ}(\underline{\sigma})$ and the functional algebra of the $\mathscr{D}_{(A)}^{WZ}(\underline{\sigma})$ we may use canonical Poisson brackets or graded commutators for the operators obtained by replacing the canonical momenta by functional derivatives with respect to the conjugate variables (e.g., $p_\mu(\underline{\sigma})$ by $i\hbar(\delta/\delta x^\mu(\underline{\sigma}))$, etc.). Thus we can look at the expressions below as computed in either way, but it is convenient to stress the *classical* origin of the modification (the 'classical anomaly', cf. section 8.6) of the original algebras (8.8.19) and (8.8.6) to be discussed below, since it is due to a Wess-Zumino term in which the only parameter is T (and where there is no $\hbar$). The only difficulty, however, is that in the presence of second class constraints, Poisson brackets have to be replaced by *Dirac* brackets. Nevertheless, as will be explicitly shown below, the left/right graded commutation $\{Q_{(A)}, \mathscr{D}_{(A)}(\underline{\sigma})\} = 0$ is maintained in the presence of the Wess-Zumino term, $\{Q_{(A)}^{WZ}, \mathscr{D}_{(A)}^{WZ}(\underline{\sigma})\} = 0$, which means that Dirac and Poisson brackets coincide and that the *algebra* of the operators can be obtained by using Poisson brackets.

The quasi-invariance of **A** is now evident, since the action of $\tilde{s}$ on **A** gives

$$\tilde{s}\mathbf{A} = \mathbf{d}\Delta \tag{8.8.21}$$

due to the quasi-invariance of $\mathscr{L}$, where $\Delta = c^A \Delta_{(A)}$ is a zero-form on loop superspace, *i.e.* a functional of z, $\Delta_{(A)} = \Delta_{(A)}[z(\underline{\sigma})]$. In fact, taking into account the way **A** has been obtained from b, the following equality holds:

$$\tilde{s}\tilde{\phi}^*\mathbf{A} = \tilde{s}\, T L_{WZ}\, d\tau = \tilde{s}\left(T\oint d^p\sigma\,\mathscr{L}_{WZ}\right)d\tau = c^A\left(\oint d^p\sigma\,\partial_i\triangle^i_{(A)}\right)d\tau$$

$$= \left(c^A\oint d^p\sigma\,\partial_I\triangle^I_{(A)}\right)d\tau + \left(c^A\oint d^p\sigma\,\partial_\tau\triangle^0_{(A)}\right)d\tau$$

$$= \left(c^A\oint d^p\sigma\,\partial_I\triangle^I_{(A)}\right)d\tau + \frac{d}{d\tau}(c^A\tilde{\phi}^*\triangle_{(A)}[z(\underline{\sigma})])\,d\tau \quad .$$

$$(8.8.22)$$

Thus, provided that $\oint d^p\sigma\,\partial_I\triangle^I_{(A)}(z(\underline{\sigma})) = 0$, $\triangle$ is given by

$$\triangle[z(\underline{\sigma})] = c^A\oint d^p\sigma\,\triangle^0_{(A)}(z(\underline{\sigma})) \qquad (8.8.23)$$

and we obtain (8.8.21). Clearly, we may assume that the bosonic coordinates are non-periodic (for a closed superstring we may take x^1, say, to be $x^1 = nR\sigma + Y(\tau,\sigma)$, where R is the string radius, n the number of times that the string wraps around the circle, and Y (as well as the other x^μ coordinates, $\mu \neq 1$) is periodic in σ). The fermionic coordinates θ, however, are assumed to be periodic, so that the rigid supersymmetry is not spoiled when they are pulled back to W by ϕ^*. In these situations, and due to the explicit form of h, it can be seen that $\partial_I\triangle^I_{(A)}$ always has a factor proportional to the derivatives of θ, so that it is always possible to extract ∂_I in such a way that $\partial_\sigma x^\mu$, but not x^μ, appears. As a result, $\triangle^I_{(A)}(z(\underline{\sigma}))$ is periodic and $\oint d^p\sigma\,\partial_I\triangle^I_{(A)} = 0$ and then the WZ Lagrangian changes, as in mechanics, by a time derivative,

$$\tilde{s}(\tilde{\phi}^*\mathbf{A}) = \tilde{s}\, T L_{WZ} = \frac{d}{d\tau}[\tilde{\phi}^*\triangle] \quad ; \qquad (8.8.24)$$

also, $0 = \tilde{s}^2\mathbf{A} = \mathbf{d}\tilde{s}\triangle$. The statements $\tilde{s}\triangle \equiv \Omega$ and $\tilde{s}\Omega = 0$ are related to a modification of the algebra of charges due to $\triangle^0_{(A)}$ that it is given by Ω_{AB}. Indeed, the equation $\tilde{s}\triangle = \Omega$, where Ω is a two-form in ghost space, gives rise to

$$\Omega = -\frac{1}{2}c^B c^A\Omega_{AB} = \left[c^A\oint d^p\sigma\,\mathscr{2}_{(A)}(\underline{\sigma}) + \frac{1}{2}c^B c^A C_{AB}{}^C\frac{\partial}{\partial c^C}\right]c^{B'}\triangle_{(B')}$$

$$= c^B c^A\left[-\oint d^p\sigma\,\mathscr{2}_{(A)}(\underline{\sigma}).\triangle_{(B)} + \frac{1}{2}C_{AB}{}^C\triangle_{(C)}\right]$$

$$= -\frac{1}{2}c^B c^A\left[\oint d^p\sigma\,\mathscr{2}_{(A)}(\underline{\sigma}).\triangle_{(B)} - (-1)^{AB}\oint d^p\sigma\,\mathscr{2}_{(B)}(\underline{\sigma}).\triangle_{(A)}\right.$$

$$\left. - C_{AB}{}^C\triangle_{(C)}\right]$$

$$(8.8.25)$$

so that

$$\Omega_{AB} = \oint d^p\sigma \oint d^p\sigma' \{ \mathcal{Q}_{(A)}(\underline{\sigma}).\Delta^0_{(B)}(\underline{\sigma}') - (-1)^{AB} \mathcal{Q}_{(B)}(\underline{\sigma}').\Delta^0_{(A)}(\underline{\sigma})$$

$$- C_{AB}{}^C \delta(\underline{\sigma} - \underline{\sigma}') \Delta^0_{(C)}(\underline{\sigma}) \}$$

$$= Q_{(A)}.\Delta_{(B)}[z(\underline{\sigma})] - (-1)^{AB} Q_{(B)}.\Delta_{(A)}[z(\underline{\sigma})] - C_{AB}{}^C \Delta_{(C)}[z(\underline{\sigma})]$$

$$(8.8.26)$$

is the term that modifies the algebra of equal time charges. From the equation $\mathbf{d}\,\Omega = 0$ and $\mathbf{d}\,c^A = 0$ we deduce that Ω is functionally constant,

$$\mathbf{d}\,\Omega_{AB} = 0 \quad, \tag{8.8.27}$$

and this means that Ω_{AB} is the integral of a divergence. Thus, in the cases where we have spaces with non-trivial homology, topological charges Ω_{AB} may appear. In the case of the closed superstring ($p=1$), for instance, its only non-zero components (θ periodic) may be seen to be given by the string winding modes, $\Omega_{\alpha\beta} \simeq \oint d\sigma \partial_\sigma x^\mu (\gamma_\mu)_{\alpha\beta}$. In the other cases it may be shown that $\Omega_{\alpha\beta}$ is proportional to

$$(\Gamma_{\mu_1...\mu_p})_{\alpha\beta} \oint d^p\sigma \epsilon^{I_1...I_p} \partial_{I_1} x^{\mu_1} ... \partial_{I_p} x^{\mu_p} \quad . \tag{8.8.28}$$

Eq. (8.8.26) shows that the operators Q that include the contribution for the WZ term,

$$(Q^{WZ})_{(A)} \equiv \oint d^p\sigma \mathcal{Q}_{(A)}(\underline{\sigma}) + \Delta_{(A)}[z(\underline{\sigma})] \quad, \tag{8.8.29}$$

close into an algebra different from that of the Q's alone, eq. (8.8.19). Using (8.8.29) and (8.8.26), we see that the modification of the supersymmetry algebra produced by the topological charges is given by the graded Poisson brackets

$$\{(Q^{WZ})_{(A)}, (Q^{WZ})_{(B)}\} = C_{AB}{}^C (Q^{WZ})_{(C)} + \Omega_{AB} \quad . \tag{8.8.30}$$

The previous discussion is expressed by the right half of Diagram 8.8.1

$$\begin{array}{ccccc}
& [2] & [1] & [0] & \text{degree of form} \\
\end{array}$$

$$\begin{array}{ccccc}
[0] & 0 \xleftarrow{\ \mathbf{d}\ } \mathbf{F} \xleftarrow{\ \mathbf{d}\ } \mathbf{A} & & & \\
& \tilde{s}\downarrow \quad\ \tilde{s}\downarrow & & & \\
[1] & 0 \xleftarrow{\ \mathbf{d}\ } \mathbf{d}\Delta \xleftarrow{\ \mathbf{d}\ } \Delta & & \Delta = c^A \Delta_{(A)}(\underline{\sigma}) \\
& \tilde{s}\downarrow \quad\ \tilde{s}\downarrow & & & \\
[2] & 0 \xleftarrow{\ \mathbf{d}\ } \Omega & & \Omega = -\tfrac{1}{2}c^B c^A \Omega_{AB} \\
& \tilde{s}\downarrow & & & \\
\text{ghost} & 0 & & & \\
\text{number} & & & &
\end{array}$$

Diagram 8.8.1

which here replaces Diagram 8.6.1; in it, $\mathbf{F} = \mathbf{dA}$ is invariant under the group action. It remains to be shown that the equation $\mathbf{dA} = \mathbf{F}$ also has the meaning of a Lie algebra modification given by the components $\mathscr{F}_{AB}$ of $\mathbf{F}$, but in this case of the algebra of the 'left' functional vector fields $\mathscr{D}_{(A)}(\underline{\sigma})$. Adding $\mathscr{A}_{(A)}(\underline{\sigma})$ to the generators of (8.8.5), we construct

$$(\mathscr{D}^{WZ})_{(A)}(\underline{\sigma}) \equiv \mathscr{D}_{(A)}(\underline{\sigma}) + \mathscr{A}_{(A)}(\underline{\sigma}) \,, \tag{8.8.31}$$

and obtain that their (equal τ, graded) Poisson brackets are given by

$$\{(\mathscr{D}^{WZ})_{(A)}(\underline{\sigma}), (\mathscr{D}^{WZ})_{(B)}(\underline{\sigma}')\} =$$
$$\delta(\underline{\sigma} - \underline{\sigma}')(-C_{AB}{}^C)(\mathscr{D}^{WZ})_{(C)}(\underline{\sigma}') + \mathscr{F}_{AB}(\underline{\sigma}, \underline{\sigma}') \,, \tag{8.8.32}$$

$$\mathscr{F}_{AB}(\underline{\sigma}, \underline{\sigma}') = \mathscr{D}_{(A)}(\underline{\sigma}).\mathscr{A}_{(B)}(\underline{\sigma}') - (-1)^{AB}\mathscr{D}_{(B)}(\underline{\sigma}').\mathscr{A}_{(A)}(\underline{\sigma})$$
$$- (-C_{AB}{}^C)\mathscr{A}_{(C)}(\underline{\sigma})\,\delta(\underline{\sigma} - \underline{\sigma}') \,; \tag{8.8.33}$$

notice that for $\mathbf{A}=0$ ($\Rightarrow \Delta=0$) eqs. (8.8.32) and (8.8.30) are just left and (integrated) right copies of the same functional algebra. The $\mathscr{F}_{AB}$ in (8.8.32) are the components of $\mathbf{F}$ on loop space, i.e.,

$$\mathbf{F} = -\frac{1}{2}\oint d^p\sigma d^p\sigma'\, \mathbf{E}^{(B)}(\underline{\sigma}) \wedge \mathbf{E}^{(A)}(\underline{\sigma}')\,\mathscr{F}_{AB}(\underline{\sigma}, \underline{\sigma}') \,. \tag{8.8.34}$$

To see this, let us write $\mathbf{F} = \mathbf{dA}$ in coordinates using (8.8.4). Then $\mathbf{F}$ is given by

$$\oint d^p\sigma d^p\sigma' \Big\{ \frac{1}{2}\mathbf{E}^{(B)}(\underline{\sigma}') \wedge \mathbf{E}^{(A)}(\underline{\sigma})(-C_{AB}{}^C)\mathscr{A}_{(C)}(\underline{\sigma})\delta(\underline{\sigma}-\underline{\sigma}')$$

$$- \mathbf{E}^{(A)}(\underline{\sigma}) \wedge \mathbf{E}^{(B)}(\underline{\sigma}')\mathscr{D}_{(B)}(\underline{\sigma}')\mathscr{A}_{(A)}(\underline{\sigma}) \Big\}$$

$$= \frac{1}{2}\oint d^p\sigma d^p\sigma' \, \mathbf{E}^{(B)}(\underline{\sigma}') \wedge \mathbf{E}^{(A)}(\underline{\sigma})\Big\{ - C_{AB}{}^C\mathscr{A}_{(C)}(\underline{\sigma})\delta(\underline{\sigma}-\underline{\sigma}')$$

$$+ (-1)^{AB}\mathscr{D}_{(B)}(\underline{\sigma}')\mathscr{A}_{(A)}(\underline{\sigma}) - \mathscr{D}_{(A)}(\underline{\sigma})\mathscr{A}_{(B)}(\underline{\sigma}') \Big\} \quad .$$

$$(8.8.35)$$

The explicit form of $\mathscr{F}_{AB}$ is deduced from the fact that $\mathbf{A}$ is constructed from b, eqs. (8.8.10) and (8.8.14). Then, since $h = db$, it is not difficult to show that $\mathscr{F}_{AB}(\underline{\sigma},\underline{\sigma}')$ must be proportional to

$$\delta(\underline{\sigma}-\underline{\sigma}')\epsilon^{I_1...I_p}\omega_{I_p}^{(A_p)}(\underline{\sigma})...\omega_{I_1}^{(A_1)}(\underline{\sigma})h_{A_1...A_pAB} \quad . \qquad (8.8.36)$$

The fact that $\mathscr{F}_{AB}(\underline{\sigma},\underline{\sigma}')$ is proportional to the δ function and not to its derivatives is often referred to as the *ultralocality* property of $\mathscr{F}_{AB}$.

When $p=0$ the expressions (8.8.32) [(8.8.30)] are simply the left [right] mutually commuting copies of the supertranslation algebra extended by a central charge, the mass, that are provided by the LI [RI] vector fields $\bar{X}^L$ [$\bar{X}^R$] on the appropriate graded supertranslation group. Let us now calculate in general the Poisson brackets of the $(\mathscr{D}^{WZ})_{(A)}(\underline{\sigma})$ with the extended generators $(Q^{WZ})_{(B)}$ (Noether charges). From (8.8.29) and (8.8.31) we get

$$\{(\mathscr{D}^{WZ})_{(A)}(\underline{\sigma}), (Q^{WZ})_{(B)}\} = \mathscr{D}_{(A)}(\underline{\sigma})\Delta_{(B)}$$

$$- (-1)^{AB}\oint d^p\sigma' \, \mathscr{Q}_{(B)}(\underline{\sigma}').\mathscr{A}_{(A)}(\underline{\sigma}) = 0 \quad . \qquad (8.8.37)$$

This is a consequence of the equation $\check{s}\mathbf{A} = \mathbf{d}\Delta$, as can be seen by writing it in coordinates:

$$c^B\oint d^p\sigma' \, \mathscr{Q}_{(B)}(\underline{\sigma}'). \oint d^p\sigma \, \mathbf{E}^{(A)}(\underline{\sigma})\mathscr{A}_{(A)}(\underline{\sigma}) = \oint d^p\sigma \mathbf{E}^{(A)}(\underline{\sigma})\mathscr{D}_{(A)}(\underline{\sigma}).c^B\Delta_{(B)} \quad ,$$

$$(8.8.38)$$

$$c^B\oint d^p\sigma \, \mathbf{E}^{(A)}(\underline{\sigma})\Big\{ \mathscr{D}_{(A)}(\underline{\sigma}).\Delta_{(B)} - (-1)^{AB}\oint d^p\sigma' \, \mathscr{Q}_{(B)}(\underline{\sigma}').\mathscr{A}_{(A)}(\underline{\sigma}) \Big\} = 0 \quad ,$$

$$(8.8.39)$$

which means that the (graded) commutator in (8.8.37) is zero. In (8.8.37), however, the commutators contain functional differential operators $(\mathscr{D}^{WZ})$ and integrated ones (Q^{WZ}). For the supersymmetric extended objects, it can be seen that the first ones come from the fermionic constraints, and the second are of course the Noether charges; the fact that (8.8.37) is zero simply states that, as anticipated, the Noether supersymmetry charges commute with the constraints.

Notice that, since the forms $\Pi^{(A)}$ are left-invariant, $\tilde{s}E^{(A)}(\sigma) = 0$. Taking into account that $\tilde{s}F = 0$, we arrive at $\tilde{s}\mathscr{F}_{AB} = 0$, which is tantamount to saying that $\mathscr{F}_{AB}$ is invariant. Note that neither does $\mathscr{F}_{AB}$ give rise to a topological extension $(\mathbf{d}E^A(\underline{\sigma}) \neq 0, \ \mathbf{d}F = 0 \Rightarrow \mathbf{d}\mathscr{F}_{AB} \neq 0$ in general) nor is Ω_{AB} invariant in general $(\tilde{s}c^A \neq 0, \ \tilde{s}\Omega = 0 \Rightarrow \tilde{s}\Omega_{AB} \neq 0)$. All the above results can be simply summarized by defining, in the presence of a WZ term, the operators

$$\mathbf{D} = \mathbf{d} + \mathbf{A} \quad , \quad S = \tilde{s} + \Delta \quad . \tag{8.8.40}$$

It can be immediately verified that $[\mathbf{D}, S] = 0$ on account of the fact that $[\tilde{s}, \mathbf{d}] = 0$ and eq. (8.8.21). The new covariant exterior derivative and BRST operators are, however, no longer nilpotent. Instead, $\mathbf{D}^2 = \mathbf{d}\mathbf{A} = \mathbf{F}$ and $S^2 = \tilde{s}\Delta = \Omega$. Thus, the additional terms (8.8.33) and (8.8.26) in the original 'left' (functional) and 'right' (Noether charges) algebras appear, as one might expect, as *curvatures* associated with the above covariant operators.

Summarizing, from Diagram 8.8.1 we deduce that the components of Ω give topological extensions of the algebra of supersymmetric charges and that those of $\mathbf{F}$ give terms that modify the algebra of constraints. These two algebras (anti)commute due to the fact that $\tilde{s}\mathbf{A} = \mathbf{d}\Delta$, as we already know from mechanics. For $p = 0$ – the massive superparticle case – these modified algebras are finite-dimensional and Ω_{AB} and $\mathscr{F}_{AB}$ define the RI and LI algebras of the supersymmetry generators and covariant derivatives extended by a central charge by arguments the same as those of section 8.6.

8.9 A Lagrangian description of the magnetic monopole

The above example concludes our discussion of Wess-Zumino terms in the sense of Definition 8.2.1 and their effects, the *classical* nature of which is stressed by the fact that they can be computed entirely in terms of Poisson brackets; neither the algebras previously described, nor the Wess-Zumino term that originates them, contain the Planck constant $\hbar$. Before moving to the gauge anomalies in the next chapter, and as a preparatory example, it is worth discussing here a Lagrangian for the magnetic monopole as an illustration of the *quantum* effects that appear when the action is not invariant under a *gauge* transformation. The Lagrangian for a particle in the field of a monopole is given by a term that shares with the Wess-Zumino terms previously considered the property of being quasi-invariant (*now*, under a *non-constant* parameter transformation). It will be seen that, due to the non-trivial topology of the problem, this quasi-invariance gives the Dirac quantization condition.

As we saw in section 2.7, Wu and Yang circumvented the Dirac string by showing that the potential of the magnetic monopole could be understood as a connection on a *non-trivial* $U(1)$-principal bundle, the $U(1)$ being the gauge group ('*electromagnetism is the gauge-invariant manifestation of a non-integrable phase factor*'). The compatibility condition for the solutions (wavesections) of the Schrödinger equation on each of the two charts reproduced Dirac's quantization condition, $2gq/\hbar c \in Z$. Let us consider now an equivalent description of the magnetic monopole, which uses a (non-relativistic) Lagrangian defined on the Hopf bundle the structure group of which is the $U(1)$-gauge group, as a first example of a Wess-Zumino term in theories with a gauge symmetry.

Since we cannot define singularity-free potentials describing the monopole on ordinary space, and we know that the gauge group of electromagnetism is a $U(1)$ group, let us add a new variable $\zeta = \exp i\psi$ to $\mathbf{x} \in R^3$. Consider now the Hopf bundle, $SU(2)(U(1), S^2)$. Locally, it can be parametrized by $(\zeta, \hat{\mathbf{x}})$, where ζ is the element of the phase group and $\hat{\mathbf{x}} \in S^2$ is a unit vector. Let us now show that the space $R^+ \times SU(2)$, where R^+ is parametrized by the radius (so that r and $\hat{\mathbf{x}}$ characterize a point in R^3) is the appropriate configuration space for a singularity-free description of a particle in the field of a magnetic monopole.

The canonical LI form on $SU(2)$ is given by $ig^\dagger dg$ [(1.9.7)], where the i has been added to obtain a hermitian expression. Since it is $su(2)$-valued, $ig^\dagger dg$ can be written as $\omega^{(1)}(\sigma_1/2) + \omega^{(2)}(\sigma_2/2) + \omega^{(3)}(\sigma_3/2)$. Thus, the third or 'vertical' component of the canonical form on $SU(2)$ is the LI one-form

$$\omega^{(3)} = \frac{i}{2} \text{Tr} \left(\sigma_3 g^\dagger dg \right) \quad . \tag{8.9.1}$$

Explicitly, and with the conventions used in section 1.3,

$$\omega^{(3)} = i(z_1^* dz_1 + z_2^* dz_2) = i\mathbf{z}^* d\mathbf{z} = -\frac{1}{2}(d\psi + \cos\theta d\phi) \quad . \tag{8.9.2}$$

Now let $s(t)$ be a possible trajectory of the particle in the manifold $R^+ \times SU(2)$. Using $s(t)$ to pull back (8.9.1) to R (time) the (Balachandran *et al.*) Lagrangian for a non-relativistic particle in the presence of a magnetic monopole is written as

$$L = \frac{1}{2}m[\frac{d}{dt}(r\hat{\mathbf{x}})]^2 + \lambda i \text{Tr} \left(\sigma_3 g^\dagger \dot{g} \right) \equiv L_0 + L_{WZ} \quad , \tag{8.9.3}$$

where $\dot{g} = dg/dt$ and λ includes the monopole charge. The first term may be rewritten as

$$L_0 = \frac{1}{2}m\dot{r}^2 + \frac{1}{2}mr^2\dot{\hat{\mathbf{x}}}^2 = \frac{1}{2}m\dot{r}^2 + \frac{1}{4}mr^2\text{Tr}(\dot{\sigma}\dot{X})^2 \tag{8.9.4}$$

using $\hat{\mathbf{x}} \cdot \dot{\hat{\mathbf{x}}} = 0$ and (1.3.12). To find the Lagrange equations we apply the ordinary variational principle. Taking variations with respect to r we get

$$\ddot{r} = r\dot{\hat{\mathbf{x}}}^2 \quad . \tag{8.9.5}$$

The arbitrary variations with respect to the other variables can be jointly written as

$$\delta g = i(\alpha \cdot \sigma)g \quad . \tag{8.9.6}$$

Eq. (8.9.6) implies

$$\delta X = i[\sigma \cdot \alpha, X] \quad , \quad \delta \text{Tr}(\dot{g}\sigma_3 g^\dagger) = \text{Tr}\ (i\sigma \cdot \dot{\alpha} X) \quad , \tag{8.9.7}$$

$$\delta \text{Tr}\ \dot{X}^2 = 2i \text{Tr}(\sigma \cdot \dot{\alpha}[X, \dot{X}]) \quad . \tag{8.9.8}$$

Inserting (8.9.7), (8.9.8) into the *last* terms of (8.9.4) and (8.9.3), the condition $\delta I = 0$ leads to

$$\text{Tr}\left\{(\sigma \cdot \alpha)\frac{d}{dt}\left[\frac{i}{2}mr^2[X, \dot{X}] - \lambda X\right]\right\} = 0 \quad ,$$

i.e., with $\lambda = qg/c$,

$$\frac{d}{dt}(\mathbf{x} \wedge m\dot{\mathbf{x}} - \frac{qg}{c}\frac{\mathbf{x}}{r}) = 0 \quad . \tag{8.9.9}$$

Eq. (8.9.9) may be written as $\dot{\mathbf{J}} = 0$; the second term in (8.9.9), which was introduced by Poincaré in 1896, is the contribution of the e.m. field to the angular momentum. Eqs. (8.9.5), (8.9.9) combine into the Lorentz force equation

$$m\ddot{\mathbf{x}} = \frac{q}{c}\dot{\mathbf{x}} \wedge \mathbf{B}$$

for $\mathbf{B} = g\mathbf{x}/r^3$, which shows that (8.9.3) is indeed the Lagrangian for a particle in the field of a magnetic monopole of charge g. Notice that, again, no equations of motion have been obtained for ψ in (8.9.2).

No singularities have appeared in the above treatment. In fact, the closed and invariant two-form

$$\Omega = i d\mathbf{z}^* \wedge d\mathbf{z} \tag{8.9.10}$$

is *exact on* $SU(2)$, and we may write $\Omega = d\omega^{(3)}$ where the one-form $\omega^{(3)}$ is given by (8.9.2) (in fact, $\Omega = d\omega^{(3)}$ just expresses that Ω is the curvature of the connection $\omega^{(3)}$). This is also in accord with the discussion of the Chevalley-Eilenberg Lie algebra cohomology: $SU(2)$ is compact and, according to Theorem 6.7.2, the CE and the de Rham cohomology classes are the same. Since $SU(2) \sim S^3$ and $H^p_{DR}(S^n, R) = 0$ for $p \neq n, p \neq 0$ (section 1.5), $H^2_{DR}(S^3) = 0$; Ω is trivial in both cohomologies and so it is the exterior differential of the LI form (8.9.2). However, if we try to define a Lagrangian on ordinary space, *i.e.*, if we remove the additional variable

ψ, the singularities appear because $SU(2) \sim S^2 \times U(1)$ only *locally*. Let us take (8.9.1) as L_{WZ}; ignoring the other constants, this corresponds to $\lambda = \frac{1}{2}$ in (8.9.3). Using the local trivializations of section 1.3 we find on S_+^2, S_-^2

$$L_{WZ+} = \frac{1}{2}(1 - \cos\theta)d\phi \equiv \frac{1}{2}\frac{1 - \cos\theta}{r\sin\theta}\omega_\phi \quad , \tag{8.9.11}$$

$$L_{WZ-} = -\frac{1}{2}(1 + \cos\theta)d\phi \equiv -\frac{1}{2}\frac{1 + \cos\theta}{r\sin\theta}\omega_\phi \quad , \quad L_+ - L_- = d\phi \quad , \tag{8.9.12}$$

which exhibit a string of singularities for $\theta = \pi, 0$ respectively. (8.9.11), (8.9.12) also give the potentials A_+, A_- on the two charts S_+, S_- of S^2, and $A_+ - A_- = \nabla\phi$, which is eq. (2.2.21) for the abelian case ($\nabla = (\partial/\partial r, (1/r)\partial/\partial\theta, (1/r\sin\theta)\partial/\partial\phi)$). In terms of cartesian coordinates $\mathbf{r} = (x, y, z)$, (8.9.11) for instance reads

$$L_{WZ+} = \frac{x\,dy - y\,dx}{2r(r + z)} \quad , \tag{8.9.13}$$

and corresponds to the potential one-form describing a monopole of strength $\frac{1}{2}$; L_{WZ+} is singular along the negative z axis. The singularities in (8.9.11), (8.9.12) had to be expected: $H_{DR}^2(S^2) = R \neq 0$, and on S^2 the two-form (8.9.10) cannot be exact since its integral has to be non-zero (and equal to the magnetic charge inside S^2 by Gauss' theorem).

Under $U(1)$ *gauge* (or *local*) transformations

$$g \rightarrow g \exp i\sigma_3\psi(t) \quad , \quad \mathbf{z} \rightarrow \mathbf{z} \exp i\psi(t) \quad , \quad \mathbf{z}^* \rightarrow \mathbf{z}^* \exp{-i\psi(t)} \tag{8.9.14}$$

(see (1.3.10)), the L_0 term (8.9.4) of the action ($\text{Tr}(\vec{\sigma}\dot{X})^2 = [\frac{d}{dt}(\mathbf{z}^\dagger\vec{\sigma}\mathbf{z})]^2$) is of course invariant but, clearly,

$$\delta\omega^{(3)} = -\dot{\psi} \quad , \tag{8.9.15}$$

i.e., is a total derivative. As a result, if we look at L_{WZ} as the effective Lagrangian, its variation

$$\delta L_{WZ} = -2\lambda\dot{\psi} \tag{8.9.16}$$

implies that when ψ makes a round turn the phase ψ of the wavefunction will experience a change equal to

$$\frac{i}{\hbar}\int_0^T (-2\lambda\dot{\psi})dt = -\frac{i}{\hbar}2\lambda(2\pi) \quad . \tag{8.9.17}$$

Consequently, the *quantum* version of the theory restricts the values of λ to satisfy

$$2\lambda/\hbar = n \in Z \quad , \quad \lambda = n\hbar/2 \quad (\frac{2qg}{c\hbar} = n) \quad , \tag{8.9.18}$$

which is none other than the Dirac quantization condition, an example of quantization of classical parameters due to the topology of the problem. We thus see that we can interpret $(2\lambda/\hbar)$ as the *winding number* which tells us how many complete turns the quantum phase undergoes when ψ moves from 0 to 2π; as we already know, it is the same number which classifies the $U(1)$-bundles over S^2 and which for $n = 1$ corresponds to the Hopf bundle.

We have used repeatedly that, when a Lagrangian one-form is quasi-invariant, then the Noether theorem states that the expression for the conserved charges acquires an additional term. Gauge symmetries do not correspond to Noether currents, but it is possible to derive an identity independent of the equations of motion, which simply shows that for every gauge symmetry there is a constraint. This constraint is modified when the theory is only quasi-invariant under local transformations. The analysis of this constraint for (8.9.3) provides an equivalent way to derive (8.9.18).

Under the transformations (8.9.14), $\delta z_j = i\psi(t)z_j$, $\delta z_j^* = -i\psi(t)z_j^*$, $j = 1, 2$, the Lagrangian $L = L_0 + L_{WZ}$ is only quasi-invariant, i.e. $\delta L = -2\lambda\dot\psi$. In terms of δz, δz^*, δL is given by

$$\frac{\partial L}{\partial z_j}\delta z_j + \frac{\partial L}{\partial z_j^*}\delta z_j^* + \frac{\partial L}{\partial \dot z_j}\delta \dot z_j + \frac{\partial L}{\partial \dot z_j^*}\delta \dot z_j^* = -2\lambda\dot\psi \quad , \tag{8.9.19}$$

$$\left(z_j\frac{\partial L}{\partial z_j} - z_j^*\frac{\partial L}{\partial z_j^*} + \dot z_j\frac{\partial L}{\partial \dot z_j} - \dot z_j^*\frac{\partial L}{\partial \dot z_j^*}\right)i\psi + \left(z_j\frac{\partial L}{\partial \dot z_j} - z_j^*\frac{\partial L}{\partial \dot z_j^*} - 2i\lambda\right)i\dot\psi = 0 . \tag{8.9.20}$$

Since the function ψ is arbitrary, both parentheses have to vanish separately. The first term equal to zero means that the variation of the Lagrangian is zero for rigid ($\dot\psi = 0$) transformations. The term proportional to $\dot\psi$ gives

$$z_j\pi_{z_j} - z_j^*\pi_{z_j^*} - 2i\lambda = 0 \quad , \tag{8.9.21}$$

in terms of the canonical momenta. This is a constraint of the theory, and has a term $-2i\lambda$ due to the quasi-invariance of the Lagrangian.

In the quantum theory, (8.9.21) is a condition to be imposed on the wavefunction $\Psi(z, z^*)$ after the replacement $\pi_{z_j} \mapsto -i\hbar\frac{\partial}{\partial z^i}$, which is

$$\left(iz_j\frac{\partial}{\partial z_j} - iz_j^*\frac{\partial}{\partial z_j^*} + \frac{2i\lambda}{\hbar}\right)\Psi = (\Xi + \frac{2i\lambda}{\hbar})\psi = 0 \quad , \tag{8.9.22}$$

after identifying the first two terms with the generator of the transformations of $U(1)$ as it is immediately read from (8.9.14). This equation states that the wavefunction is not invariant under the gauge transformations but gets a phase:

$$\Psi(e^{i\psi}z, e^{-i\psi}z^*) = \exp\left(-\frac{2i\lambda}{\hbar}\psi(t)\right)\Psi(z, z^*) \quad . \tag{8.9.23}$$

Again, the quantization condition results from the fact that ψ has to be a univalued function which implies $\lambda = \frac{n\hbar}{2}$.

To conclude, we mention that this example exhibits another feature which is present for *topologically non-trivial* Wess-Zumino terms: the quantization of the physical parameter which constitutes the coefficient of the Wess-Zumino term.

Bibliographical notes for chapter 8

The reader who wants additional information on jet bundles and variational principles may look at Hermann (1970), Goldschmidt and Sternberg (1973), García-Pérez (1974), Aldaya and de Azcárraga (1978, 1980) and the book of Saunders (1989); these contain further references.

The geometric approach to mechanics has been covered in many books and articles. A very readable presentation is the book of Godbillon (1969). For more detailed accounts the reader may wish to consult the treatise of Abraham and Marsden (1978) and the books of Arnold (1978), Thirring (1978), Guillemin and Sternberg (1984), and Marsden (1992). The pioneering article in discussing quasi-invariance and central extensions in mechanics (section 8.4) is by Lévy-Leblond (1969; 1971), where more details can be found. See also, *e.g.*, the papers of Houard (1977), Aldaya and de Azcárraga (1980, 1985a, 1987), Marmo *et al.* (1985, 1988), and Cariñena and Ibort (1988). The analysis of the possible kinematical groups mentioned in section 8.3 is given by Bacry and Lévy-Leblond (1968); see also Cariñena *et al.* (1981).

On geometric quantization in general, see the references in chapter 3.

There is a large amount of references on Wess-Zumino terms, a name which now goes beyond its original meaning. They were introduced by Wess and Zumino (1971) in the context of chiral theories, and further discussed by Witten (1983a,b), Knizhnik and Zamolodchikov (1984), Braaten *et al.* (1985), Krichever *et al.* (1986) and others; see also the book by Mickelsson (1989). All these contain further information. More references are also given in chapter 10.

The BRST transformations were introduced by Becchi *et al.* (1976) and Tyutin (1975). For a recent account see the book of Henneaux and Teitelboim (1992), which contains many references; see also the bibliography in chapter 10.

Superstring theory is discussed at length in the two-volume book of Green *et al.* (1987), which contains further references to the earlier papers; see also Brink and Henneaux (1988). The interpretation of the second term in the string action as a Wess-Zumino term was given in Henneaux and Mezincescu (1985). The massive superparticle actions with a group Wess-Zumino term may be found in de Azcárraga and Lukierski (1982); ·for their fermionic (Siegel) symmetry see de Azcárraga and Lukierski (1983) and Siegel (1983). The mathematics of supermanifold cohomology were discussed by Rabin (1987). For supersymmetric extended objects the reader may look at Achúcarro et al. (1987), in which their classification was given, and the lectures by Townsend (1989, 1993) which contain further references. The relevance of the Chevalley-Eilenberg cohomology for supersymmetric extended objects is given in de Azcárraga and Townsend (1989); see also de Azcárraga et al. (1989). The properties of spinors in arbitrary dimensions are discussed, e.g., in Van Nieuwenhuizen (1984) and Sohnius (1985).

Since 1994, superstring theory is undergoing a 'second revolution'. The interested reader may look at the reviews of Duff (1996), Schwarz (1977) and Townsend (1977), and to the references therein, for details on super-membranes, M-theory and related topics.

For the fibre bundle description of the monopole see the references given in chapter 2; see also Álvarez (1985) for its analysis in terms of Čech cohomology. For additional information on geometrical and topological structures in general the reader may consult e.g. Coleman (1985) which contains many of his Erice lectures, Boya et al. (1978), Goddard and Mansfield (1986), Leinaas (1980), Madore (1981), Jackiw (1984) and Isham (1984); see also the collection of articles in Shapere and Wilczek (1989). The monopole Lagrangian of section 8.9 is due to Balachandran et al. (1980); see also Balachandran (1988).

9

Infinite-dimensional Lie
groups and algebras

This chapter is devoted to introducing infinite-dimensional Lie groups and algebras and cohomology (in contrast with the finite-dimensional case) as a preparation for chapter 10. Special attention is given to gauge groups and current algebras (an example of the same kind of generalization was already given in chapter 8 in the context of supersymmetric extended objects).

A second set of examples, in which the extension properties of the Lie algebras involved are studied, is provided by the Virasoro and Kac-Moody algebras, as well as the two-dimensional conformal group. It is also shown that Polyakov's induced two-dimensional gravity (which itself is not discussed) provides yet another example of a Wess-Zumino term, obtained from the group $Diff\, S^1$ of the diffeomorphisms of the circle.

9.1 Introduction

Most of the infinite-dimensional groups that appear in physics are defined by endowing the space of mappings of a (finite-dimensional) manifold M into another finite-dimensional manifold Q with a group structure. For instance, this is the case for the groups $Map(M, G) \equiv G(M)$ that correspond to the smooth mappings $M \to G$ and which (in physics) are sometimes referred to as 'local' groups; this is also the case of the gauge groups, as well as of the *loop groups* LG or $Map(S^1, G)$ given by the smooth mappings $S^1 \to G$. The loop groups can be generalized to groups of mappings $S^n \to G$; these *generalized loop groups* or *sphere groups* may be denoted by $L^n G$ (with $L^1 G \equiv LG$ and $L^0 G = G$), and were already encountered in section 8.8 in a particular case in which G was the (graded) supertranslation group. In these examples, the 'target' manifold Q is a Lie (or even a graded Lie) group G, but if $Q = M$ then $Map(M, M)$ is the group $Diff\, M$ of the diffeomorphisms of a manifold M, the Lie algebra of which is $\mathscr{X}(M)$. An important subgroup of *Diff M* is the group of volume-preserving diffeomorphisms; its Lie algebra is given by the space of vector fields on M of divergence identically zero (see eq. (1.4.50)). If M is endowed with a symplectic metric form Ω_S, another subgroup is that of the symplectic diffeomorphisms; its Lie algebra is generated by the (locally) Hamiltonian vector fields on M (section 3.6). A special example of *Diff M* is *Diff S*^{1}; the central extension of its algebra is the Virasoro algebra to be discussed in section 9.5.

To endow these infinite-dimensional spaces with a Lie group structure, it is necessary to extend the property that characterizes a finite-dimensional Lie group (*i.e.*, that the group law $G \times G \to G$ and $g \mapsto g^{-1}$ are smooth mappings) to the case where G is an infinite-dimensional manifold. Infinite-dimensional manifolds are defined in a way similar to the finite-dimensional ones. In the finite-dimensional case the open sets of the local covering $\{U_i\}$ of M are set into one-to-one correspondence with open sets of R^n; it is said that the local charts (U_i, φ_i) are *modelled'* on R^n. If, in particular, M is a finite-dimensional Lie group G, we know that the model space is the vector space R^r (r=dim G) which constitutes its own Lie algebra $\mathscr{G}$, and that the canonical local coordinates of an element g *near* the identity element $e \in G$ are determined by the exponential mapping. If the manifold is infinite-dimensional, R^n is replaced by some infinite-dimensional topological space E such as a Hilbert space, Banach space, locally convex* space, etc. For an infinite-dimensional Lie group the model space is also its associated Lie algebra. For instance, the unitary group of all the isometries of a Hilbert space is an example of a group modelled on a Banach space, but this is not always so: as a vector space, the Lie algebra $\mathscr{G}(M)$ of the group $G(M)$ of smooth mappings $M \to G$ is complete and locally convex, but it is not Banach. Thus, it turns out that it is better to use locally convex vector spaces rather than Banach spaces to model infinite-dimensional Lie groups in general. This permits including many interesting groups in physics which cannot be given a Banach structure, and motivates

(9.1.1) Definition (*General Lie group*)

A Lie group is a group which is also a manifold modelled on a complete, locally convex vector space, such that the group operations $G \times G \to G$ and $g \mapsto g^{-1}$ are smooth mappings.

---

* A *Banach space* is a complete normed vector space. A convex set in a vector space contains the line segment joining any two points in it. A topological vector space is *locally convex* if every neighbourhood U of zero contains a neighbourhood V of zero which is convex.

The Lie algebra $\mathcal{G}(M)$, as we shall see below, can be considered as the space of smooth mappings of $M \to \mathcal{G}$ (in particular, the algebra $\mathcal{L}\mathcal{G}$ of LG is the space of smooth mappings $S^1 \to \mathcal{G}$); as mentioned, section 8.8 already provided the expression for the commutation relations of a Lie algebra of the type $\mathcal{G}(M)$. The Lie algebra of the group of diffeomorphisms of a compact manifold *Diff M* is given by the algebra of the smooth vector fields X on M, $\mathcal{X}(M)$, which is a locally convex vector space. However, although with the previous definitions there are many similarities between the finite-dimensional and the infinite-dimensional cases, there are also strong differences. For instance, the simplest example of infinite-dimensional groups, the groups $G(M)$, can be shown to be analytic (the analytic structure comes completely from G), with canonical coordinate system given by the exponential mapping and, as such, the product of elements near the identity is given by the Campbell-Baker-Hausdorff formula (groups with this property may be referred to as *Campbell-Baker-Hausdorff Lie groups*). But in contrast, and although it is of course smooth, *Diff M* (*Diff S¹* in particular) is not analytic and does not have a canonical coordinate chart at the identity element: the exponential mapping is not locally one-to-one since it is possible to encounter elements arbitrarily *close* to the identity that are not in any one-parameter subgroup*. This is traced to the fact that *Diff M* cannot be given a *Banach* Lie group structure. This implies that the inverse function theorem (which is automatically valid for Banach spaces) is lost, and that in particular it is not valid for the mapping $exp : \mathcal{X}(M) \to \textit{Diff } M$ at the origin. Another important fact is the failure of the converse to Lie's third theorem in the infinite-dimensional case, which for finite r-dimensional algebras $\mathcal{G}$ states that there is a simply connected r-dimensional Lie group whose Lie algebra is isomorphic with $\mathcal{G}$: even in the Banach case, there

---

* For example (Freifeld), the element $f \in \textit{Diff } S^1$, $f(\varphi) = \varphi + \pi/n + \epsilon \sin^2 n\varphi$, n large, ϵ small, is arbitrarily close to the identity but does not belong to a one-parameter subgroup: since it is not the square of another diffeomorphism it cannot be a pure exponential (if it were, we could write $f = exp\, v = g \circ g$ with $g = exp(v/2)$). For a finite-dimensional Lie group such as $GL(2, R)$ or $SL(2, R)$ this may also happen but *not* for elements arbitrarily close to the identity element since the exponential mapping $exp : \mathcal{G} = T_e(G) \to G$ is *locally* a diffeomorphism. For instance, the group elements $\begin{pmatrix} -2 & 0 \\ 0 & -1 \end{pmatrix}$ and $\begin{pmatrix} -e^\rho & 0 \\ 0 & -e^{-\rho} \end{pmatrix}$ ($\rho \neq 0$; if $\rho=0$, it is a rotation of angle π) cannot be obtained by exponentiating a certain Lie algebra element. If G is a connected and compact group, then the mapping $exp : \mathcal{G} \to G$ is onto (if G is just compact, exp is obviously onto on the component of the identity element), but this property is not inherited by $G(M)$.

When G is not compact, an element can be written as the product of only a small number of exponentials. For instance, coming back to the $SL(2, R)$ example above, we find $\begin{pmatrix} -e^\rho & 0 \\ 0 & -e^{-\rho} \end{pmatrix} =$

$exp\left[\pi \begin{pmatrix} 0 & 1 \\ -1 & 0 \end{pmatrix}\right] . exp\left[\rho \begin{pmatrix} 1 & 0 \\ 0 & -1 \end{pmatrix}\right]$. A similar statement is valid for *DiffM*, M compact: any element of *DiffM* can at least be written as a finite product $(exp\, X_1)...(exp\, X_l)$, $X_1, ..., X_l \in \mathcal{X}(M)$.

are infinite-dimensional algebras for which this is not true. Finally, there are no simple classifying theorems for infinite-dimensional Lie algebras analogous to the powerful theorems of the finite case. We shall not dwell on these differences any further, and limit ourselves in this chapter to discussing some specific examples of physical interest.

Let us first turn to gauge groups associated with compact finite-dimensional groups G. In the previous chapter it has been shown in detail how a non-zero variation of the action in mechanics under a symmetry transformation is related to an extension of the original symmetry algebra, the modified structure constants of which are determined by the cohomological descent process. In the next chapter we shall consider an analogous situation for the case of gauge theories. There are two important differences, however, from the cases previously considered: first, gauge groups are infinite-dimensional, and, secondly, the breaking of gauge invariance is, as we shall see, an inherent property of the quantum (path integral) formulation of the anomalous theory, since its Lagrangian is gauge-invariant. The infinite-dimensional character of these groups appears because their parameters are functions $g_i(x)$ of the coordinates of the finite-dimensional manifold M; this reflects the freedom of choosing 'locally' (at each $x \in M$) the group transformation, *i.e.* g_i becomes $g_i(x)$, $x \in M$, and thus, *at least locally*,[*] they are of the type $Map(M, G)$.

In spite of their infinite dimensionality, the structure of $Map(M, G)$ still 'remembers' that of the original finite-dimensional Lie groups as will be seen in next section. As a result, it is possible to study *e.g.* the associated algebra cohomology problem for central extensions in much the same way as in the finite-dimensional case. Nevertheless, the cohomology structure turns out to be richer as exhibited *e.g.* by the central extension of the $su(2)$ loop algebra, the $su(2)$ *Kac-Moody algebra* (section 9.4), which exists despite the fact that the Whitehead lemma establishes that there are no central extensions of the finite-dimensional $su(2)$ algebra. Moreover, having a Lie algebra extension does not necessarily imply the existence of a corresponding group extension. For instance, both the loop algebra $\mathscr{L}\mathscr{G} = \mathscr{G}(S^1)$ and the loop group $LG = G(S^1)$ of a compact group G

---

[*] It may be worth while to comment here on a terminology which is potentially confusing from a mathematical point of view. In physics a symmetry transformation g is frequently called 'global' when it does not depend on $x \in M$, $g \neq g(x)$, where M is the spacetime manifold; it is called 'local' when $g = g(x)$. These 'local' transformations are often referred to as *local gauge transformations*. But, as we shall see in section 10.1, a more precise definition of the gauge transformations will show that the mappings $g : x \mapsto g(x)$ which define the 'local' group $G(M) = Map(M, G)$ correspond to the group of gauge transformations only when the principal bundle $P(G, M)$ on which the (Yang-Mills) connections are defined is trivial, $P = G \times M$ (section 2.2(b)). In general, gauge transformations are elements of $Map(G, M)$ only locally, *i.e.* on $U_\alpha \subset M$. As for the 'global' transformations (elements of G), they are now often called *rigid* to avoid the overlap with the mathematical terminology.

can be (non-trivially) centrally extended by R and $U(1)$ respectively. But this is not the case for generalized loop algebras and groups as, for instance, for $Map(S^2, su(2)) = \mathscr{L}^2(su(2))$ and $Map(S^2, SU(2)) = L^2(SU(2))$: the $\mathscr{L}^2(su(2))$ algebra may be centrally extended, but not so the $SU(2)$ 'sphere' group. It follows that a Lie algebra extension may correspond to a trivial group extension: the infinite-dimensional Lie algebra has a larger cohomology group. Consider now the simple case of the loop group LG of a compact and simply connected group G. As was mentioned in section 6.7 the de Rham cohomology of G is isomorphic with the Chevalley-Eilenberg Lie algebra cohomology in the finite-dimensional case. This is not true for infinite-dimensional Lie groups (and fails in particular for $Map(M, G)$, dim $M \geq 2$) but for $M = S^1$ the loop group and loop algebra cohomologies are the same. To a central extension $\widetilde{\mathscr{L}\mathscr{G}}$ of the loop algebra $\mathscr{L}\mathscr{G}$ by R defined by the two-cocycle ω corresponds an extension $\widetilde{LG}$ of the loop group LG by S^1 iff $\omega/2\pi$ is of integer cohomology class on LG. This is so because the extension $\widetilde{LG}$ may be viewed as a principal bundle $\widetilde{LG}(U(1), LG)$ which is characterized by its Chern class $c_1(\omega)$; if $\omega/2\pi$ is not of integer class, there is no group extension corresponding to $\overline{\mathscr{L}\mathscr{G}}$.

Let us now address the problem of extending some of the expressions given in chapter 1 to the case where the r-dimensional Lie group is replaced by the 'local' group $G(M)$ defined on a manifold M. In so doing, vector fields will become *functional* differential operators and the finite sums over the indices of the Lie algebra $\mathscr{G}$ will be accompanied by integrals over M. As mentioned, G will be compact, as is the case in theories of Yang-Mills type. This mathematically convenient restriction is physically motivated by the fact that fields transform under unitary representations of G and that the field multiplets (which determine the basis of the representation space) should have a finite number of components.

9.2 The group of mappings G(M) associated with a compact Lie group G, and its Lie algebra 𝒢(M)

Let us start by introducing the local group $G(M)$ associated to G. In a gauge theory in which the potentials are connections on a trivial bundle $P(G, M)$, the group $G(M)$ is the gauge group (in general, this is not the case; a precise definition will be given in section 10.1).

(9.2.1) DEFINITION (*The group G(M) or 'local' group associated with G*)

Let G be a compact (finite-dimensional) Lie group. The group $G(M)$ (or $Map(M, G)$) associated with G over the manifold M is the Lie group of the smooth mappings $g : M \to G$, $g : x \in M \mapsto g(x) \in G$, with the pointwise *composition group law given by*

$$g''(x) = g'(x) * g(x) , \quad g''(x), g'(x), g(x) \in G(M) ; \qquad (9.2.1)$$

$g^{-1}(x) = [g(x)]^{-1}$ *and* $e(x) = e.$

Although $G(M)$ is not finite-dimensional, its Lie algebra $\mathscr{G}(M)$ (which clearly is infinite-dimensional) can be obtained from the finite-dimensional algebra $\mathscr{G}$ by replacing the ordinary derivatives by *functional* derivatives. Consider for example the generators (in the physics terminology) of the right action. For the finite-dimensional Lie group G the infinitesimal generators were given by (1.1.14),

$$X_{(i)}^L = X_{(i)}^{Lj}(g)\frac{\partial}{\partial g^j} , \quad X_{(i)}^{Lj} = \left.\frac{\partial g''^j(g',g)}{\partial g^i}\right|_{\substack{g'=g \\ g=e}} , \quad i,j = 1,...,r = \dim(G) .$$

$$(9.2.2)$$

Similarly, the generators of $\mathscr{G}(M)$ associated with the action of $G(M)$ on itself can be derived from (9.2.1):

$$X_{(i)}^L(x) = \int d^n y \left.\frac{\delta g''^j(y)}{\delta g^i(x)}\right|_{\substack{g(x)=e \\ g'(x)=g(x)}} \frac{\delta}{\delta g^j(y)}$$

$$= \int d^n y \, \delta^n(x-y)\left.\frac{\partial g''^j}{\partial g^i}(x)\right|_{\substack{g(x)=e \\ g'(x)=g(x)}} \frac{\delta}{\delta g^j(y)} , \qquad (9.2.3)$$

and thus the expression for a generator of $\mathscr{G}(M)$ (cf. (9.2.2)) is given by

$$X_{(i)}^L(x) = X_{(i)}^{Lj}(x)\frac{\delta}{\delta g^j(x)} , \qquad (9.2.4)$$

where the dependence of $X_{(i)}^L$ and its components $X_{(i)}^{Lj}$ on $x \in M$ is through $g(x)$, and will usually be omitted. In (9.2.3), the coordinates x, y on M can be regarded as additional continuous indices which label the elements of the infinite-dimensional group, and $\frac{\delta g^i(x)}{\delta g^j(y)} = \delta_j^i \delta^n(x-y)$. The integral over y plays the same rôle as the sum over j for the continuous indices, for which the Dirac delta $\delta^n(x-y)$ is the analogue of the Kronecker delta δ_j^i for the finite-dimensional Lie group parameters. It is now easy to check that if the structure constants of $\mathscr{G}$ as defined by the LI vector fields (9.2.2) are C_{ij}^k, then the generators of $\mathscr{G}(M)$ satisfy the following algebra commutation relations:

$$[X_{(i)}^L(x), X_{(j)}^L(y)] = \delta^n(x-y)C_{ij}^k X_{(k)}^L(y) . \qquad (9.2.5)$$

Explicitly,

$$[X_{(i)}^L(x), X_{(j)}^L(y)] = X_{(i)}^{Lr}(g(x))\frac{\delta X_{(j)}^{Ls}}{\delta g^r(x)}(g(y))\frac{\delta}{\delta g^s(y)} - \begin{pmatrix} x \leftrightarrow y \\ i \leftrightarrow j \end{pmatrix}$$

$$= X_{(i)}^{Lr}(g(x))\frac{\partial X_{(j)}^{Ls}}{\partial g^r}(g(x))\delta^n(x-y)\frac{\delta}{\delta g^s(y)} - \begin{pmatrix} x \leftrightarrow y \\ i \leftrightarrow j \end{pmatrix}$$

$$= [X_{(i)}^{Lr}(g(x))\frac{\partial X_{(j)}^{Ls}}{\partial g^r}(g(x)) \qquad\qquad (9.2.6)$$

$$- X_{(j)}^{Lr}(g(x))\frac{\partial X_{(i)}^{Ls}}{\partial g^r}(g(x))] \cdot \delta^n(x-y)\frac{\delta}{\delta g^s(x)}$$

$$= C_{ij}^k X_{(k)}^{Ls}(g(x))\delta^n(x-y)\frac{\delta}{\delta g^s(x)} = \delta^n(x-y)C_{ij}^k X_{(k)}^L(x).$$

Notice that since $X_{(i)}^{Lj}$ depends only on $g(x)$ and not on its derivatives, $\frac{\delta X_{(i)}^{Lj}(y)}{\delta g^k(x)} = \frac{\partial X_{(i)}^{Lj}}{\partial g^k}(x)\delta^n(x-y)$; as a result, the calculation proceeds as in the finite-dimensional case except for the extra factor $\delta^n(x-y)$. Eqs. (9.2.4) and (9.2.5) define the infinite-dimensional Lie algebra $\mathscr{G}(M)$.

We know that an arbitrary element ζ of the finite-dimensional Lie algebra $\mathscr{G}$ can be written as $\zeta = \zeta^i T_i \equiv \zeta \cdot T$, $\zeta^i \in R$, where T_i is a basis of $\mathscr{G}$. Then the commutator of two elements is given by $[\zeta, \zeta'] = [\zeta \cdot T, \zeta' \cdot T] = C_{ij}^k \zeta^i \zeta'^j T_k \equiv [\zeta, \zeta'] \cdot T$ if we introduce the notation $[\zeta, \zeta']^k \equiv C_{ij}^k \zeta^i \zeta'^j$. The R-linearity of $\mathscr{G}$ is expressed by saying that $[T_i, T_j] = C_{ij}^k T_k$ implies that $[\zeta, \zeta'] = [\zeta, \zeta']^k T_k$. Similarly, an arbitrary element ζ of the infinite-dimensional Lie algebra $\mathscr{G}(M)$ can be expressed (omitting the superscript L) as

$$\zeta \in \mathscr{G}(M) \quad , \quad \zeta := \zeta \cdot X \equiv \int d^n y\, \zeta^i(y) X_{(i)}(y), \qquad (9.2.7)$$

where the functions $\zeta^i(x)$, $i = 1, ..., r$, are the coordinates of ζ, i.e. the coefficients of the expression for ζ in the basis $X_{(i)}(x)$. Obviously, this involves a sum over i and an integral over the n continuous labels $y \in R^n$ that determine the points of M. In particular, the coordinates of the basis elements $X_{(i)}(x)$ themselves are $\zeta_{(i)}^j(y) = \delta_{(i)}^j \delta^n(x-y)$, which concentrate all the weight in $j = i$ and $y = x$. The commutator of two algebra elements $\zeta = \zeta \cdot X$, $\zeta' = \zeta' \cdot X$ is given by

$$[\zeta \cdot X, \zeta' \cdot X] = \int d^n y \int d^n z\, \zeta^i(y) \zeta'^j(z)[X_{(i)}(y), X_{(j)}(z)]$$

$$= \int d^n y \int d^n z\, \zeta^i(y) \zeta'^j(z) C_{ij}^k \delta^n(y-z) X_{(k)}(y) \qquad (9.2.8)$$

$$= \int d^n y\, \zeta^i(y) \zeta'^j(y) C_{ij}^k X_{(k)}(y) \quad \forall \zeta, \zeta' \in \mathscr{F}(M).$$

Using the same notation as used for $\mathscr{G}$, the above commutator of two elements of $\mathscr{G}(M)$ can be written as

$$[\zeta, \zeta'] \equiv [\zeta \cdot X, \zeta' \cdot X] = [\zeta, \zeta'] \cdot X . \tag{9.2.9}$$

Thus, in much the same way that the Lie bracket of $\mathscr{G}$ is R-linear, in the case of the $\mathscr{G}(M)$ Lie algebra the following lemma holds:

(9.2.1) LEMMA ($\mathscr{F}(M)$-*linearity in* $\mathscr{G}(M)$)

The bracket in $\mathscr{G}(M)$ *is* $\mathscr{F}(M)$-*linear.*

Another equivalent way of expressing this $\mathscr{F}(M)$-linearity is by saying that

$$\left[\int d^n x \, \alpha(x) X_{(i)}(x), \int d^n y \, \beta(y) X_{(j)}(y) \right]$$

$$= C_{ij}^k \int d^n z \, \alpha(z)\beta(z) X_{(k)}(z) \quad \forall \alpha(x), \beta(x) \in \mathscr{F}(M) . \tag{9.2.10}$$

The above can be summarized in the following statement:

$$\mathscr{G}(M) \approx \mathscr{G} \otimes \mathscr{F}(M) . \tag{9.2.11}$$

The identification $\mathscr{G}(M) = T_e(G(M))$ and the previous construction show that $\mathscr{G}(M)$ itself can be looked at as the Lie algebra of smooth mappings $M \to \mathscr{G}$; this explains why, formally, the structures of $\mathscr{G}(M)$ and $\mathscr{G}$ look so similar.

9.3 Current algebras as infinite-dimensional Lie algebras

Infinite-dimensional Lie algebras like (9.2.5) are quite old in physics. In fact, the algebras $Map(M, \mathscr{G}) = \mathscr{G}(M)$ are called *current* or *local algebras*. They appeared first as the algebra of the (equal time) charge *densities* $Q_{(i)}(x)$ associated with the internal symmetry generators of the Lie group G of a field theory, $[Q_{(i)}(x), Q_{(j)}(y)]_{x^0=y^0} = i\delta^{n-1}(x-y)C_{ij}^k Q_{(k)}(y)$, where the expression of the hermitian charge density operators $Q_{(i)}(x)$ is given in terms of quantized fields.

Let us recall how they arise in quantum field theory. Consider, in the absence of gauge fields, a Lagrangian density $\mathscr{L}(\varphi(x), \partial_\mu \varphi(x))$, $\mu = 0, ..., (D-1)$, which depends on fields φ^s over D-dimensional spacetime and assume that, under the action of the group $G(M)$, they transform as

$$\varphi'(x) = e^{ig^j \tau_j} \varphi(x) \quad , \quad \delta\varphi^s = (ig^j \tau_j)^s_{\,t} \varphi^t , \quad [\tau_l, \tau_j] = iC_{lj}^k \tau_k \quad , \tag{9.3.1}$$

where the indices s, t (omitted in the first and last expressions) are the representation indices that determine a certain multiplet φ^t (*i.e.*, a basis for the representation space of G) and $\tau = \tau^\dagger$. Using the Euler-Lagrange equations $\frac{d}{dx^\mu}\left(\frac{\partial \mathscr{L}}{\partial(\partial_\mu\varphi)}\right) - \frac{\partial \mathscr{L}}{\partial \varphi} = 0$, the variation of the Lagrangian is found to be

$$\delta \mathscr{L} = \partial_\mu \left(\frac{\partial \mathscr{L}}{\partial(\partial_\mu \varphi^s)}(ig^j \tau_j)^s_{\cdot t}\varphi^t \right) \ . \tag{9.3.2}$$

Now, if we define the internal currents $j^\mu_{(l)}(x)$ by

$$j^\mu_{(l)} := -\frac{\partial \mathscr{L}}{\partial(\partial_\mu\varphi^s)}(i\tau_l)^s_{\cdot t}\varphi^t \quad , l = 1, ..., (\dim G) \, , \ \mu = 0, 1, ..., (D-1) \, , \tag{9.3.3}$$

the variation of the Lagrangian can be expressed as

$$\delta \mathscr{L} = -(\partial_\mu j^\mu_{(i)})g^i - j^\mu_{(i)}\partial_\mu g^i \ . \tag{9.3.4}$$

Eqs. (9.3.4) lead to the so-called *Gell-Mann–Lévy equations*,

$$j^\mu_{(i)} = -\frac{\partial(\delta\mathscr{L})}{\partial(\partial_\mu g^i)} \quad , \quad \partial_\mu j^\mu_{(i)} = -\frac{\partial(\delta\mathscr{L})}{\partial g^i} \quad , \tag{9.3.5}$$

which tell us that if $\mathscr{L}$ is invariant under a transformation of G (*i.e.*, under a 'global', or *rigid* transformation, $g^i \neq g^i(x)$) the current $j^\mu_{(i)}$ given by the first equation in (9.3.5) is conserved as a consequence of (9.3.4) or of the second equation in (9.3.5).

In terms of the canonical momenta $\pi_s(x) = \frac{\partial \mathscr{L}}{\partial \dot{\varphi}^s(x)}$ associated with the fields $\varphi^s(x)$ the charge densities read

$$Q_{(i)}(x) \equiv j^0_{(i)}(x) = -i\pi_s(x)(\tau_i)^s_{\cdot t}\varphi^t(x) \ . \tag{9.3.6}$$

Assuming the canonical commutation relations (φ bosonic, $\hbar = 1$),

$$[\varphi^s(x), \pi_t(y)]_{x^0=y^0} = i\delta^s_t\delta(\mathbf{x}-\mathbf{y}) \quad , \tag{9.3.7}$$

which allow us to represent $\pi_t(x)$ by

$$\pi_s(x) = -i\frac{\delta}{\delta\varphi^s(x)} \quad , \quad \left.\frac{\delta\varphi(x)}{\delta\varphi(y)}\right|_{x^0=y^0} = \delta(\mathbf{x}-\mathbf{y}) \ . \tag{9.3.8}$$

We can immediately find the equal-time commutation relations of the (hermitian) charge densities. They are given by

$$[Q_{(i)}(x), Q_{(j)}(y)]_{x^0=y^0} = iC^k_{ij}\delta(\mathbf{x}-\mathbf{y})Q_{(k)}(y) \ . \tag{9.3.9}$$

Two comments are in order here. The first is that the derivation of (9.3.9) proceeds in a similar way if the fields φ^s, instead of being bosonic, are assumed to be fermionic; it suffices to replace the commutators of (9.3.7) by anticommutators. The second is that (9.3.9) has been obtained independently of the conservation of the currents $j^\mu_{(i)}(x)$ or, in other words, of the invariance of the Lagrangian under the finite-dimensional

Lie group G: the commutator does not depend on the detailed structure of the Lagrangian. When the currents are conserved, $\partial_\mu j^\mu_{(i)} = 0$, the charges

$$Q_{(i)} = \int d^3x\, Q_{(i)}(x) \tag{9.3.10}$$

do not depend on time and the algebra of conserved charges $[Q_{(i)}, Q_{(j)}] = iC^k_{ij}Q_{(k)}$ reproduce the Lie algebra $\mathscr{G}$ of G. But even if the currents are not conserved, the double integration of (9.3.9) over the space variables shows that $\mathscr{G}$ (eq. (9.3.1)) is then reproduced by the algebra of equal-time charges

$$[Q_{(i)}(t), Q_{(j)}(t)] = iC^k_{ij}Q_{(k)}(t) \quad . \tag{9.3.11}$$

The above reasonings constitute the basis of the old current algebra hypothesis, which was formulated in the early sixties by Gell-Mann in a quark model field-theoretical context. In it, the expression for the currents $j^\mu_{(i)}$ is given by bilinears of the fermionic quark fields $q(x)$. Within this explicit model it is possible to compute not only the charge–charge densities commutation relations (9.3.9), but also the charge–current commutation relations. A naïve calculation leads to

$$[j^0_{(i)}(x), j^\mu_{(j)}(y)]_{x^0=y^0} = iC^k_{ij}\delta(\mathbf{x}-\mathbf{y})j^\mu_{(k)}(y) \quad . \tag{9.3.12}$$

However, such a result ignores the problem that arises from the fact that the local currents are given by products of (quark) fields taken at the same spacetime point. When the commutators (9.3.12) are calculated by expressing the currents as bilinears of the quark fields in different, spatially separated points, and then the limit leading to zero separation is taken, those for $\mu = 1, 2, 3$ are modified by additional terms that contain the gradients of the delta function, as first shown by Schwinger in 1959. These are the original Schwinger terms; by extension, the name refers nowadays to any additional term including the delta distribution or its derivatives that modifies the r.h.s. of a local algebra's commutation relations. This is the case for the central term of the Kac-Moody algebra, which we now discuss.

9.4 The Kac-Moody or untwisted affine algebra

Let $G(M)$ now be $G(S^1) = Map(S^1, G)$, the group of smooth mappings (loops) $z \mapsto g(z)$ of the circle $S^1 = \{z \in C|\ |z| = 1\}$ into a simple, compact and connected finite-dimensional Lie group G. The group structure is defined by the pointwise multiplication of functions,

$$(g'g)(z) = g'(z)g(z) \quad . \tag{9.4.1}$$

As previously discussed, $Map(S^1, G)$ is an infinite-dimensional group, the loop group LG, the elements of which can be represented by

$$g(z) = exp\, \alpha^a(z) T_a \quad , \quad a = 1, ..., r = dim\, G \quad , \tag{9.4.2}$$

where $T_a = -T_a^\dagger$ are the generators of the finite-dimensional Lie algebra $\mathscr{G}$, $[T_a, T_b] = C_{ab}^c T_c$. For elements near the identity,

$$g(z) \simeq 1 + \alpha^a(z) T_a \quad . \tag{9.4.3}$$

Making a Laurent expansion of $\alpha^a(z)$ on the circle,

$$\alpha^a(z) = \sum_{n=-\infty}^{\infty} \alpha^a_{-n} z^n \quad , \tag{9.4.4}$$

expression (9.4.3) reads

$$g(z) \simeq 1 + \sum_{n=-\infty}^{\infty} \alpha^a_{-n} T_a z^n = 1 + \sum_{n=-\infty}^{\infty} \alpha^a_{-n} T_a^n \quad , \quad T_a^n \equiv T_a z^n \quad , \tag{9.4.5}$$

where T_a^n are the generators of the algebra $\mathscr{G}(S^1)$. We may now write the commutation relations of the Lie algebra in terms of the generators T_a^m. The commutators of the finite-dimensional $\mathscr{G}$ then imply

$$[T_a^m, T_b^n] = C_{ab}^c T_c^{m+n} \quad . \tag{9.4.6}$$

Eqs. (9.4.6) are the defining relations of the loop algebra associated with $\mathscr{G}$, that is the algebra $\mathscr{LG} = Map(S^1, \mathscr{G})$ of the loop group LG. The original finite-dimensional Lie algebra $\mathscr{G}$ is reproduced by the generators T_a^0; they correspond to the generators of the group of the constant maps $S^1 \to G$, which is isomorphic to G. With the previous conventions, $T_a^{m\dagger} = -T_a^{-m}$ since, z being of unit modulus, $z^* = z^{-1}$. The reader will notice that the integer superscript is conserved, suggesting that it labels some sort of momentum. This is indeed the case: if we start from the equal-time current algebra commutators (9.3.9) in two dimensions

$$[j_a^0(x), j_b^0(y)] = i\hbar C_{ab}^c \delta(x - y) j_c^0(y) \quad , \tag{9.4.7}$$

then the momenta defined by

$$j_a^m = \int dx\, x^m j_a^0(x) \tag{9.4.8}$$

close into (*i.e.*, realize) the Lie algebra (9.4.6) (with an i in the r.h.s. These expressions are also valid for higher $(D-1)$ spatial dimensions: $x \to \mathbf{x}$, $x^m \to |\mathbf{x}|^m$ and $dx \to d^{(D-1)}x$).

It turns out that the loop algebra may be centrally extended; the result is the *Kac-Moody* (or *untwisted affine*) algebra. The cocycle of the extension satisfies the same general properties used in chapter 6 for the finite-dimensional case. To find the extension two-cocycle here, let us write the commutation relations of the Kac-Moody algebra in the form

$$[\bar{T}_a^m, \bar{T}_b^n] = C_{ab}^c \bar{T}_c^{m+n} + i\omega(T_a^m, T_b^n)\Xi \qquad (9.4.9)$$

where ω is the two-cocycle of the extension to be determined and the central generator Ξ is antihermitian. This implies (see, e.g., eq. (6.3.9))

$$\omega([T_a^m, T_b^n], T_c^s) + \omega([T_b^n, T_c^s], T_a^m) + \omega([T_c^s, T_a^m], T_b^n) = 0 \qquad (9.4.10)$$

which in turn implies

$$C_{ab}^d \omega(T_d^{m+n}, T_c^s) + C_{bc}^d \omega(T_d^{n+s}, T_a^m) + C_{ca}^d \omega(T_d^{s+m}, T_b^n) = 0 ; \qquad (9.4.11)$$

as we know, this relation guarantees that the extended Lie algebra (9.4.9) satisfies the Jacobi identity.

It is not difficult to find that

$$\omega(T_a^m, T_b^n) \propto \delta_{ab}\delta^{m+n,0}m \qquad (m, n \in Z) \quad , \qquad (9.4.12)$$

is a solution of (9.4.11). Indeed, for a semisimple group eq. (1.8.12) holds, so that taking $s = m = 0$ in (9.4.11) and multiplying it by $C^{ba}{}_e$, it is found that

$$\omega(T_a^n, T_b^0) + [C^{ce}{}_a C_{cb}{}^d - C^{dc}{}_a C_{bc}{}^e]\omega(T_d^n, T_e^0) = 0 \qquad (9.4.13)$$

and, using the Jacobi identity,

$$\omega(T_a^n, T_b^0) = C_{ab}{}^c C_c{}^{de}\omega(T_d^n, T_e^0) \quad . \qquad (9.4.14)$$

This means that, with the coboundary generated by $b(T_a^n) = C_a{}^{de}\omega(T_d^n, T_e^0)$, it is possible to eliminate $\omega(T_a^n, T_b^m)$ whenever one of the indices m, n is zero (cf. section 6.4) since

$$\omega'(T_a^n, T_b^0) = \omega(T_a^n, T_b^0) - C_{ab}{}^c b(T_c^n)$$
$$= \omega(T_a^n, T_b^0) - C_{ab}{}^c C_c{}^{de}\omega(T_d^n, T_e^0) = 0 \quad , \qquad (9.4.15)$$

by virtue of (9.4.14). Using this fact we find, setting $s = 0$ in (9.4.11), that

$$C_{bc}^d \omega(T_d^n, T_a^m) + C_{ca}^d \omega(T_d^m, T_b^n) = 0 \quad , \qquad (9.4.16)$$

i.e., that $\omega(T_a^m, T_b^n)$ is an invariant tensor under the adjoint representation of G. Since G is compact and simple, we conclude that

$$\omega(T_a^m, T_b^n) = \delta_{ab}\omega^{m,n} \quad , \qquad (9.4.17)$$

where $\omega^{m,n} = -\omega^{n,m}$ is such that

$$\omega^{m+n,s} + \omega^{n+s,m} + \omega^{s+m,n} = 0 \qquad (9.4.18)$$

and $\omega^{m,0} = 0$. Taking $s = 1$, the following two equations are easily deduced:

$$\omega^{m,n} = \omega^{m-1,n+1} + \omega^{1,m+n-1} \quad , \tag{9.4.19}$$
$$\omega^{m,n} = \omega^{m+1,n-1} - \omega^{1,m+n-1} \quad .$$

Iterating either of these equations to end up with $\omega^{0,n+m} = 0$ and summing up the expressions obtained we find

$$\omega^{m,n} = m\omega^{1,m+n-1} \quad . \tag{9.4.20}$$

Re-inserting this into (9.4.18), it is found that $(m + n + s)\omega^{1,m+n+s-1} = 0$, so that $q\omega^{1,q-1} = 0$. Thus, $\omega^{1,q-1} = \delta^{q,0}\omega^{1,-1}$, and then

$$\omega^{m,n} = m\delta^{m+n,0}\omega^{1,-1} \quad , \tag{9.4.21}$$

together with (9.4.17), reproduces (9.4.12). Moreover, eq. (9.4.12) defines a non-trivial two-cocycle since a redefinition of T_a^m of the form

$$\bar{T}'^m_a = \bar{T}^m_a + \eta^m_a \Xi \tag{9.4.22}$$

does not permit us to remove the central term in (9.4.9). Thus the elements of the second cohomology group of $Map(S^1, G)$ are characterized by a number k; $H^2_0(LG, R)$ is a one-dimensional space, and the *Kac-Moody algebra* commutation relations may be written replacing $i\Xi$ by 1 as

$$[\bar{T}^m_a, \bar{T}^n_b] = C^c_{ab}\bar{T}^{m+n}_c - k\delta_{ab}\delta^{m+n,0}m \quad . \tag{9.4.23}$$

Clearly, once there is no risk of confusing the generators of (9.4.23) with those of the unextended Lie algebra (9.4.6), the bars may be dropped. The central term is zero if both $m, n \in N$. Expression (9.4.23) also checks Whitehead's lemma (Theorem 6.5.2): if $m = n = 0$, the algebra is that of the simple Lie algebra $\mathscr{G}$ and the central term becomes zero. Following the terminology of previous chapters we could denote the above extension (9.4.23) by $\mathscr{G}(S^1)_{(k)}$, the subscript k making reference to the parameter of the specific extension[*].

The present expressions for both the Kac-Moody algebra (9.4.23) and the unextended (9.4.6) loop algebra do not seem to correspond to the general structure of a current algebra, as one might expect from the original definition. This may be made explicit by introducing a field (a

---

[*] Notice that if G were abelian (and hence not semisimple) the second term in (9.4.23) would still satisfy the appropriate properties to define a central extension. The resulting extended algebra

$$[T^m_a, T^n_b] = k\delta_{ab}\delta^{m+n,0}m$$

is sometimes called the *infinite-dimensional Weyl-Heisenberg* algebra.

'current') J_a by

$$J_a(x) = \frac{\hbar}{L} \sum_{n=-\infty}^{\infty} T_a^{-n} e^{i2\pi nx/L} \quad , \tag{9.4.24}$$

which instead of depending on z (the unit circle) is defined on the real line, being periodic with period L, $J_a(x) = J_a(x + L)$ (we could also use $\varphi = 2\pi x/L$). If we now compute the commutation relations of $J_a(x)$ and $J_b(y)$ (it is convenient to set $y = 0$ to simplify) we obtain, using (9.4.23),

$$[J_a(x), J_b(0)] = \left(\frac{\hbar}{L}\right)^2 [\sum_{m=-\infty}^{\infty} T_a^{-m} e^{i2\pi mx/L}, \sum_{n=-\infty}^{\infty} T_b^{-n}]$$

$$= \left(\frac{\hbar}{L}\right)^2 \sum_{m,n} \exp(i2\pi mx/L)[C_{ab}^c T_c^{-(m+n)} - k\delta_{ab}\delta^{m+n,0}(-m)]$$

$$= \left(\frac{\hbar}{L}\right)^2 \sum_{m} \exp(i2\pi mx/L) \sum_{l} \exp(i2\pi lx/L) C_{ab}^c T_c^{-l}$$

$$+ \left(\frac{\hbar}{L}\right)^2 k\delta_{ab} \sum_{m,n} \exp(i2\pi mx/L)\delta^{m+n,0} m$$

$$= \left(\frac{\hbar}{L}\right) \sum_{m} \exp(i2\pi mx/L) C_{ab}^c J_c(x) - \left(\frac{\hbar}{L}\right)^2 k\delta_{ab} \sum_{m} m\exp(i2\pi mx/L) .$$

$$\tag{9.4.25}$$

From the Poisson summation formula,

$$\frac{1}{L} \sum_{m=-\infty}^{\infty} \exp(i2\pi mx/L) = \sum_{m=-\infty}^{\infty} \delta(x - mL) = \delta(x) \quad , \tag{9.4.26}$$

we get, differentiating with respect to x, that

$$\frac{2\pi i}{L^2} \sum_{m} m\exp(i2\pi mx/L) = \delta'(x) \quad . \tag{9.4.27}$$

Hence, using this in (9.4.25) we obtain

$$[J_a(x), J_b(0)] = \hbar C_{ab}^c J_c(0)\delta(x) - i\frac{\hbar^2}{2\pi}k\delta_{ab}\delta'(x) \quad , \tag{9.4.28}$$

where the fact that the (constant, or c-number) Schwinger term is of order $\hbar^2$ should be noticed. In this form (but with $\hbar = 1$) the Kac-Moody algebra will appear again in section 10.11.

In a field theory it is convenient to have hermitian current densities $j_a^0 \equiv j_a = iJ_a$. This is easily remedied by taking hermitian rather than anti-hermitian T_a and has the effect of adding an i in the C_{ab}^c term and changing the sign of the other. Restoring the argument y, eq. (9.4.28) finally reads

$$[j_a(x), j_b(y)] = i\hbar C_{ab}^c j_c(x)\delta(x - y) + i\frac{\hbar^2}{2\pi} k\delta_{ab}\partial_x\delta(x - y) \quad . \qquad (9.4.29)$$

We may consider eq. (9.4.29) as an example of current algebra relations in a two-dimensional spacetime theory in which space is periodic. The use of eq. (9.4.26) implied that the delta function and its derivative are being treated as periodic functions with period L. Nevertheless, since the period does not appear in the space arguments (x, y), these expressions can also be considered for $L \to \infty$, which corresponds to the infinite radius limit.

9.5 The Virasoro algebra

Another interesting example which illustrates both the techniques of Lie algebra cohomology (chapter 6) and the peculiarities of the infinite-dimensional case is provided by the Virasoro algebra. This Lie algebra plays an important rôle in physics where it appears associated with the conformal group in two dimensions. The Virasoro algebra is a central extension of the Lie algebra of the group $Diff\, S^1$ or 'Witt algebra'. The extensions are characterized by a parameter c, so that the name Virasoro algebra actually refers to a class of isomorphic Lie algebras corresponding to the different values of c. We wish to derive the Virasoro algebra starting from the Witt algebra.

The Lie algebra associated with the infinite-dimensional group $Diff\, S^1$ of the orientation preserving diffeomorphisms of S^1 coincides with the algebra of vector fields on the circle $\mathscr{X}(S^1)$ (the notation $Vec\, S^1$ is also used), which is the model space for $Diff\, S^1$. The elements of $Diff\, S^1$ are the diffeomorphisms $z \mapsto G(z)$, $z \in S^1$, with composition law

$$G''(z) = (G' \circ G)(z) \equiv G'(G(z)) \quad . \qquad (9.5.1)$$

As in the Kac-Moody case, we can write $z = e^{i\varphi}$ and consider instead the mappings $\varphi \mapsto G(\varphi)$ such that $G(\varphi) = G(\varphi + 2\pi)$. Then, again extending eq. (1.1.14) to the infinite-dimensional case, the generators $X^L(\varphi)$ are obtained as follows:

$$
\begin{aligned}
X^L(\varphi) &= \oint d\tilde{\varphi} \frac{\delta G''(\tilde{\varphi})}{\delta G(\varphi)} \bigg|_{\substack{G(\varphi)=\varphi \\ G'(\varphi)=G(\varphi)}} \frac{\delta}{\delta G(\tilde{\varphi})} \\
&= \oint d\tilde{\varphi} \frac{\partial G'(G(\tilde{\varphi}))}{\partial G(\tilde{\varphi})} \frac{\delta G(\tilde{\varphi})}{\delta G(\varphi)} \bigg|_{\substack{G(\varphi)=\varphi \\ G'(\varphi)=G(\varphi)}} \frac{\delta}{\delta G(\tilde{\varphi})} \\
&= \oint d\tilde{\varphi} \frac{\partial G'(\tilde{\varphi})}{\partial \tilde{\varphi}} \delta(\varphi - \tilde{\varphi}) \bigg|_{G'(\varphi)=G(\varphi)} \frac{\delta}{\delta G(\tilde{\varphi})} \\
&= \partial_\varphi G(\varphi) \frac{\delta}{\delta G(\varphi)} \quad .
\end{aligned}
\tag{9.5.2}
$$

It is simple to derive the algebra commutators for the generators $X^L(\varphi)$ of $Diff\,S^1$:

$$
\begin{aligned}
[X^L(\varphi), X^L(\tilde{\varphi})] &= \partial_\varphi G(\varphi) \partial_{\tilde{\varphi}} \delta(\varphi - \tilde{\varphi}) \frac{\delta}{\delta G(\tilde{\varphi})} - \varphi \leftrightarrow \tilde{\varphi} \\
&= X^L(\varphi) \partial_{\tilde{\varphi}} \delta(\varphi - \tilde{\varphi}) - X^L(\tilde{\varphi}) \partial_\varphi \delta(\varphi - \tilde{\varphi}) \quad ,
\end{aligned}
\tag{9.5.3}
$$

where in the second equality of (9.5.3), the distributional identity

$$
a(x)\partial_y \delta(x - y) = -a(y)\partial_x \delta(x - y) + a'(x)\delta(x - y)
\tag{9.5.4}
$$

has been used.

It is customary to take advantage of the fact that the functions $G(\varphi)$ can be expanded in Fourier series and write the Lie algebra (9.5.3) in a discrete basis. Let us define

$$
L^n = -i \oint d\varphi z^n X^L(\varphi) \quad .
\tag{9.5.5}
$$

In terms of the L^n, the algebra $\mathscr{X}(S^1)$ is written

$$
\begin{aligned}
[L^m, L^n] &= -\oint d\varphi \oint d\tilde{\varphi} z^m \tilde{z}^n [X^L(\varphi), X^L(\tilde{\varphi})] \\
&= -\oint d\varphi \oint d\tilde{\varphi} z^m \tilde{z}^n \{ X^L(\varphi) \partial_{\tilde{\varphi}} \delta(\varphi - \tilde{\varphi}) - X^L(\tilde{\varphi}) \partial_\varphi \delta(\varphi - \tilde{\varphi}) \} \\
&= -i \oint d\varphi \{ -z^m n z^n X^L(\varphi) + z^n m z^m X^L(\varphi) \} \\
&= -i \oint d\varphi z^{m+n} X^L(\varphi) (m - n) \quad ,
\end{aligned}
$$

i.e.,
$$
\tag{9.5.6}
$$

$$
[L^m, L^n] = (m - n) L^{m+n} \quad , \quad L^{m\dagger} = L^{-m} \quad .
\tag{9.5.7}
$$

This is the most common form of $\mathscr{X}(S^1)$. The realization of the generators L^n acting on functions f on the circle is given by

$$L^n = -z^{n+1}\frac{d}{dz} = ie^{in\phi}\frac{d}{d\phi} \quad , \quad n \in Z \quad . \tag{9.5.8}$$

The central extensions of (9.5.7) appear as the result of quantization in many physical examples. Proceeding in the usual manner, a central extension of (9.5.7) will have the structure

$$[L^m, L^n] = (m-n)L^{m+n} + \omega(L^m, L^n) \equiv (m-n)L^{m+n} + C_{m,n} \quad , \tag{9.5.9}$$

where the additional central generator is represented by 1, and $C_{m,n}$ are the additional structure constants (where the superscript corresponding to the central generator is omitted). The two-cocycle condition now reads (see again eq. (6.3.9))

$$(m-n)C_{m+n,p} + (n-p)C_{n+p,m} + (p-m)C_{m+p,n} = 0 \quad . \tag{9.5.10}$$

The solution of this equation may be simplified by adding a coboundary to $C_{m,n}$. Specifically, a coboundary generated by $b(m)$ is given by (see eq. (6.4.7))

$$C_{m,n}^{cob} = -b(m+n)(m-n) \quad . \tag{9.5.11}$$

Taking this into account, we see that, with $b(m) = \frac{1}{m}C_{m,0}$ when $m \neq 0$ and $b(0) = \frac{1}{2}C_{1,-1}$, $C'_{m,n} = C_{m,n} + C_{m,n}^{cob}$ is given by

$$C'_{0,n} = C_{0,n} + \frac{1}{n}C_{n,0}\, n = 0 \quad (n \neq 0),$$

$$C'_{1,-1} = C_{1,-1} - \frac{1}{2}C_{1,-1}\, 2 = 0 \quad . \tag{9.5.12}$$

Thus, eq. (9.5.10) can be solved setting $C_{0,n} = 0 = C_{1,-1}$ without loss of generality. Particularizing eq. (9.5.10) to the case $p = 0$, the condition

$$(n+m)C_{m,n} = 0 \tag{9.5.13}$$

is obtained. This means that $C_{m,n} = 0$ unless $m = -n$, so that

$$C_{m,n} = c(m)\delta_{m,-n} \quad , \quad c(-m) = -c(m) \quad , \tag{9.5.14}$$

where the last equation follows from the antisymmetry of $C_{m,n}$.
   To find the expression for $c(m)$ it suffices to put $p = -(m+1)$, $n = 1$ in (9.5.10), and use (9.5.14):

$$(m-1)c(m+1)\delta_{m+1,m+1} + (2+m)c(-m)\delta_{-m,-m} = 0 \quad , \tag{9.5.15}$$

$$(m-1)c(m+1) - (m+2)c(m) = 0 \quad .$$

This is a difference equation, which is not difficult to solve. Since $c(0) = 0$ and $c(1) = 0$ because $C_{0,0} = 0 = C_{1,-1}$, the first non-zero value is $c(2)$, so that

$$c(m+1) = \frac{m+2}{m-1}c(m) \quad , \quad m \geq 2 \quad , \tag{9.5.16}$$

and

$$c(m+1) = \frac{(m+2)!}{3!(m-1)!}c(2) = \frac{1}{12}(m+2)(m+1)mc \quad , \tag{9.5.17}$$

with $c(2) \equiv c/2$. Then, finally,

$$c(m) = \frac{1}{12}(m+1)m(m-1)c = \frac{1}{12}m(m^2 - 1)c \quad . \tag{9.5.18}$$

By substituting (9.5.18) in (9.5.10), it can be checked that this is indeed a cocycle.

Using (9.5.18), the Virasoro algebra is given by

$$[L^m, L^n] = (m-n)L^{m+n} + \frac{c}{12}m(m^2 - 1)\delta_{m,-n} \quad ; \tag{9.5.19}$$

the central charge in the Virasoro algebra is referred to as the *conformal anomaly* (section 9.7).

It remains to be shown that the extension defined by (9.5.18) is non-trivial. If the extension were trivial, $C_{m,n}$ would be a coboundary, and there would exist a $b(m)$ for which

$$\frac{c}{12}m(m^2 - 1)\delta_{m,-n} = -b(m+n)(m-n) \quad . \tag{9.5.20}$$

But this is not possible because for e.g. $m+n=0$ it would imply $\frac{c}{12}m(m^2 - 1)=-b(0)2m$, a condition that obviously cannot be fulfilled for all $m \in N$. Notice, however, that the term $-\frac{c}{12}m$ in (9.5.20) is a coboundary, since it can be generated by $b(m) = \frac{c}{24}\delta_{m,0}$. This means that the elements of the cocycle class of a given c are of the form $\frac{1}{12}(cm^3 - c'm)$; the above is the representative obtained by setting $c' = c$. The fact that the second cohomology group of $Diff\,S^1$ is finite- (in fact one-) dimensional is a particular case of a more general result, due to Gelfand and Fuks: Let M be a compact oriented manifold, $\mathscr{X}(M)$ the algebra of the smooth vector fields on M, and let $H^q(\mathscr{X}(M), R)$ be the q-th cohomology group. Then $H^q(\mathscr{X}(M), R)$ is finite-dimensional for any q.

The Virasoro algebra contains the finite-dimensional subalgebra generated by L^0, L^1, and L^{-1},

$$[L^0, L^1] = -L^1 \quad , \quad [L^0, L^{-1}] = L^{-1} \quad , \quad [L^1, L^{-1}] = 2L^0 \quad . \tag{9.5.21}$$

This subalgebra corresponds to the $SU(1,1)$ subgroup of $Diff\,S^1$ of the Möbius transformations

$$z \mapsto G(z) = \frac{az+b}{b^*z+a^*} \quad , \quad |a|^2 - |b|^2 = 1 \tag{9.5.22}$$

($su(1,1) \approx sl(2,R) \approx so(2,1)$), which cover S^1 once positively as z goes around S^1 positively. The fact that $C_{1,-1}$ and $C_{0,1}$ can be eliminated by a coboundary is again consistent with Whitehead's lemma because the subalgebra (9.5.21) admits no non-trivial central extensions.

In the basis of the $X^L(\varphi)$ the Virasoro algebra is given by

$$[X^L(\varphi), X^L(\tilde{\varphi})] = X^L(\varphi)\partial_{\tilde{\varphi}}\delta(\varphi - \tilde{\varphi}) - X^L(\tilde{\varphi})\partial_\varphi\delta(\varphi - \tilde{\varphi})$$
$$+ \frac{i}{24\pi}(c\partial_\varphi^3\delta(\varphi - \tilde{\varphi}) + c'\partial_\varphi\delta(\varphi - \tilde{\varphi})) \quad , \tag{9.5.23}$$

so that the r.h.s. of (9.5.3) acquires a central term for which

$$C(\varphi, \tilde{\varphi}) = \frac{i}{24\pi}(c\partial_\varphi^3\delta(\varphi - \tilde{\varphi}) + c'\partial_\varphi\delta(\varphi - \tilde{\varphi})) \tag{9.5.24}$$

(this expression will be derived in the next section). As a consistency check, let us here recover $C_{m,n}$ from it using (9.5.5):

$$C_{m,n} = \frac{i}{24\pi}(-1)\oint d\varphi \oint d\tilde{\varphi}[c\partial_\varphi^3\delta(\varphi - \tilde{\varphi}) + c'\partial_\varphi\delta(\varphi - \tilde{\varphi})]z^m\tilde{z}^n$$
$$= \frac{-i}{24\pi}\oint d\varphi((-i)(-1)cm^3 + (-1)ic'm)z^{m+n} \tag{9.5.25}$$
$$= \frac{1}{12}(cm^3 - c'm)\delta_{m+n,0} \quad .$$

Note that, if instead of considering the generators $X^L(\varphi)$ of (9.5.23) we consider elements of the Virasoro algebra of the form $\eta_1 \cdot X$, $\eta_2 \cdot X$ as in (9.2.7), we immediately find that the central c-term in the commutator is proportional to

$$\int d\varphi[(\partial_\varphi^3\eta_1)\eta_2 - \eta_1(\partial_\varphi^3\eta_2)] \quad \text{or} \quad \int d\varphi \begin{vmatrix} \partial_\varphi\eta_1 & \partial_\varphi\eta_2 \\ \partial_\varphi^2\eta_1 & \partial_\varphi^2\eta_2 \end{vmatrix} \quad , \tag{9.5.26}$$

where the determinant is obtained by a partial integration (in this last form, the Gelfand-Fuks Lie algebra two-cocycle (9.5.26) generalizes easily to the three-cocycle by using a third order determinant).

Let us comment on the relationship between the Virasoro algebra and the Kac-Moody algebra for a Lie group G. They can be combined into a semidirect product, which arises naturally in some physical examples. Let $G(S^1)$ be the group of mappings $S^1 \to G$ with elements $g(z)$ and composition law (9.4.1). The semidirect product of $Diff\,S^1$ by $G(S^1)$ is then given by the elements $\tilde{g}(z) = (g(z), G(z))$ with the group law

$$\tilde{g}''(z) = \tilde{g}'(z)\tilde{g}(z) = [g'(G(z))g(z), (G' \circ G)(z)] \quad . \tag{9.5.27}$$

To compute its Lie algebra, it suffices to note that the generators $X^L(\varphi)$ are now given by

$$X^L(\varphi) = \partial_\varphi G(\varphi)\frac{\delta}{\delta G(\varphi)} + \partial_\varphi g^a(\varphi)\frac{\delta}{\delta g^a(\varphi)} \tag{9.5.28}$$

(cf. (9.5.2)). Since, by eq. (9.2.4), the generators of the Kac-Moody algebra $X_a(\varphi)$ are given by $X^L_{(a)}(\varphi) = X^{Lb}_{(a)}(g(\varphi))\frac{\partial}{\partial g^b(\varphi)}$, it is simple to find, by using the identity

$$b(y)\partial_y\delta(x - y) = b(x)\partial_y\delta(x - y) - \partial_x b(x)\delta(x - y) \quad, \tag{9.5.29}$$

the mixed commutator

$$[X^L(\varphi), X^L_a(\tilde{\varphi})] = -\partial_\varphi\delta(\varphi - \tilde{\varphi})X^L_a(\varphi) \quad. \tag{9.5.30}$$

In the discrete bases L^m (eq. (9.5.5)), $T^n_a = \frac{L}{2\pi\hbar}\int_0^L d\varphi\, X^L_a(x)e^{i2\pi nx/L}$ (we use here $\varphi = 2\pi x/L$ rather than x), (9.5.30) is written

$$
\begin{aligned}
[L^m, T^n_a] &= [-i\oint d\varphi\, z^m X^L(\varphi), \left(\frac{L}{2\pi\hbar}\right)\oint d\tilde{\varphi}X^L_a(\tilde{\varphi})\tilde{z}^n] \\
&= (-i)\left(\frac{L}{2\pi\hbar}\right)\oint d\varphi\oint d\tilde{\varphi}z^m\tilde{z}^n(-\partial_\varphi\delta(\varphi - \tilde{\varphi})X^L_a(\varphi)) \\
&= \frac{-iL}{2\pi\hbar}\oint d\varphi z^n\partial_\varphi(z^m X^L_a(\varphi)) = -\frac{L}{2\pi\hbar}\oint d\varphi nz^{n+m}X^L_a(\varphi) \\
&= -nT^{m+n}_a \quad.
\end{aligned}
\tag{9.5.31}
$$

On the space of functions $f : S^1 \to V$, where V is a vector space, a realization of the Lie algebra (9.4.6), (9.5.7) and (9.5.31) is obtained by adding to the generators of (9.5.8) the T^n_a of (9.4.5).

The extensions of the Kac-Moody and $\mathcal{X}(S^1)$ algebras are consistent with the above semidirect structure: due to the central nature of the additional generators, the only Jacobi identity that has to be checked is

$$[[T^m_a, T^n_b], L^p] + [[L^p, T^m_a], T^n_b] + [[T^n_b, L^p], T^m_a] = 0 \quad; \tag{9.5.32}$$

this is the only combination that gives a central term after computing the two successive Lie brackets. For an extension $\omega(T^m_a, T^n_b)$ of the Kac-Moody algebra to be consistent, it is then necessary that

$$-m\omega(T^{m+p}_a, T^n_b) + n\omega(T^{n+p}_b, T^m_a) = 0 \quad, \tag{9.5.33}$$

a condition that is fulfilled by the solutions (9.4.12). Thus, the *semidirect product of the Virasoro by the extended Kac-Moody algebra* reads, for T^m_a

satisfying $T_a^m = T_a^{-m\dagger}$ (rather than $-T_a^{-m\dagger}$ (eq. (9.4.23)),

$$[T_a^m, T_b^n] = iC_{ab}^c T_c^{m+n} + k\delta_{ab}\delta^{m+n,0}m \quad ,$$

$$[L^m, T_a^n] = -nT_a^{m+n} \quad , \qquad\qquad (9.5.34)$$

$$[L^m, L^n] = (m-n)L^{m+n} + \frac{c}{12}m(m^2-1)\delta_{m,-n} \quad .$$

In physical problems, the Virasoro and Kac-Moody algebras appear in such a way that they are not independent. For instance, in the Wess-Zumino-Witten (WZW) model, to be mentioned in section 10.11, the currents realize two Kac-Moody algebras, and the components of the stress-energy tensor close into two Virasoro algebras; this is the case in any conformal field theory in two dimensions. The Virasoro algebra is obtained from the Kac-Moody algebra generators in terms of (normal-ordered) bilinears of the currents. In fact starting from a Kac-Moody algebra it is possible to construct the semidirect product with a Virasoro algebra in this way. This procedure is called the Sugawara construction, which we now sketch briefly.

Consider a Kac-Moody algebra for a simple group G. In its enveloping algebra, take the following normal ordering:

$$: T_a^n T_b^m : \quad = \begin{cases} T_a^n T_b^m & \text{for } n \leq m \quad , \\ \frac{1}{2}(T_a^n T_b^m + T_b^m T_a^n) & \text{for } n = m \quad , \\ T_b^m T_a^n & \text{for } n > m \quad , \end{cases} \qquad (9.5.35)$$

and then define

$$L^n = \frac{1}{2k+Q} \sum_{a=1}^{\dim G} \sum_{k\in Z} : T_a^k T_a^{n-k} : \quad , \qquad (9.5.36)$$

where Q is the eigenvalue of the Casimir operator in the adjoint representation of $\mathscr{G}$ ($C_{ib}^c C_{jc}^b = Q\delta_{ij}$) and k is the constant defining the extension for (9.4.12). Then it may be seen that the operators L^n and T_a^n obey (9.5.31) and the L^n's alone close into a Virasoro algebra with c given by

$$c = \frac{2k\dim G}{Q+2k} \quad . \qquad (9.5.37)$$

9.6 Chevalley-Eilenberg cohomology on
Diff S^1 and two-dimensional gravity

We have seen in previous chapters the relevance of the Chevalley-Eilenberg approach to Lie algebra cohomology in the construction of physical actions. As a further example, we consider the case of *Diff S^1* and construct first the CE two-cocycle (see section 6.7) which determines the (Virasoro) extension of its Lie algebra. We shall see that this CE cocycle is given by a closed invariant functional two-form on *Diff S^1*. The corresponding potential one-form gives a field theory model (as *e.g.* in section 8.8), which here corresponds to the induced two-dimensional quantum gravity action of Polyakov. In other words, the action is given by a Wess-Zumino term on $Diff\,S^1$.

The first step in the formal derivation of this result is to calculate the invariant functional differential forms on $Diff\,S^1$. We choose the right-invariant ones because this corresponds in the field theory to invariance under reparametrizations of the φ variable (the fields will depend on t and φ, $G = G(t, \varphi)$, see below). Alternatively we might change (9.5.1) to read $G''(\varphi) = G(G'(\varphi))$; then the generators below would be the left-invariant ones. The functional vector fields $X^R(\varphi)$ are calculated as in (9.5.2), but now it is convenient to write $G'(G(\tilde{\varphi}))$ as

$$G'(G(\tilde{\varphi})) = \oint d\bar{\varphi}\, G'(\bar{\varphi})\delta(\bar{\varphi} - G(\tilde{\varphi})) \quad, \tag{9.6.1}$$

so that $\delta G'(G(\tilde{\varphi}))/\delta G'(\varphi) = \delta(\varphi - G(\tilde{\varphi}))$. Then

$$X^R(\varphi) = \oint d\tilde{\varphi}\, \frac{\delta G'(G(\tilde{\varphi}))}{\delta G'(\varphi)}\bigg|_{G'(\varphi)=\varphi} \frac{\delta}{\delta G(\tilde{\varphi})} = \oint d\tilde{\varphi}\, \delta(\varphi - G(\tilde{\varphi}))\frac{\delta}{\delta G(\tilde{\varphi})} \quad. \tag{9.6.2}$$

It is now possible to obtain the forms $\omega^R(\varphi)$ by duality, $\omega^R(\varphi)[X^R(\tilde{\varphi})] = \delta(\varphi - \tilde{\varphi})$; this now involves the functional exterior derivative $\mathbf{d}$, $\mathbf{d}G(\varphi)[\frac{\delta}{\delta G(\tilde{\varphi})}] = \delta(\varphi - \tilde{\varphi})$. The result is

$$\omega^R(\varphi) = \oint d\tilde{\varphi}\, \delta(\tilde{\varphi} - G^{-1}(\varphi))\mathbf{d}G(\tilde{\varphi})$$
$$= \oint d\tilde{\varphi}\, \delta(G(\tilde{\varphi}) - \varphi)\partial_{\tilde{\varphi}}G(\tilde{\varphi})\mathbf{d}G(\tilde{\varphi}) \quad, \tag{9.6.3}$$

where in the second line use has been made of the fact that $\delta(\tilde{\varphi}-G^{-1}(\varphi)) = \frac{1}{\partial_\varphi G^{-1}(\varphi)}\delta(\varphi - G(\tilde{\varphi})) = \partial_{G^{-1}(\varphi)}G(G^{-1}(\varphi))\delta(\varphi - G(\tilde{\varphi})) = \partial_{\tilde{\varphi}}G(\tilde{\varphi})\delta(\varphi - G(\tilde{\varphi}))$.

The next task is to find a closed RI two-form that is a non-trivial element of the second CE cohomology group. Such a form will have to be expressed as a combination of the $\omega^R(\varphi)$,

$$\omega = \oint d\varphi \oint d\tilde{\varphi}\, N(\varphi, \tilde{\varphi})\omega^R(\varphi) \wedge \omega^R(\tilde{\varphi}) \quad, \tag{9.6.4}$$

where N can be taken to be antisymmetric and its dependence on φ, $\tilde{\varphi}$ is *not* through G; ω will have to be closed, $\mathbf{d}\omega = 0$. In searching for ω, it is useful to take into account the analogue of the Maurer-Cartan equations, which can be deduced from the algebra of the X^R (the one given in (9.5.3) but with a minus sign in the structure constants), or directly from (9.6.3).

These are given by[*]

$$\mathbf{d}\omega^R(\varphi) = \partial_\varphi \omega^R(\varphi) \wedge \omega^R(\varphi) \quad . \tag{9.6.5}$$

The formula for $\mathbf{d}\omega$ is then easily found to be

$$\mathbf{d}\omega = 2 \oint d\varphi \oint d\tilde{\varphi} \, N(\varphi, \tilde{\varphi}) \partial_\varphi \omega^R(\varphi) \wedge \omega^R(\varphi) \wedge \omega^R(\tilde{\varphi}) \quad . \tag{9.6.6}$$

Let us now look for the combinations $N(\varphi, \tilde{\varphi})$ for which the expression (9.6.6) vanishes. First of all, for $\mathbf{d}\omega$ to vanish, $N(\varphi, \tilde{\varphi})$ has to be local, *i.e.*, a combination of $\delta(\varphi - \tilde{\varphi})$ and its derivatives. The possibility $N(\varphi, \tilde{\varphi}) = \delta(\varphi - \tilde{\varphi})$ is excluded because it is not antisymmetric. The next possibility is

$$N(\varphi, \tilde{\varphi}) = \partial_\varphi \delta(\varphi - \tilde{\varphi}) \quad , \tag{9.6.7}$$

which is antisymmetric. Inserting this into (9.6.6) it is found that the result is zero, so that it gives a closed invariant form. However, this is a trivial CE cocycle since by (9.6.5) it can be written as $\omega = -\mathbf{d}\left(\oint d\varphi \, \omega^R(\varphi) \right)$ which means that ω is the differential of an invariant form. The next possibility involves second derivatives of $\delta(\varphi - \tilde{\varphi})$, and would be of the form $\partial_\varphi^2 \delta(\varphi - \tilde{\varphi}) - \partial_{\tilde{\varphi}}^2 \delta(\varphi - \tilde{\varphi}) = 0$. We thus have to look for non-trivial cocycles using expressions for the antisymmetric $N(\varphi, \tilde{\varphi})$ which contain third order derivatives of $\delta(\varphi - \tilde{\varphi})$. These may be written as

$$N(\varphi, \tilde{\varphi}) \propto \partial_\varphi^3 \delta(\varphi - \tilde{\varphi}) \quad . \tag{9.6.8}$$

Then (9.6.8) gives

$$\omega \propto \oint d\varphi \, \partial_\varphi^2 \omega^R(\varphi) \wedge \partial_\varphi \omega^R(\varphi) \quad , \tag{9.6.9}$$

and again $\mathbf{d}\omega = 0$ ($\mathbf{d}\omega$ is proportional to $\oint d\varphi \partial_\varphi (\partial_\varphi^2 \omega \wedge \partial_\varphi \omega \wedge \omega) = 0$); we shall see immediately that ω gives a non-trivial CE cocycle. If we allow ourselves the possibility of adding the two-coboundary (9.6.7), ω can be written in the form

$$\omega = -\frac{1}{48\pi}\left\{ c \oint d\varphi \, \partial_\varphi^2 \omega^R(\varphi) \wedge \partial_\varphi \omega^R(\varphi) - c' \oint d\varphi \, \partial_\varphi \omega^R(\varphi) \wedge \omega^R(\varphi) \right\} ; \tag{9.6.10}$$

---

[*] Notice that the algebra (9.5.3) may be written as $[X^L(\varphi), X^L(\tilde{\varphi})] = \int d\bar{\varphi} \, C(\varphi, \tilde{\varphi}, \bar{\varphi})X^L(\bar{\varphi})$ with (cf. (9.5.3)) $C(\varphi, \tilde{\varphi}, \bar{\varphi}) = (\partial_{\tilde{\varphi}}\delta(\varphi - \tilde{\varphi}))\delta(\varphi - \bar{\varphi}) - (\partial_\varphi \delta(\varphi - \tilde{\varphi}))\delta(\tilde{\varphi} - \bar{\varphi})$. Then the Maurer-Cartan equation (cf. (1.7.5)) is written $\mathbf{d}\omega(\bar{\varphi}) = \frac{1}{2}\int d\varphi d\tilde{\varphi} C(\varphi, \tilde{\varphi}, \bar{\varphi})\omega(\varphi) \wedge \omega(\tilde{\varphi})$, which gives the next formula for $\mathbf{d}\omega(\varphi)$ once $\bar{\varphi}$ is relabelled as φ.

and will be a non-trivial CE two-cocycle for $c \neq 0$. A potential form β for ω ($\omega = \mathbf{d}\beta$) is

$$\beta = -\frac{1}{48\pi}\left\{ c \oint d\varphi \, \frac{\mathbf{d}\partial_\varphi^2 G(\varphi)}{\partial_\varphi G(\varphi)} - c' \oint d\varphi \, \omega^R(\varphi) \right\} \, . \tag{9.6.11}$$

As expected, β cannot be written in terms of the ω^R alone, so it is not invariant and ω is indeed non-trivial. The structure constants (cf. (9.5.24)) which give the Virasoro extension of the Witt algebra may be recovered from the CE two-cocycle ω as usual:

$$\omega(X^R(\varphi), X^R(\tilde{\varphi}))$$

$$= \frac{-1}{48\pi}\left\{ 2c \oint d\bar{\varphi} \, \partial_{\bar{\varphi}}^2 \delta(\bar{\varphi} - \varphi) \partial_{\bar{\varphi}} \delta(\bar{\varphi} - \tilde{\varphi}) - 2c' \oint d\bar{\varphi} \, \partial_{\bar{\varphi}} \delta(\bar{\varphi} - \varphi) \delta(\bar{\varphi} - \tilde{\varphi}) \right\}$$

$$= \frac{-1}{24\pi}\{ c\partial_\varphi^3 \delta(\varphi - \tilde{\varphi}) + c'\partial_\varphi \delta(\varphi - \tilde{\varphi}) \} \, . \tag{9.6.12}$$

Let us come back to the form β of (9.6.11). Consider a mapping $\phi : R \to Diff\,S^1$, $\phi : t \in R \mapsto G(t) \in Diff\,S^1$, so that $G(t) : \varphi \mapsto G(t)(\varphi) \equiv G(t,\varphi)$. In this way the pull-back $\phi^*(\beta)$ of β to R defines the action

$$\int \phi^* \beta = -\frac{1}{48\pi} \int dt \oint d\varphi \left\{ c \left(2\frac{(\partial_\varphi^2 G)^2 \partial_t G}{(\partial_\varphi G)^3} - \frac{\partial_\varphi^3 G \, \partial_t G}{(\partial_\varphi G)^2} \right) - c' \partial_\varphi G \, \partial_t G \right\} \tag{9.6.13}$$

(where two integrations by parts have been performed) of a field theory model, which is given by a Wess-Zumino term on the group $Diff\,S^1$. If, instead of considering S^1, one takes the variable φ to be defined in R, and reinterprets t and φ as the light-cone coordinates $x^\pm = \frac{1}{\sqrt{2}}(x^0 \pm x^1)$, the action (9.6.13) takes the form, for $c' = 0$,

$$\frac{c}{48\pi} \int d^2x \, \frac{\partial_+ G}{\partial_- G} \left(\frac{\partial_-^3 G}{\partial_- G} - 2\frac{(\partial_-^2 G)^2}{(\partial_- G)^2} \right) \, ; \tag{9.6.14}$$

we shall just add here that this is the Polyakov-Wiegmann action of the two-dimensional induced gravity in the light-cone gauge.

9.7 The conformal algebra

Let us conclude this chapter with a few comments on the conformal group. Consider a flat D-dimensional Minkowski spacetime, with metric $\eta_{\mu\nu}$. The *conformal group* is then defined as the group of coordinate transformations $f : x \mapsto x' = x'(x)$ such that the metric $(f^*g')_{\mu\nu}$, where $g' = \eta_{\mu\nu}dx'^\mu \otimes dx'^\nu$, is given by

$$g_{\mu\nu}(x) = \lambda(x)\eta_{\mu\nu} \, . \tag{9.7.1}$$

Condition (9.7.1) states that

$$(f^*g')_{\rho\sigma} = \frac{\partial x'^\mu}{\partial x^\rho}\frac{\partial x'^\nu}{\partial x^\sigma}\eta_{\mu\nu} = \lambda(x)\eta_{\rho\sigma} \quad . \tag{9.7.2}$$

The physical meaning of (9.7.1) is tantamount to saying that the conformal group preserves the light-like structure. This excludes the conformal group as a symmetry of massive particle equations, the classical trajectories of which are time-like worldlines. If massive particles are included, the condition $\lambda(x) = 1$ must be imposed which restricts the symmetry to the Poincaré subgroup: Maxwell equations are conformally invariant as already proved by Bateman and Cunningham in 1910, but the massive Klein-Gordon equation is not.

For spacetime dimension $D > 2$, the conformal group is finite-dimensional. This is checked by solving eq. (9.7.2) for general infinitesimal transformations of spacetime, i.e. for $x'^\mu = x^\mu + \epsilon^\mu(x)$, which leads to $\partial_\rho\epsilon_\sigma + \partial_\sigma\epsilon_\rho = \lambda'\eta_{\rho\sigma}$ ($\lambda \simeq 1 + \lambda'$) or, taking the trace to obtain $\lambda' = \frac{2}{D}\partial^\mu\epsilon_\mu$,

$$\partial_\rho\epsilon_\sigma + \partial_\sigma\epsilon_\rho = \frac{2}{D}\partial_\mu\epsilon^\mu\eta_{\rho\sigma} \quad . \tag{9.7.3}$$

Eq. (9.7.3) defines ϵ_μ as a *conformal Killing vector* (cf. eq. (1.4.49)). Now, putting $\psi = \partial_\mu\epsilon^\mu$, and differentiating (9.7.3) with respect to x^ν and x^σ, the following equations are obtained:

$$\left(1 - \frac{2}{D}\right)\partial_\nu\partial_\rho\psi + \partial^\sigma\partial_\sigma\partial_\nu\epsilon_\rho = 0 \quad ,$$
$$\left(1 - \frac{2}{D}\right)\partial_\nu\partial_\rho\psi + \frac{1}{D}\eta_{\rho\nu}\partial^\sigma\partial_\sigma\psi = 0 \quad , \tag{9.7.4}$$

where in the second line we have symmetrized in ρ, ν and used (9.7.3). By contracting the indices ρ, ν, we arrive at

$$\left(2 - \frac{2}{D}\right)\partial_\nu\partial^\nu\psi = 0 \quad , \tag{9.7.5}$$

which means that $\partial_\nu\partial^\nu\psi = 0$ except for $D = 1$; in the case $D = 1$, any coordinate transformation is conformal, eq. (9.7.3). Inserting this into (9.7.4), it is seen that $(1 - \frac{2}{D})\partial_\nu\partial_\rho\psi = 0$ for $D > 1$. When $D > 2$, this implies $\partial_\nu\partial_\rho\psi = 0$, i.e. ψ is of the form $\psi = D\rho + 2Dc_\mu x^\mu$. Differentiating (9.7.3) with respect to x^ν, substituting the value of ψ and using (9.7.3) again with the indices ν, σ, we obtain $\partial_\sigma(\partial_\nu\epsilon_\rho - \partial_\rho\epsilon_\nu) = 4(c_\nu\eta_{\rho\sigma} - c_\rho\eta_{\nu\sigma})$ or, after integration,

$$\partial_\nu\epsilon_\rho - \partial_\rho\epsilon_\nu = 4(c_\nu x_\rho - c_\rho x_\nu) + 2a_{\rho\nu} \tag{9.7.6}$$

for some constant antisymmetric tensor $a_{\nu\rho}$. Adding (9.7.6) to (9.7.3) and integrating again we find

$$\epsilon^\mu = a^\mu + a^\mu{}_\nu x^\nu + \rho x^\mu + 2c_\nu x^\nu x^\mu - c^\mu x^2 \quad . \tag{9.7.7}$$

From this expression it is easy to deduce that the generators of the conformal group in $D > 2$ are

$$P_{(\mu)} = \frac{\partial}{\partial x^\mu} \quad , \quad M_{(\mu\nu)} = x_\mu \frac{\partial}{\partial x^\nu} - x_\nu \frac{\partial}{\partial x^\mu} \quad ,$$

$$D = x^\mu \frac{\partial}{\partial x^\mu} \quad , \quad K_{(\mu)} = 2x_\mu x^\nu \frac{\partial}{\partial x^\nu} - x^2 \frac{\partial}{\partial x^\mu} \quad , \qquad (9.7.8)$$

where $P_{(\mu)}$, $M_{(\mu\nu)}$ are the generators of the Poincaré group, D is the generator of the dilatations and $K_{(\mu)}$ generate the special conformal transformations. They close into the conformal Lie algebra

$$[M_{(\mu\nu)}, M_{(\rho\sigma)}] = \eta_{\nu\rho} M_{(\mu\sigma)} + \eta_{\mu\sigma} M_{(\nu\rho)} - \eta_{\mu\rho} M_{(\nu\sigma)} - \eta_{\nu\sigma} M_{(\mu\rho)} \ ,$$

$$[P_{(\mu)}, M_{(\rho\sigma)}] = \eta_{\mu\rho} P_{(\sigma)} - \eta_{\mu\sigma} P_{(\rho)} \quad , \quad [P_{(\mu)}, P_{(\sigma)}] = 0 \quad ,$$

$$[D, P_{(\mu)}] = -P_{(\mu)} \quad , \quad [D, K_{(\mu)}] = K_{(\mu)} \quad , \quad [D, M_{(\mu\nu)}] = 0 \quad ,$$

$$[K_{(\mu)}, P_{(\nu)}] = -2(\eta_{\mu\nu} D + M_{(\mu\nu)}) \quad ,$$

$$[K_{(\mu)}, M_{(\rho\sigma)}] = \eta_{\mu\rho} K_{(\sigma)} - \eta_{\mu\sigma} K_{(\rho)} \quad , \quad [K_{(\mu)}, K_{(\nu)}] = 0 \ .$$

$$(9.7.9)$$

The first two lines of (9.7.9) are the Poincaré subalgebra and the rest correspond to the dilatations and special conformal transformations which in their finite form are given by

$$x'^\mu = \rho x^\mu \quad , \quad x'^\mu = \frac{x^\mu - c^\mu x^2}{1 - 2c \cdot x + c^2 x^2}, \quad \rho \in R \quad , \quad c^\mu \in R^D \quad . \quad (9.7.10)$$

The special conformal transformations map certain points of spacetime to infinity which is not a point of Minkowski space; thus, the conformal group is defined on its compactification. It depends on $\binom{D+2}{2}$ parameters: one dilatation, D special conformal transformations, D spacetime translations and $D(D-1)/2$ Lorentz transformations. Thus, the conformal group has the structure of a (pseudo-) orthogonal group in $D+2$ dimensions. Notice that $[P^{(\mu)} P_{(\mu)}, D] = 2P^2$; this shows that P^2 is not a Casimir of the conformal group and explains why the massive relativistic equations do not have conformal symmetry.

In $D = 2$, however, $(1 - \frac{2}{D})\partial_\nu \partial_\rho \psi = 0$ does not imply $\partial_\nu \partial_\rho \psi = 0$, which was crucial for the finiteness of the group in the $D > 2$ case. In this case every harmonic function ψ, $\partial^\nu \partial_\nu \psi = 0$, determines a solution: the conformal group becomes infinite-dimensional. To see this, it is convenient to write the metric tensor field in terms of light-cone coordinates:

$$g' = \eta_{\mu\nu} dx'^\mu \otimes dx'^\nu = dx'^0 \otimes dx'^0 - dx'^1 \otimes dx'^1 = dx'^+ \otimes dx'^- + dx'^- \otimes dx'^+ \ .$$

$$(9.7.11)$$

The transformation $x \to x'$ is a conformal one if

$$
dx'^+ \otimes dx'^- + dx'^- \otimes dx'^+ = \left\{ \left(\frac{\partial x'^+}{\partial x^+} dx^+ + \frac{\partial x'^+}{\partial x^-} dx^- \right) \right.
$$

$$
\otimes \left(\frac{\partial x'^-}{\partial x^+} dx^+ + \frac{\partial x'^-}{\partial x^-} dx^- \right) + x'^+ \leftrightarrow x'^- \right\}
$$

$$
= \left\{ \frac{\partial x'^+}{\partial x^+} \frac{\partial x'^-}{\partial x^+} dx^+ \otimes dx^+ + \frac{\partial x'^+}{\partial x^+} \frac{\partial x'^-}{\partial x^-} dx^+ \otimes dx^- \right.
$$

$$
+ \frac{\partial x'^+}{\partial x^-} \frac{\partial x'^-}{\partial x^+} dx^- \otimes dx^+ + \frac{\partial x'^+}{\partial x^-} \frac{\partial x'^-}{\partial x^-} dx^- \otimes dx^- + x'^+ \leftrightarrow x'^- \right\}
$$

$$
(9.7.12)
$$

is equal to $\lambda(x)(dx^+ \otimes dx^- + dx^- \otimes dx^+)$. This is satisfied if

$$
\frac{\partial x'^+}{\partial x^-} = 0 = \frac{\partial x'^-}{\partial x^+} \quad . \tag{9.7.13}
$$

We then read from (9.7.12) and (9.7.13) that $\lambda(x^+, x^-) = \frac{\partial x'^+}{\partial x^+} \frac{\partial x'^-}{\partial x^-}$ and that the most general transformations of the conformal group in two dimensions take the form

$$
x'^+ = f(x^+) \quad , \quad x'^- = \tilde{f}(x^-) \quad , \tag{9.7.14}
$$

for *two* independent smooth functions f and $\tilde{f}$ on the variables x^+, x^-: the conformal group can be written as the product of the groups of functions of the light-cone coordinates. For the same reason, the Lie algebra can be written as the direct sum of two commuting infinite-dimensional Lie algebras, each of which turns out to be equivalent to the Witt algebra. This is proved using the same procedure as in (9.5.1)–(9.5.3), due to the similarity between (9.7.14) and the law $x' = G(x)$.

In physical situations such as two-dimensional statistical models with conformal invariance, the WZW model or string theory (which is a conformal field theory on the worldsheet), the algebra obtained through the commutators of the T_{++} and T_{--} components of the stress-energy tensor is not – once more – the two-dimensional conformal algebra that we have described above. Instead of obtaining two copies of the Witt algebra the calculation leads to two copies of the Virasoro algebra, each labelled by a real parameter c. This is why the usual terminology refers to such extensions as *conformal anomalies*. In general, both algebras are extended with different values of c although in conformal field theory they are taken to be the same. Setting them equal to zero, the 'classical' Lie algebra (that of the two-dimensional conformal group) is recovered.

To conclude, we mention that it is possible to study the conformal group by considering complex coordinates $z, \bar{z} \in C$ instead of $x^+, x^- \in R$ and treating them as independent complex variables. Then the metric tensor is

$g = d\bar{z} \otimes dz + dz \otimes d\bar{z}$ and the transformations of the group are $z' = f(z)$ and $\bar{z}' = f(\bar{z})$. An infinitesimal transformation can be written (e.g. for z) as $z' = z - \varepsilon(z)$ and, expanding $\varepsilon(z)$ in a Laurent series around $z = 0$, it is seen that the generators of the action of the group on the functions of S^1 are given by

$$L^n = -z^{n+1}\frac{d^n}{dz^n} \quad , \quad \bar{L}^n = -\bar{z}^{n+1}\frac{d^n}{d\bar{z}^n} \quad , \tag{9.7.15}$$

from which we recover the statement that the two-dimensional conformal algebra is the direct sum of two independent Witt algebras. The generators (9.7.15) are written as in (9.5.8) and the equivalence with the Witt algebra is obvious. To recover the Minkowski (resp. Euclidean) space it is sufficient to restrict the results to the real section of the complex plane $z = x^+$, $\bar{z} = x^-$ (resp. $z = (x^1 + ix^2)/\sqrt{2}$, $\bar{z} = z^*$).

Bibliographical notes for chapter 9

For infinite-dimensional manifolds see, e.g., the book of Lang (1972) and Choquet-Bruhat *et al.* (1982). In section 9.1 we have used the review of Milnor (1984) and the book of Pressley and Segal (1986) to which we refer for a more detailed account. For the structure of *Banach-Lie* groups see Ebin and Marsden (1970). Additional general information about infinite-dimensional Lie algebras and groups can be found in the books of Kac (1985), Cornwell (1989) and in the paper by Kobayashi *et al.* (1985); they contain further references. A collection of relevant articles is given in Goddard and Olive (1988). For the cohomology of infinite-dimensional Lie algebras in general see the book of Fuks (1986), where the references to the original papers of Gelfand, Feigin and Fuks on infinite-dimensional Lie algebras can be found. For the projective representations of $Map(S^1, G)$ and $Diff\, S^1$ see, in particular, Segal (1981). The case of loop groups is also treated in Pressley and Segal (1986) (see also Coquereaux and Pilch (1989)); that of sphere groups is given in Nash *et al.* (1993).

The physical aspects of the Kac-Moody and Virasoro algebras are especially discussed *e.g.* in Dolan (1984) and in the review article of Goddard and Olive (1986) which is included in the reprint volume of the same authors (1988). They are also treated in the book of Mickelsson (1989) and in the review of Ragoucy and Sorba (1992). The geometry of the circle reparametrizations with a fixed point ($Diff\, S^1/S^1$) is described in the lecture of Zumino (1988). The representations of the Virasoro group are discussed by Witten (1988); see also Aldaya and Navarro-Salas (1990). An introduction to current algebras in particle physics can be found in the reprint volume of Adler and Dashen (1968); Schwinger terms were introduced in Schwinger (1959). The modification of the current algebra commutators due to the presence of a Wess-Zumino-Witten term on a compact group is given in de Azcárraga *et al.* (1990).

For an introduction to the ($D=4$) conformal group and field theory see *e.g.* Fulton *et al.* (1962), Kastrup (1966), Mack and Salam (1969) and Ferrara *et al.* (1973). Basic papers in two-dimensional conformal invariance are, *e.g.*, those of Belavin *et al.* (1984), Knizhnik and Zamolodchikov (1984) and Zamolodchikov (1986). A collection of reprints on two-dimensional conformal invariance with a short introduction is provided in Itzykson *et al.* (1988). See also the paper by Jackiw (1990) and the review articles of Olive (1990), Álvarez-Gaumé *et al.* (1990), Ginsparg (1990) and Asorey (1992); these contain further references. A personal account on the history of conformal anomalies is given in Duff (1994). For a complete textbook reference on conformal field theory see Di Francesco *et al.* (1997).

The two-dimensional gravity action was given by Polyakov (1987, 1990); the form given in section 9.6 is that of Alekseev and Shatashvili (1989) which they derived by the coadjoint orbit method. See also Rai and Rodgers (1990), Delius *et al.* (1990) and Aldaya *et al.* (1991).

10

Gauge anomalies

This chapter is devoted to the topological and cohomological properties of abelian and non-abelian chiral anomalies in Yang-Mills theories.

First, the Gribov ambiguity and the appearance of anomalies are related to the non-trivial topology of the *configuration or Yang-Mills orbit space*. This is followed by the explicit path integral calculation of the abelian chiral anomaly in $D = 2p$ dimensions and the non-abelian gauge anomalies (for $D = 2$) by using Fujikawa's method. It is seen how these results may be interpreted in terms of suitable index theorems on spaces of adequate dimensions ($D = 2p$ and $(D + 2)$ respectively). The consistency conditions for the anomalies and the Schwinger terms are interpreted in terms of a cohomology in which the cocycles are valued in $\mathscr{F}[\mathscr{A}]$, the space of functionals of the gauge fields. Then it is shown how the cohomological descent procedure starting from the Chern character forms provides a method for obtaining non-trivial candidates for both the non-abelian anomalies and the Schwinger terms.

The question of the ambiguity of the cohomological descent procedure, which gives rise to different (but cohomologous) expressions for the Schwinger terms, the BRST formulation of the gauge cohomology and the Wess-Zumino-Witten terms are also discussed. At the end, some comments on the possible consistency of anomalous gauge theories are made.

10.1 The group of gauge transformations and the orbit space of Yang-Mills potentials

The requirement that field theories invariant under rigid transformations ($g_i \neq g_i(x)$) of a group G should also be invariant under local ($g_i = g_i(x)$) transformations constitutes the *gauge invariance principle*. This symmetry principle (a result of requiring consistency with the localized field concept which underlies physical theories) is a dynamical one because, besides including the rigid symmetry, it also gives us information about the way the interaction with the gauge fields has to be described. As we know (see the first footnote in section 2.2(b)), in Yang-Mills theories this gauge invariance is achieved by introducing the gauge or Yang-Mills potentials $A^i_\mu(x)$ (i = group index, μ = spacetime index) which transform under the adjoint representation of the (compact) Lie group G, and by replacing the ordinary derivatives ∂_μ by the covariant derivatives $\mathscr{D}_\mu$.

Since the Yang-Mills potentials are themselves subjected to gauge transformations $A \mapsto A^g$, it is important to determine the independent degrees of freedom to avoid overcounting; gauge symmetries are, after all, the expression of redundancies in the description of the Yang-Mills fields. Let $\mathscr{A}$ be the (infinite-dimensional) space of all Yang-Mills potentials A. The action of the gauge group on $\mathscr{A}$ determines orbits which contain the Yang-Mills fields that are connected by a gauge transformation. Two potentials A and A' are in the same orbit if there is a gauge transformation g for which $A' = A^g$; this transformation law also induces an action of the gauge group on the space $\mathscr{F}[A]$ of functionals of the gauge fields. The superfluous degrees of freedom are eliminated by fixing a gauge, *i.e.* by finding a solution A^{g_0} to the gauge fixing equation $f(A^g) = 0$; the A^{g_0} of a given orbit is the potential satisfying the condition prescribed by f. The statement that this can be found uniquely is tantamount to saying that the gauge condition $f(A^g) = 0$ has exactly one solution g_0 for any $A \in \mathscr{A}$; in this case, the generating functional $W[A]$ of the theory projects to the independent degrees of freedom by introducing the Faddeev-Popov determinant and the corresponding ghost fields. In *non-abelian* gauge theories, however, the solution to the equation $f(A^g)$ is not unique: for example, the Coulomb gauge condition does not determine A uniquely (in abelian theories this might also appear to be the case, since $\nabla \cdot \mathbf{A}' = \nabla \cdot \mathbf{A}$ still allows $\mathbf{A}$ and $\mathbf{A}'$ to be gauge-related, $\mathbf{A}' = \mathbf{A} + \nabla\lambda$, with $\Delta\lambda = 0$, but with zero boundary conditions at spatial infinity, the harmonic λ is zero everywhere). This phenomenon, known as the *Gribov ambiguity*, implies that there is no unique intersection between the 'surface' $f(A^g) = 0$ and the orbits A^g. Nevertheless, since the intersections of the orbit with the 'surface' $f(A^g) = 0$ are separated by finite gauge transformations, this ambiguity does not affect the perturbative description of the theory, which relies on an expansion about a classical configuration.

The geometrical picture behind the Gribov ambiguity, as pointed out by Singer, is that of a non-trivial principal bundle, the *bundle of Yang-Mills potentials*, for which there is no global section. The structure group of this bundle is the (infinite-dimensional) group of gauge transformations, the fibres are the orbits of gauge-related potentials and the base is determined by the independent gauge degrees of freedom. Selecting a gauge would correspond to choosing a vector potential in each orbit in a continuous manner, *i.e.* to selecting a particular trivializing section of the Yang-Mills bundle. As we shall see, this is not possible in non-abelian theories. Moreover, this mathematical obstruction also exists for gauges other than the Coulomb gauge. No global gauge fixing is possible on compactified spacetimes; trying to extend the local chart to the whole manifold has the effect that, beyond a certain distance, the gauge condition does not fix the gauge uniquely.

To see this in more detail, let us characterize the *gauge group*. As it was described in section 2.2, a Yang-Mills potential A comes from a connection ω on a principal bundle $P(G, M)$ over spacetime M, and a gauge transformation $g(x)$ relates *e.g.* the local representatives $\sigma^*(\omega) = A$, $\sigma'^*(\omega) = A'$ determined by the pull-back of ω by two local sections σ and σ' over $U \subset M$, eq. (2.1.5). Alternatively, it is also possible to look at a gauge transformation as the change produced in A by a vertical bundle automorphism f of P (f is trivial on the base, $f_b = \mathbf{1}_{|M}$). If $f : P \to P$ is an automorphism (we drop the subscript of f_{tm}) $f^*(\omega)$ is the (gauge) transformed connection. Thus, if $A = \sigma^*(\omega)$ on U, then the new A' is given by $\sigma^*(f^*\omega) = (f \circ \sigma)^*(\omega) = A'$ where, in analogy with the previous situation, we may denote $f \circ \sigma$ by σ'. Both descriptions are equivalent, so that we can adopt the following

(10.1.1) DEFINITION (*Group of gauge transformations*)

The group of gauge transformations is the group $\mathscr{C} = \mathrm{Aut}_v(P)$ *of vertical automorphisms of* $P(G, M)$, *eq. (1.3.25).*

We constructed in section 1.3 the bundle of Lie groups $\widetilde{Ad}(P) = P \times_{\widetilde{AdG}} G$ associated to a principal bundle $P(G, M)$. By extending Proposition 1.3.1 to apply to this associated bundle, we find

(10.1.1) PROPOSITION

The group of gauge transformations is in one-to-one correspondence with

a) *the set of G-functions* $\varphi : P \to G$ (*in this case, equivariant with respect to the adjoint action of G on itself*),

b) *the set of sections* $\Gamma(P \times_{\widetilde{AdG}} G, M)$ *of the associated bundle* $\widetilde{Ad}(P)$ *over M.*

Proof: By virtue of Proposition 1.3.1, we only need to prove the first part. Let f be a gauge transformation, $f \in \mathrm{Aut}_v(P)$, and let its associated G-function φ be defined by $\varphi(p) = g$ where g is uniquely defined by the condition that $f(p) = p\varphi(p) = pg$ (since $f \in \mathrm{Aut}_v(P)$, $f_b = \mathbf{1}_{|M}$, $f(p)$ and

p are necessarily in the same fibre). Moreover, since $f(pg_0) = f(p)g_0 = p\varphi(p)g_0$ is also equal to $pg_0\varphi(pg_0)$, it follows that $\varphi(pg_0) = g_0^{-1}\varphi(p)g_0$ and φ is equivariant under $\widetilde{Ad}\, g_0^{-1}$. Thus, $f \mapsto \varphi$. Conversely, given φ, the mapping $\varphi \mapsto f$ is established by defining $f(p) = p\varphi(p)$, q.e.d.

If $\sigma_\alpha^*(\omega) = A_\alpha$, and $\sigma'_\alpha(x) = f(\sigma_\alpha(x)) = \sigma_\alpha(x)\varphi(\sigma_\alpha(x)) \equiv \sigma_\alpha(x)\varphi_\alpha(x)$, the gauge transformation is given by

$$A'_\alpha(x) = Ad(\varphi_\alpha^{-1}(x))A_\alpha(x) + \varphi_\alpha^{-1}(x)d\varphi_\alpha(x) \quad . \tag{10.1.1}$$

If two local sections are given, $\sigma_\beta(x) = \sigma_\alpha(x)g_{\alpha\beta}(x)$, then $\sigma'_\beta(x) = f(\sigma_\beta(x)) = \sigma_\beta(x)\varphi_\beta(x) = f(\sigma_\alpha(x)g_{\alpha\beta}(x)) = \sigma_\alpha(x)\varphi_\alpha(x)g_{\alpha\beta}(x)$ so that $\varphi_\beta(x) = g_{\alpha\beta}^{-1}(x)\varphi_\alpha(x)g_{\alpha\beta}(x)$ for $x \in U_\alpha \cap U_\beta$.

Having defined the elements of the group $\mathscr{C}$ of gauge transformations as sections in $\Gamma(\widetilde{Ad}(P), M)$ it may be shown that $\mathscr{C}$ is a Baker-Campbell-Hausdorff group, the Lie algebra of which is given by the sections $\Gamma(Ad(P), M)$ of the associated bundle of Lie algebras $Ad(P) = P \times_{Ad\, G} \mathscr{G}$. Thus, we may summarize the above by saying that a gauge transformation may be seen as a section of $\widetilde{Ad}(P)$, and that an infinitesimal gauge transformation is a section of $Ad(P)$. If the bundle $P(G, M)$ is trivial, the group of gauge transformations is given by the group $G(M)$ of the smooth mappings $M \to G$, as was the case in (2.2.6); in the general case, $G(M)$ describes the gauge group locally.

Clearly, inequivalent (*i.e.*, gauge-unrelated) Yang-Mills potentials are projected onto different points of the quotient $\mathscr{A}/\mathscr{C}$; $\mathscr{A}/\mathscr{C}$ is called the *orbit space* of the (Yang-Mills) connections on $P(G, M)$. In general, the action of $\mathscr{C}$ on $\mathscr{A}$ is not free, but with certain technical restrictions it is possible to obtain a free action. This may be achieved, for instance, by restricting $\mathscr{C}$ to being the group of automorphisms preserving a point $p \in P$ or to the group of based gauge transformations (leaving infinity – a point of $M^m = S^m$- fixed); this will be assumed in what follows without changing the notation. Then $\mathscr{A}$ has a principal bundle structure $\mathscr{A}(\mathscr{C}, \mathscr{A}/\mathscr{C})$, the *bundle of Yang-Mills connections*. Fixing a gauge in this bundle (*i.e.*, selecting a representative in each fibre) would imply that there is a global section and that the bundle is trivial. Thus Gribov's result for $G = SU(2)$ may be reformulated by saying that the Coulomb gauge is not a global section.

It is not difficult to see that the bundle of connections is not trivial in general. If it were, we could write

$$\mathscr{A} = \mathscr{C} \times \mathscr{A}/\mathscr{C} \quad . \tag{10.1.2}$$

The functional space of connections $\mathscr{A}$ is an affine space (see below Theorem 2.1.2), and hence contractible; thus, it has no topological invariants.

But (10.1.2) implies that

$$0 = \pi_j(\mathscr{A}) = \pi_j(\mathscr{C}) + \pi_j(\mathscr{A}/\mathscr{C}) \quad , \tag{10.1.3}$$

a relation which cannot be fulfilled since in general $\mathscr{C}$ possesses non-zero homotopy groups (see below): thus, the bundle may be non-trivial and in this case no *continuous* gauge fixing is possible. Since $\mathscr{A}$ is topologically trivial, the topology of $\mathscr{A}/\mathscr{C}$ comes entirely from $\mathscr{C}$.

Let us now look briefly at the topology of the orbit space $\mathscr{A}/\mathscr{C}$, the space of measurable physical fields or configuration space of the Yang-Mills theory. This depends on the topology of the manifold M. The usual asymptotic conditions allow us to replace the m-dimensional Euclidean M by its conformal compactification S^m and we shall consider only this case when doing topological considerations. To have a finite Yang-Mills action it is also assumed that the potentials $A(x)$ become pure gauge at infinity, $A(x) \mapsto g^{-1}(x)dg(x)$, where $g : S^{m-1} \to G$ and S^{m-1} is the sphere at infinity (cf. section 2.8). These mappings fall into homotopy classes $\pi_{m-1}(G)$ which classify the bundles $P_k(G, S^m)$ over S^m (see the third remark after (1.3.19)). If m is even, as will later on be the case, the class k is obtained by computing the Chern class $c_{(m/2)}$ on S^m (cf. section 2.8). This k-dependence is then reflected in the other constructions: $\mathscr{A}$ splits into spaces $\mathscr{A}_k$ of connections on the bundles $P_k(G, S^m)$ with gauge groups $\mathscr{C}_k$ (base point preserving automorphisms of P_k). For a given k, $\mathscr{A}_k(\mathscr{C}_k, \mathscr{A}_k/\mathscr{C}_k)$ is a principal bundle with structure group $\mathscr{C}_k$. The important point, as discussed by Atiyah and Jones, is that the homotopy type of $\mathscr{A}_k/\mathscr{C}_k$ does not make reference to k, $\pi_i(\mathscr{A}_k/\mathscr{C}_k) = \pi_i(\mathscr{A}_0/\mathscr{C}_0)$, and thus we may take $k = 0$. This means that, as far as the homotopy properties are concerned, we may take the trivial bundle $P_0(G, S^m)$ for which $P_0 = G \times S^m$, so that $\mathscr{C}_0 = Map\,(S^m, G) = G(S^m)$ and ignore the subscript k. Since by selecting a $g(x)$ on S^{m-1} by a normalization condition $(g(s_0) = e$ for some fixed point $s_0 \in S^{(m-1)}$ as mentioned) g is uniquely determined by A, the net outcome of their analysis is that there is a mapping

$$\mathscr{A}/\mathscr{C} \to G(S^{m-1}) \tag{10.1.4}$$

which is a homotopy equivalence (the notation $\Omega^n(G)$ for $G(S^n)$, here denoting the based loops or mappings $g : S^n \to G$ with $g(s_0) = e$ is commonly used).

In fact, from the homotopy sequence induced by $\mathscr{A}(\mathscr{C}, \mathscr{A}/\mathscr{C})$ we find that

$$0 = \pi_n(\mathscr{A}) \to \pi_n(\mathscr{A}/\mathscr{C}) \to \pi_{n-1}(\mathscr{C}) \to \pi_{n-1}(\mathscr{A}) = 0 \quad , \tag{10.1.5}$$

from which we read that

$$\pi_j(\mathscr{A}/\mathscr{C}) = \pi_{j-1}(\mathscr{C}) \quad . \tag{10.1.6}$$

On the other hand, the property

$$\pi_j(G(S^m)) = \pi_{j+m}(G) \qquad (10.1.7)$$

means that

$$\pi_j(\mathscr{C}) = \pi_j(G(S^m)) = \pi_{j+m}(G) \quad , \qquad (10.1.8)$$

which checks the homotopy equivalence (10.1.4) since

$$\pi_j(\mathscr{A}/\mathscr{C}) = \pi_j(G(S^{m-1})) = \pi_{j+m-1}(G) \quad . \qquad (10.1.9)$$

In the two- and four-dimensional cases ($M = S^2, S^4$) this gives, respectively,

$$\pi_0(\mathscr{C}) = \pi_1(\mathscr{A}/\mathscr{C}) = \pi_2(G) = 0 \quad , \quad \pi_1(\mathscr{C}) = \pi_2(\mathscr{A}/\mathscr{C}) = \pi_3(G) \quad ,$$
$$\pi_0(\mathscr{C}) = \pi_1(\mathscr{A}/\mathscr{C}) = \pi_4(G) \quad , \quad \pi_1(\mathscr{C}) = \pi_2(\mathscr{A}/\mathscr{C}) = \pi_5(G) \quad .$$
$$(10.1.10)$$

These homotopy groups can be found in section 1.11; in particular $\pi_3(G) = Z$ for a simple Lie group. Other cases are, for instance, $\pi_4(SU(n)) = (Z_2 \ (n = 2), 0 \ (n > 2))$; $\pi_5(SU(n)) = (Z_2 \ (n = 2), Z \ (n \geq 3))$; $\pi_5(O(n)) = 0 \ (n > 6)$. The previous discussion shows that the non-triviality of the orbit space is tied to the non-triviality of the homotopy groups of G. In particular, for an abelian gauge theory in four dimensions $\pi_j(\mathscr{A}/\mathscr{C}) = \pi_{j+3}(U(1)) = 0$: the $U(1)$-gauge bundle of QED is trivial and so there is no Gribov ambiguity in quantum electrodynamics. We shall see in section 10.6 that the non-abelian anomaly of chiral gauge theories in four dimensions may be extracted topologically by looking for non-contractible two-spheres in orbit spaces for which $\pi_2(\mathscr{A}/\mathscr{C}) = \pi_5(G) = Z$. In general, in (even) dimension $m = 2p$ a *sufficient* condition for the existence of a perturbative non-abelian anomaly is that $\pi_2(\mathscr{A}/\mathscr{C}) = \pi_1(\mathscr{C}) = \pi_{2p+1}(G) = Z$ (examples can be found that show that the condition is not always necessary). In four dimensions the only *unsafe* simple Lie algebras are $su(n)$, $n \geq 3$, as follows from (1.11.3) (recall that $so(6) \sim su(4)$, $A_3 \sim D_3$).

If $j = 0$ above, we find $\pi_0(\mathscr{C}) = \pi_{2p}(G)$. For $G = SU(2)$ and $D = 2p = 4$, for instance, we find $\pi_0(\mathscr{C}) = \pi_4(SU(2)) = Z_2$. If, as in this case, $\pi_0(\mathscr{C}) \neq 0$, the gauge group $\mathscr{C}$ is not connected: it contains 'pieces' besides that including the identity element. This produces another type of anomalous behaviour under gauge transformations which cannot be continuously reached from the identity, and results in another type of anomaly, Witten's global (or non-perturbative) anomaly. In contrast with the anomalies to be discussed below, this anomaly is not reflected in the non-conservation of currents of any sort; it leads to a non-perturbative failure of gauge symmetry, and will not be considered further.

10.2 Theory with Dirac fermions: the abelian anomaly and the index theorem

In general, a quantum theory is called *anomalous* if there is an exact symmetry of the classical action which is not preserved as a symmetry of its path integral quantum formulation: there is an obstruction to the 'lifting' of the classical symmetry to the quantum case. Although we shall consider only Yang-Mills theories in this chapter, there are also other types of anomalies such as gravitational or conformal anomalies. In gauge theories there are many types of anomalies depending on the symmetry which is being broken and the properties they satisfy. For instance, in a theory with massless fermions the axial vector Noether current j_5^μ associated with the rigid chiral symmetry of the classical action turns out to be not conserved in the quantum theory. In this case, the (Adler-Bell-Jackiw) anomaly contributes to physical processes ($\pi^0 \to 2\gamma$) and to the solution of the $U(1)$ 'problem' of QCD and, as such, it is not a 'bad' anomaly. When the anomalies affect the gauge symmetry of a theory, however, they make it inconsistent: gauge invariance is an essential ingredient of Yang-Mills theories, since it is the only way known at present to achieve unitarity and renormalizability (in $D = 4$). As a result, quantum theories that are anomalous in this sense are discarded and the anomaly cancellation condition is a useful constraint in the construction of physical models. Anomalies of this type appear in gauge theories in the presence of chiral (Weyl) fermionic fields and satisfy the Wess-Zumino consistency condition (section 10.7). We shall restrict ourselves to these two types of anomalies: the axial abelian $U(1)$ anomaly and the non-abelian anomalies for chiral fermions.

We shall first give an example of the first type of anomalies, namely the $U(1)$ (or *abelian, chiral, or axial*) anomaly in Dirac theory in arbitrary even D dimensions (the restriction to D even is due to the fact that only in even dimensions may chirality be defined). The reasons for doing this, apart from completeness, are the following: first, the abelian anomaly constitutes an immediate application of the index theorem for the twisted spin complex (eq. (2.11.19)) or, equivalently, a field theory proof for it. Secondly, it gives an introductory example of the Fujikawa method that we shall also use to calculate the non-abelian gauge anomaly. Finally, the derivation of abelian anomaly involves a generalization of the Noether argument that was given in section 9.3 to the quantum case which is worth discussing.

In *classical* theories all that matters is the Lagrangian (*e.g.* $\mathcal{L}(A, \psi)$ for a theory involving gauge and fermionic fields ψ). In the *quantum case,* however, the relevant quantity is the non-local functional

$$W[A] = \int D\psi D\bar{\psi} e^{iI[A,\psi]} = e^{iZ[A]} \qquad (10.2.1)$$

(or $Z[A] \equiv -i \log W[A]$, $\hbar = 1$), where $D\psi D\bar{\psi}$ is the fermionic functional integration measure,

$$I[A,\psi] = \int d^D x \bar{\psi} i \mathscr{D} \psi \quad , \quad \mathscr{D}_\mu = \partial_\mu + A_\mu \quad , \quad A_\mu = A_\mu^i T_i \quad , \qquad (10.2.2)$$
$$\mathscr{D} = \gamma^\mu \mathscr{D}_\mu \quad ,$$

ψ and $\bar{\psi}$ are independent Grassmann variables and T_i are the matrices of a representation of the generators of the Lie algebra $\mathscr{G}$; they act on the group representation index in ψ. We shall set $D = 2p$ in general. This may look unphysical but, apart from their mathematical interest, higher- (and lower-) dimensional spacetimes are often considered in physics and it is worth doing the discussion for any even dimension though in general the resulting theories are not renormalizable. In (10.2.2), ψ is a Dirac (complex) spinor and the Minkowski metric is $\eta^{\mu\nu} = (+,-,...,-)$; the gamma matrices satisfy $\{\gamma^\mu, \gamma^\nu\} = 2\eta^{\mu\nu}$, $\gamma^{0\dagger} = \gamma^0$, $\gamma^{i\dagger} = -\gamma^i$, $i = 1, 2, ..., (D-1)$, and $\gamma_{D+1} = (i)^{D/2-1}\gamma^0...\gamma^{D-1}$ (cf. section 2.11). Due to the oscillatory character of the exponential of (10.2.1), the path integral is not well defined. As is known, this is remedied by defining it in Euclidean space, which means replacing (10.2.1) by

$$W_E[A] = \int D\psi D\bar{\psi} e^{-I_E(A,\psi)} = e^{-Z_E[A]} \quad , \qquad (10.2.3)$$

where the action reads

$$I_E(A,\psi) = \int d^D x \, \bar{\psi} i \mathscr{D} \psi \quad . \qquad (10.2.4)$$

The Euclidean Dirac matrices entering in $\mathscr{D}$ are all now hermitian,

$$\{\gamma^\mu, \gamma^\nu\} = 2\delta^{\mu\nu} \quad , \quad \gamma^{\mu\dagger} = \gamma^\mu \quad , \quad \gamma_{D+1} = i^{D/2}\gamma^0...\gamma^{D-1} \quad , \qquad (10.2.5)$$
$$\gamma_{D+1}^2 = 1 \quad , \quad \gamma_{D+1}^\dagger = \gamma_{D+1} \quad , \quad \gamma_{D+1}\gamma^\mu = -\gamma^\mu\gamma_{D+1} \quad .$$

Indeed, the Wick rotation that changes $W[A]$ into $W_E[A]$ is given by the substitutions $x_M^0 \mapsto -ix_E^0$, $x_M^i \mapsto x_E^i = x_{Ei}$, $A_0^M \mapsto iA_0^E$, $A_i^M \mapsto A_i^E = A^{Ei}$, $i = 1, ..., (D-1)$, $\psi_M \mapsto \psi_E$, $\bar{\psi}_M \mapsto i\bar{\psi}_E$, and the identification $\gamma_M^i = i\gamma_E^i$, $\gamma_M^0 = \gamma_E^0$, where the index M (E) stands here for quantities in Minkowski (Euclidean) space. From this, and our definition of the γ_{D+1} matrices, we see that $\gamma_{D+1}^M = (-1)^{\frac{D}{2}-1}\gamma_{D+1}^E$.

The action $I[A,\psi]$ (or $I_E[A,\psi]$) is invariant under the rigid chiral transformations of the Dirac fields

$$\psi' = e^{i\alpha\gamma_{D+1}}\psi \;,\; \bar{\psi}' = \bar{\psi}e^{i\alpha\gamma_{D+1}} \;,\; \delta\psi = i\alpha\gamma_{D+1}\psi \;,\; \delta\bar{\psi} = i\alpha\gamma_{D+1} \;. \qquad (10.2.6)$$

Classically, this means that there is a conserved Noether *chiral* or *axial current* j_{D+1}^μ (j_5^μ for short) which, from the point of view of section 9.3, can be deduced by writing the variation of $I[A, \psi]$ under a non-constant parameter transformation $\alpha = \alpha(x)$ as

$$\delta I = -\int d^D x\, \partial_\mu \alpha j_5^\mu \quad, \qquad j_5^\mu = \bar\psi \gamma^\mu \gamma_{D+1} \psi \quad, \qquad (10.2.7)$$

and a formally identical expression for δI_E and j_{5E}^μ. But it turns out that quantum mechanically the axial current is no longer conserved, the anomalous divergence being given by

$$\partial_\mu J_5^\mu := \frac{1}{W[A]} \int D\psi D\bar\psi\, \partial_\mu j_5^\mu e^{iI[A,\psi]} = \mathcal{B} \qquad (10.2.8)$$

(see below). The quantity $\mathcal{B}$, which depends only on the fields A_μ, is called the *abelian* or *axial anomaly*, the name 'abelian' referring to the fact that the rigid symmetry is an axial $U(1)$ transformation (note that the gauge fields A_μ do not have to be $U(1)$-valued connections). As pointed out by Fujikawa, the non-conservation of the axial current j_5^μ in the quantum case may be tied to the transformation properties of the fermionic functional measure in (10.2.3). To see this, let us try to rederive the current conservation for this case by taking transformations of the type (10.2.6) with $\alpha = \alpha(x)$. If we use them to perform a change of variables in (10.2.3), the path integral has to remain the same and then, expanding in terms of α the difference between the two integrals after and before the change, we obtain

$$0 = \int \delta(D\psi D\bar\psi) e^{-I_E[A,\psi]} + \int D\psi D\bar\psi e^{-I_E[A,\psi]} \int d^D x\, \partial_\mu \alpha j_{5E}^\mu \quad, \qquad (10.2.9)$$

where use has been made of eq. (10.2.7) for the Euclidean action. If the fermionic measure were invariant, then we would deduce that

$$\int D\psi D\bar\psi e^{-I_E[A,\psi]} \int d^D x\, \alpha \partial_\mu j_{5E}^\mu = 0 \quad, \qquad (10.2.10)$$

or $\partial_\mu J_{5E}^\mu = 0$. But $D\psi D\bar\psi$ is not invariant, hence the $\mathcal{B}$ ($\mathcal{B}_E$ in the Euclidean case; the subscript E will be dropped in most places) in (10.2.8), which is related to eq. (2.11.19) as we now see.

The value of $\delta(D\psi D\bar\psi)$ in (10.2.9) is minus twice the trace (in the sense of the space of functions) of the operator $i\alpha\gamma_{D+1}$:

$$D\psi' D\bar\psi' = exp(-2\mathrm{Tr}(i\alpha\gamma_{D+1}))D\psi D\bar\psi \,,$$
$$\delta(D\psi D\bar\psi) = -2\mathrm{Tr}(i\gamma_{D+1}\alpha)D\psi D\bar\psi \qquad (10.2.11)$$

(note the appearance of the *inverse* of the usual Jacobian due to the fact that we are dealing with Grassmann odd variables). Let us now compactify the Euclidean space to the sphere S^D and expand ψ and $\bar\psi$ in terms of the

(Grassmann even) eigenfunctions ψ_n of the hermitian operator $i\slashed{D}$ with (real) eigenvalues λ_n, $i\slashed{D}\psi_n = \lambda_n\psi_n$,

$$\psi = \sum_n a_n\psi_n \quad, \quad \bar{\psi} = \sum_n \bar{b}_n\psi_n \quad, \quad (\psi_i, \psi_j) = \int d^D x\, \psi_i^\dagger \psi_j = \delta_{ij} \quad;$$

(10.2.12)

the odd character of ψ, $\bar{\psi}$ is carried by the coefficients a_n, $\bar{b}_n$. This is possible because as already remarked in section 2.11, $i\slashed{D}$, being elliptic in Euclidean space, is a Fredholm operator and then it has discrete spectrum. Using (10.2.12) we see that the Jacobian is given by $2\mathrm{Tr}(i\alpha\gamma_{D+1}) = 2\int d^D x \sum_n \alpha\psi_n^\dagger i\gamma_{D+1}\psi_n$, and we finally obtain

$$0 = \int D\psi D\bar{\psi}\left[-2\int d^D x\, \alpha \sum_n \psi_n^\dagger i\gamma_{D+1}\psi_n - \int d^D x\, \partial_\mu j_{5E}^\mu \alpha\right]e^{-I_E[A,\psi]} \quad,$$

(10.2.13)

or, since this is true for any $\alpha(x)$,

$$\partial_\mu J_{5E}^\mu \equiv \frac{1}{W_E[A]}\int D\psi D\bar{\psi}\,\partial_\mu j_{5E}^\mu e^{-I_E[A,\psi]} = -2i\sum_n \psi_n^\dagger\gamma_{D+1}\psi_n \equiv \mathscr{B}_E$$

(10.2.14)

(cf. (10.2.8)); thus *the anomalous divergence of the current is the result of the non-trivial Jacobian associated with the transformation.*

The above expression for $\mathscr{B}_E$ is divergent, but may be regularized using a Gaussian cut-off (heat kernel regularization) as

$$\mathscr{B}_E = \lim_{M\to\infty}(-2i)\sum_n e^{-\lambda_n^2/M^2}\psi_n^\dagger\gamma_{D+1}\psi_n \quad.$$

(10.2.15)

Let us now consider the quantity $\int d^D x\, \mathscr{B}_E$. Only the ψ_n corresponding to zero modes contribute to it, since if $i\slashed{D}\psi_n = \lambda_n\psi_n$ then $i\slashed{D}\gamma_{D+1}\psi_n = -\lambda_n(\gamma_{D+1}\psi_n)$ and thus, for $\lambda_n \neq 0$, ψ_n and $(\gamma_{D+1}\psi_n)$ are eigenvectors corresponding to different eigenvalues of a hermitian operator and hence orthogonal. Since $\gamma_{D+1} = P_+ - P_-$, $P_\pm = \frac{1}{2}(1 \pm \gamma_{D+1})$, these contributions are just the difference between the positive and negative chirality zero modes, i.e.

$$\int d^D x\, \mathscr{B}_E = -2i[\dim\ker(i\slashed{D}_+) - \dim\ker(i\slashed{D}_-)] = -2i\,\mathrm{index}(i\slashed{D}_+) \quad,$$

(10.2.16)

where the elliptic non-self-adjoint $((i\slashed{D}_+)^\dagger = i\slashed{D}_-)$ operators for each chirality are given by

$$i\slashed{D}_\pm = i\slashed{D}P_\pm \quad, \quad i\slashed{D} = i\slashed{D}_+ + i\slashed{D}_- \quad.$$

(10.2.17)

Thus, the integration of the abelian anomaly gives the index for the twisted spin complex associated with the operator $i\slashed{D}_+$. In fact, it is possible to deduce eq. (2.11.19) by computing (10.2.15) directly. This can be done by writing $\mathscr{B}_E$ as

$$\mathscr{B}_E = -2i \lim_{M\to\infty} \lim_{x\to y} \sum_n \psi_n^\dagger(y)\gamma_{D+1}e^{-(i\slashed{\mathscr{D}})^2(x)/M^2}\psi_n(x)$$

$$= -2i \lim_{M\to\infty} \lim_{x\to y} \mathrm{Tr}(\gamma_{D+1}e^{-(i\slashed{\mathscr{D}})^2(x)/M^2} \sum_n \psi_n(x)\psi_n^\dagger(y)) \qquad (10.2.18)$$

$$= -2i \lim_{M\to\infty} \lim_{x\to y} \mathrm{Tr}(\gamma_{D+1}e^{-(i\slashed{\mathscr{D}})^2(x)/M^2})\delta(x-y) \quad,$$

where here the trace is the ordinary trace involving the group and the spinorial indices and in the last step we have used the completeness relation for ψ_n, $\delta^D(x-y) = \sum_n \psi_n(x)\psi_n^\dagger(y)$. Expressing $\delta^D(x-y)$ in a plane wave basis, $\delta^D(x-y) = \frac{1}{(2\pi)^D}\int d^D k\, e^{ik(x-y)}$, (10.2.18) gives

$$\mathscr{B}_E = -2i \lim_{M\to\infty} \frac{1}{(2\pi)^D}\int d^D k\, e^{-ikx}\mathrm{Tr}\left(\gamma_{D+1}e^{-(i\slashed{\mathscr{D}})^2/M^2}\right)e^{ikx} \quad. \qquad (10.2.19)$$

From this expression it is not difficult to calculate the terms that contribute to the $M \to \infty$ limit. Note first that $-(i\slashed{\mathscr{D}})^2 = \slashed{\mathscr{D}}^2$ is given by

$$\slashed{\mathscr{D}}^2 = \mathscr{D}_\mu \mathscr{D}_\nu \gamma^\mu \gamma^\nu = \frac{1}{2}\mathscr{D}_\mu \mathscr{D}_\nu \{\gamma^\mu, \gamma^\nu\} + \frac{1}{2}\mathscr{D}_\mu \mathscr{D}_\nu [\gamma^\mu, \gamma^\nu]$$

$$= \mathscr{D}^\mu \mathscr{D}_\mu + \frac{1}{4}[\mathscr{D}_\mu, \mathscr{D}_\nu][\gamma^\mu, \gamma^\nu] = \mathscr{D}^\mu \mathscr{D}_\mu + \frac{1}{4}F_{\mu\nu}[\gamma^\mu, \gamma^\nu] \quad. \qquad (10.2.20)$$

Making the change of variables $s^\mu = k^\mu/M$ eq. (10.2.19) reads

$$\mathscr{B}_E = -2i \lim_{M\to\infty} \frac{1}{(2\pi)^D}\int d^D s\, M^D \mathrm{Tr}\left(\gamma_{D+1}e^{-(\frac{i\slashed{\mathscr{D}}}{M}-\slashed{s})^2}\right) \quad. \qquad (10.2.21)$$

The exponential in (10.2.21) can be written, using (10.2.20),

$$e^{-(\frac{i\slashed{\mathscr{D}}}{M}-\slashed{s})^2} = \exp\left(\frac{\mathscr{D}^2}{M^2} + \frac{1}{4M^2}F_{\mu\nu}[\gamma^\mu, \gamma^\nu] + \frac{2i}{M}s^\mu \mathscr{D}_\mu\right)e^{-s^2} \quad. \qquad (10.2.22)$$

Since $\mathrm{Tr}(\gamma_{D+1}\gamma^{\mu_1}...\gamma^{\mu_k}) = 0$ for $k < D$, all potentially divergent terms in (10.2.21) vanish, and the result of the limit is the contribution to the exponential in (10.2.22) that is proportional to $1/M^D$ and contains enough gamma matrices at the same time. This contribution is the one that comes entirely from the $F_{\mu\nu}[\gamma^\mu, \gamma^\nu]$ term,

$$
\begin{aligned}
\mathscr{B}_E &= \frac{1}{(D/2)!} \frac{-2i}{(4\pi)^D} \int d^D s \, e^{-s^2} \operatorname{Tr}(\gamma_{D+1} F_{\mu_1\mu_2} \ldots F_{\mu_{D-1}\mu_D} [\gamma^{\mu_1}, \gamma^{\mu_2}] \ldots [\gamma^{\mu_{D-1}}, \gamma^{\mu_D}]) \\
&= \frac{-2i}{(4\pi)^D} \frac{1}{(D/2)!} (2\pi)^{D/2} \operatorname{Tr}(F_{\mu_1\mu_2} \ldots F_{\mu_{D-1}\mu_D}) \operatorname{Tr}(\gamma_{D+1} \gamma^{\mu_1} \ldots \gamma^{\mu_D}) \\
&= -2i \left(\frac{-i}{4\pi} \right)^{D/2} \frac{1}{(D/2)!} \epsilon^{\mu_1 \cdots \mu_D} \operatorname{Tr}(F_{\mu_1\mu_2} \ldots F_{\mu_{D-1}\mu_D}) \quad ,
\end{aligned}
$$

$$ (10.2.23) $$

where we have used that $\int d^D s \, e^{-s^2} = \pi^{D/2}$, $\operatorname{Tr}(\gamma_{D+1} \gamma^{\mu_1} \ldots \gamma^{\mu_D}) = (-2i)^{\frac{D}{2}} \epsilon^{\mu_1 \cdots \mu_D}$.

The abelian anomaly in D-dimensional Minkowski space $\mathscr{B}$ is obtained from $\mathscr{B}_E$ by adding a factor $(-i)$ coming from the trace part $(\partial_0^E \rightarrow -i\partial_0^M$, $A_0^E \mapsto -iA_0^M)$, a factor $(-1)^{D/2}$ since $(\partial_\mu J_5^\mu)_E$ goes to $(-1)^{D/2} \partial_\mu J_5^\mu$ and a minus sign since in Minkowski spacetime we take $\epsilon^{0\ldots D-1} = (-1)^{D-1}$. The result is

$$
\mathscr{B} = \left(\frac{i}{4\pi} \right)^{D/2} \frac{2}{(D/2)!} \epsilon^{\mu_1 \cdots \mu_D} \operatorname{Tr}(F_{\mu_1\mu_2} \ldots F_{\mu_{D-1}\mu_D}) \quad . \tag{10.2.24}
$$

This agrees with the Feynman diagram expression for the $D = 4$ triangle anomaly,

$$
\partial_\mu J_5^\mu = -\frac{1}{16\pi^2} \epsilon^{\mu\nu\rho\sigma} \operatorname{Tr}(F_{\mu\nu} F_{\rho\sigma}) = -\frac{1}{4\pi^2} \epsilon^{\mu\nu\rho\sigma} \partial_\mu \operatorname{Tr}(A_\nu \partial_\rho A_\sigma + \frac{2}{3} A_\nu A_\rho A_\sigma) \quad . \tag{10.2.25}
$$

Clearly, *the abelian anomaly is gauge-invariant.* Let us check that this calculation gives the correct formula ((2.11.19)) for the index of $i\mathscr{D}_+$. Using (10.2.16), (10.2.23) and (2.5.12) we have *

$$
\begin{aligned}
\text{index } i\mathscr{D}_+ &= \frac{i}{2} \int d^D x \, \mathscr{B}_E \\
&= \left(\frac{-i}{2\pi} \right)^{D/2} \frac{1}{(D/2)!} \frac{1}{2^{D/2}} \int d^D x \, \epsilon^{\mu_1 \cdots \mu_D} \operatorname{Tr}(F_{\mu_1\mu_2} \ldots F_{\mu_{D-1}\mu_D}) \\
&= \left(\frac{-i}{2\pi} \right)^{D/2} \frac{1}{(D/2)!} \frac{1}{2^{D/2}} \int_{S^D} dx^{\mu_1} \wedge \ldots \wedge dx^{\mu_D} \operatorname{Tr}(F_{\mu_1\mu_2} \ldots F_{\mu_{D-1}\mu_D})
\end{aligned}
$$

---

* In this chapter the symbol $\wedge$ for the exterior product will be usually omitted and the dimension of the manifold M will be specified in each case.

$$= (-1)^{D/2} \int_{S^D} \left(\frac{i}{2\pi}\right)^{D/2} \frac{1}{(D/2)!} \mathrm{Tr}(F^{D/2})$$

$$= (-1)^{D/2} \int_{S^D} ch_{(D/2)}(F) \quad , \tag{10.2.26}$$

as we wanted to show (since $M \sim S^D$, the spin connection does not enter and the index density for the Dirac operator reduces to the Chern character, section 2.11). Thus, the expectation value $(\partial_\mu J_5^\mu)_E$ is given by the index of the Dirac operator.

10.3 The action of gauge transformations on the space of functionals

Let us now consider the gauge group action on a functional $F[A]$. Since locally a gauge transformation is an element of $G(M)$, we shall consider this group without worrying about the possible non-triviality of $P(G, M)$. As we know, the Yang-Mills potentials A^i can be viewed as the components of $\mathcal{G}$-valued forms A over M, and may be written as

$$A(x) = A^i(x)T_i , \quad A^i(x) \equiv A^i_\mu(x)dx^\mu , \tag{10.3.1}$$

where T_i is a basis of the algebra $\mathcal{G}$ of G. The corresponding curvature tensor F and dF are given by

$$F = dA + A^2 , \quad dF = [F, A] . \tag{10.3.2}$$

The space $\mathcal{F}[\mathcal{A}]$ of functionals has an obvious abelian group and vector space structure under the sum of functionals. Let the functional $F[A]$ be an element of $\mathcal{F}[\mathcal{A}]$. The action of an element $g(x)$ of $G(M)$ on A is given by the gauge transformation law of the Yang-Mills potentials,

$$A^g(x) = g^{-1}(x)Ag(x) + g^{-1}(x)dg(x) \quad ; \tag{10.3.3}$$

since $(A^{g'})^g = A^{g'g}$, a gauge transformation defines a *right* action of $G(M)$ on A. $G(M)$ defines an action $\sigma(g)$ on $\mathcal{F}[\mathcal{A}]$ as follows:

$$(\sigma(g)F)[A] \equiv F[A^g] \equiv F[g^{-1}Ag + g^{-1}dg] , \quad \forall F[A] \in \mathcal{F}[\mathcal{A}] , \tag{10.3.4}$$

where $g(x) = exp(g^i(x)T_i)$. This is a left action on F, as can be seen (cf. section 5.4) from the fact that $(\sigma(g'g)F) = \sigma(g')(\sigma(g)F)$ since

$$(\sigma(g'g)F)[A] = F[A^{g'g}] = F[(A^{g'})^g] = (\sigma(g')(\sigma(g)F))[A] \quad . \tag{10.3.5}$$

The following proposition gives the infinitesimal version ρ of this action:

(10.3.1) PROPOSITION

The generators of the action on $\mathscr{F}[\mathscr{A}]$ are given by

$$Y_{(i)}(x) = -\frac{\partial}{\partial x^\mu}\frac{\delta}{\delta A^i_\mu(x)} - C^k_{ij}A^j_\mu(x)\frac{\delta}{\delta A^k_\mu(x)} := -(\delta^k_i\partial_\mu + C^k_{ij}A^j_\mu(x))\frac{\delta}{\delta A^k_\mu}$$

(10.3.6)

and their algebra is that of (9.2.6).

Proof: To prove the first part, consider an infinitesimal gauge transformation of parameters $\zeta^i(x)$. Then, taking $g(x) = \exp g^i(x)T_i$ to first order (for infinitesimal $g^i(x) = \zeta^i(x)$, $g(x) \simeq 1 + \zeta^i(x)T_i$), we find

$$(A^g)^i_\mu T_i = g^{-1}A_\mu g + g^{-1}\partial_\mu g \simeq A_\mu - [\zeta^i T_i, A_\mu] + \partial_\mu\zeta^i T_i \quad . \quad (10.3.7)$$

Thus, the generator of the gauge transformation (dim $M = m$)

$$Y_{(i)}(x) = \int d^m y\,\frac{\delta(A^g)^k_\mu(y)}{\delta g^i(x)}\bigg|_{g=e}\frac{\delta}{\delta A^k_\mu(y)}$$

(10.3.8)

is given by

$$Y_{(i)}(x) = -\int d^m y\, C^k_{ij}A^j_\mu(y)\delta^m(x-y)\frac{\delta}{\delta A^k_\mu(y)} + \int d^m y\,\partial_{y^\mu}\delta^m(x-y)\frac{\delta}{\delta A^i_\mu(y)} \quad ,$$

(10.3.9)

which reproduces (10.3.6). To calculate the commutators, let us first write $Y_{(i)}(x)$ as

$$Y_{(i)}(x) = -\left[\int d^m z\,\partial_{x^\mu}\delta^m(x-z)\frac{\delta}{\delta A^i_\mu(z)}\right] - C^k_{ij}A^j_\mu\frac{\delta}{\delta A^k_\mu(x)} \quad . \quad (10.3.10)$$

In this way, we have

$$[Y_{(i)}(x), Y_{(j)}(y)]$$
$$= \int d^m z\,\partial_{x^\mu}\delta^m(x-z)C^k_{ji}\delta^m(y-z)\frac{\delta}{\delta A^k_\mu(y)}$$
$$+ C^s_{ir}A^r_\mu(x)C^t_{js}\delta^m(x-y)\frac{\delta}{\delta A^t_\mu(y)} - \binom{x \leftrightarrow y}{i \leftrightarrow j}$$
$$= \int d^m z(\partial_{x^\mu}\delta^m(x-z)\delta^m(y-z)C^k_{ji} - \partial_{y^\mu}\delta^m(y-z)\delta^m(x-z)C^k_{ij})\frac{\delta}{\delta A^k_\mu(z)}$$
$$+ (C^s_{ir}C^t_{js} - C^s_{jr}C^t_{is})A^r_\mu\delta^m(x-y)\frac{\delta}{\delta A^t_\mu(y)} \quad .$$

(10.3.11)

Now, using in (10.3.11) the distributional identity

$$\partial_{x^\mu}\delta^m(x-z)\delta^m(y-z) + \partial_{y^\mu}\delta^m(y-z)\delta^m(x-z) = \delta^m(x-y)\partial_{x^\mu}\delta^m(x-z)$$

(10.3.12)

and the Jacobi identity $C_{ir}^s C_{sj}^t + C_{ji}^s C_{sr}^t + C_{rj}^s C_{si}^t = 0$, we finally obtain

$$[Y_{(i)}(x), Y_{(j)}(y)] = -C_{ij}^k \delta^m(x-y) \int d^m z \, \partial_{x^\mu} \delta^m(x-z) \frac{\delta}{\delta A_\mu^k(z)}$$

$$- C_{ij}^s C_{sr}^t A_\mu^r \delta^m(x-y) \frac{\delta}{\delta A_\mu^t(y)} = \delta^m(x-y) C_{ij}^k Y_{(k)}(y),$$

q.e.d.
$$(10.3.13)$$

It is easy to see that, writing $\rho(\zeta \cdot X)$ for $\zeta \cdot Y$,

$$\rho(\zeta \cdot X) A_\mu^k(x) = \left[\int d^m y \zeta^i(y) Y_{(i)}(y) \right] A_\mu^k(x)$$

$$= - \int d^m y \zeta^i(y) [\partial_{y^\mu} \delta^m(x-y) \delta_i^k + C_{ij}^k A_\mu^j(y) \delta^m(x-y)]$$

$$= \int d^m y [\partial_\mu \zeta^k(y) - C_{ij}^k \zeta^i(y) A_\mu^j(y)] \delta^m(x-y)$$

$$= \partial_\mu \zeta^k(x) - C_{ij}^k \zeta^i(x) A_\mu^j(x) \equiv (\mathscr{D}_\mu)_{\cdot l}^k \zeta^l(x);$$

$$(10.3.14)$$

which is the same as (10.3.7) and (2.2.8). Thus, *an infinitesimal gauge transformation is given locally by an element of* $\mathscr{G}(M)$. In the rest of the chapter we shall write

$$Y_{(i)}(x) \equiv \rho(X_{(i)}(x)), \quad \rho(\zeta \cdot X) \equiv \zeta \cdot Y \equiv \int d^m y \zeta^i(y) Y_{(i)}(y), \quad (10.3.15)$$

to express that $Y_{(i)}(x)$ is the generator of the infinitesimal action ρ on $\mathscr{F}[\mathscr{A}]$ associated with the element $X_{(i)}(x)$ of the canonical basis of the gauge algebra $\mathscr{G}(M)$. Then $[\zeta \cdot Y, \zeta' \cdot Y] = [\zeta, \zeta'] \cdot Y$ as for (9.2.9).

10.4 Cohomology on $G(M)$, $\mathscr{G}(M)$ and its BRST formulation

Having introduced the infinite-dimensional group $G(M)$, its Lie (gauge) algebra $\mathscr{G}(M)$ and its action on $\mathscr{F}[\mathscr{A}]$, we extend the techniques described in chapters 5 and 6 to the present case and obtain the cochains and the operators which define the corresponding Lie group and algebra cohomologies. We shall study the group cohomology associated with the action $\sigma(g)$ (for $G(M)$) on the space of local functionals of A, i.e. the space of functionals constructed as integrals of A_μ^i and their derivatives up to a finite order, $\mathscr{F}_L[\mathscr{A}]$. The reason is that chiral anomalies, the cohomology of which we want to study, are given by local functionals; in fact, they can be computed in perturbation theory, where all the expressions are local. Eq. (10.3.6) also defines how $\mathscr{G}(M)$ acts on $\mathscr{F}_L[\mathscr{A}]$, which in fact becomes a $\rho(\mathscr{G}(M))$-module. This enables us to consider the cohomology of the gauge algebra $\mathscr{G}(M)$ by following essentially the same lines as used in chapter 6. The corresponding cohomologies will be relevant in the geometrical description of anomalies, as we shall see.

(a) $H_\sigma(G(M), \mathscr{F}_L[\mathscr{A}])$.

In this case, the cochains have arguments that are elements of $G(M)$ i.e., functions $g(x)$ and, at the same time, they belong to $\mathscr{F}_L[\mathscr{A}]$. Thus, they are mappings $\Omega : G(M) \times \overset{n}{\cdots} \times G(M) \to \mathscr{F}_L[\mathscr{A}]$, $\Omega : (g_1, ..., g_n) \mapsto \Omega_n[\ \cdot \ ; g_1, ..., g_n]$, which is a functional of A, $\Omega_n[\ \cdot \ ; g_1, ..., g_n] : A \mapsto \Omega_n[A; g_1, ..., g_n] \in R$. We shall write them as $\Omega_n[A; g_1, ..., g_n]$, using a square bracket to indicate that Ω is a functional of both A and $g_1, ..., g_n$, and a semicolon to separate the vector potentials A from the true arguments of the n-cochain, the group elements $g_a(x) \in G(M)$, $a = 1, ..., n$.

Taking this into account, a simple extension of formula (5.4.7) provides the definition of δ for this case,

$$\delta\Omega_n[A; g_1, ..., g_{n+1}] \equiv (\sigma(g_1)\Omega_n)[A; g_2, ..., g_{n+1}]$$

$$+ \sum_{a=1}^{n} (-1)^a \Omega_n[A; g_1, ..., g_{a-1}, g_a g_{a+1}, g_{a+2}, ..., g_{n+1}] \qquad (10.4.1)$$

$$+ (-1)^{n+1}\Omega_n[A; g_1, ..., g_n] , \qquad \delta^2 = 0 \quad ,$$

where in the first term the operator $\sigma(g_1)$ only acts on the argument A through (10.3.4). The first three examples of (10.4.1) are

$$(\delta\Omega_0)[A; g] = \Omega_0[A^g] - \Omega_0[A] , \qquad (10.4.2)$$

$$(\delta\Omega_1)[A; g_1, g_2] = \Omega_1[A^{g_1}; g_2] - \Omega_1[A; g_1 g_2] + \Omega_1[A; g_1] , \qquad (10.4.3)$$

$$(\delta\Omega_2)[A; g_1, g_2, g_3] = \Omega_2[A^{g_1}; g_2, g_3] + \Omega_2[A; g_1, g_2 g_3]$$
$$- \Omega_2[A; g_1 g_2, g_3] - \Omega_2[A; g_1, g_2] . \qquad (10.4.4)$$

(b) $H_\rho(\mathscr{G}(M), \mathscr{F}_L[\mathscr{A}])$

The cochains are skewsymmetric multilinear mappings $\omega_n : \mathscr{G}(M) \times \overset{n}{\cdots} \times \mathscr{G}(M) \to \mathscr{F}_L[\mathscr{A}]$. Then $\omega_n : (X_1, ..., X_n) \mapsto \omega_n[\ \cdot \ ](X_1, ..., X_n)$, where $\omega_n[\ \cdot \ ](X_1, ..., X_n)$ is a functional of A, $\omega_n[\ \cdot \ ](X_1, ..., X_n) : A \mapsto \omega_n[A](X_1, ..., X_n) \in R$. To simplify the notation in the case where the X's are the basis elements $X_{(i)}(x)$ of $\mathscr{G}(M)$, this will be written as

$$\omega_n[A](X_{(i_1)}(x_1), ..., X_{(i_n)}(x_n)) \equiv \omega_{i_1...i_n}[A](x_1, ..., x_n) , \qquad (10.4.5)$$

omitting the index n since the order of the cochain is indicated by the number of subscripts. When the arguments of the n-cochains depend on n arbitrary elements ζ_a of the algebra, $\zeta_a = \zeta_a \cdot X$ ($a = 1, ..., n$), it is convenient to introduce, using (10.4.5), the notation

$$\omega_n[A](\zeta_1, ..., \zeta_n) \equiv$$

$$\omega_n[A]\left(\int d^m y_1 \, \zeta_1^{i_1}(y_1)X_{(i_1)}(y_1), ..., \int d^m y_n \, \zeta_n^{i_n}(y_n)X_{(i_n)}(y_n)\right)$$

$$= \int d^m y_1 ... \int d^m y_n \, \zeta_1^{i_1}(y_1)...\zeta_n^{i_n}(y_n)\omega_{i_1...i_n}[A](y_1, ..., y_n) \quad . \tag{10.4.6}$$

The notation in formula (10.4.5) will be used to give the explicit expressions for the cocycles, whereas that of formula (10.4.6) is more suitable for calculations. Clearly, it is easy to obtain $\omega_{i_1...i_n}[A](x_1, ..., x_n)$ from $\omega_n[A](\zeta_1, ..., \zeta_n)$ just by taking $\zeta_l^{j_l}(y_l) = \delta_{i_l}^{j_l}\delta^m(y_l - x_l)$, $l = 1, ..., n$.

The gauge algebra coboundary operator will be again denoted by s in the infinite-dimensional case. Its action on an n-cochain, defined on general elements ζ_a of $\mathcal{G}(M)$, is given by an extension of the formula (6.10.4) for the Lie algebra cohomology:

$$s\omega_n[A](\zeta_1, ..., \zeta_{n+1})$$

$$= \sum_{a=1}^{n+1}(-1)^{a+1}(\zeta_a \cdot Y)\omega_n[A](\zeta_1, ..., \hat{\zeta}_a, ..., \zeta_{n+1})$$

$$+ \sum_{\substack{a,b \\ a<b}}^{n+1}(-1)^{a+b}\omega_n[A]([\zeta_a, \zeta_b], \zeta_1, ..., \hat{\zeta}_a, ..., \hat{\zeta}_b, ..., \zeta_{n+1}), \quad s^2 = 0,$$

$$\tag{10.4.7}$$

where, we recall, $[\zeta_a, \zeta_b]^k$ is $\zeta_a^i \zeta_b^j C_{ij}^k$. Similarly,

$$(s\omega_n)_{i_1...i_{n+1}}[A](x_1, ..., x_{n+1})$$

$$= \sum_{a=1}^{n+1}(-1)^{a+1}\rho(X_{(i_a)}(x_a))\omega_{i_1...\hat{i}_a...i_{n+1}}[A](x_1, ..., \hat{x}_a, ..., x_{n+1})$$

$$+ \sum_{\substack{a,b \\ a<b}}^{n+1}(-1)^{a+b}C_{i_a i_b}^l \delta^m(x_a - x_b)\omega_{l i_1...\hat{i}_a...\hat{i}_b...i_{n+1}}[A](x_a, x_1, ..., \hat{x}_a, ..., \hat{x}_b, ..., x_{n+1})$$

$$\tag{10.4.8}$$

when the cochain is defined by its action on the elements of the canonical basis of $\mathcal{G}(M)$. It is simple to obtain (10.4.7) from (10.4.8) just by multiplying the latter by $\zeta_1^{i_1}(x_1)...\zeta_{n+1}^{i_{n+1}}(x_{n+1})$, integrating over $x_1, ..., x_{n+1}$ and using the definitions (10.4.5) and (10.4.6). Explicitly,

$$(s\omega_n)[A](\zeta_1, ..., \zeta_{n+1})$$

$$= \int d^m x_1 ... \int d^m x_{n+1} \, \zeta_1^{i_1}(x_1)...\zeta_{n+1}^{i_{n+1}}(x_{n+1})(s\omega_n)_{i_1...i_{n+1}}[A](x_1, ..., x_{n+1})$$

$$= \sum_{n=1}^{n+1}(-1)^{a+1}\int d^m x_1 ... \int d^m x_{n+1} \, \zeta_a^{i_a}\rho(X_{(i_a)}(x_a))\zeta_1^{i_1}(x_1)...\hat{\zeta}_a^{i_a}(x_a)...\zeta_{n+1}^{i_{n+1}}(x_{n+1})$$

$$. \, \omega_{i_1...\hat{i}_a...i_{n+1}}[A](x_1, ..., \hat{x}_a, ..., x_{n+1})$$

$$+ \sum_{\substack{a,b \\ a<b}}^{n+1}(-1)^{a+b}\int d^m x_1 ... \int d^m x_{n+1} \, \zeta_a^{i_a}(x_a)\zeta_b^{i_b}(x_b)\delta^m(x_a - x_b)C^l_{i_a i_b}$$

$$.\zeta_1^{i_1}(x_1)...\hat{\zeta}_a^{i_a}(x_a)...\hat{\zeta}_b^{i_b}(x_b)...\zeta_{n+1}^{i_{n+1}}(x_{n+1})$$

$$. \, \omega_{l i_1...\hat{i}_a...\hat{i}_b...i_{n+1}}[A](x_a, x_1, ..., \hat{x}_a, ..., \hat{x}_b, ..., x_{n+1})$$

$$= \sum_{n=1}^{n+1}(-1)^{a+1}(\zeta_a \cdot Y)\omega_n[A](\zeta_1 \cdot X, ..., \hat{\zeta}_a \cdot X, ..., \zeta_{n+1} \cdot X)$$

$$+ \sum_{\substack{a,b \\ a<b}}^{n+1}(-1)^{a+b}\omega_n[A]([\zeta_a \cdot X, \zeta_b \cdot X], \zeta_1 \cdot X, ..., \hat{\zeta}_a \cdot X, ..., \hat{\zeta}_b \cdot X, ..., \zeta_{n+1} \cdot X),$$

$$(10.4.9)$$

and the last expression (see (9.2.9)) is the same as (10.4.7).

The first examples for the action of s on the cochains of the canonical basis are

$$(s\omega_0)_i[A](x) = Y_{(i)}(x).\omega_0[A],$$ $$(10.4.10)$$

$$(s\omega_1)_{ij}[A](x, y) = Y_{(i)}(x).\omega_j[A](y) - Y_{(j)}(y).\omega_i[A](x)$$
$$- \delta^m(x - y)C^k_{ij}\omega_k[A](x),$$ $$(10.4.11)$$

$$(s\omega_2)_{ijk}[A](x, y, z) = Y_{(i)}(x).\omega_{jk}[A](y, z) - Y_{(j)}(y).\omega_{ik}[A](x, z)$$
$$+ Y_{(k)}(z).\omega_{ij}[A](x, y)$$
$$- \delta^m(x - y)C^l_{ij}\omega_{lk}[A](x, z) + \delta^m(x - z)C^l_{ik}\omega_{lj}[A](x, y)$$
$$- \delta^m(y - z)C^l_{jk}\omega_{li}[A](y, x),$$ $$(10.4.12)$$

and, for a general element $\zeta \cdot X \in \mathscr{G}(M)$ (cf. (10.3.14)),

$$(s\omega_0)[A](\zeta) = \frac{d}{dt}\omega_0[A_\mu + t\mathscr{D}_\mu\zeta]_{t=0},$$ $$(10.4.13)$$

$$(s\omega_1)[A](\zeta_1,\zeta_2) = \frac{d}{dt}\omega_1[A_\mu + t\mathscr{D}_\mu\zeta_1]_{t=0}(\zeta_2) - \frac{d}{dt}\omega_1[A_\mu + t\mathscr{D}_\mu\zeta_2]_{t=0}(\zeta_1)$$
$$- \omega_1[A]([\zeta_1,\zeta_2]) \,,$$

(10.4.14)

$$(s\omega_2)[A](\zeta_1,\zeta_2,\zeta_3) = \frac{d}{dt}\omega_2[A_\mu + t\mathscr{D}_\mu\zeta_1]_{t=0}(\zeta_2,\zeta_3)$$
$$- \frac{d}{dt}\omega_2[A_\mu + t\mathscr{D}_\mu\zeta_2]_{t=0}(\zeta_1,\zeta_3)$$
$$+ \frac{d}{dt}\omega_2[A_\mu + t\mathscr{D}_\mu\zeta_3]_{t=0}(\zeta_1,\zeta_2)$$
$$- \omega_2[A]([\zeta_1,\zeta_2],\zeta_3) + \omega_2[A]([\zeta_1,\zeta_3],\zeta_2)$$
$$- \omega_2[A]([\zeta_2,\zeta_3],\zeta_1) \,.$$

(10.4.15)

In the above formulae we had to introduce the rather cumbersome expression $\frac{d}{dt}\omega[A_\mu + t\mathscr{D}_\mu\zeta]_{t=0}$ since $\zeta \cdot Y$ is a derivation and accordingly $(\zeta \cdot Y)\omega[A] \neq \omega[((\zeta \cdot Y)A]$; for instance $(\zeta \cdot Y)A^2 = [(\zeta \cdot Y)A]A + A[(\zeta \cdot Y)A] = \frac{d}{dt}[A + t\mathscr{D}\zeta]^2_{t=0}$.

It is also possible to use in the case of gauge algebras the BRST formalism of section 6.8. Looking at (6.8.2), we see that the corresponding definition of the BRST operator $\tilde{s}$ for the infinite-dimensional Lie algebras is

$$\tilde{s} = \int d^m x \left\{ c^i(x) Y_{(i)}(x) - \frac{1}{2}c^i(x)\,c^j(x)\,C^k_{ij}\frac{\delta}{\delta c^k(x)} \right\} \,,$$

(10.4.16)

where apart from the contraction of the Lie algebra indices there is an integration that effects the summation over the 'continuous indices'. Notice that for this reason the parameters c^i (which can be identified with the Faddeev-Popov anticommuting ghost fields) depend now on x. From a Lie algebra cochain given by $\omega_{i_1 \ldots i_n}[A](x_1, \ldots, x_n)$ we construct the cochain of ghost number n (cf. (6.8.7))

$$\tilde{\omega}_n = \int d^m x_1 \ldots d^m x_n \, c^{i_1}(x_1) \ldots c^{i_n}(x_n)\omega_{i_1 \ldots i_n}[A](x_1, \ldots, x_n) \,.$$

(10.4.17)

A natural extension of Theorem 6.8.1 shows that the action of s on an n-cochain ω can be implemented by the operator $\tilde{s}$ acting on $\tilde{\omega}$. The action of $\tilde{s}$ on $c^i(x)T_{(i)} \equiv c(x)$ and $A_\mu(x)$ is the geometric contents of the *BRST transformations*

$$\tilde{s}c(x) = -\frac{1}{2}[c(x),c(x)] \,,$$

(10.4.18)

$$\tilde{s}A_\mu(x) = \partial_\mu c(x) + [A_\mu(x),c(x)] = \mathscr{D}_\mu c(x) \,.$$

(10.4.19)

As is known in gauge theories, when the gauge group is $U(1)$ (the familiar case of electromagnetism) the bracket in (10.4.19) is absent and the

Faddeev-Popov ghost 'decouples'. As in (6.8.14), eq. (10.4.18) for the ghost fields $c^i(x)$ has the structure of a Maurer-Cartan equation (cf. (1.7.15)); *ghosts in gauge theories can be interpreted as the Maurer-Cartan forms for the infinite-dimensional group $G(M)$.* Eqs. (10.4.18) and (10.4.19) can be used to obtain $\check{s}\tilde{\omega}$ for any cochain.

As we shall see later on, the non-trivial elements of the cohomology groups just described will turn out to be relevant in the analysis of non-abelian gauge anomalies and their consequences (Schwinger terms). We shall devote the rest of this chapter to analysing their geometrical description, particularly stressed by Stora, Zumino, Faddeev, and others, but, before doing so, let us consider some physical and topological aspects of non-abelian anomalies.

10.5 Theory with Weyl fermions: non-abelian gauge anomalies and their path integral calculation

Non-abelian *gauge* anomalies arise in theories with gauge fields coupled to chiral (Weyl) fermions; thus, they appear, like the abelian ones, in even dimensions. As in the previous case we shall start in arbitrary even dimension, although at the end of the section we shall consider the simplest case ($D = 2$) for an explicit calculation.

Clearly, the complete generating functional of the theory with dynamical gauge fields will include their kinetic term, the functional measure DA_μ, etc., but these are not important for the problem of the gauge invariance of the theory that we want to analyse and which is concentrated in (10.2.1) since the other contributions are gauge-invariant. In fact, since the presence of gauge anomalies makes the gauge theory inconsistent, the problem may be split into two stages, the first of which is to quantize the fermion degrees of freedom keeping A_μ as an external field to investigate the gauge transformation behaviour of the resulting theory.

The functional $Z[A]$ is as (10.2.1) but now ψ is a *Weyl* spinor which we take with positive chirality, $\gamma_{D+1}\psi = \psi$, $P_+\psi = \psi$ (thus, on ψ, $\mathcal{D}$ can be replaced by $\mathcal{D}_+$). The variation of $Z[A]$ under a gauge transformation $\zeta^i(x)$ is given by

$$(\zeta \cdot Y)Z[A] = \int d^D x \zeta^i(x) Y_{(i)}(x) Z[A] \quad . \tag{10.5.1}$$

Since

$$\frac{\delta Z[A]}{\delta A^i_\mu(x)} = \frac{1}{W[A]} \int D\psi D\bar\psi\, j^\mu_{(i)}(x) \exp iI[A,\psi] =: J^\mu_{(i)}(x) \quad , \tag{10.5.2}$$

where $j^{\mu}_{(i)} = \bar{\psi}\gamma^{\mu}(iT_{(i)})\psi$, $(iT_{(i)})^{\dagger} = (iT_{(i)})$, is the current coupled to $A^i_{\mu}(x)$, and $J^{\mu}_{(i)}$ is defined by (10.5.2), we find

$$(\zeta \cdot Y)Z[A] = \int d^D x \zeta^i(x)\mathscr{D}_{\mu}J^{\mu}_{(i)}(x) \equiv \int d^D x \zeta^i(x)\mathscr{V}_{(i)}[A](x) \quad . \quad (10.5.3)$$

Thus, if the current $J^{\mu}_{(i)}$ is not *covariantly* conserved, the functional will not be gauge-invariant and there will be an anomaly: the classical co-variant conservation of the source (matter) current $\mathscr{D}_{\mu}j^{\mu}_{(i)} = 0$ may not be maintained in the quantum case.

Similarly to the abelian case, the non-invariance of $Z[A]$ may be tied to the fact that, although the exponential in (10.2.1) is gauge-invariant (it is given by the classical action) the fermionic functional measure $D\psi D\bar{\psi}$ is not. This implies the appearance of a Jacobian determinant. Formally, this means that

$$\frac{D(g^{-1}\psi)D(\bar{\psi}g)}{D\psi D\bar{\psi}} = e^{i\alpha_1[A;g]} \quad . \quad (10.5.4)$$

If $W[A^g] = e^{i\alpha_1[A;g]}W[A]$, $W[A] = e^{iZ[A]}$, it follows that $Z[A^g] = Z[A] + \alpha_1[A,g]$. Then $Z[A^g] - Z[A] \simeq (\zeta \cdot Y)Z[A] = \int d^D x\, \zeta^i \mathscr{V}_{(i)}[A](x)$, and

$$\alpha_1[A,g] = \int d^D x(\zeta^i \mathscr{V}_{(i)}[A](x) + ...) \quad . \quad (10.5.5)$$

Before discussing the geometric structure of the anomaly, it is worth describing in detail a simple example of the calculations leading to (10.5.4). The starting point will be the functional $W[A]$ given in eqs. (10.2.1), (10.2.2) where ψ is now a Weyl spinor. Apart from the Euclidean formulation, some additional modification is required to define properly the path integral due to the fact that now Weyl spinors are involved.

The standard path integral formulation of the Euclidean generating functional (10.2.3) requires evaluating the determinant of the 'matrix' $i\mathscr{D}$. For this, it is natural to expand ψ in terms of the eigenfunctions of the Dirac operator $i\mathscr{D}$. The problem is that, since $i\mathscr{D}$ changes the chirality (γ^{D+1} anticommutes with the gamma matrices), there is no well-defined eigenvalue problem for chiral fermion fields. The equation $\mathscr{D}\psi = \psi$ does not make sense for a Weyl spinor (or, equivalently, there is no well-posed eigenvalue problem for a Weyl operator $\mathscr{D}_+$). A way of solving this difficulty is considering complex (Dirac) fermions instead of chiral (Weyl) ones and replacing the operator $i\mathscr{D}$ by $i\tilde{\mathscr{D}}$, where

$$\tilde{\mathscr{D}} = \mathscr{D}_+ + \partial_- \equiv \mathscr{D}P_+ + \partial P_- = \partial + AP_+ \quad , \quad (10.5.6)$$

so that now the generating functional becomes $\int D\psi D\bar{\psi}exp(-\int d^D x\, \bar{\psi}i\tilde{\mathscr{D}}\psi)$ $\equiv \det(i\tilde{\mathscr{D}})$. This is possible because, with such a $\tilde{\mathscr{D}}$, the negative chirality fermions do not couple to A, and therefore they just give a proportionality factor in (10.2.3), independent of A. This means that the chiral gauge theory defined by $\tilde{\mathcal{D}}$ is the same as the one based on $\mathcal{D}$ up to this proportionality factor. The new operator has a well-defined eigenvalue problem, although its eigenvalues are not real ($i\tilde{\mathcal{D}}$ is not selfadjoint). It will be useful to have the expression for the square of $\tilde{\mathcal{D}}$:

$$\tilde{\mathcal{D}}^2 = \mathcal{D}\partial P_- + \partial\mathcal{D}P_+ = \partial^2 + \partial A P_+ + A\partial P_- \quad . \tag{10.5.7}$$

Note that the new action

$$\tilde{I}_E(A,\psi) = \int d^D x\, \bar{\psi} i\tilde{\mathcal{D}}\psi \tag{10.5.8}$$

is invariant under the infinitesimal gauge transformations

$$\delta_\zeta \psi = -\zeta P_+ \psi \quad , \quad \delta_\zeta \bar{\psi} = \bar{\psi} P_- \zeta \quad , \quad \delta_\zeta A_\mu = \partial_\mu \zeta + [A_\mu, \zeta] \quad , \tag{10.5.9}$$

with $\zeta(x) = \zeta^i(x) T_i$, which are nothing but the ordinary gauge transformations for the positive chirality components of ψ, the negative ones being unaffected. Under such transformations, however, the measure is not invariant, and produces a consistent anomaly (the meaning of the word consistent in the present context will be given in section 10.7). Indeed, denoting by ψ' and A' the transformed fields

$$\tilde{W}_E[A'] = \int D\psi D\bar{\psi} e^{-\tilde{I}_E(A',\psi)} = \int D\psi' D\bar{\psi}' e^{-\tilde{I}_E(A',\psi')}$$

$$= \int D\psi' D\bar{\psi}' e^{-\tilde{I}_E(A,\psi)} = \int D\psi D\bar{\psi} J e^{-\tilde{I}_E(A,\psi)} \quad , \tag{10.5.10}$$

where in the second step the integration variables have simply been relabelled and in the third one the invariance of the tilded action (10.5.8) has been used. The Jacobian determinant J (cf. eq. (10.5.4)) for the transformation (10.5.9) (again, the inverse of the usual one due to the Grassmann character of the fields ψ) is given by the product of the determinants of the operators $(1 + \zeta P_+)$ and $(1 - \zeta P_-)$, which are the respective contributions of the $\mathcal{D}\psi$ and $\mathcal{D}\bar{\psi}$ parts of the measure. To write these determinants it is convenient to consider the eigenvalue problem for the operator $i\tilde{\mathcal{D}}$. Since the operator $i\tilde{\mathcal{D}}$ is an elliptic one, it has a discrete spectrum on a compact manifold, again Euclidean space compactified to S^D. But, since $i\tilde{\mathcal{D}}$ is not selfadjoint, its eigenvalues are complex and this implies that the left and right eigenfunctions

$$i\tilde{\mathcal{D}}\phi_n = \lambda_n \phi_n \quad , \quad \chi_n^\dagger i\tilde{\mathcal{D}} = \lambda_n \chi_n^\dagger \quad , \quad \int d^D x\, \chi_m^\dagger \phi_n = \delta_{mn} \tag{10.5.11}$$

have to be considered. The fields ψ, $\bar{\psi}$ may be expanded in terms of these eigenfunctions,

$$\psi = \sum_n a_n \phi_n \quad , \quad \bar{\psi} = \sum_n \bar{b}_n \chi_n^\dagger \quad . \tag{10.5.12}$$

Each factor of the Jacobian can be formally written as the exponential of the trace of $\zeta P_+, -\zeta P_-$ (ζ small). Using the fields ϕ_n and $\chi_n^\dagger$ the trace is computed by summing over n the diagonal 'matrix elements' $\chi_n^\dagger \zeta P_+ \phi_n$ and $\chi_n^\dagger (-\zeta P_-)\phi_n$, respectively, so that

$$
\begin{aligned}
J &= exp\{ \int d^D x \sum_n \chi_n^\dagger(x)(\zeta(x)P_+ - \zeta(x)P_-)\phi_n(x)\} \\
&= exp\{ \int d^D x \sum_n \chi_n^\dagger(x)\gamma_{D+1}\zeta\phi_n(x)\} \quad ,
\end{aligned}
\tag{10.5.13}
$$

where the summation over the group indices in ζ and in ϕ_n, χ_n is understood.

Since (10.5.13) is divergent it is necessary to introduce a cut-off to regulate it. Although the eigenvalues of $(i\tilde{\slashed{\mathcal{D}}})$ are not gauge-invariant[*], they may still be used to introduce a Gaussian regulator (the gauge transformation properties of the fermion determinant $det(i\tilde{\slashed{\mathcal{D}}})$ are discussed in the next section). The Jacobian J is computed as the limit

$$
J = exp\{ \lim_{M\to\infty} \int d^D x \sum_n (\chi_n^\dagger(x)\gamma_{D+1}\zeta\phi_n(x))e^{-\lambda_n^2/M^2}\}
\tag{10.5.14}
$$

which means that, in evaluating (10.5.14), it is only necessary to retain the term independent of M. The regulator $exp(-\lambda_n^2/M^2)$ may now be seen as the result of taking the trace of an operator involving $\tilde{\slashed{\mathcal{D}}}^2$:

$$
\begin{aligned}
J &= exp\{ \lim_{M\to\infty} \int d^D x \, \mathrm{Tr}[\gamma_{D+1}\zeta(x)\sum_n \phi_n(x)\chi_n^\dagger(x)]e^{-\lambda_n^2/M^2}\} \\
&= exp\{ \lim_{M\to\infty} \int d^D x \lim_{x\to y} \mathrm{Tr}[\gamma_{D+1}\zeta(x)e^{-(i\tilde{\slashed{\mathcal{D}}})^2(x)/M^2}\sum_n \phi_n(x)\chi_n^\dagger(y)]\} \\
&= exp\{ \lim_{M\to\infty} \int d^D x \lim_{x\to y} \mathrm{Tr}[\gamma_{D+1}\zeta(x)e^{-(i\tilde{\slashed{\mathcal{D}}})^2(x)/M^2}\delta(x-y)]\} \quad ,
\end{aligned}
\tag{10.5.15}
$$

where Tr indicates trace on the spinorial as well as on the gauge group indices, and for the last equality we have used the completeness relation for the eigenvectors $\phi_n(x)$, $\chi_n^\dagger(x)$, $\sum_n \phi_n(x)\chi_n^\dagger(y) = \delta^D(x-y)$. The actual calculation of the expression (10.5.15) in arbitrary even dimension involves a large amount of algebra. It is possible to calculate only the leading term of the anomaly and then use the Wess-Zumino consistency condition (see section 10.7) to determine the remaining terms; this can be done because the Wess-Zumino condition relates different orders in the gauge field. Since

---

[*] If they were, $i\tilde{\slashed{\mathcal{D}}}(A)\phi_n = \lambda_n\phi_n$ would imply $i\tilde{\slashed{\mathcal{D}}}(A^g)g^{-1}\phi_n = \lambda_n g^{-1}\phi_n$ and hence $g\tilde{\slashed{\mathcal{D}}}(A^g)g^{-1} = \tilde{\slashed{\mathcal{D}}}(A)$ which is not the case as may be checked using (10.5.6) and $A_\mu^g = g^{-1}(A_\mu + \partial_\mu)g$.

even this procedure is rather long, we shall perform the direct calculation in the simplest case, *i.e.* for $D = 2$.

The first step is to write the Dirac δ as the Fourier transform of unity. Then (10.5.15) can be written as

$$J = exp\{ \lim_{M \to \infty} \int d^2x \lim_{x \to y} \text{Tr}[\gamma_{D+1}\zeta(x)e^{\tilde{\not{\mathcal{D}}}^2(x)/M^2}\frac{1}{(2\pi)^D}\int d^2k\, e^{ik(x-y)}]\}$$

$$= exp\{\frac{1}{(2\pi)^D}\lim_{M \to \infty}\int d^2x \text{Tr}[\gamma_{D+1}\zeta(x)\int d^2k\, e^{-ikx}e^{\tilde{\not{\mathcal{D}}}^2(x)/M^2}e^{ikx}]\} \quad .$$

$$(10.5.16)$$

It is convenient to use the Baker-Campbell-Hausdorff formula (see footnote in section 3.8 for $A = \tilde{\not{\mathcal{D}}}^2/M^2$ and $B = -\partial^2/M^2$) and the second equality of eq. (10.5.7) to write

$$e^{\tilde{\not{\mathcal{D}}}^2/M^2} \simeq exp\{\frac{1}{M^2}(\not{\partial}\!AP_+ + A\not{\partial}P_-) + \frac{1}{2M^4}[\partial^2, \not{\partial}\!AP_+ + A\not{\partial}P_-] + ...\}e^{\partial^2/M^2},$$

$$(10.5.17)$$

where

$$[\partial^2, \not{\partial}\!AP_+ + A\not{\partial}P_-] = \not{\partial}(\partial^2 A)P_+ + (\partial^2 A)\not{\partial}P_- + 2\{\not{\partial}(\partial^\mu A)P_+ + (\partial^\mu A)\not{\partial}P_-\}\partial_\mu\,.$$

$$(10.5.18)$$

The operator e^{∂^2/M^2}, when acting on e^{ikx}, gives $e^{-k^2/M^2}e^{ikx}$. To look for the terms that contribute in the limit $M \to \infty$, we again make the change $k^\mu \to s^\mu = k^\mu/M$, $e^{-k^2/M^2} \to e^{-s^2}$ and $d^2k \to d^2sM^2$. Then it is sufficient to look for the terms in the first exponential of the r.h.s. of (10.5.18) with a factor $\frac{1}{M^2}$, because only these terms survive in the limit $M \to \infty$. Each time that a ∂ outside brackets in (10.5.18) acts on e^{iMsx}, it produces a factor M. The potentially divergent terms do not contribute: the term proportional to M^2 comes from the 1 in the expansion of (10.5.17) and it is zero since $\text{Tr}(\zeta\gamma_3) = 0$; the term proportional to M comes from the linear term in that expansion (the last one in (10.5.18)) and $\text{Tr}[\zeta\gamma_3(\not{s}\!AP_+ + A\not{s}P_-)] = \text{Tr}[\zeta(\not{s}\!AP_+ - A\not{s}P_-)] = \text{Tr}[\zeta(\not{s}\!AP_+ - A P_+\not{s})] = 0$. This means that the terms we are looking for are

$$J = exp\{\frac{1}{(2\pi)^2}\int d^2x\, \text{Tr}[\gamma_3\zeta \int d^2s\, e^{-s^2}[(\not{\partial}\!A)P_+$$

$$(10.5.19)$$

$$- \frac{1}{2}(\not{s}\!A)^2P_+ - \frac{1}{2}(A\not{s})^2P_- - (\not{s}(\partial^\mu A)P_+ + (\partial^\mu A)\not{s}P_-)s_\mu]]\} \,.$$

There are no more contributions to (10.5.19) because the number of derivatives in the higher order terms from (10.5.17) available to compensate the powers of $1/M$ in such a way that only $1/M^2$ is left is not large enough (we see immediately that for $D \geq 2$ the calculation becomes much more involved: higher dimensions require more terms in the Baker-Campbell-Hausdorff expansion). In the expression (10.5.19), all derivatives act on the A fields only, contrary to the expression (10.5.17). The second and third terms in (10.5.19) as well as the fourth and fifth terms cancel between themselves due to the presence of Tr.

Since $\int d^2s\, e^{-s^2} = \pi$, the remaining first term gives the exponential we are looking for:

$$
\begin{aligned}
J &= exp\{\frac{\pi}{(2\pi)^2}\int d^2x\, \text{Tr}[\gamma_3\zeta(\slashed\partial\slashed A)P_+]\}\\
&= exp\{-\frac{i}{4\pi}\int d^2x\,\text{Tr}(\zeta\epsilon_{\mu\nu}\partial^\mu A^\nu) + \frac{1}{4\pi}\int d^2x\,\text{Tr}(\zeta\partial_\mu A^\mu)\}\ ,
\end{aligned}
$$

(10.5.20)

since $\text{Tr}(\gamma_3\gamma_\mu\gamma_\nu) = -2i\epsilon_{\mu\nu}$. Nevertheless, this expression may be replaced by a simpler one because the second term in the exponential of (10.5.20) is not an essential part of the anomaly. In fact, it is not difficult to see that a redefinition of $Z[A]$ by means of a local functional can produce a gauge variation compensating this term. Such a redefinition is just proportional to $\int d^2x\,\text{Tr}(A^\mu A_\mu)$, since

$$
\begin{aligned}
\delta_\zeta\int d^2x\,\text{Tr}(A^\mu A_\mu) &= 2\int d^2x\,\text{Tr}(A^\mu\partial_\mu\zeta + A^\mu[A_\mu,\zeta])\\
&= -2\int d^2x\,\text{Tr}(\partial_\mu A^\mu\zeta)\ .
\end{aligned}
$$

(10.5.21)

As will be shown later on, the anomaly is given by a non-trivial one-cocycle; the second term in (10.5.20) then corresponds to a one-coboundary. If we only consider the other term, the result is the integral of a differential form which has the same expression when going back to Minkowski spacetime. Thus, looking at (10.5.20) and dividing by i (see (10.5.4)), it is seen that the anomaly in $D = 2$ is given, to first order in ζ, by

$$
\int d^2x\,\zeta^i\mathcal{V}_{(i)}[A](x) = -\frac{1}{4\pi}\int\text{Tr}(\zeta\, dA)
$$

(10.5.22)

(see eq. (10.5.3)). Proceeding as in the two-dimensional case, it can be found that the expression for the anomaly for $D = 4$ is

$$
\frac{-i}{24\pi^2}\int\text{Tr}[\zeta(d(AdA + \frac{1}{2}A^3)]\ ,
$$

(10.5.23)

which means that

$$
(\mathcal{D}_\mu J^\mu)_{(j)} = \frac{-i}{24\pi^2}\epsilon^{\mu\nu\rho\sigma}\text{Tr}[T_{(j)}\partial_\mu(A_\nu\partial_\rho A_\sigma + \frac{1}{2}A_\nu A_\rho A_\sigma)]\ .
$$

(10.5.24)

We note in passing that the anomaly (10.5.23) is of order q^2; we reabsorbed the coupling constant q in the definition of the gauge field (see the second footnote in section 2.2). In contrast with the abelian anomaly, *the non-abelian anomaly is not gauge-invariant* and it is given by the *covariant* divergence of a non-abelian current. Expressions (10.5.22) and (10.5.23) divided by 2π will be recovered later on by purely geometric arguments based on the Chern forms (alternatively, we may multiply the Chern forms by 2π since they are normalized to produce integers).

Eq. (10.5.23) may be rewritten in the form

$$d_{ijk}\frac{i}{(24\pi)^2}\int \text{Tr}[d\zeta^i A^j (dA^k + \frac{1}{4}C^k_{st}A^s A^t)] \qquad (10.5.25)$$

where the purely group theoretic factor d_{ijk} is given by

$$d_{ijk} = \frac{1}{2}\text{Tr}(T_i\{T_j, T_k\}) := \text{sTr}(T_i, T_j, T_k) \quad (i,j,k = 1,..,\dim, G) \quad (10.5.26)$$

where sTr is the symmetric trace. Thus, a $D=4$ theory will not be anomalous if the factor (10.5.26) is zero. The Lie algebra matrices appearing in (10.5.26) are the generators of the representation of G on the chiral fermions; they were contained in the expression for the covariant derivative $\mathscr{D}_\mu = \partial_\mu + A_\mu$ on the spinors. Thus, this expression depends on the representation of the Weyl fermions. In arbitrary $D = 2p$ dimensions the anomaly will not be present when

$$d_{i_1...i_{p+1}} = 0 \quad \Longleftrightarrow \quad \text{Tr}(T_{i_1}\{T_{i_2}, T_{i_3}, ..., T_{i_{p+1}}\}) = 0 \qquad (10.5.27)$$

where $\{T_{i_2}, T_{i_3}, ..., T_{i_{p+1}}\}$ is the symmetrized product of the generators. Groups or algebras for which this is zero irrespective of the representation used are called *safe*; in four dimensions, all simple algebras are safe with the exception of $su(n)$, $n \geq 3$. Safe groups are groups that do not admit totally symmetric invariant tensors of rank $(p + 1)$ in the adjoint index. When they are irreducible, these tensors are proportional to the generalized Casimir invariants* of $\mathscr{G}$. Looking at Table 6.7.1 and at eq. (10.5.27), it is seen that the dimensions for which anomalies appear are given by $D = 2(m_i - 1)$ $(i = 1,...,l)$, i.e. $2, 4, ..., 2l$ (A_l); $2, 6, 10, ..., 2(2l - 1)$ $(B_l$ and $C_l)$, $2, 6, 10, ..., 2(2l - 3)$ and $2(l - 1)$ (D_l), etc. In particular, since no third rank antisymmetric tensor exists for a simple algebra but for $su(n)$, $n \geq 3$, only these algebras are unsafe in four dimensions.

When the group is not safe, the anomaly may still vanish for some representations. This is what happens in the Weinberg-Salam standard model for which the gauge group, $SU(3) \otimes SU(2) \otimes U(1)$, is not safe. In

---

* The ring $I(G)$ of AdG-invariant polynomials of a compact simple Lie group is isomorphic to the centre of the enveloping algebra of $\mathscr{G}$; if G has rank l, $I(G)$ has l independent generators of order $m_1, ..., m_l$. This means that there are l polynomials P_r, $r = 1, ..., l$, or symmetric tensors d of order $m_1, ..., m_l$ (see section 6.7). For instance, for the simple algebra A_l ($su(n)$) the order of the $l = n - 1$ generalized Casimir invariants is $2, ..., n$. Thus, for $su(3)$ and writing still T_i for the matrices $D(T_i)$ in a given representation, we find $\text{Tr}(T_i T_j) = C_2(D)\delta_{ij}$ and $\frac{1}{2}\text{Tr}(T_i\{T_j, T_k\}) = C_3(D)d_{ijk}$, where $C_2(D)$ and $C_3(D)$ are the values of the second and third Casimir operators for the representation D of $\mathscr{G}$.

the one-family approximation, the standard model contains a left-handed $SU(2)$ doublet of leptons (ν_L, e_L) and another one of quarks (u_L, d_L) plus the right-handed singlets e_R, u_R, d_R; $SU(3)$ is the colour group, which gives three copies of each quark flavour, and $U(1)$ is generated by the weak hypercharge Y; $Q = T_3 + Y/2$ so that $Y_R = 2Q$ and $Y_L = 2(Q - T_3)$. The electroweak gauge group $SU(2) \otimes U(1)$ initially contains two gauge coupling constants because it is not simple. From the previous calculation, it is clear that the expression for the anomaly for right-handed fermions is the same but for a minus sign (changing the chirality is equivalent to the transformation $\gamma_5 \mapsto -\gamma_5$) and, since different chiralities do not mix, when there are spinors of both chiralities the $D = 4$ anomaly is proportional to

$$\text{Tr}(\lambda_i^L \{\lambda_j^L, \lambda_k^L\}) - \text{Tr}(\lambda_i^R \{\lambda_j^R, \lambda_k^R\}) \tag{10.5.28}$$

(thus, a possibility for anomaly cancellation is having the same number of L and R degrees of freedom, but this would not be a chiral theory). In the present case, the $\lambda_i^{R,L}$ can be organized into three mutually commuting sets, $\lambda_{3i}^{R,L}$, $\lambda_{2i}^{R,L}$, $\lambda_1^{R,L}$, which are the representation matrices of the $su(3)$, $su(2)$, $u(1)$ algebras respectively ($\lambda_1^R = 2Q$, $\lambda_1^L = 2(Q - T_3)$, $\lambda_{2i}^R = 0$, $\lambda_{2i}^L = \frac{\sigma_i}{2}$). For instance, $\lambda_{2i}^L = \mathbf{1} \otimes (\sigma_i/2) \otimes \mathbf{1}$. Let us now consider the different possibilities for eq. (10.5.28):

a) Only one matrix is a generator of $su(2)$ or $su(3)$, i.e. there is only a λ_{2i} or a λ_{3i}. Then the result is zero because it contains $\text{Tr}(\lambda_{2,3}^{R,L}) = 0$ and the Pauli and Gell-Mann matrices λ_{3i} are traceless.
b) All three matrices belong to $su(3)$. The result is zero because $\lambda_{3i}^R = \lambda_{3i}^L \equiv \lambda_{3i}$.
c) Two matrices are in $su(3)$ and the other is the $U(1)$ generator. Eq. (10.5.28) is then the difference of the traces of tensor product matrices which is proportional to $\text{Tr}(Q - T_3) - \text{Tr}(Q) = 0$.
d) All generators belong to $su(2)$. This gives zero result because $SU(2)$ is a safe group (there is no d_{ijk} for $SU(2)$ since $\text{Tr}(\sigma_i \{\sigma_j, \sigma_k\}) = 0$).
e) Two generators are in $su(2)$ and the third one belongs to $u(1)$. Then the result is proportional to $\text{Tr}(\lambda_1^L) = 2\text{Tr}(Q)$ because the right part of (10.5.28) does not contribute.
f) All generators belong to $u(1)$. Eq. (10.5.28) is proportional to

$$\text{Tr}((\lambda_1^L)^3) - \text{Tr}((\lambda_1^R)^3) \propto \text{Tr}(Q^3 - \frac{3}{2}\sigma_3 Q^2 + \frac{3}{4}Q - \frac{1}{8}\sigma_3) - \text{Tr}(Q^3) \propto \text{Tr}(Q) \ . \tag{10.5.29}$$

From the above analysis, it is clear that the total anomaly in the standard model vanishes if $\text{Tr}(Q) = 0$, i.e., if the sum of the charges of all fermions is zero. But this is indeed the case for *each* family because, in each, there are two leptons of charge -1 (e_L, e_R say, one for each chirality), one lepton of charge 0, ν_L, six quarks with charge 2/3 (two chiralities u_R, u_L in three colours) and six quarks (d_R, d_L in three colours) with charge $-1/3$, which gives a total charge of $-2 + 6(\frac{2}{3} - \frac{1}{3}) = 0$. Thus, the cancellation of the anomaly in the standard model is achieved with three colours irrespective of the number of families due to their suitable (*i.e.*, anomaly cancelling) structure.

10.6 The non-abelian anomaly as a probe for non-trivial topology

The reader will have noticed the similarity between the expression for the $D = 4$ $U(1)$ anomaly and the non-abelian gauge anomaly. They are given by the divergence and covariant divergence, respectively, of the currents J_5^μ and $J_{(i)}^\mu$, eqs. (10.2.25), (10.5.24); ignoring the overall coefficient, they are similar apart from the numerical factor of the A^3 term. In fact, as shown by Atiyah and Singer and by Álvarez-Gaumé and Ginsparg, *the expression for the non-abelian anomaly in $D = 2p$ dimensions may be obtained from that for the abelian anomaly in a $(D + 2)$-dimensional space* by using a suitable $(D + 2)$-dimensional index theorem (a two-higher-dimensional index theorem is also a convenient tool to investigate the gravitational anomalies, which exist only for $D = 4k - 2$). We shall devote this section to sketching how this may be done, and to relating the non-abelian anomaly to the non-trivial topology of the orbit space as was anticipated in section 10.1.

Let us consider again the action of a Yang-Mills theory with a chiral fermion on an even-dimensional Euclidean space compactified to S^{2p}, and let G be a simple, simply connected group as *e.g.* $SU(n)$. The effective action $Z[A]$ is given by

$$e^{-Z[A]} = \int D\psi D\bar{\psi} e^{-\int d^{2p}x\, \bar{\psi} i \slashed{D} P_+ \psi} \quad . \tag{10.6.1}$$

As we saw in the previous section, the identification of (10.6.1) with $\det(i\slashed{D}P_+)$ is not possible since, apart from the regularization problems, the operator $i\slashed{D}P_+$ does not have a well-defined eigenvalue problem. This difficulty was avoided there by introducing a new operator $\tilde{\slashed{D}}$ (eq. (10.5.6)) and by using a Dirac spinor instead of a Weyl one. When this is done, the new action (10.5.8) permits expressing (10.6.1) as a determinant

$$e^{-Z[A]} = \int D\psi D\bar{\psi} e^{-\int d^{2p}x\, \bar{\psi} i \tilde{\slashed{D}} \psi} \equiv \det(i\tilde{\slashed{D}}) \quad . \tag{10.6.2}$$

As we know, the functional integral (or, equivalently, $\det(i\tilde{\slashed{D}})$) is not gauge-invariant,

$$e^{-Z[A^g]} = e^{i\alpha_1[A,g]} e^{-Z[A]} \quad ; \tag{10.6.3}$$

this lack of invariance has the effect that (10.6.2) cannot be expressed as an integral over the orbit space $\mathscr{A}/\mathscr{C}$.

By eq. (10.6.3), the fermion determinant may itself serve to probe the non-trivial topology of the orbit space. It is easy to check that its variation must be imaginary since its modulus is gauge-invariant. Using (10.5.6) it is seen that

$$(\det i\tilde{\mathscr{D}})(\det i\tilde{\mathscr{D}}^\dagger) = \det(i\tilde{\mathscr{D}}i\tilde{\mathscr{D}}^\dagger) = \det[(i\mathscr{D}_+ + i\partial_-)(i\mathscr{D}_- + i\partial_+)]$$

$$= \det(i\mathscr{D}_+ i\mathscr{D}_- + i\partial_- i\partial_+) = \det \begin{bmatrix} i\partial_- i\partial_+ & \\ & i\mathscr{D}_+ i\mathscr{D}_- \end{bmatrix} \qquad (10.6.4)$$

where the Weyl realization for the gamma matrices (which has zero diagonal boxes, eq. (2.11.9)), has been used. Ignoring the unimportant factor $\det(i\partial_- i\partial_+)$, (10.6.4) gives $|\det(i\tilde{\mathscr{D}})|^2 \propto \det(i\mathscr{D}_+ i\mathscr{D}_-)$. But, since $\det(i\mathscr{D}_+ i\mathscr{D}_-)^2 = \det(i\mathscr{D})^2$, where $\mathscr{D}$ is the ordinary Dirac operator, it is found that, up to a constant factor,

$$|\det i\tilde{\mathscr{D}}| = (\det i\mathscr{D})^{1/2} \qquad (10.6.5)$$

which, as we already know, can be regulated in a gauge-invariant manner. This shows that only the imaginary part of $\det(i\tilde{\mathscr{D}})$ may change as implied by (10.6.3). In fact, general arguments (with some qualifications in the $D = 2$ case) may be put forward that show that only the imaginary part of Z may change since the real part always admits a gauge-invariant definition.

To gain some insight into the topological nature of the non-abelian gauge anomaly let us look at the connection between the D-dimensional non-abelian anomaly and the $D + 2$ index theorem which describes, as we saw in section 10.2, the abelian anomaly in $D + 2$ dimensions. Let us consider now a family of gauge transformations $g(\theta, x)$ depending on a parameter $\theta \in S^1$ satisfying the boundary conditions

$$g(0, x) = g(2\pi, x) = e \quad . \qquad (10.6.6)$$

These transformations define mappings $g : S^1 \times S^{2p} \to G$ that are classified, on account of (10.6.6), by the homotopy classes $\pi_{2p+1}(G)$. This may be seen by noticing that with these conditions on $g(\theta, x)$ the product of S^1 and S^{2p} is topologically equivalent to S^{2p+1}. To see this, consider two topological spaces A and B with base points a and b, and let $A \times B$ be their Cartesian product. Then, if in this set we identify (or 'smash' to a point) the points of the *one-point union* $A \vee B \equiv (A \times \{b\}) \cup (\{a\} \times B)$, the result (*i.e.*, the quotient $A \times B / A \vee B$) is the *smash product* $A \wedge B$ (sometimes called *reduced join*). For instance, if both $A = B = S^1$, the result is the two-dimensional sphere S^2 considered as a 'loop of loops': it may be obtained (Figure 10.6.1) by gliding a circle over another while keeping a

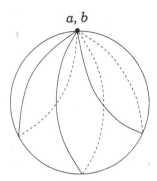

Figure 10.6.1   The smashed product $S^1 \wedge S^1$ or loop of loops

common point (*e.g.* the north pole) fixed (similarly, a loop of S^m spheres is S^{m+1}, so that $S^1 \wedge S^m = S^{m+1} = S^m \wedge S^1$; in general, $S^p \wedge S^q = S^{p+q}$).

Let $A(x)$ now be a 'reference' gauge field in the zero instanton sector (it corresponds to a connection on a trivial bundle) and let

$$A^\theta = g^{-1}(\theta, x)(d + A)g(\theta, x) \quad , \tag{10.6.7}$$

where $d := \sum_{\mu=0}^{D-1} dx^\mu (\partial/\partial x^\mu)$, be the resulting one-parameter family of group-transformed Yang–Mills configurations. It may be assumed that $i\tilde{\mathscr{D}}(A)$ has no zero modes; then, since only the phase can pick up an anomalous change under a gauge transformation, the operator $i\tilde{\mathscr{D}}(A^\theta)$ does not have them either $\forall \theta \in U(1)$ since the modulus of the determinant is gauge-invariant, $|\det(i\tilde{\mathscr{D}}(A^\theta))| = |\det(i\tilde{\mathscr{D}}(A))|$. We may then write

$$e^{-Z[A^\theta]} = \det(i\tilde{\mathscr{D}}(A^\theta)) = \det(i\tilde{\mathscr{D}}(A))e^{i\alpha[A,\theta]} \quad ; \tag{10.6.8}$$

θ enters in α and in A^θ through $g(\theta, x)$. The functional $\det(i\tilde{\mathscr{D}}(A^\theta))$ defines, through the exponent in $\alpha[A, \theta]$, a mapping $\theta \in S^1 \mapsto e^{i\alpha[A,\theta]} \in U(1)$. These mappings are characterized by an integer winding number

$$k = \frac{1}{2\pi} \int_0^{2\pi} d\theta \frac{\alpha[A, \theta]}{\partial \theta} \quad . \tag{10.6.9}$$

The writing of the integrand of (10.6.9) is formal since it is not exact; otherwise k would always be zero. To obtain this winding number, and to find a local density for it, let us now extend A^θ to a *two-parameter* family $A^{t,\theta}$ of gauge fields. If A^θ describes a circle S^1 in the gauge fields space $\mathscr{A}$, $A^{t,\theta}$ is defined on a two-dimensional disk D^2, $\partial D^2 = S^1$, by

$$A^{t,\theta} \equiv tA^\theta = tg^{-1}(\theta, x)(d + A)g(\theta, x) \tag{10.6.10}$$

where $d = \sum dx^\mu (\partial/\partial x^\mu)$ as before and (t, θ) are the polar coordinates of the disc D^2, $0 \le t \le 1$. At its boundary, $A^{t,\theta}$ becomes the one-parameter family of gauge-related configurations (10.6.7). Geometrically, if A^θ describes a circle in $\mathscr{A}$, $A^{t,\theta}$ determines a two-dimensional disk in the space of all gauge fields on S^{2p} that belong to the trivial topological class in $\mathscr{A}$ (as was briefly discussed in section 10.1, this triviality is not important in the discussion of the topology of the orbit space $\mathscr{A}/\mathscr{C}$). The Dirac determinant $\det(i\tilde{\mathscr{D}}(A^{t,\theta}))$ becomes a complex functional of the gauge fields on this disk, and (10.6.9) measures the winding number of the phase of $\det(i\tilde{\mathscr{D}}(A^{t,\theta}))$ as we move along the circle of gauge fields at its boundary. We shall not enter into the details required to compute the winding number determined by $\det(i\tilde{\mathscr{D}}(A^{t,\theta}))$. We shall just mention that it may be shown to be equal to the difference of the positive and negative chirality zero modes of the ordinary Dirac operator $i\mathscr{D}_{2p+2}$ in $(D+2)$ dimensions, *i.e.* that

$$index(i\mathscr{D}_{2p+2}) = n_+ - n_- = k = \frac{1}{2\pi} \int_0^{2\pi} d_\theta \alpha[A,\theta] \qquad (10.6.11)$$

where $\mathscr{D}_{2p+2} = \sum_{\mu=0}^{2p+1} \gamma^\mu \mathscr{D}_\mu$ (the γ's being now those of a $(D+2)$-dimensional space with coordinates t, θ, x^μ) and we have introduced the notation $d\theta\frac{\partial}{\partial\theta} = d_\theta$. This is accomplished by relating the zero modes of $i\tilde{\mathscr{D}}(A^{t,\theta})$ to those of $i\mathscr{D}_{2p+2}$ (note that, although $i\tilde{\mathscr{D}}(A)$ was assumed to have no zero modes so that $i\tilde{\mathscr{D}}(A^\theta)$ is also free of them, $A^{t,\theta}$ is not related to A^θ by a gauge transformation and so $i\tilde{\mathscr{D}}(A^{t,\theta})$ may have zero modes in the interior of the disk D^2). We shall admit (10.6.11) without further discussion and check now that the winding number (10.6.9) of the phase of the D-dimensional Weyl determinant is measured by the homotopy class in $\pi_{2p+1}(G)$ of the mapping $g(\theta, x)$.

Let us then evaluate $index(i\mathscr{D}_{2p+2})$. The gauge field $A^{t,\theta}(x)$ is defined on the manifold $D^2 \times S^{2p}$ with boundary $S^1 \times S^{2p}$, and so is the operator $i\mathscr{D}_{2p+2}$. In order to use the index theorem for manifolds without boundary, *i.e.* to avoid the boundary corrections, it is convenient to work with the boundaryless manifold $S^2 \times S^{2p}$. Then the Yang-Mills fields $\mathbf{A}$ on this larger manifold $S^2 \times S^{2p}$ are the pull-backs of connections on a principal bundle over $S^2 \times S^{2p}$ with structure group G. Thus, as far as the two-sphere is concerned, we can use to avoid a singular parametrization two local charts $D_\pm^2 \times S^{2p}$ where the disks D_+ (D_-) are parametrized by (t, θ) $((s, \theta))$, D_+ being the previous D^2. The overlapping region at the equator corresponds to taking $t = 1 = s$; the north (south) poles of S^2 are given by $t = 0$ ($s = 0$). Let us now take for the two local gauge fields

$$(0, 0, \mathbf{A}_\pm) \qquad (10.6.12)$$

on the two charts D_+ $(0 \leq t \leq 1)$ and D_- $(0 \leq s \leq 1)$ the expressions

$$\mathbf{A}_+(t,\theta,x) = t[g^{-1}(A+d+d_\theta)g] = A^{t,\theta} + tg^{-1}d_\theta g \quad, \quad \mathbf{A}_-(s,\theta,x) = A \quad,$$
$$(10.6.13)$$

the lower disk being trivial. If $\tilde{d}$ is given by

$$\tilde{d} = d + d_\theta + d_t + d_s \tag{10.6.14}$$

with obvious notation, we see that $\mathbf{A}_\pm$ at the equator $t = 1 = s$ are gauge-related,

$$\mathbf{A}_+ = g^{-1}(\tilde{d} + \mathbf{A}_-)g \tag{10.6.15}$$

since $d_t g(\theta, x) = 0 = d_s g(\theta, x)$ so that $\mathbf{A}_\pm$ do define a connection on the principal bundle over $S^2 \times S^{2p}$ with group G. Thus, $g(\ ,x) : S^1 \to \mathscr{C}$ (which by the previous considerations also defines a mapping $g(\theta, x)$ homotopically equivalent to a mapping $g : S^{2p+1} \to G$) is just the transition function between the local expressions $\mathbf{A}_\pm$ at the boundary $S^1 \times S^{2p}$ of the two patches. These mappings (or transition functions) define loops in $\mathscr{C}$ that are classified by $\pi_1(\mathscr{C})$. As the reader will have recognized, this picture is analogous to the geometric description of the magnetic monopole in section 2.7.

Since the expression of the $\hat{A}$-roof genus of the manifold $S^2 \times S^{2p}$ is one (see section 2.11), the expression of the index that we need to compute is taken to be $(p = D/2)$

$$index(i\mathscr{D}_{2p+2}(\mathbf{A})) = \int_{S^2 \times S^{2p}} ch_{p+1}(\mathbf{F}) = \int_{D_+ \times S^{2p}} ch_{p+1}(\mathbf{F}_+) + \int_{D_- \times S^{2p}} ch_{p+1}(\mathbf{F}_-) \quad .$$
$$(10.6.16)$$

As we know, the Chern character $(2p + 2)$-form $ch_{p+1}(\mathbf{F})$ (eq. (2.5.12)) is closed, and the local potential $(2p + 1)$-forms are given by the Chern-Simons forms $Q^{(2p+1)}$ for the gauge fields (10.6.13) on the corresponding charts. Proceeding in the same way as in previous calculations (see, *e.g.*, section 2.7), we find that (10.6.16) is given by integrating over the common boundary $S^1 \times S^{2p}$ the *difference*

$$Q^{(2p+1)}(\mathbf{A}_+, \mathbf{F}_+)|_{t=1} - Q^{(2p+1)}(\mathbf{A}_-, \mathbf{F}_-)|_{s=1} \quad . \tag{10.6.17}$$

But, on account of (10.6.15), this expression is the variation of $Q^{(2l+1)}$ under a gauge transformation, and is given by eq. (2.6.15) once l is replaced there by $(p + 1)$,

$$Q_+^{(2p+1)} - Q_-^{(2p+1)} = Q^{(2p+1)}(g^{-1}\tilde{d}g, 0) + \tilde{d}\alpha^{(2p)} \quad, \tag{10.6.18}$$

where the $\tilde{d}$ appearing in (10.6.18) is now the appropriate exterior derivative $\tilde{d} = d + d_\theta$. Taking into account that the second term in (10.6.18) is an exact form and that accordingly it does not contribute to the integration over $S^1 \times S^{2p}$ we find, using (2.6.19),

$$index(i\mathscr{A}_{2p+2}) = (-1)^p \left(\frac{i}{2\pi}\right)^{(p+1)} \frac{p!}{(2p+1)!} \int_{S^1 \times S^{2p}} \mathrm{Tr}(g^{-1}\tilde{d}g)^{2p+1} \quad .$$

(10.6.19)

By recalling previous arguments and eq. (2.6.21) we see that (10.6.19) is the number of times $g(\theta, x)$ wraps around S^{2p+1}; this winding number is in $\pi_{2p+1}(G) \equiv \pi_{D+1}(G)$.

Let us now identify the local density $id_\theta\alpha[A, \theta]$, which describes the non-abelian anomaly in a D-dimensional spacetime. For this, we have to relate both sides of (10.6.11). First, we notice that in terms of $Q^{(2p+1)}$, $index(i\mathscr{A}_{2p+1})$ reduces to

$$index(i\mathscr{A}_{2p+2}) = \int_{S^1 \times S^{2p}} Q^{(2p+1)}(A^\theta + \hat{v}, F^\theta) \tag{10.6.20}$$

where $\hat{v}$ is the one-form $\hat{v} \equiv g^{-1}d_\theta g \equiv v d\theta$ and $F^\theta = \tilde{d}(A^\theta + \hat{v}) + (A^\theta + \hat{v})^2 = g^{-1}Fg$ since the term in Q_-, evaluated at $s = 1$, does not have a $d\theta$ component (see (10.6.13)) and cannot contribute to the integral of $S^1 \times S^{2p}$. The only term contributing in (10.6.20) with the required $d\theta$ is the term linear in $\hat{v}$. By (10.6.11) we know that the local density for the anomaly is the local density for $index(i\mathscr{A}_{2p+2})$. Thus, what is needed is the first order variation

$$Q_1^{(2p)}(\hat{v}, A^\theta, F^\theta) = Q^{(2p+1)}(A^\theta + \hat{v}, F^\theta) - Q^{(2p+1)}(A^\theta, F^\theta) \tag{10.6.21}$$

($\hat{v}^2 \equiv 0$). Since the expression for $Q^{(2p+1)}$ is known, this is not difficult to compute. For $D = 2p = 4$ this gives using (10.6.11)

$$\frac{1}{2\pi}d_\theta\alpha[A, \theta] = \int_{S^4} Q_1^4(\hat{v}, A^\theta, F^\theta) = \frac{-i}{48\pi^3} \int_{S^4} \mathrm{Tr}\hat{v}d(A^\theta dA^\theta + \frac{1}{2}(A^\theta)^3) \quad .$$

(10.6.22)

Factoring out $d\theta$ and setting $\theta = 0$ we find the expression for the anomaly as

$$\frac{\partial\alpha}{\partial\theta}[A, g] = \frac{-i}{24\pi^2} \int_{S^4} \mathrm{Tr}[vd(AdA + \frac{1}{2}A^3)] \quad . \tag{10.6.23}$$

From eqs. (10.5.3) and (10.5.5), we find

$$\mathscr{V}_{(i)}[A](x) = \mathscr{D}_\mu J_{(i)}^\mu = \frac{-i}{24\pi^2} \mathrm{Tr}[T_{(i)}\epsilon^{\nu\rho\sigma\kappa}\partial_\nu(A_\rho\partial_\sigma A_\kappa + \frac{1}{2}A_\rho A_\sigma A_\kappa)] \tag{10.6.24}$$

as given in eq. (10.5.24).

We may now look at the topological implications of these results in connection with the comments in section 10.1. The previous arguments may be considered as a check of the equality $\pi_1(\mathscr{C}) = \pi_5(G)$ (we did not prove (10.6.11), but we did find the anomaly (10.6.24) from it). What about $\pi_2(\mathscr{A}/\mathscr{C})$? Consider again the disk of Yang-Mills potentials $A^{t,\theta}$ in $\mathscr{A}$. Since the potentials A^θ on the boundary $t = 1$ of the disk are gauge-related, the projection of this disk on the orbit space is a two-sphere, since all A^θ are projected onto the reference potential A.

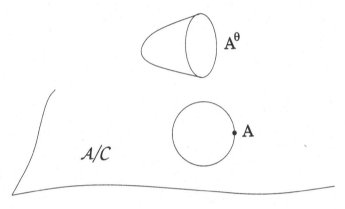

Figure 10.6.2

Since the 't part' of $A^{t,\theta}$ is topologically trivial, the two-sphere in $\mathscr{A}/\mathscr{C}$ will always be contractible ($\pi_2(\mathscr{A}/\mathscr{C}) = 0$) if and only if the loop A^θ in $\mathscr{A}$ is trivial, *i.e.* if $\pi_1(\mathscr{C}) = 0$. The homotopic non-triviality of the gauge group is equivalent to that of the two-sphere (otherwise the homotopy used to contract the sphere could be used to deform the loop $g(\theta, x)$ to a point $g(x)$), and $\pi_1(\mathscr{C}) = \pi_2(\mathscr{A}/\mathscr{C})$). When this non-triviality is present, there is a global topological obstruction to removing the phase factor in the definition of the determinant. As a result, the determinant cannot be consistently restricted to the orbit space $\mathscr{A}/\mathscr{C}$ since it cannot be defined in a gauge-invariant manner. Instead, if det A is computed by using some regularization procedure, it is found that

$$\det(A^g) = e^{i\alpha_1[A,g]} \det(A) \quad , \tag{10.6.25}$$

where α_1 may be looked at as a mapping of $\mathscr{A} \times \mathscr{C}$ in the multiplicative group of complex numbers (alternatively, it associates a functional of A to g) hence the subscript 1. The action of two transformations requires that

$$\alpha_1[A^g, g'] - \alpha_1[A, gg'] + \alpha_1[A, g] = 0 \pmod{2\pi} \quad , \tag{10.6.26}$$

which is a one-cocycle condition (see lemma 10.7.1 below); it relates topological properties with cohomology since the variation of the phase of the determinant is an obstruction to projecting gauge orbits in $\mathscr{A}$ onto points of $\mathcal{A}/\mathcal{C}$. This picture defines a complex line bundle over $\mathcal{A}/\mathcal{C}$, the associated *determinant bundle*. It is characterized by its first Chern class which determines an element in $\pi_1(\mathcal{C})$ $(= \pi_{2p+1}(G)) = \pi_2(\mathcal{A}/\mathcal{C}) = Z$: *the determinant bundle over a non-contractible two-sphere in orbit space with winding number in* $\pi_1(\mathcal{C})$ *is identical to the bundle describing a monopole of the same number of units of magnetic charge.* Giving a representative of the first Chern class in a topologically non-trivial configuration corresponds to giving a specific form of the anomaly.

Summarizing, in $D = 4$ the abelian anomaly is governed by $\pi_0(\mathcal{A}/\mathcal{C}) = \pi_3(G)$ and the non-abelian anomaly by $\pi_2(\mathcal{A}/\mathcal{C}) = \pi_5(G)$ (Witten's global anomaly depends on $\pi_1(\mathcal{A}/\mathcal{C}) = \pi_4(G)$). It must be stressed, however, that having $\pi_5(G) = Z$ $(\pi_{2p+1}(G) = Z)$ is a *sufficient* but not a *necessary* condition for the existence of an anomalous variation of $Z[A]$ (an example is provided by a $U(1)$ gauge theory coupled to a Weyl fermion in any even dimension, which is perturbatively anomalous despite the fact that $\pi_{2p+1}(U(1)) = 0$, $p \neq 0$). After all, the determinant may have a local variation even if the topology does not force it to acquire a non-trivial topological phase. Thus, we may look at the previous formulae as a practical way of extracting the local phase variation (and hence, the anomaly) not necessarily related to the existence of a non-zero integer winding number (in $D = 4$, nevertheless, all anomaly free groups have $\pi_5(G) = 0$). Hence, even if the variation of the phase functional (10.6.22) does not have a global topological meaning for a compactified spacetime S^D (although it may appear as a topological obstruction in less topologically trivial spacetimes), it still provides the physical (perturbative) expression for the non-abelian anomaly. As a result, we may look at (10.6.23) and ask ourselves whether such an expression might be given by the variation of a local functional (an integral over base space of polynomials in the fields), in which case it could be removed. This point of view allows us to look at anomalies as defining non-trivial classes in the *local* cohomology of gauge fields by using the techniques of section 10.4. This approach, particularly emphasized by Stora and Zumino, results in the cohomological descent procedure, already hinted at when writing (10.6.21).

10.7 The geometry of consistent non-abelian gauge anomalies

Let us now discuss non-abelian anomalies making emphasis on their local properties. The implication of (10.5.4) or (10.5.3) is that, under a gauge transformation, the generating functional $Z[A]$ of an anomalous theory behaves as (see (10.5.5))

$$Z[A^g] = Z[A] + \alpha_1[A;g], \quad Y_{(i)}(x).Z[A] = \mathcal{V}_{(i)}[A](x). \qquad (10.7.1)$$

The terms α_1 and $\mathscr{V}_{(i)}$ are, respectively, the finite and infinitesimal versions of the *anomaly*: the right hand sides in eqs. (10.7.1) contain an unexpected term indicating that the gauge invariance of the quantum theory described by such functionals has been lost. Both expressions for the anomaly fulfil a condition which results from the fact that the finite and infinitesimal gauge transformations have a group and algebra structure respectively. Explicitly, the relations

$$Z[(A^g)^{g'}] - Z[A^{gg'}] = 0 \quad (\text{mod } 2\pi) \quad , \tag{10.7.2}$$

$$(Y_{(i)}(x)Y_{(j)}(y) - Y_{(j)}(y)Y_{(i)}(x) - [Y_{(i)}(x), Y_{(j)}(y)]).Z[A] = 0 ,$$

have to be satisfied. This leads to the simple but powerful Wess-Zumino consistency condition

(10.7.1) LEMMA (*Wess-Zumino consistency conditions*)

The finite and infinitesimal anomalies must satisfy

$$\alpha_1[A^g; g'] - \alpha_1[A; gg'] + \alpha_1[A; g] = 0 \quad (\text{mod } 2\pi) \quad , \tag{10.7.3}$$

$$Y_{(i)}(x).\mathscr{V}_{(j)}[A](y) - Y_{(j)}(y).\mathscr{V}_{(i)}[A](x) - \delta^D(x - y)C_{ij}^k \mathscr{V}_{(k)}[A](x) = 0 . \tag{10.7.4}$$

Comparing with eqs. (10.6.26) ((10.4.3)) and (10.4.11), it is seen that the consistency conditions[*] (10.7.3) and (10.7.4) are just the one-cocycle condition expressions for $\alpha_1[A; g]$ and $\mathscr{V}_{(i)}[A]$ respectively. Thus, the problem of finding the non-abelian anomalies is a cohomology problem; they are called *consistent anomalies* since they satisfy (10.7.3), (10.7.4). These conditions again indicate that the presence of anomalies can be traced to general principles which do not make reference to the perturbation theory arguments that originally led to them. The cohomological method described below allows us to determine the chiral anomalies without needing to compute any Feynman diagram; in fact, in this treatment anomalies appear as purely differential geometric objects which can be separated from their original quantum context. Its normalization cannot be given by the (linear) Wess-Zumino condition, but it is naturally incorporated into the formalism as we shall see.

It is possible to obtain a functional equivalent to $Z[A]$ by adding to it a *local* functional $\Omega_0[A] = \omega_0[A]$. This will give rise to new anomalies α'_1

---

[*] Note that, in (8.2.6), (8.2.7), the consistency condition was a condition on $d\Delta$ because the variation of the Lagrangian was given by an *exact* form, while here this is not the case. Here the cocycles will be obtained by integration of certain differential forms, which are themselves not cocycles but cocycles modulo d.

and $\mathcal{V}'_{(i)}$ equivalent to α_1 and $\mathcal{V}_{(i)}$. Taking into account (10.7.1) they are related by

$$\alpha'_1[A;g] = \alpha_1[A;g] + \Omega_0[A^g] - \Omega_0[A] \,,$$
$$\mathcal{V}'_{(i)}[A](x) = \mathcal{V}_{(i)}[A](x) + Y_{(i)}(x).\omega_0[A] \,. \tag{10.7.5}$$

Again, these equations have a cohomological interpretation, because they can be rewritten by (10.4.2) and (10.4.10) as

$$\alpha'_1[A;g] = \alpha_1[A;g] + \delta\Omega_0[A;g] \,,$$
$$\mathcal{V}'_{(i)}[A](x) = \mathcal{V}_{(i)}[A](x) + \delta\omega_{0(i)}[A](x) \,. \tag{10.7.6}$$

Thus, equivalent anomalies differ by a coboundary; they are given by cohomologous non-trivial one-cocycles (we saw an example of this in the previous $D = 2$ calculation). A true, non-trivial anomaly cannot be obtained from a local functional $\omega_0[A]$ as $\mathcal{V}_{(i)}(x) = Y_{(i)}(x).\omega_0[A]$ since, if this were the case, it could be compensated by adding a counterterm to the original action. These results can be summarized in the following form:

(10.7.2) LEMMA

Every class of equivalent consistent non-abelian anomalies in D dimensions associated to the compact group G defines a non-trivial element of $H^1_\sigma(G(M^D), \mathcal{F}_L[\mathcal{A}])$ (for the group transformation) and $H^1_\rho(\mathcal{G}(M^D), \mathcal{F}_L[\mathcal{A}])$ (for the infinitesimal case).

We have shown in previous chapters how a non-zero variation of the action in mechanics under a symmetry transformation is related to an extension of the original symmetry algebra, the structure constants of which are determined by the cohomological descent process (cf. sections 8.5 and 8.6). In the case of gauge theories, the presence of anomalies has an effect that again can be related to the *second* cohomology group. It is a known fact that in gauge theories there are certain constraints, due precisely to the gauge invariance of the classical theory, which are called Gauss law constraints. They are the quantities that multiply the spurious variables $A^0_i(x)$ (which can be viewed as Lagrange multipliers) inside the action $I[A; \psi]$. Thus, they are given by

$$G_{(i)}(x) = \frac{\delta I[A; \psi]}{\delta A^i_0(x)} \,. \tag{10.7.7}$$

Since in I the gauge fields are coupled to the fermions, the Gauss law constraints include a term which is bilinear in the spinors ψ. These constraints become operators when ψ is quantized, but one has to be careful with the definition of the bilinears of operators ψ because they are ill-defined and have to be regularized, for instance via the point splitting method. It turns out that, while for non-anomalous theories the Gauss law constraint operators close under commutation reproducing the original algebra $\mathscr{G}(M^{D-1})$, for the anomalous theories this is not the case. Thus, the anomaly prevents the use of the Gauss law to constrain the physical states. The commutators are modified by the presence of a term, the 'Schwinger term' (cf. section 9.3). In general, the Schwinger terms will depend on the gauge field, but in $1+1$ spacetime dimensions the Schwinger term may be given by a c-number and determines the central extension which characterizes the Kac-Moody algebra. Explicitly, the *equal-time* commutators of the operators $G_{(i)}(x)$ are

$$[G_{(i)}(x), G_{(j)}(y)]_{x^0=y^0} = i\delta(\mathbf{x} - \mathbf{y})C^k_{ij}G_{(k)}(x) + iS_{ij}[A](x, y), \qquad (10.7.8)$$

where the i multiplying the structure constants is due to the fact that $G_{(i)}$ are hermitian operators. The additional term S_{ij} in (10.7.8) modifies the original structure of the Lie algebra and comes from the fermionic term in $G_{(i)}$. The gauge part of $G_{(i)}$ acts on local functionals of A as $G_{(i)}(x).F[A] = iY_{(i)}(x).F[A]$. We now show that the terms $S_{ij}[A](x, y)$ satisfy a consistency condition which implies that they are given by two-cocycles on the space gauge algebra.

Since the Schwinger terms S_{ij} appear in an algebra commutator, the Jacobi identity imposes on them the relation

$$\begin{aligned}
0 &= [[G_{(i)}(x), G_{(j)}(y)], G_{(k)}(z)] + \text{cycl. perm.} \\
&= -\delta(\mathbf{x} - \mathbf{y})C^l_{ij}S_{lk}[A](x, z) - \delta(\mathbf{y} - \mathbf{z})C^l_{jk}S_{li}[A](y, x) \\
&\quad - \delta(\mathbf{z} - \mathbf{x})C^l_{ki}S_{lj}[A](z, y) \\
&\quad + Y_{(k)}(z).S_{ij}[A](x, y) + Y_{(i)}(x).S_{jk}[A](y, z) + Y_{(j)}(y).S_{ki}[A](z, x) .
\end{aligned}$$
$$\qquad (10.7.9)$$

Eq. (10.7.9) is just the two-cocycle condition (10.4.12) for $\omega_{ij} = S_{ij}$ since $S_{ij}[A](x, y) = -S_{ji}[A](y, x)$. If, making use of the freedom of regularization for $G_{(i)}(x)$, we had considered an equivalent Gauss-law operator given by $G'_{(i)}(x) = G_{(i)}(x) + \omega_{(i)}[A](x)$, then substituting in (10.7.8) we would have obtained

$$\begin{aligned}
[G'_{(i)}(x), G'_{(j)}(y)] &= [G_{(i)}(x), G_{(j)}(y)] + [G_{(i)}(x), \omega_{(j)}[A](y)] \\
&\quad + [\omega_{(i)}[A](x), G_{(j)}(y)] \\
&= i\delta(\mathbf{x} - \mathbf{y})C^k_{ij}G_{(k)}(x) + iS_{ij}[A](x, y) \\
&\quad + iY_{(i)}(x).\omega_{(j)}[A](y) - iY_{(j)}(y).\omega_{(i)}[A](x) \qquad (10.7.10) \\
&= i\delta(\mathbf{x} - \mathbf{y})C^k_{ij}G'_{(k)}(x) \\
&\quad + i(S_{ij}[A](x, y) + Y_{(i)}(x).\omega_{(j)}[A](y) \\
&\quad - Y_{(j)}(y).\omega_{(i)}[A](x) - \delta(\mathbf{x} - \mathbf{y})C^k_{ij}\omega_{(k)}[A](x)) .
\end{aligned}$$

From (10.7.10) we deduce that, if

$$[G'_{(i)}(x), G'_{(j)}(y)] = i\delta(\mathbf{x} - \mathbf{y})C_{ij}^k G'_{(k)}(x) + iS'_{ij}[A](x, y),\qquad(10.7.11)$$

then

$$S'_{ij}[A](x, y) = S_{ij}[A](x, y) + Y_{(i)}(x).\omega_{(j)}[A](y)$$
$$- Y_{(j)}(y).\omega_{(i)}[A](x) - \delta(\mathbf{x} - \mathbf{y})C_{ij}^k \omega_{(k)}[A](x),\qquad(10.7.12)$$

i.e., S'_{ij} and S_{ij} are cohomologous two-cocycles. The above may be stated as follows:

(10.7.3) LEMMA (*Classes of Schwinger terms*)

Every class of equivalent Schwinger terms for a D-dimensional anomalous gauge theory associated to a compact group G defines a non-trivial element of $H^2_\rho(\mathscr{G}(M^{D-1}), \mathscr{F}_L[\mathscr{A}])$.

Thus, summarizing, the anomaly is an element of H^1 in the cohomology of the gauge group on the D-dimensional spacetime and the Schwinger term belongs to H^2 in the cohomology of the gauge algebra on the $(D-1)$-dimensional space.

10.8 The cohomological descent in the trivial $P(G, M)$ case: cochains and coboundary operators

Let us apply the cohomological descent procedure to find the form of both the consistent non-abelian anomalies and the Schwinger terms. For simplicity in the presentation, only potentials with trivial topological quantum number will be considered in what follows or, in other words, *it will be assumed that the principal bundle $P(G, M)$ is trivial*. As in chapter 8, the method will rely on the existence of two commuting nilpotent operators, δ (or s) and d, the main difference being that here s has been designed to take into account the gauge transformations associated with $G(M)$ (section 10.3) and that the group G is always compact. From lemmas 10.7.2, 10.7.3 it follows that to obtain candidates for both the anomaly and the Schwinger term it is sufficient to find non-trivial elements of the cohomology groups $H^1_\sigma(G(M^D), \mathscr{F}_L[\mathscr{A}])$ (or $H^1_\rho(\mathscr{G}(M^D), \mathscr{F}_L[\mathscr{A}]))$ and $H^2_\rho(\mathscr{G}(M^{D-1}), \mathscr{F}_L[\mathscr{A}])$ respectively. In the cohomological descent method, the cochains are obtained as integrals of differential forms depending on A and n arguments, $\Omega_n^m(A; g_1, ..., g_n)$ and $\omega_n^m(A; \zeta_1, ..., \zeta_n)$, over an m-dimensional compact oriented manifold M^m which will be taken as S^m, $M^m \sim S^m$. The discussion of section 10.6 indicates that these forms will be obtained by descending from a $2l = D + 2$ form, the Chern character form $ch_l(F)$. These forms are differential polynomials Ω_n^m, ω_n^m which can be integrated over the corresponding manifold M^m giving the cochains

$$\Omega_n[A; g_1, ..., g_n] = \int_{M^m} \Omega_n^m(A; g_1, ..., g_n) ,$$

$$\omega_n[A](\zeta_1, ..., \zeta_n) = \int_{M^m} \omega_n^m(A; \zeta_1, ..., \zeta_n) . \tag{10.8.1}$$

The action of the group δ and algebra s cohomology operator on the differential m-forms Ω_n^m and ω_n^m respectively, is defined by (cf. (10.4.1) and (10.4.7))

$$(\delta\Omega_n^m)(A; g_1, ..., g_{n+1}) \equiv \Omega_n^m(A^{g_1}; g_2, ..., g_{n+1})$$
$$+ \sum_{a=1}^{n} (-1)^a \Omega_n^m(A; g_1, ..., g_{a-1}, g_a g_{a+1}, g_{a+2}, ..., g_{n+1}) \tag{10.8.2}$$
$$+ (-1)^{n+1} \Omega_n^m(A; g_1, ..., g_n) ,$$

$$(s\omega_n^m)(A; \zeta_1, ..., \zeta_{n+1}) \equiv \sum_{a=1}^{n+1} (-1)^{a+1} \frac{d}{dt} \omega_n^m(A + t\mathcal{D}\zeta_a; \zeta_2, ..., \hat{\zeta}_a, ..., \zeta_{n+1})_{t=0}$$
$$+ \sum_{\substack{a,b \\ a<b}}^{n+1} (-1)^{a+b} \omega_n^m(A; [\zeta_a, \zeta_b], \zeta_1, ..., \hat{\zeta}_a, ..., \hat{\zeta}_b, ..., \zeta_{n+1}) . \tag{10.8.3}$$

These definitions are formally the same as (10.4.1) and (10.4.7) but now the 'cochains' are differential forms. Then it follows from (10.8.2) and (10.8.3) on account of the linearity of the integral with respect to sums of forms that

$$\delta\Omega_n = \int_{M^m} \delta\Omega_n^m , \quad s\omega_n = \int_{M^m} s\omega_n^m . \tag{10.8.4}$$

In addition to this, we have the following.

(10.8.1) LEMMA

The nilpotent operators δ and s acting on forms commute with the exterior differential d, i.e.

$$[\delta, d] = 0 , \quad [s, d] = 0 . \tag{10.8.5}$$

Proof: The differential d acts on the variables of the manifold M^m. Since we can look at δ (s) as a kind of differential which acts on A and on the group (algebra) variables, but not on x, the commutativity follows, *q.e.d.*

Let us now turn to the cohomological descent. Its basic ingredients are the commuting*, nilpotent ($s^2 = 0 = d^2$) operators d and s, and the existence of a class of closed gauge-invariant differential forms the potentials of which are not gauge-invariant. These forms are the Chern character forms $ch_l(\bar{\Omega})$ of order $2l$ discussed in section 2.5, here given by

$$ch_l(F) = b_l \text{Tr}(F^l) , \quad b_l = \frac{1}{l!}\left(\frac{i}{2\pi}\right)^l , \tag{10.8.6}$$

where F is given by (10.3.2), and $F^l \equiv F \overset{l}{\ldots} F$. Since ch_l depends on A through F, and F transforms by $g^{-1}Fg$, the Chern character form is gauge-invariant by the cyclic property of the trace. The proportionality constant b_l is not important in the reasonings below and it will not be made explicit.

First of all, we rewrite eqs. (2.6.10) and (2.6.11) in the form of the following.

(10.8.1) PROPOSITION

For $P(G,M)$ trivial, the form $ch_l(F)$ is closed and exact[†], $ch_l(F) = d\Omega_0^{2l-1}(A) \equiv d\omega_0^{2l-1}(A)$, and its potential $(2l-1)$-form is read from

$$ch_l(F) = b_l d\left[\int_0^1 \delta t \, l\text{Tr}(AF_t^{l-1})\right] \equiv b_l d\Omega_0^{2l-1}(A) , \tag{10.8.7}$$

where $F_t \equiv tF + (t^2 - t)A^2 = tdA + t^2 A^2$.

Proof: This is a particular case of the proof in section 2.6. It is also easy to check (10.8.7) directly using that

$$dF_t^k = t[F_t^k, A] , \tag{10.8.8}$$

a relation that can be proved by induction in k, q.e.d.

Eq. (10.8.7) is the starting point for obtaining solutions to the gauge cohomology problem stated above. It leads to a chain of equations in which each member of the chain is a form that satisfies the cocycle condition modulo an exact form. Indeed, we can apply the operator δ or s to the first equation in (10.8.7) to obtain, on account of the gauge invariance of $ch_l(F)$,

$$d[(\delta\Omega_0^{2l-1})(A;g)] = 0 , \quad d[(s\omega_0^{2l-1})(A;\zeta)] = 0 , \tag{10.8.9}$$

---

* It is also possible to have an anticommuting $\tilde{s}$, see section 10.10.
† We recall that, for a general non-trivial bundle $P(G,M)$, this is only true on the local chart over which the bundle is trivial (see eq. (2.4.17)). Thus, if $P(G,M)$ is non-trivial all the following expressions should be understood as *local* expressions.

which means that $\delta\Omega_0^{2l-1}$ and $s\omega_0^{2l-1}$ are closed forms. Thus, $\delta\Omega_0^{2l-1}$ and $s\omega_0^{2l-1}$ are, locally, the exterior derivative of forms Ω_1^{2l-2}, ω_1^{2l-2}. Let us now see whether they are exact.

First, we note that the forms of the type ω that appear in the *algebra* cohomology are constructed out of the one-forms A, the two-forms F and the zero- and one-forms ζ, $d\zeta \equiv \mu$. They fulfil the relations

$$dA = F - A^2, \quad dF = [F, A], \quad d\zeta \equiv \mu, \quad d\mu = 0, \qquad (10.8.10)$$

and the differential of any expression containing them can be calculated from these relations. It is said that the ω's are elements of the *free differential algebra* generated by A, F, ζ, μ. Now, whenever a free differential algebra is generated by pairs of differential forms G^p, G^{p+1} ($p, p+1$ being the order of the forms) such that

$$dG^p = G^{p+1}, \quad dG^{p+1} = 0, \qquad (10.8.11)$$

the algebra is called *contractible*. For these algebras the de Rham cohomology is trivial, *i.e.*, every closed form can be written as the differential of a form that depends on the generators. In order to show this, it is convenient to use the Cartan homotopy formula. This formula has already been derived in section 2.4 (cf. eqs. (2.4.24), (2.4.25)) as another way of obtaining the transgression formula, but since its validity relies on the fact that the forms can be treated as purely *algebraic* objects that are polynomials in G^p and G^{p+1} and their differentials (without looking at their commutation or symmetry properties), it will be derived again here from this point of view.

(10.8.2) LEMMA (*Cartan homotopy operator*)

Let $G_t^p = tG^p$ and $G_t^{p+1} = tG^{p+1}$ and let the Cartan homotopy operator be given by

$$k_{01} = \int_0^1 \delta t \, k_t \quad , \quad k_t G_t^p = 0 \quad , \quad k_t G_t^{p+1} = G^p = \frac{d}{dt} G_t^p \qquad (10.8.12)$$

($\delta t k_t G_t^{p+1} = \delta G_t^p$) *and the fact that k_t is an antiderivation on forms. Then* $k_{01}d + dk_{01} = \int_0^1 \delta t \frac{d}{dt}$.

Proof: Since k_t (d) is an antiderivation of degree -1 ($+1$), $k_t d + dk_t$ is a derivation of degree zero. Moreover, since

$$(dk_t + k_t d)G_t^p = k_t G_t^{p+1} = \frac{d}{dt} G_t^p \quad ,$$

$$(dk_t + k_t d)G_t^{p+1} = dk_t G_t^{p+1} = d(\frac{d}{dt} G_t^p) = \frac{d}{dt} G_t^{p+1} \quad , \qquad (10.8.13)$$

it follows that

$$dk_t + k_t d = \frac{d}{dt} \qquad (10.8.14)$$

on forms constructed out of G_t^p and G_t^{p+1}, and we find

$$\int_0^1 \delta t \frac{d}{dt} = \int_0^1 \delta t (dk_t + k_t d) = dk_{01} + k_{01} d \quad , \qquad (10.8.15)$$

q.e.d.

We can now prove the following.

(10.8.2) Proposition

Every contractible free differential algebra has trivial de Rham cohomology.

Proof: Let β be a form in the free differential algebra, and let β_t be the form which is given by the same polynomial constructed now with G_t^p, G_t^{p+1}. Then from (10.8.15) it follows that

$$\int_0^1 \delta t \frac{d\beta_t}{dt} = \beta = \int_0^1 \delta t (dk_t + k_t d)\beta_t . \qquad (10.8.16)$$

Let β now be closed. Then the potential form of β is given by

$$\int_0^1 \delta t \, k_t \beta_t \quad , \qquad (10.8.17)$$

q.e.d.

Proposition 10.8.1 can be seen as a consequence of Proposition 10.8.2 for the differential algebra defined by A, dA; it suffices to make the identifications $A = G^1$, $dA = G^2$, $A_t = tA$, $dA_t = tdA$. Then $k_t G_t^1 = k_t A_t = 0$, $k_t G_t^2 = k_t dA_t = (-dk_t + \frac{d}{dt})A_t = A$ and also $k_t F_t = k_t (dA_t - A_t^2) = A$ (cf. section 2.4). Then eq. (10.8.7) follows from (10.8.17) since $k_t \text{Tr}(F^l) = l \, \text{Tr}((k_t F_t)F_t^{l-1}) = l \, \text{Tr}(AF_t^{l-1})$ (cf. (2.6.11)).

Proposition 10.8.2 also applies to the case of the *algebra* generated by (10.8.10), because it is generated by the pairs (ζ, μ), (A, B):

$$d\zeta \equiv \mu , \quad d\mu = 0 , \quad dA = F - A^2 \equiv B , \quad dB = 0 . \qquad (10.8.18)$$

Therefore, every closed form ω constructed from A and ζ is exact. However, in the case of the forms of the type Ω that appear in the *group* cohomology we do not have the same result: while one might think of writing every closed form as the differential of a form that depends on A

and the group element coordinates g_i in the exponent of $g = exp(g^i T_i)$, for compact groups such an expression is not exact. The situation is already exhibited clearly by the simple example of the group $U(1)$, of elements $\eta = exp\,i\varphi$, $\varphi \in [0, 2\pi]$. The one-form $\eta^{-1}d\eta$ is obviously closed, but it is not exact: $\eta^{-1}d\eta = id\varphi$ only *locally*.

This argument shows that, for compact groups, the de Rham cohomology for forms which depend on A and $g \in G$ (G compact) is not trivial. For instance, consider as another example the group $SU(2)$ and its volume three-form. Let g be an element of $SU(2)$. The volume form on $SU(2)$ (cf. (2.6.20)) is defined as

$$\Omega = \lambda \mathrm{Tr}(g^{-1}dg\,g^{-1}dg\,g^{-1}dg)\,, \tag{10.8.19}$$

where λ is a numerical factor. It is easy to see that Ω is a closed non-exact form defined on $SU(2)$. It is obviously closed since it is a volume form. On the other hand, Ω cannot be the differential of a two-form on the group manifold because its integral over $SU(2)$ (the volume of the group, $16\pi^2$) is non-zero and the sphere S^3 ($\sim SU(2)$) has no boundary. Moreover, this property is preserved under mappings, since d is natural with respect to them.

10.9 The descent equations. Cocycles and coboundaries

Let us now go back to the forms $\Omega_n^m(A; g_1, ..., g_n)$ and $\omega_n^m(A; \zeta_1, ..., \zeta_n)$. From the previous discussion it can be concluded, in general locally for $\delta\Omega$ and globally for $s\omega$, that

$$(\delta\Omega_0^{2l-1})(A; g) = d[\Omega_1^{2l-2}(A; g)]\,, \tag{10.9.1}$$
$$(s\omega_0^{2l-1})(A, \zeta) = d[\omega_1^{2l-2}(A; \zeta)]$$

(in short, $\delta\Omega_0^{2l-1} = d\Omega_1^{2l-2}$, $s\omega_0^{2l-1} = d\omega_1^{2l-2}$), where Ω_1^{2l-2} (ω_1^{2l-2}) are certain $(2l-2)$-forms depending on one gauge group (algebra) argument. Applying these operators again to (10.9.1) we conclude that both $\delta\Omega_1^{2l-2}$ and $s\omega_1^{2l-2}$ are closed. It is obvious that this can be continued until the order of the differential forms is reduced to zero. In the group case, the result is a chain of local equalities for the forms Ω_n^m, $m + n = 2l - 1$,

$$ch_l(F) = d\Omega_0^{2l-1} \;(= d\omega_0^{2l-1})\,,$$

$$\dots\dots\dots\dots$$

$$\delta\Omega_{k-1}^{2l-k} = d\Omega_k^{2l-k-1}\,, \quad k = 1, ..., 2l - 1\,, \tag{10.9.2}$$

$$\dots\dots\dots\dots$$

$$\delta\Omega_{2l-1}^0 = 0\,.$$

The double differential complex (a consequence of lemma 10.8.1) can be expressed in terms of the following commutative diagram:

$$
\begin{array}{ccccc}
ch_l(F) & \xleftarrow{d} & \Omega_0^{2l-1} & & \\
\downarrow{\scriptstyle\delta} & & \downarrow{\scriptstyle\delta} & & \\
0 & \xleftarrow{d} & \delta\Omega_0^{2l-1} & \xleftarrow{d} & \Omega_1^{2l-2} \\
& & \downarrow{\scriptstyle\delta} & & \downarrow{\scriptstyle\delta} \\
& & 0 & \xleftarrow{d} & \delta\Omega_1^{2l-2} \quad \ldots \\
& & & & \vdots
\end{array}
$$

Diagram 10.9.1

(the equation $\delta ch_l(F) = 0$ is due to the original gauge invariance of the Chern form. We stress that the above group expressions are local even if, as is assumed here, the bundle $P(G, M)$ is trivial). In the algebra case the corresponding chain leads to the global relations

$$
ch_l(F) = d\omega_0^{2l-1}(= d\Omega_0^{2l-1}),
$$

$$
\cdots\cdots\cdots\cdots
$$

$$
s\omega_{k-1}^{2l-k} = d\omega_k^{2l-k-1}, \quad k = 1, \ldots, 2l-1, \qquad (10.9.3)
$$

$$
\cdots\cdots\cdots\cdots
$$

$$
s\omega_{2l-1}^0 = 0,
$$

or, again diagrammatically,

$$
\begin{array}{ccccc}
ch_l(F) & \xleftarrow{d} & \omega_0^{2l-1} & & \\
\downarrow{\scriptstyle s} & & \downarrow{\scriptstyle s} & & \\
0 & \xleftarrow{d} & s\omega_0^{2l-1} & \xleftarrow{d} & \omega_1^{2l-2} \\
& & \downarrow{\scriptstyle s} & & \downarrow{\scriptstyle s} \\
& & 0 & \xleftarrow{d} & s\omega_1^{2l-2} \quad \ldots \\
& & & & \vdots
\end{array}
$$

$$
\begin{array}{ccccc}
& & \vdots & & \\
\ldots & & s\omega_{k-1}^{2l-k} & \xleftarrow{d} & \omega_k^{2l-k-1} \\
& & \downarrow{\scriptstyle s} & & \downarrow{\scriptstyle s} \\
& & 0 & \xleftarrow{d} & s\omega_k^{2l-k-1} \quad \ldots \\
& & & & \vdots
\end{array}
$$

Diagram 10.9.2

The first two squares of these diagrams have the same structure as those of the Wess-Zumino terms given by non-trivial elements of the Chevalley-Eilenberg cohomology (Diagram 8.6.1).

The elements Ω_{k-1}^{2l-k} and ω_{k-1}^{2l-k} in (10.9.2) and (10.9.3) do not yet provide the cocycles for the cohomology groups $H_\sigma^{k-1}(G(M^{2l-k}), \mathscr{F}_L[A])$, $H_\rho^{k-1}(\mathscr{G}(M^{2l-k}), \mathscr{F}_L[A])$ directly: $\delta\Omega_{k-1}^{2l-k} = d\Omega_k^{2l-k-1}$ and $s\omega_{k-1}^{2l-k} = d\omega_k^{2l-k-1}$ and these are not zero. Since there is still an integration to be performed [(10.8.1)], the cocycles are, really, 'cocycles modulo d'. To obtain the true cocycles, we have to integrate the cocycle forms over M^{2l-k} ($\sim S^{2l-k}$). In this case, writing $\omega_{k-1} = \int_{M^{2l-k}} \omega_{k-1}^{2l-k}$ and using that $\delta \int \omega = \int \delta\omega$,

$$(s\omega_{k-1})[A](\zeta_1, ..., \zeta_k) \equiv \int_{M^{2l-k}} (s\omega_{k-1}^{2l-k})(A; \zeta_1, ..., \zeta_k)$$

$$= \int_{M^{2l-k}} d(\omega_k^{2l-k-1}(A; \zeta_1, ..., \zeta_k)) = 0 . \tag{10.9.4}$$

For the group forms Ω, however, the last equality is not true because the analogous expression for $\delta\Omega$ contains terms of the type (10.8.19) (cf. (2.6.15), (2.6.19)) which are not exact and give non-vanishing integrals. We shall see this explicitly in section 10.11, where we shall find (as in section 2.6) that $d\Omega_1^{2l-2}$ contains a term proportional to $\mathrm{Tr}(g^{-1}dg)^{2l-1}$, the integral of which gives the winding number of the mapping $g : S^{2l-1} \to G$ ($M^{2l-1} \sim S^{2l-1}$), i.e. a multiple of 2π once the appropriate factor is included. Thus, for the Ω's, the zero in the r.h.s. of the cocycle condition is replaced by 'multiple of 2π' as we have already seen in (10.7.3). For the rest of this section we shall concentrate only on the Lie algebra cohomology, for which global expressions can be found.

In every step $s\omega_{k-1}^{2l-k} = d\omega_k^{2l-k-1}$ of the cohomological descent method, there is an ambiguity due to the freedom of adding an exact differential form $d\beta_k^{2l-k-2}$ to ω_k^{2l-k-1} since, in that case, the equation remains unchanged. This has an effect in the next step of the chain, which is given by

(10.9.1) PROPOSITION

If $\omega'^{2l-k-1}_k = \omega_k^{2l-k-1} + d\beta_k^{2l-k-2}$ then $\omega'_{k+1} = \omega_{k+1} + s\beta_k$, i.e. the ambiguity in the choice of ω_k^{2l-k-1} leads in the next step of the chain to elements ω_{k+1} that belong to the same s-cohomology class.

Proof: We wish to show that, although $\omega'_k = \omega_k$, the presence of β leads to different, but cohomologous, ω'_{k+1}, ω_{k+1}. Acting with s on both sides of the equation $\omega'^{2l-k-1}_k = \omega_k^{2l-k-1} + d\beta_k^{2l-k-2}$ and then using (10.9.3), we get

$$d\omega'^{2l-k-2}_{k+1} = d\omega_{k+1}^{2l-k-2} + sd\beta_k^{2l-k-2} .$$

Since d and s commute, this means that $\omega'^{2l-k-2}_{k+1} - \omega^{2l-k-2}_{k+1} - s\beta^{2l-k-2}_k$ is zero up to a differential that disappears after integration over M^{2l-k-2}, so that the above statement follows with $\beta_k = \int_{M^{2l-k-2}} \beta^{2l-k-2}_k$, q.e.d.

Finally, there is the question of the non-triviality of the cocycles provided by the cohomological descent method. This requires

(10.9.2) PROPOSITION

If ω_{k-1} is a trivial $(k-1)$-cocycle (i.e., ω^{2l-k}_{k-1} is trivial 'modulo d'), then all the higher cocycles in the chain (and hence in lower positions in Diagram 10.9.2) are trivial.

Proof: The s-triviality of ω_{k-1} means that $\omega^{2l-k}_{k-1} = s\alpha^{2l-k}_{k-2} + d\beta^{2l-k-1}_{k-1}$ where we have included a possible exact $(2l - k - 1)$-form, irrelevant for the integrated cocycle ω_{k-1}. The action of s on this equality implies, using (10.9.3) once more, that

$$\omega^{2l-k-1}_k = s\beta^{2l-k-1}_{k-1} + d\gamma^{2l-k-2}_k \; , \tag{10.9.5}$$

where γ^{2l-k-2}_k is a certain $(2l - k - 2)$-form. This means that ω^{2l-k-1}_k is trivial 'modulo d'. We can apply the same reasoning several times to conclude that $\omega^{2l-k}_{k-1}, ..., \omega^0_{2l-1}$ are all trivial 'modulo d'. This implies that $\omega_{k-1}, ..., \omega_{2l-1}$ are trivial, q.e.d.

Using this result it will be shown at the end of section 10.11 that all the cocycles that are obtained by descending from the Chern form are non-trivial.

Finally, let us recall that the starting point in all the previous reasonings was eq. (10.8.7), which assumed the triviality of $P(G, M)$. If this is not the case, eq. (10.8.7) has to be replaced by the corresponding transgression formula for $P(F) - P(F_0)$, where F_0 is the curvature associated with another (background) connection A_0 (cf. (2.4.13)). In this way, the above results, which would just have a local character for a non-trivial $P(G, M)$, can be extended to new expressions that are globally valid for non-trivial bundles. Starting from

$$P(F^l) - P(F^l_0) = d\omega^{2l-1}_0(A, A_0) \; , \tag{10.9.6}$$

the descent equations (where the coboundary operator does not act on A_0) produce the chain of forms $\omega^{2l-k}_{k-1}(A, A_0; \zeta_1, ..., \zeta_{k-1})$, also subject to ambiguities of the type already discussed in the absence of A_0. Among these forms, $\omega^{2l-2}_1(A, A_0; \zeta)$ satisfies the Wess-Zumino consistency condition and may be used to describe the chiral anomaly in the presence of the background field A_0.

10.10 A compact form for the gauge algebra descent equations

The descent equations (10.9.3) may be expressed in a more compact form. This expression makes use of a different cohomology operator $\not{s}$ constructed from both the de Rham and the BRST cohomology operators. In order to introduce $\not{s}$, it is convenient to define the action of $\tilde{s}$ on differential forms in such a way that it now *anticommutes* with d,

$$\{\tilde{s}, d\} = 0 \quad , \tag{10.10.1}$$

and (cf. (10.4.18), (10.4.19))

$$\tilde{s}c(x) = -c(x)^2 \quad , \quad \tilde{s}A = -dc(x) - [A(x), c(x)] = -\mathcal{D}c(x) \quad , \tag{10.10.2}$$

where $c(x)$ is the odd Lie algebra-valued ghost zero form. To have $\tilde{s}$ anticommuting with d it suffices to define its action in such a way that, on a p-form α, $\tilde{s}(\text{new})\alpha = (-1)^p \tilde{s}(\text{old})\alpha$, where $\tilde{s}(\text{old})$ commutes with d. This justifies the additional global minus sign in $\tilde{s}A$, eq. (10.10.2). Using the new BRST operator, we now define the nilpotent operator

$$\not{s} := d + \tilde{s} \quad , \quad \not{s}^2 = 0 \quad . \tag{10.10.3}$$

This is a cohomology operator on the set of cochains that are non-homogeneous sums (in ghost number and form degree) of terms $\tilde{\omega}_n^m$. The fact that $\not{s}$ is not a homogeneous operator on the cochains $\tilde{\omega}_n^m$ is not a difficulty: the Hodge-de Rham operator $\not{d} = d + \delta$ of eq. (1.5.30), is also not homogeneous (however, the parallel stops there since, in contrast with $\not{s}$ above, $\not{d}^2$ is the Laplace-de Rham operator and not zero).

Using $\not{s}$, the equations in (10.9.3) can be summarized in a single equation. In terms of the BRST operator $\tilde{s}$ of (10.10.1), (10.10.2), the corresponding cohomological descent equations from $ch_l(F)$ can be rewritten in the form

$$ch_l(F) = -d\tilde{\omega}_0^{2l-1} \quad ,$$

$$\cdots\cdots\cdots$$

$$\tilde{s}\tilde{\omega}_{k-1}^{2l-k} = -d\tilde{\omega}_k^{2l-k-1} \quad , \quad k = 1, ..., 2l - 1 \quad , \tag{10.10.4}$$

$$\cdots\cdots\cdots$$

$$\tilde{s}\tilde{\omega}_{2l-1}^0 = 0 \quad ,$$

where it is convenient to introduce minus signs in the r.h.s. Adding all the equations of the chain we get

$$ch_l(F) + \sum_{k=1}^{2l} \tilde{s}\tilde{\omega}_{k-1}^{2l-k} = -\sum_{k=1}^{2l} d\tilde{\omega}_{k-1}^{2l-k} \quad , \tag{10.10.5}$$

and, using the definition (10.10.3) of $\not{s}$,

$$ch_l(F) = -\not{s}\left(\sum_{k=1}^{2l} \tilde{\omega}_{k-1}^{2l-k}\right) , \qquad (10.10.6)$$

which expresses the fact that $ch_l(F)$ is a trivial cocycle ($\not{s}ch_l(F) = 0$) for the $\not{s}$ cohomology. Conversely, given a solution $\mathcal{W}$ of the equation $ch_l(F) = \not{s}\mathcal{W}$ we can obtain, separating the terms of $\mathcal{W}$ homogeneous in the ghost number, the equations (10.10.4). Thus, the equation

$$ch_l(F) = \not{s}\mathcal{W} \qquad (10.10.7)$$

summarizes all the descent equations. It always has a solution because the free differential algebra generated by A, dA, c and dc (in terms of which all the forms in the chain are written), is contractible relative to the operator $\not{s}$. Indeed, this algebra can be generated by A, $\not{s}A$, c and $\not{s}c$: since

$$\not{s}A = dA - dc - [A, c] , \quad \not{s}c = dc - c^2 , \qquad (10.10.8)$$

dA and dc can be written in terms of A, $\not{s}A$, c and $\not{s}c$. Explicitly, $dc = \not{s}c + c^2$ and $dA = \not{s}A + \not{s}c + c^2 + [A, c]$.

10.11 Specific results in $D = 2, 4$ spacetime dimensions

We now show a few examples of the use of the cohomological descent method for $D = 2, 4$ to obtain candidates for both the anomalies and the associated Schwinger terms. This requires extracting in each step the potential form that provides the next element of the chain. For each spacetime dimension, the appropriate value of l has to be taken: the anomaly is defined in $D = 2l - 2$ dimensions and the Schwinger term in $D - 1 = 2l - 3$ dimensions, which determines the degree $2l$ of the Chern character form which is the first element of the chain: $l = \frac{D}{2} + 1$, D even.

(a) $D = 2$ ($l = 2$)
In this case $l = 2$. The anomaly is given by an algebra (group) cohomology element of the type ω_1^2 (Ω_1^2) (a spacetime two-form which is a 'group one-form') and the Schwinger term in the algebra by an ω_2^1. First of all, we have, using (10.8.7)

$$ch_2(F) = b_2 \text{Tr}(F^2) = d[b_2 \text{Tr}(AdA + \frac{2}{3}A^3)] = d\omega_0^3(A) = d\Omega_0^3(A) . \quad (10.11.1)$$

Let us consider first the *algebra* cohomology by applying s and using (10.3.14) or (2.2.8). Due to the ambiguity mentioned before Proposition 10.9.1, for the next step we can have either $\omega_1'^2 = b_2 \text{Tr}(Ad\zeta)$

or $\omega_1^2 = b_2 \mathrm{Tr}(\zeta dA)$, which differ by the exact form* $d\beta_1^1 = -b_2 d\mathrm{Tr}(A\zeta)$. These two forms originate two different but cohomologous chains denoted by ω' and ω, and they are given by

$$ch_2(F) = b_2 \mathrm{Tr}(F^2) ,$$

$$\omega'^3_0 = b_2 \mathrm{Tr}(AdA + \frac{2}{3}A^3) ,$$

$$\omega'^2_1 = b_2 \mathrm{Tr}(Ad\zeta) , \qquad (10.11.2)$$

$$\omega'^1_2 = b_2 \mathrm{Tr}(\zeta_{[1}d\zeta_{2]}) ,$$

$$\omega'^0_3 = \frac{1}{3}b_2 \mathrm{Tr}(\zeta_{[1}\zeta_2\zeta_{3]}) ,$$

where $b_2 = -\frac{1}{8\pi^2}$, and by

$$ch_2(F) = b_2 \mathrm{Tr}(F^2) ,$$

$$\omega_0^3 = b_2 \mathrm{Tr}(AdA + \frac{2}{3}A^3) \; (= \omega'^3_0) ,$$

$$\omega_1^2 = b_2 \mathrm{Tr}(\zeta dA) , \qquad (10.11.3)$$

$$\omega_2^1 = b_2 \mathrm{Tr}(A[\zeta_1, \zeta_2]) ,$$

$$\omega_3^0 = \frac{1}{3}b_2 \mathrm{Tr}(\zeta_{[1}\zeta_2\zeta_{3]}) = \omega'^0_3 ,$$

where the square brackets for indices indicate antisymmetrization (without factorials) in the enclosed indices. The differential forms obtained are not yet the cocycles; to obtain them, they have to be integrated. This means that, for instance, the integral of $2\pi\omega'^2_1$ or $2\pi\omega_1^2$ reproduces the anomaly in D=2 (eq. (10.5.22)); the integral $2\pi \int d^2x\, \omega_1^2$ gives the first term in the expansion of $\alpha_1[A, g]$ for $D = 2$. Similarly, $\omega'^1_2 = b_2 \mathrm{Tr}(\zeta_{[1}d\zeta_{2]})$ gives (see (10.8.1))

$$\omega'_2[A](\zeta_1, \zeta_2) = b_2 \int_{M^1} \mathrm{Tr}(\zeta_{[1}d\zeta_{2]}) = b_2 \int dx\, \mathrm{Tr}\{\zeta_{[1}(\partial\zeta_{2]}/\partial x)\}(x)$$

$$= b_2 \int dxdy\, \delta(x-y)\mathrm{Tr}\{\zeta_{[1}(x)(\partial\zeta_{2]}(y)/\partial y)\} \qquad (10.11.4)$$

$$= -b_2 \int dxdy\, \partial_y\delta(x-y)\mathrm{Tr}(\zeta_{[1}(x)\zeta_{2]}(y)) ;$$

factoring out $\zeta_1(x)\zeta_2(y)$ and using (10.4.6), we deduce that the two-cocycle in the canonical basis

$$\omega'_{ij}[A](x, y) = -2b_2\partial_y\delta(x-y)\mathrm{Tr}(T_iT_j) \qquad (10.11.5)$$

---

* This ambiguity is clearly not restricted to the first step of the chain.

gives a possible Schwinger term $S_{ij} = 2\pi\omega'_{ij}$ (and that it satisfies $S_{ij}(x, y) = -S_{ji}(y, x)$). Note that this term is independent of A, and so in this case it defines a central extension; in fact this central extension is the *Kac-Moody algebra* (section 9.4),

$$[X_{(i)}(x), X_{(j)}(y)] = \delta(x - y)C_{ij}^k X_{(k)}(x) + i2\pi b_2 \delta_{ij} \partial_y \delta(x - y) . \quad (10.11.6)$$

In general, however, the Schwinger term depends on the gauge fields. The unprimed form of the anomaly and the Schwinger term are calculated in the same way:

$$\omega_i[A](x) = \omega'_i[A](x) = b_2 \epsilon^{\mu\nu} \mathrm{Tr}(T_i \partial_\mu A_\nu) ,$$
$$\omega_{ij}[A](x, y) = b_2 \mathrm{Tr}(A[T_i, T_j]) \delta(x - y) . \quad (10.11.7)$$

For the *group* cohomology, the same ambiguity is also present. Nevertheless, in this cohomology only the one-cocycle is really interesting since it gives the anomaly. It is obtained by integrating the differential form Ω_1^2, which is obtained from Ω_0^3 in (10.11.1) as we now show. Using that, under a gauge transformation (10.3.3),

$$A \mapsto g^{-1}(A + a)g ,$$
$$dA \mapsto -g^{-1}aAg + g^{-1}dAg - g^{-1}Adg - (g^{-1}dg)^2 \quad (10.11.8)$$
$$= g^{-1}(-aA + dA - Aa - a^2)g ,$$

where $a \equiv dgg^{-1}$, a simple calculation shows that $\delta\Omega_0^3 = d\Omega_1^2$ with

$$\Omega_1^2(A, g) = -b_2 \mathrm{Tr}[aA + \frac{1}{3}d^{-1}a^3] , \quad (10.11.9)$$

where $d^{-1}a^3$ is used since $\mathrm{Tr}(a^3)$ does not admit a global expression for its potential form; to first order, Ω reduces to the corresponding algebra cocycle ω. Nevertheless, it is possible to give an expression for the cocycle $\Omega_1[A; g]$, provided we modify the domain of integration. First we recall that $M^2 \sim S^2$, due to the standard boundary conditions on $g(x)$. Then, if D^3 is a three-dimensional manifold with boundary such that $\partial D^3 = S^2$ (the boundary of D^3 is the two-dimensional spacetime), we can write, using Stokes' theorem,

$$\Omega_1[A; g] = \int_{S^2} \Omega_1^2 = \int_{D^3} d\Omega_1^2 = -b_2 \int_{D^3} d\mathrm{Tr}\{aA\} - \frac{b_2}{3} \int_{D^3} \bar{a}^3$$
$$= -b_2 \int_{S^2} \mathrm{Tr}\{aA\} - \frac{b_2}{3} \int_{D^3} \bar{a}^3 . \quad (10.11.10)$$

In the second term of (10.11.10) we use the bar on $\bar{a}$ to indicate that the group elements are defined on D^3. This is possible because for any Lie group $\pi_2(G) = 0$, and so the mappings $g : S^2 \to G$ are homotopically trivial and can be extended to mappings $g : D^3 \to G$; the first term in (10.11.10) is standard in the sense that it is defined on ∂D^3. As a result, Ω_1 is given by

$$\Omega_1[A;g] = -b_2 \int_{S^2} d^2 x \epsilon^{\mu\nu} \mathrm{Tr}\{a_\mu A_\nu\} - \frac{b_2}{3} \int_{D^3} d^3 \bar{x} \epsilon^{\mu\nu\rho} \mathrm{Tr}\{\bar{a}_\mu \bar{a}_\nu \bar{a}_\rho\} \quad (10.11.11)$$

where $\bar{a}_\mu$ reduces to a_μ on S^2.

The Wess-Zumino-Witten term
The cocycle Ω_1, viewed as a term for a sigma model action with target space G, may be used to define the *Wess-Zumino-Witten term* in two dimensions, and it is worth discussing its presence in an action since it is topologically non-trivial. The last part of $\Omega_1[A, g]$ does not depend on A, and including a global factor λ it is given by the term

$$\alpha = \lambda \frac{1}{3} b_2 \int_{D^3} \mathrm{Tr}(\bar{a}^3) \quad . \qquad (10.11.12)$$

Classically, this term is well defined because then all that matter are the Euler-Lagrange equations. To derive them one has to compute the variation of α induced by δg, extending it to D in such a way that it coincides with δg on ∂D^3. This does not produce an ambiguity because $\delta \mathrm{Tr}(a^n) = n d \mathrm{Tr}(\delta g g^{-1} a^{n-1})$ so that, in the present case,

$$\delta\alpha = \lambda \frac{1}{3} b_2 \int_{D^3} \delta \mathrm{Tr}(\bar{a}^3) = \lambda b_2 \int_{D^3} d \mathrm{Tr}(\delta g g^{-1} \bar{a}^2) = \lambda b_2 \int_{S^2} \mathrm{Tr}(\delta g g^{-1} a^2),$$
$$(10.11.13)$$

which is independent of the particular form chosen for δg and of the particular disk such that $\partial D^3 = S^2$. As a result, the classical Euler-Lagrange equations (which we shall not write because we are not considering a kinetic sigma model term for g) are not ambiguous.

In the *quantum* theory, however, the action itself is important because it appears as an exponential in a Feynman path integral. This means that there is a potential source of ambiguity in the election of the disk D^3. But since the action appears in the exponential $e^{\frac{i}{\hbar}I}$, the ambiguity disappears if the coefficient of I is fixed in such a way that two different choices for D^3 give rise to the same exponential. As we shall now see, the requirement that the quantum theory is well defined leads to the quantization of the coefficient of the Wess-Zumino-Witten term. This fact, pointed out by Witten, is analogous to the monopole charge quantization which follows from the requirement that the Dirac string should be unobservable. Let

D^3 and D'^3 be two possible choices with a common boundary S^2. The difference between them is

$$\alpha - \alpha' = \lambda \frac{b_2}{3} \left[\int_{D^3} \mathrm{Tr}(\bar{a}^3) - \int_{D'^3} \mathrm{Tr}(\bar{a}^3) \right] = \lambda \frac{b_2}{3} \int_{S^3} \mathrm{Tr}(\bar{a}^3) \qquad (10.11.14)$$

where $S^3 \in G$ is the 'sphere' made by joining the disks D^3 and D'^3 by their common boundary. The last member of (10.11.14) is a topological invariant that measures $\pi_3(G)$, the winding number. For any simple compact Lie group, $\pi_3(G) = Z$ or, in other words, every three-sphere in G is a 'multiple' of a basic three-sphere. Thus, if the quantum theory has to be independent of the choice of D, the exponential $\exp\{\frac{i}{\hbar}\lambda\frac{b_2}{3}\int_{S^3}\mathrm{Tr}(\bar{a}^3)\}$ has to be equal to one, and the coefficient λ (see eq. (2.6.21)) must be quantized. In complete analogy with the monopole case, *topologically non-trivial WZW terms require the quantization of the coefficient that accompanies them.*

In field-theoretical calculations, all the above examples of anomalies and Schwinger terms have been found. The reason why a $(D=2)$-dimensional field theory leads to an algebra depending on one continuous (space) index is that the calculation of the commutators with canonically quantized fields is performed by keeping the time coordinate fixed and thus they depend on D-1 continuous indices. This is also true in general; in the $D = 4$ case below, the algebra commutators depend on three spatial indices.

(b) $D = 4$ $(l = 3)$

Let us again consider first the case of the *algebra* cohomology. As before, two chains associated to the choice of $\omega_1^4(x)$ can be given:

$$ch_3(F) = b_3 \mathrm{Tr}(F^3),$$

$$\omega'^5_0 = b_3 \mathrm{Tr}[(dA)^2 A + \frac{3}{2}dAA^3 + \frac{3}{5}A^5],$$

$$\omega'^4_1 = -\frac{1}{2}b_3 \mathrm{Tr}[d\zeta(AdA + dAA + A^3)],$$

$$\omega'^3_2 = b_3 \mathrm{Tr}([d\zeta_1, d\zeta_2]A),$$

$$\omega'^2_3 = b_3 \mathrm{Tr}(d\zeta_{[1}d\zeta_2\zeta_{3]}), \qquad (10.11.15)$$

$$\omega'^1_4 = -\frac{1}{2}b_3 \mathrm{Tr}(d\zeta_{[1}\zeta_2\zeta_3\zeta_{4]}),$$

$$\omega'^0_5 = \frac{1}{10}b_3 \mathrm{Tr}(\zeta_{[1}\zeta_2\zeta_3\zeta_4\zeta_{5]}),$$

where $b_3 = -\frac{i}{48\pi^3}$, eq. (10.8.6), and

$$ch_3(F) = b_3 \text{Tr}(F^3),$$

$$\omega_0^5 = b_3 \text{Tr}[(dA)^2 A + \frac{3}{2} dAA^3 + \frac{3}{5} A^5] \, (= \omega_0'^5),$$

$$\omega_1^4 = \frac{1}{2} b_3 \text{Tr}[\zeta d(AdA + dAA + A^3)] = b_3 \text{Tr}[\zeta d(AdA + \frac{1}{2} A^3)],$$

$$\omega_2^3 = \frac{1}{2} b_3 \text{Tr}\{[\zeta_1, \zeta_2](dAA + AdA + A^3) + \zeta_1 dA\zeta_2 A - \zeta_2 dA\zeta_1 A\},$$

$$\omega_3^2 = \frac{1}{2} b_3 \text{Tr}\{[\zeta_{[1}\zeta_2\zeta_{3]}dA + A\zeta_{[1}A\zeta_2\zeta_{3]}\},$$

$$\omega_4^1 = \frac{1}{2} b_3 \text{Tr}(\zeta_{[1}\zeta_2\zeta_3\zeta_{4]}A),$$

$$\omega_5^0 = \frac{1}{10} b_3 \text{Tr}(\zeta_{[1}\zeta_2\zeta_3\zeta_4\zeta_{5]}) = \omega'_5^0.$$

$$(10.11.16)$$

We again recognize that e.g. $2\pi\omega_1^4$ is the integrand of (10.5.23) with the exact factor; its integral is the first term in the expansion of $\alpha_1[A, g]$ for $D = 4$. The difference between ω_1^4 and ω'_1^4 is given by the differential of $\beta_3^1 = -\frac{1}{2}\text{Tr}[\zeta(AdA + dAA + A^3)]$. The corresponding expressions for the anomaly and the two possible Schwinger terms can be calculated as in (10.11.4) to obtain

$$\omega'_i[A](x)$$
$$= \omega_i[A](x) = \frac{1}{2} b_3 \epsilon^{\mu\nu\rho\sigma} \text{Tr}(T_i \partial_\mu (A_\nu \partial_\rho A_\sigma + \partial_\nu A_\rho A_\sigma + A_\nu A_\rho A_\sigma)),$$

$$\omega'_{ij}[A](x, y) = b_3 \epsilon^{abc} \text{Tr}(\{T_i, T_j\} T_k) \partial_a A_b^k(y) \partial_{x^c} \delta(\mathbf{x} - \mathbf{y}),$$

$$\omega_{ij}[A](x, y) = \frac{1}{2} b_3 \epsilon^{abc} \text{Tr}\{[T_i, T_j](\partial_a A_b A_c + A_a \partial_b A_c$$
$$+ A_a A_b A_c) + \partial_a (T_i A_b T_j A_c)\} \delta(\mathbf{x} - \mathbf{y}),$$

$$(10.11.17)$$

which are also obtained in field theoretical calculations*. We calculate explicitly, as an example, the Schwinger term $\omega'_{ij}[A](x, y)$ starting from the corresponding differential form in (10.11.15):

$$\omega'_2[A](\zeta_1, \zeta_2)$$
$$= b_3 \int d^3x \epsilon^{abc} \text{Tr}(\{\partial_a \zeta_1, \partial_b \zeta_2\} A_c)$$
$$= b_3 \int d^3x d^3y \epsilon^{abc} \text{Tr}(\{\partial_a \zeta_1(y), \partial_b \zeta_2(x)\} A_c(x)) \delta(\mathbf{x} - \mathbf{y})$$

---

* Note that, as in the $D = 2$ case, there are Schwinger terms proportional to $\delta(x - y)$ and also proportional to $\partial\delta(x - y)$. Terms of the first kind are usually called *ultralocal*.

$$= b_3 \int d^3x d^3y \epsilon^{abc} \text{Tr}(\{T_i, T_j\} T_k) \partial_a \zeta_1^i(y) \partial_b \zeta_2^j(x) A_c^k(x) \delta(\mathbf{x} - \mathbf{y})$$

$$= -b_3 \int d^3x d^3y \epsilon^{abc} \text{Tr}(\{T_i, T_j\} T_k) \zeta_1^i(y) \partial_b \zeta_2^j(x) A_c^k(x) \partial_{y^a} \delta(\mathbf{x} - \mathbf{y})$$

$$= b_3 \int d^3x d^3y \epsilon^{abc} \text{Tr}(\{T_i, T_j\} T_k) \zeta_1^i(y) \zeta_2^j(x) \partial_b A_c^k(x) \partial_{y^a} \delta(\mathbf{x} - \mathbf{y}) .$$

$$(10.11.18)$$

Now, looking at (10.4.6), we get

$$\omega'_{ij}[A](x, y) = b_3 \epsilon^{abc} \text{Tr}(\{T_i, T_j\} T_k) \partial_b A_c^k(y) \partial_{x^a} \delta(\mathbf{x} - \mathbf{y}) . \qquad (10.11.19)$$

One may check that this expression is antisymmetric under the simultaneous interchange of (i, x) and (j, y) as it corresponds to a term modifying a commutator:

$$\omega'_{ji}[A](y, x) = b_3 \epsilon^{abc} \text{Tr}(\{T_j, T_i\} T_k) \partial_a A_b^k(x) \partial_{y^c} \delta(\mathbf{x} - \mathbf{y})$$

$$= -b_3 \epsilon^{abc} \text{Tr}(\{T_i, T_j\} T_k) \partial_a A_b^k(y) \partial_{x^c} \delta(\mathbf{x} - \mathbf{y}) \qquad (10.11.20)$$

$$= \omega'_{ij}[A](x, y) ,$$

where the identity

$$f(x) \partial_{y^a} \delta(\mathbf{x} - \mathbf{y}) = -f(y) \partial_{x^a} \delta(\mathbf{x} - \mathbf{y}) + \partial_{x^a} f(x) \delta(\mathbf{x} - \mathbf{y}) \qquad (10.11.21)$$

has been used, together with the fact that the term proportional to $\delta(\mathbf{x} - \mathbf{y})$ vanishes because second derivatives are symmetric in their indices.

As an example of the *group* cohomology, we give here the expression for the one-cocycle (modulo d) four-form $\Omega_1^4(A, g)$. Using (10.11.8) we find that

$$\delta\Omega_0^5 = \frac{b_3}{2} \text{Tr}[d(dAAa + AdAa + A^3 a - \frac{1}{2}(Aa)^2 - Aa^3) + \frac{1}{5}a^5] \quad (10.11.22)$$

so that, up to a total derivative, Ω_1^4 is given by

$$\Omega_1^4 = b_3 \frac{1}{2} \text{Tr}[dAAa + AdAa + A^3 a - \frac{1}{2}(Aa)^2 - Aa^3 + d^{-1}\frac{1}{5}a^5] \quad . \quad (10.11.23)$$

Upon integration, this gives the one-cocycle $\Omega_1[A; g]$:

$$\Omega_1[A; g] = \frac{1}{2}b_3 \left[\int_{S^4} d^4x \epsilon^{\mu\nu\rho\sigma} \text{Tr}\{\partial_\mu A_\nu A_\rho a_\sigma + A_\mu \partial_\nu A_\rho a_\sigma + A_\mu A_\nu A_\rho a_\sigma \right.$$

$$\left. - \frac{1}{2}A_\mu a_\nu A_\rho a_\sigma - A_\mu a_\nu a_\rho a_\sigma\} + \frac{1}{5} \int_{D^5} d^5\bar{x} a^5 \right] ,$$

$$(10.11.24)$$

where $\partial D^5 = S^4$. It is obtained from a potential form for $(\delta\Omega_0^5)(A, g)$ and, as in the two-dimensional case, is only locally defined. If Ω_1 is used to construct the $D = 4$ *Wess-Zumino-Witten term* of the action, the term that depends on $d^{-1}\text{Tr}(a^5)$ leads to a quantization of the proportionality constant, by an argument identical to that of the $D = 2$ case, only that now $\pi_4(G) = 0$ and $\pi_5(G) = Z$ are required. This condition is fulfilled, e.g., by $SU(n)$ for $n \geq 3$ (and also by $U(n)$ since $\pi_k(U(n)) = \pi_k(SU(n))$ for $k \geq 2$, $n \geq 2$, section 1.11).

It is noticeable from (10.11.2), (10.11.3), (10.11.15) and (10.11.16) that the last element of the chain is always proportional to $\text{Tr}(\zeta_{[1}\zeta_2\zeta_{3]})$ or $\text{Tr}(\zeta_{[1}\zeta_2\zeta_3\zeta_4\zeta_{5]})$. Also, at this last stage of the chain, the ζ's do not depend on any spacetime variable and are just constants. Then the following proposition holds:

(10.11.1) PROPOSITION

The cochain of the Lie algebra $\mathcal{G}$ given by

$$\omega(\zeta_1, \ldots, \zeta_{D+1}) = \text{Tr}(\zeta_{[1} \cdots \zeta_{D+1]}) \quad , \tag{10.11.25}$$

where $\zeta_a = \zeta_a^i T_i$ and $\zeta_a^i \in R$, defines a non-trivial cocycle of $H^{D+1}(\mathcal{G}, R)$.

Proof: The proof of this statement follows easily by using the Chevalley-Eilenberg approach to Lie algebra cohomology (section 6.7). To see it explicitly, let us construct an odd (D is even) LI $(D + 1)$-form associated with ω. The coordinates of this skew-symmetric linear mapping are given by $\omega_{i_1 \ldots i_{D+1}} = \omega(\delta_{i_1}^{j_1} T_{j_1}, \ldots, \delta_{i_{D+1}}^{j_{D+1}} T_{j_{D+1}})$, from which we get

$$\omega^{D+1}(g) = \frac{1}{(D+1)!} \omega_{i_1 \ldots i_{D+1}} \omega^{L(i_1)} \cdots \omega^{L(i_{D+1})}$$

$$= \text{Tr}(T_{(i_1)} \cdots T_{(i_{D+1})}) \, \omega^{L(i_1)} \cdots \omega^{L(i_{D+1})} \tag{10.11.26}$$

where $\omega^{L(i)}(g)$ are the elements of the basis of LI one-forms on G.

The canonical one-form on G is given by $g^{-1}dg = \omega^{L(i)}(g)T_i$, with $g = \exp g^i T_i$. Thus, the above differential form is given by

$$\omega^{D+1}(g) = \text{Tr}((g^{-1}dg)^{D+1}) \quad .$$

This is a closed form, because $d(g^{-1}dg) = -(g^{-1}dg)^2$ and then

$$d\omega = d\text{Tr}((g^{-1}dg)^{D+1}) = -\text{Tr}((g^{-1}dg)^{D+2}) = 0 \quad ,$$

since $\text{Tr}((g^{-1}dg)^{even}) = 0$. Moreover, the odd ω cannot be the differential of an invariant form because $\text{Tr}((g^{-1}dg)^D)$ is also zero, *q.e.d.*

From this proposition, we conclude that the cochain (10.11.25) is a $(D + 1)$-cocycle in $H_0^{D+1}(\mathcal{G}, R)$, which defines the cohomology groups of the final step of the chain, where the cochains do not depend on A. Since, as has been shown, every chain has this non-trivial cocycle as its last element*, all the cocycles given here are non-trivial, because if they were trivial, ω_{D+1} or ω'_{D+1} would also be trivial by Proposition 10.9.2.

10.12 On the existence of consistent anomalous theories

We have seen that eqs. (10.7.1) express that the quantum theory is anomalous, a fact that can be associated with the gauge non-invariance of the fermionic functional measure $D\psi D\psi^\dagger$. The presence of the gauge anomaly spoils the reduction of degrees of freedom required by gauge invariance, itself a manifestation of the fact that some degrees of freedom are spurious. When gauge anomalies are present, the quantum theory is not consistent. The divergence structure of the theory becomes more severe, and loss of renormalizability follows. The elimination of the anomaly, as we saw in the example of the standard model, restricts the representation contents of the fermions of the theory: the anomaly cancellation imposes severe constraints on the form of the theory and in fact may be invoked as a justification of its structure. A question immediately arises: could it be possible to make an anomalous theory consistent? We shall mention here a procedure, particularly suggested by Faddeev, because of its relation with the geometric picture of the anomalies described in previous sections. The *ansatz* consists of adding to the original action I a new piece I_{WZW} (the Wess-Zumino-Witten term) which is *not* gauge-invariant, in such a way that the complete quantum theory becomes gauge-invariant. In so doing, *new* variables (fictitious scalar fields) with adequate gauge transformation properties are added to the original theory, a fact that explains how a non-trivial anomaly (the variation of a non-local functional of the original fields of the theory) may be obtained from the variation of the local functional I_{WZW} (cf. (10.7.1)). This is reminiscent of what happened in the context of the Wess-Zumino terms of chapter 8, where the addition of a new degree of freedom with the appropriate transformation properties to the quasi-invariant Lagrangian cancelled the quasi-invariance. The way of implementing this idea is to use the cocycle property (10.7.3) of $\alpha_1[A;g]$ to define a Wess-Zumino-Witten action $I_{WZW}[A;u] \equiv \alpha_1[A;u]$ where u is a G-valued function on spacetime and defines *new* fields $u(x)$. Using this definition, eq. (10.7.3) gives, with $u^g = g^{-1}u$,

$$I_{WZW}[A^g, u^g] - I_{WZW}[A, u] = \alpha_1[A^g; g^{-1}u] - \alpha_1[A; u] = -\alpha_1[A; g] .$$
$$(10.12.1)$$

Thus, since the variation of this action is *minus* the anomaly, the WZW action can be used for its cancellation. Consider now the total action $I' =$

---

* This follows easily looking at the structure of $s\omega_D^1 = d\omega_{D+1}^0$, where the term in A, if present, has the form of $\text{Tr}(\zeta_1...[\zeta_{D+1},A])$ which is zero since D is even.

$I[A;\psi] + I_{WZW}[A;u]$ (which depends on the additional fields u) and the path integral formulation of the theory associated to it. Looking again at A as an external field, the theory is described by

$$W'[A] = \int D\psi D\psi^\dagger Du\, e^{i(I[A;\psi]+I_{WZW}[A;u])} \quad . \tag{10.12.2}$$

If now $W'[A^g]$ is calculated two distinct contributions are found: the first one comes from the non-invariance of $D\psi D\psi^\dagger$ and the second from the non-invariance of I_{WZW}. They are exponentials of equal quantities but for a sign and so they cancel. Explicitly,

$$
\begin{aligned}
W'[A^g] &= \int D\psi D\psi^\dagger Du\, e^{i(I[A^g;\psi]+I_{WZW}[A^g;u])} \\
&= \int D(g^{-1}\psi)D(g^{-1}\psi)^\dagger D(u^g) e^{i(I[A^g;g^{-1}\psi]+I_{WZW}[A^g;u^g])} \\
&= e^{i\alpha_1[A;g]} \int D\psi D\psi^\dagger Du\, e^{i(I[A^g;g^{-1}\psi]+I_{WZW}[A^g;u^g])} \qquad (10.12.3) \\
&= e^{i\alpha_1[A;g]} \int D\psi D\psi^\dagger Du\, e^{i(I[A;\psi]+I_{WZW}[A;u]-\alpha_1[A;g])} \\
&= W'[A] \quad ,
\end{aligned}
$$

or, equivalently,

$$Z'[A^g] - Z'[A] = \alpha_1[A,g] - \alpha_1[A,g] = 0 \quad . \tag{10.12.4}$$

Clearly, the price paid by the theory is the inclusion of the new scalar fields u but, formally, it is anomaly free. Thus, this new quantum theory is gauge-invariant and, provided that it is free of other problems such as power-counting non-renormalizability, it might be taken as acceptable (*i.e.*, consistent) gauge theory. However, it may be argued that by introducing the additional scalar fields what is achieved is just a gauge-invariant description of the original anomalous theory. Another possibility might be considering an anomalous theory as an *effective theory* for low energy physics: the presence of the anomaly could be interpreted as an indication that new physics should appear at high energies, much in the same way as the non-renormalizability of Fermi's theory of weak interactions indicated the need of a better description at high energy. In any case, recent analyses indicate that any sensible method of quantizing an anomalous gauge theory in four dimensions leads to a non-renormalizable theory with non-linearly realized gauge symmetry. Thus, it seems that the absence of gauge anomalies is required for a consistent theory. We shall finish at this point, and refer to the literature cited below for further discussions on the possibility of having consistent anomalous gauge theories.

10.13 Appendix: calculating Chern-Simons forms[*]

Eq. (10.8.7) is the starting point for a practical way of computing the Chern-Simons form Ω_0^{2l-1} (or $Q^{(2l-1)}$ in (2.6.2)), which can be given in terms of A and dA, or in terms of A and F. We give here the first cases for both (equivalent) expressions.

(a) *Expressions in A, dA*
Putting $F_t = tdA + t^2A^2$ inside the trace in (10.8.7), it follows that

$$\Omega_0^{2l-1} = b_l \sum_{s=0}^{l-1} \int_0^1 \delta t\, lt^{l-s-1}(t^2)^s \text{Tr}(A\,\mathscr{S}((dA)^{l-s-1},(A^2)^s)) , \quad b_l = \left(\frac{i}{2\pi}\right)^l \frac{1}{l!} ,$$

(10.13.1)

where $\mathscr{S}((dA)^{l-s-1},(A^2)^s)$ indicates sum over all possible products ('words') containing a total of $(l-s-1)$ dA's and s A^2's. The integral in t produces a factor $1/(l+s)$ and this gives, for the lowest l cases, the results

$l = 2$ $(s = 0, 1)$

$$\Omega_0^3 = b_2\left[\frac{2}{2+0}\text{Tr}(AdA) + \frac{2}{2+1}\text{Tr}(A^3)\right] = b_2\text{Tr}(AdA + \frac{2}{3}A^3) ,$$

(10.13.2)

$l = 3$ $(s = 0, 1, 2)$

$$\Omega_0^5 = b_3\left[\frac{3}{3+0}\text{Tr}(A(dA)^2) + \frac{3}{3+1}2\text{Tr}(A^3dA) + \frac{3}{3+2}\text{Tr}(A^5)\right]$$
$$= b_3\text{Tr}(A(dA)^2 + \frac{3}{2}A^3dA + \frac{3}{5}A^5) ,$$

(10.13.3)

$l = 4$ $(s = 0, ..., 3)$

$$\Omega_0^7 = b_4\left[\frac{4}{4+0}\text{Tr}(A(dA)^3) + \frac{4}{4+1}\text{Tr}(2A^3(dA)^2 + AdAA^2dA)\right.$$
$$\left. + \frac{4}{4+2}3\text{Tr}(dAA^5) + \frac{4}{4+3}\text{Tr}(A^7)\right]$$
$$= b_4\text{Tr}(A(dA)^3 + \frac{8}{5}A^3(dA)^2 + \frac{4}{5}AdAA^2dA + 2dAA^5 + \frac{4}{7}A^7) ,$$

(10.13.4)

---

[*] With J. Mateos Guilarte.

$l = 5$ $(s = 0, ..., 4)$

$$\Omega_0^9 = b_5 \left[\frac{5}{5+0} \text{Tr}(A(dA)^4) + \frac{5}{5+1} \text{Tr}(2A^3(dA)^3 + AdAA^2(dA)^2 \right.$$

$$+ A(dA)^2 A^2 dA) + \frac{5}{5+2} \text{Tr}(3A^5(dA)^2 + 2A^3 dAA^2 dA + AdAA^4 dA)$$

$$\left. + \frac{5}{5+3} 4\text{Tr}(A^7 dA) + \frac{5}{5+4} \text{Tr}(A^9) \right]$$

$$= b_5 \text{Tr}[A(dA)^4 + \frac{5}{3} A^3(dA)^3 + \frac{5}{6}(AdAA^2(dA)^2 + A(dA)^2 A^2 dA)$$

$$+ \frac{15}{7} A^5(dA)^2 + \frac{10}{7} A^3 dAA^2 dA + \frac{5}{7} AdAA^4 dA + \frac{5}{2} A^7 dA + \frac{5}{9} A^9] \quad .$$

$$(10.13.5)$$

The first three cases correspond to eqs. (2.6.5), (2.6.6).

(b) *Expressions in A, F*
In this case, $F_t = tF + (t^2 - t)A^2$ is used and (10.8.7) now gives

$$\Omega_0^{2l-1} = b_l \sum_{s=0}^{l-1} \int_0^1 \delta t \, lt^{l-s-1}(t^2 - t)^s \text{Tr}(A\mathcal{S}(F^{l-s-1}, (A^2)^s)) \quad . \quad (10.13.6)$$

The factors multiplying each polynomial in the sum are now given by

$$l b_l \int_0^1 \delta t \, t^{l-1}(t - 1)^s = b_l l \sum_{j=0}^s \binom{s}{j}(-1)^{s-j} \frac{1}{1+j} \quad . \quad (10.13.7)$$

For every value of $s (\leq l - 1)$, the factor $(-1)^{s-j} \binom{s}{j}$ can be read from a trivial modification of Pascal's triangle:

$s = 0$	1
$s = 1$	-1 1
$s = 2$	1 -2 1
$s = 3$	-1 3 -3 1
$s = 4$	1 -4 6 -4 1
$s = 5$	-1 5 -10 10 -5 1

For a given l, there are l different polynomials with a global factor lb_l. Each of them contains the same number of F's and A's, and its coefficient is the sum of the elements (times $1/(l + j)$) in a line of the above triangle (these polynomials are the same as before with dA replaced by F). It is then easy to find the expressions for the Chern-Simons forms in any dimension. For instance,

$$(l = 2) \quad \Omega_0^3 = 2b_2 \left[\frac{1}{2}\mathrm{Tr}(AF) + (-\frac{1}{2} + \frac{1}{3})\mathrm{Tr}(A^3) \right] = b_2 \mathrm{Tr}(AF - \frac{1}{3}A^3) \quad,$$

$$(10.13.8)$$

$$(l = 3) \quad \Omega_0^5 = 3b_3 \left[\frac{1}{3}\mathrm{Tr}(AF^2) + (-\frac{1}{3} + \frac{1}{4})2\mathrm{Tr}(A^3 F) + (\frac{1}{3} - \frac{2}{4} + \frac{1}{5})\mathrm{Tr}(A^5) \right]$$

$$= b_3 \mathrm{Tr}(AF^2 - \frac{1}{2}A^3 F + \frac{1}{10}A^5) \quad,$$

$$(10.13.9)$$

$$(l = 4) \quad \Omega_0^7 = 4b_4 \left[\frac{1}{4}\mathrm{Tr}(AF^3) + (-\frac{1}{4} + \frac{1}{5})\mathrm{Tr}(2A^3 F^2 + AFA^2 F) \right.$$

$$\left. + (\frac{1}{4} - \frac{2}{5} + \frac{1}{6})3\mathrm{Tr}(FA^5) + (-\frac{1}{4} + \frac{3}{5} - \frac{3}{6} + \frac{1}{7})\mathrm{Tr}(A^7) \right]$$

$$= b_4 \mathrm{Tr}(AF^3 - \frac{2}{5}A^3 F^2 - \frac{1}{5}AFA^2 F + \frac{1}{5}FA^5 - \frac{1}{35}A^7) \quad,$$

$$(10.13.10)$$

$$(l = 5) \quad \Omega_0^9$$

$$= 5b_5 \left[\frac{1}{5}\mathrm{Tr}(AF^4) + (-\frac{1}{5} + \frac{1}{6})\mathrm{Tr}(2A^3 F^3 + AFA^2 F^2 + AF^2 A^2 F) \right.$$

$$+ (\frac{1}{5} - \frac{2}{6} + \frac{1}{7})\mathrm{Tr}(3A^5 F^2 + 2A^3 FA^2 F + AFA^4 F)$$

$$+ (-\frac{1}{5} + \frac{3}{6} - \frac{3}{7} + \frac{1}{8})4\mathrm{Tr}(A^7 F) + (\frac{1}{5} - \frac{4}{6} + \frac{6}{7} - \frac{4}{8} + \frac{1}{9})\mathrm{Tr}(A^9) \right]$$

$$= b_5 \mathrm{Tr}[AF^4 - \frac{1}{3}A^3 F^3 - \frac{1}{6}(AFA^2 F^2 + AF^2 A^2 F) + \frac{1}{7}A^5 F^2$$

$$+ \frac{2}{21}A^3 FA^2 F + \frac{1}{21}AFA^4 F - \frac{1}{14}A^7 F + \frac{1}{126}A^9] \quad;$$

$$(10.13.11)$$

they coincide, of course, with those of (a).

Bibliographical notes for chapter 10

For general references on the geometric formulation of Yang-Mills theories, see the bibliographical notes for chapter 2.

The BRST transformations were introduced in Becchi *et al.* (1976) and Tyutin (1975). Further references can be found in the book of Henneaux and Teitelboim (1992). The interpretation of ghosts in gauge theories as Maurer-Cartan forms in the infinite-dimensional group of gauge transformations is given in Bonora and Cotta-Ramusino (1983) where the cohomology of the gauge group is studied; see also Thierry-Mieg (1980). Additional details in connection with the cohomology of Lie algebras can be found in Kastler and Stora (1986), Brandt *et al.* (1990a,b), and in van

Holten (1990). For the BRST transformation in topological quantum field theory (see below) see Kanno (1989). A general setting of ghost techniques in mathematical physics is given in Stasheff (1992).

The Gribov ambiguity was found by Gribov (1978); see also Takahashi and Kobayashi (1978). Its implications for the functional integral are discussed in Zwanziger (1989). Additional information on Coulomb gauge fixing can be found in Van Baal (1992), which contains further references. The structure of the infinite-dimensional bundle of Yang-Mills connections and its consequences for the Gribov ambiguity were given by Singer (1978; see also 1985), and further discussed by Atiyah and Jones (1978); see also Mitter and Viallet (1981). The reviews of Daniel and Viallet (1980), Marathe and Martucci (1989), the report by Viallet (1990) and the books by Mickelsson (1989) and Nash (1991) contain useful material and additional references on the topology of Yang-Mills bundles; see also Heil *et al.* (1990). The global anomalies mentioned in section 10.1 were found by Witten (1982b, 1985). See also the review of Ball (1989), which contains a detailed account of both perturbative and topological aspects of chiral theories and anomalies.

An excellent collection of lectures and articles on anomalies is contained in the book edited by Treiman *et al.* (1985), which includes in particular the lectures of Jackiw (1985a) and Zumino (1985a) at Les Houches. Jackiw's lectures contain a detailed list of references including the classical early papers on anomalies not quoted here. Other reviews may be found in Álvarez-Gaumé (1986), Morozov (1986) Yamagishi (1987) (which contains an extensive list of references) and Bonora (1990). The observation that the anomaly has its origin in the lack of invariance of the fermionic measure is due to Fujikawa (1980a,b; 1986) and Vergeles (1975 Landau Inst. preprint, unpublished); the application to the arbitrary dimensional case is given in Zumino *et al.* (1984). Further aspects are discussed in Balachandran *et al.* (1982), Sonoda (1986) and in Tsutsui (1989a). For the group properties relevant in anomaly cancellation see, *e.g.*, Okubo and Patera (1985) and Kephart (1985) in which earlier references can be found. General group properties are given in Slanski (1981).

The Wess-Zumino chiral effective action and the corresponding consistency conditions were introduced by Wess and Zumino (1971); the consistency conditions also are naturally incorporated in the cohomological descent description (references below). The mathematical framework of the Wess-Zumino model was further described by Witten (1983a,b) (see also Novikov (1982)) and in the paper by Dijkgraaf and Witten (1990), which provides a classification of topological actions. For a review of the topological field theory of Witten (1988b; see also 1991) and Atiyah (1989) (see also Atiyah and Jeffrey (1990)), not treated in the text, see Birmingham *et al.* (1991); a short account is given in Nash (1991). In two dimensions the WZ model has been studied by Polyakov and Wiegmann (1984) and, in connection with conformal theory, by Knizhnik and Zamolodchikov (1984). Further mathematical aspects are given in Krichever *et al.* (1986). For the structure of sigma models with a WZ term see, *e.g.*, the book by Zakrzewski (1989), Jack *et al.* (1990), Hull and Spence (1991) and Witten (1992b), from which earlier references can be traced; see also Wu (1993). Applications of topology to quantum theory in general are discussed in Schwarz (1993).

The geometric approach to anomalies and in particular the cohomological descent method is due to Stora, Zumino and Faddeev; see the lectures by Stora (1977, 1984, 1986) and Zumino (1985a), and the papers by Bardeen and Zumino (1984), Zumino *et al.* (1984), Faddeev and Shatashvili (1985) (cf. Alekseev *et al.* (1988)), Kastler and Stora (1986) and Mañes and Zumino (1986). Zumino's article (1985b) discusses the trivial bundle case and Mañes *et al.* (1985) treat the non-trivial one mentioned at the end of section 10.9. See also the lectures by Jackiw (1985a, 1987) and Wess (1986). A mathematical description of the local cohomology is given in Bonora and Cotta-Ramusino (1983). Additional information on some of the techniques used, such as the transgression formula, may be found in the bibliographical notes for chapter 2. Further analysis of the consistency conditions as a local cohomology problem, its solutions in the BRS framework as well as of the issues of non-triviality and completeness are given by Dubois-Violette *et al.* (1985), Schücker (1987), Brandt *et al.* (1990a,b) and Dubois-Violette *et al.* (1992). Early papers on the anomalous Gauss law commutators and Schwinger terms are those of Faddeev (1984), Mickelsson (1985b), Zumino (1985b), Faddeev and Shatashvili (1986) and Fujikawa (1986). For further discussion of the Schwinger terms as the curvature two-form on a line bundle and ray representations see Niemi and Semenoff (1985) and Nelson and Álvarez-Gaumé (1985). More general information can be found in the lectures cited above and in Mickelsson (1989), Nash (1991) and Ball (1989). For the finite form of two-cocycle see Mickelsson (1989), Laursen *et al.* (1986) and Bak (1993). Field theory computations are given in Jo (1985), Kobayashi *et al.* (1986), Percacci and Rajaraman (1989), Dunne and Trugenberger (1990) and Schwiebert (1990). The cohomological techniques may also be applied to the analysis and determination of the covariant anomalies (not discussed in the text). Recent articles on this subject are, *e.g.*, Tsutsui (1989b), Abud *et al.* (1990) and Kelnhofer (1991, 1993), which contain earlier references.

The Fujikawa method of computing the anomaly provides a natural connection between the axial anomaly and the index theorem (see the bibliographical notes for chapter 2 for references on the index theorem).

The application of the index theorem to the non-abelian gauge anomalies is due to Atiyah and Singer (1984) and to Álvarez-Gaumé and Ginsparg (1984, 1985) whose presentation we have followed. The relevance of topological considerations in non-abelian anomalies was also discussed by Gómez (1983). For a discussion on the capacity of the topological analysis to detect non-abelian anomalies see also Álvarez-Gaumé and Ginsparg (1985) and Álvarez-Gaumé (1986). These papers include a discussion of the application of the index theorem to gravitational anomalies (not treated in the text); see also in this context Álvarez et al. (1984) and Ginsparg (1986). The anomalies in odd dimensions (not mentioned in the book) are discussed in connection with the abelian and non-abelian anomalies in even dimensions in Álvarez-Gaumé et al. (1985). See also the review of Álvarez-Gaumé (1986) and the books of Mickelsson (1989), Nakahara (1990) and Nash (1991), for additional information and references.

For the free differential algebras mentioned in section 10.8 the reader may consult Sullivan (1977) and the lecture of Van Nieuwenhuizen (1983).

The possibility of having consistent anomalous gauge theories was particularly put forward by Faddeev (1986) and in two dimensions by Jackiw and Rajaraman (1985); see also Faddeev and Shatashvili (1986). Further discussion and references may be found in the review articles of Ball (1989) and Preskill (1991).

BRS algebras and anomalies in the context of supersymmetric and superstring theories have not been discussed in the text. The interested reader may look at the books of Gibbons et al. (eds.) (1986), Green et al. (1987) and Gieres (1988) for initial general information and references; see also Dixon (1990).

List of symbols

$\phi_t = exp\,tX$	One-parameter group generated by $X \in \mathcal{G}$
$*$	Hodge dual mapping
(α, β)	Scalar product of the p-forms α, β of $L^2(\Lambda^p(M))$
δ	Codifferential
$\displaystyle{\not{\!d}} := d + \delta$	Hodge-de Rham operator
$\text{Harm}^p(M)$	Harmonic forms on M
$\omega^{L,R}(g)$	Left-, right-invariant form on G (LIF, RIF)
$\omega^{R(k)}(g),\ \omega^{L(k)}(g)$	Right-, left-invariant one-forms dual to $X^R_{(k)}(g), X^L_{(k)}(g)$
$\mu^{L,R}$	Left-, right-invariant Haar measure
$\theta \equiv \theta^L,\ \theta^R$	Left-, right-invariant canonical form on G
$\mathcal{X}^{*L,R}(G)$	Vector space of the left-, right-invariant one-forms on G
$Adg,\ Ad^*g$	Adjoint, coadjoint representation of $g \in G$ on $\mathcal{G}$
$adX,\ ad^*X$	Adjoint, coadjoint representation of $X \in \mathcal{G}$ on $\mathcal{G}$
$\Sigma,\ A = (\mu, \alpha)$	Supertranslation group, superspace indices
$\Pi^{(A)},\ D_{(B)}$	Left-invariant forms, vector fields on Σ
$E_{(A)}{}^B,\ E_B{}^{(A)}$	Rigid superspace supervielbein, inverse supervielbein

Chapter 2

$H_p(P),\ V_p(P)$	Horizontal, vertical subspace of $T_p(P)$
$Y^v(p),\ Y^h(p)$	Vertical, horizontal component of $Y(p)$
$\mathcal{P}(P)$	$\mathcal{F}(M)$-module of projectable vector fields on P
Y^A	Fundamental vector field on P associated with $A \in \mathcal{G}$
$\omega,\ \Omega$	Connection, curvature on $P(G,M)$
$\mathcal{D}$	Exterior covariant derivative
$\rho(X)$	Representation of $X \in \mathcal{G}$ on a vector space V
$A = A_\mu dx^\mu,\ F = \frac{1}{2}F_{\mu\nu}dx^\mu \wedge dx^\nu$	$\mathcal{G}$-valued Yang-Mills field, field strength
Θ	Left-invariant connection
$\Omega,\ \omega^s$	Left-equivariant form, components
$D(g)$	Representation of $g \in G$ on a vector space V
$I^l(G),\ I(G)$	Vector space of AdG-invariant polynomials of order $l,\ \sum_{l=0}^{\infty} I^l(G)$
$P(\Omega)$	Differential $2l$-form induced by the polynomial P
$P(\bar{\Omega})$	Projection of $P(\Omega)$ on M
$\omega_t,\ \Omega_t$	One-parameter family of connections, curvatures
$\omega_{t_1 \ldots t_r}$	Simplex of interpolating connections
$Q^{(2l-1)}(\bar{\omega}, \bar{\omega}_0) = TP(\bar{\omega}, \bar{\omega}_0)$	Transgression of $P(\bar{\Omega})$
$Q^{(2l-r),r}$	Chern-Simons-like $(2l-r)$-forms of type r
k_{01}	Homotopy operator
$c_l(\bar{\Omega}),\ c(\bar{\Omega})$	Chern $2l$-forms on M, total Chern form
$ch_l(\bar{\Omega}),\ ch(\bar{\Omega})$	Chern character $2l$-forms, total Chern character
$Q^{(2l-1)}(A, F)$	Chern-Simons form in the Y-M case
p_{2k}	Pontrjagin classes
e	Euler class
$\sigma_x(\xi, \mathcal{D})$	Leading symbol of $\mathcal{D}$
$E^\alpha_\mu(x)$	Vielbein
ω_μ	Spin connection

Chapter 3

$B, \vec{A}, \vec{V}, R$	Galilei group elements
$\Delta(t, \vec{x}; g),\ \Delta(x, g)$	$\mathcal{F}(M)$-valued one-cochains on G
$U(g),\ \tilde{U}(g)$	Unitary operator representing $g \in G$, ray operator
$\xi(g', g),\ \omega(g', g)$	Local factor (additive two-cocycle), factor system
$\xi_{cob}(g', g),\ \omega_{cob}(g', g)$	Two-coboundaries

$\tilde{G}_{(m)}, \bar{G}_{(m)}$	Galilei group centrally extended by $U(1)$, R
ζ, θ	Central parameter in $\tilde{G}_{(m)}, \bar{G}_{(m)}$
$\bar{\mathscr{G}}_{(m)}$	Lie algebra of the extended Galilei group
$\bar{X}, \bar{\omega}$	Vector fields, forms for a central extension
$\Omega, (S;\Omega)$	Symplectic form, symplectic manifold
X_S	Vector field on S
$j(\vec{u})$	Cross product operator relative to the three-vector $\vec{u}$
$\tilde{\mathscr{P}}$	Universal covering group of the Poincaré group
$L_+^\uparrow$	Proper (restricted) Lorentz group
G_c	Contracted group of a group G

Chapter 4

$\tilde{G}$	Extension of a group G by a group K
$\lessdot$	Invariant or normal subgroup symbol
$K \otimes G$	Direct product of groups G, K
$\mathrm{Aut}\, K$	Group of automorphisms of a group K
$\mathrm{Int}\, K$	Group of the inner automorphisms of K
$\mathrm{Out}\, K$	$\mathrm{Aut}\,K/\mathrm{Int}\,K$ (classes of outer automorphisms)
$[\tilde{g}]$	Element of $\mathrm{Aut}\,K$ induced by $\tilde{g} \in \tilde{G}$
f	Homomorphism $f : \tilde{G} \to \mathrm{Aut}\,K$, $f(\tilde{g}) = [\tilde{g}]$
σ	Homomorphism $\sigma : G \to \mathrm{Out}\,K$
$\phi_{\tilde{G}}(K)$	Centralizer of K in $\tilde{G}$
C_K	Centre of K
H	Quotient $\phi_{\tilde{G}}(K)/C_K$
(K,σ)	G-kernel with group K and homomorphism σ
(C_K, σ_0)	Centre of the kernel (K,σ)
$s(g), u(g)$	Trivializing sections
$ks(g), (k,g)$	Elements of an extension of G by K
ⓢ	Semidirect product of groups for the action σ of G on K
$F(g)$	System of automorphisms $F : G \to \mathrm{Aut}\,K$

Chapter 5

A	Abelian group
$\alpha_n(g_1, ..., g_n)$	A-valued n-cochain on G
$C^n(G, A)$	Abelian group of the n-cochains on G
$\delta : C^n \to C^{n+1}$	Coboundary operator for the group cohomology
$\sigma(g)$	Action of $g \in G$ on A
$Z_\sigma^n(G, A), B_\sigma^n(G, A)$	Abelian group of n-cocycles, n-coboundaries on G for the action σ
Σ_2	$N = 2$ superspace ($N = 2$ supertranslation group)
$\Omega_n(x; g_1, ...g_n)$	$\mathscr{F}(M)$-valued cocycle on G

Chapter 6

$\omega : \mathscr{G} \times ... \times \mathscr{G} \to V$	V-valued n-cochain on $\mathscr{G}$
$C^n(\mathscr{G}, V)$	Abelian group of V-valued n-cochains on $\mathscr{G}$
$s : C^n(\mathscr{G}, V) \to C^{n+1}(\mathscr{G}, V)$	Coboundary operator for the Lie algebra cohomology
$Z_\rho^n(\mathscr{G}, V), B_\rho^n(\mathscr{G}, V)$	Abelian group of n-cocycles, n-coboundaries on $\mathscr{G}$ for the representation ρ of $\mathscr{G}$
$H_\sigma^n(G, A), H_\rho^n(\mathscr{G}, V)$	n-th cohomology group of G, $\mathscr{G}$ for the action ρ, σ
ω_{cob}	Lie algebra coboundary
$\mathscr{A}$	Abelian Lie algebra
$\tilde{\mathscr{G}}$	Lie algebra extension of $\mathscr{G}$ by $\mathscr{A}$
ⓢ	Semidirect product of Lie algebras, for the action ρ

C_{ij}^θ	Additional structure constants for the left generators of a central extension
μ_{ij}	Coboundary of a central extension of $\mathcal{G}$
$E^h(G)$, $E(G)$	n-th Chevalley-Eilenberg cohomology group, CE ring
c^i	Anticommuting ghost variable
$\tilde{s}$	BRST nilpotent operator
$\tilde{\omega}$	BRST cochain
$\text{Aut}(\mathcal{A})$	Group of automorphisms of the abelian Lie algebra $\mathcal{A}$
$s_n[g_1, ..., g_n]$	n-'simplex' in group space of vertices
	$e, g_1, g_1 g_2, ..., g_1 \cdots g_n$
$s_n(t_1, ..., t_n; g_1, ..., g_n)$	Parametrization of the points of an n-'simplex' in group space
$\hat{\omega}^n(g)$	Left-equivariant form associated with the cochain ω_n
$\omega_n(x, X_1, ... X_n)$	$\mathscr{F}(M)$-valued Lie algebra n-cochain

Chapter 7

$t : \text{Out}\,K \to \text{Aut}\,K$	Trivializing section in $\text{Aut}\,K$ as an extension
$\alpha : G \to \text{Aut}\,K$	System of automorphisms of K
$\wedge$	C_K-product of G-kernels, C_K-product of extensions
V	Group of spacetime reflections ('Vierergruppe')
(μ_1, μ_2)	Elements of V
K	Complex conjugation operator
Q, D_4	Quaternionic, dihedral group
$A(a, b, c)$	Associator of a, b, c,

Chapter 8

N	Manifold of coordinates q^i
E	Evolution space, coordinates (t, q^i)
$J^1(E)$	1-jet bundle ($= R \times T(N)$)
$\bar{s}^1 \equiv j^1(s)$	1-jet prolongation of the section s of $E \overset{\pi}{\to} R$
$\bar{X}^1$	1-jet prolongation of the vector field X on E
$L(t, q^i, \dot{q}^i)dt$	Lagrangian one-form on $J^1(E)$
Θ_{PC}	Poincaré-Cartan form
$\theta^i := dq^i - \dot{q}^i dt$	Structure one-forms
$Tr_3, B_3, \mathscr{R}$	Space translations, boosts, rotations
$(t, \mathbf{x}, \dot{\mathbf{x}})$, $(\chi, t, \mathbf{x}, \dot{\mathbf{x}})$	Evolution space, extended evolution space variables of Galilean mechanics
$\bar{\Theta}_{PC}$	Poincaré-Cartan form on the extended evolution space
$\Delta(t, q; g)$, $\Delta_{cob}(t, q; g)$	Gauge function, trivial gauge function
$\Delta_{(i)}(q)$	Infinitesimal gauge function for the parameter g^i
$\Omega_n(q, \dot{q}; g_1, ..., g_n)$	$\mathscr{F}(J^1(E))$-valued n-cochain on G
$\omega_n(q, \dot{q}; X_1, ... X_n)$	$\mathscr{F}(J^1(E))$-valued n-cochain on $\mathcal{G}$
ω_n^m	Algebra n-cochain, m-form
$\tilde{\omega}_n^m(q, \dot{q}')$	m-form, ghost number n BRST cochain
$\bar{s}$	BRST operator for the extended group
$\bar{d}$	Differential on the extended group manifold
h	CE non-trivial form
b	Potential form of h or Wess-Zumino form
$\omega^\mu(\tau)d\tau$	One-form on the worldline induced by Π^μ
$\mathscr{L}_0, \mathscr{L}_{WZ}$	Kinetic and Wess-Zumino Lagrangian densities
$\xi^i = (\tau, \underline{\sigma})$	Worldvolume coordinates
m_{ij}, M_{ij}	Induced metric on the worldvolume W
$L[z^A]$	Lagrangian functional for an extended object ($p \geq 1$)

$L\Sigma = L^1\Sigma,\ L^p\Sigma$	Loop superspace, generalized p-loop superspace
$\mathbf{d}$	Exterior functional derivative on $L^p\Sigma$
$\mathbf{E}^{(A)}(\sigma)$	One-forms on $L^p\Sigma$
$\mathscr{D}_{(A)}(\sigma)$	Functional vector fields associated with $D_{(A)}$
$\mathbf{A}$	Lagrangian one-form on $L^p\Sigma$ for I_{WZ}
$\mathscr{A}_{(A)}(\underline{\sigma})$	Components of $\mathbf{A}$ in the $\mathbf{E}^{(A)}(\sigma)$ basis
$Q_{(A)}$	Supersymmetry generator
$\mathscr{Q}_{(A)}(\underline{\sigma})$	Supersymmetry generator density on $L^1\Sigma$
$\mathscr{Q}_{(A)}^{WZ}(\underline{\sigma}),\ \mathscr{D}_{(A)}^{WZ}(\underline{\sigma})$	WZ-modified form of $\mathscr{Q}_{(A)}(\underline{\sigma}),\ \mathscr{D}_{(A)}(\underline{\sigma})$
c^A	Graded ghosts on superspace
Ω_{AB}	Schwinger term in the algebra $\{Q^{WZ}, Q^{WZ}\}$
Ω	Two-form in ghost space of coordinates Ω_{AB}
$\mathscr{F}_{AB}(\underline{\sigma}, \underline{\sigma}')$	Schwinger term in the algebra $\{\mathscr{D}^{WZ}(\underline{\sigma}), \mathscr{D}^{WZ}(\underline{\sigma}')\}$
$\mathbf{F}$	Loop-superspace two-form, $\mathbf{F} = \mathbf{dA}$

Chapter 9

$LG,\ \mathscr{L}\mathscr{G}$	Loop group of G, Lie algebra of LG
Diff M	Group of diffeomorphisms of M
$Map(M, G) = G(M)$	Group of smooth mappings $M \to G$
$g_i(x)$	Element of $G(M)$
$\mathscr{G}(M)$	Lie algebra of $G(M)$
$\zeta \cdot X = \zeta$	Element of $\mathscr{G}(M)$ of coordinates $\zeta^i(x)$
$j_{(i)}^\mu(x)$	Currents in a field theory
$Q_{(i)}(x) \equiv j_{(i)}^0(x)$	Charge densities

Chapter 10

A^g	Transformed gauge field by g
$Y_{(i)}(x)$	Generators of the gauge group action on $\mathscr{F}[\mathscr{A}]$
$F[A],\ \mathscr{F}[\mathscr{A}]$	Functional of the gauge fields A, space of functionals
$\mathscr{F}_L[\mathscr{A}]$	Space of the local functionals of the gauge potential A
$\zeta \cdot Y$	Generator of a gauge transformation $\zeta(x)$
$\mathscr{A}$	Space of the Yang-Mills potentials, A
$\mathscr{C}$	Gauge group
$\mathscr{A}/\mathscr{C}$	Orbit space of Yang-Mills potentials
$\Omega_n[A; g_1, ..., g_n]$	n-cochain $\in C_\sigma^n(G(M), \mathscr{F}_L[\mathscr{A}])$
$\omega_n[A](X_1, ..., X_n),\ \omega_n[A](\zeta_1, ..., \zeta_n)$	n-cochain $\in C_\rho^n(\mathscr{G}(M), \mathscr{F}_L[A])$
$\omega_{i_1...i_n}[A](x_1, ...x_n)$	n-cochain $\in C_\rho^n(\mathscr{G}(M), \mathscr{F}_L[\mathscr{A}])$ evaluated for the basis elements $X_{(i)}(x)$ of $\mathscr{G}(M)$
$c^i(x)$	Ghost parameters for $G(M)$
$\mathscr{D}$	Dirac operator in the presence of a Y-M field
$\mathscr{V}_{(i)}[A](x)$	Infinitesimal form of the gauge anomaly
$\alpha_1[A; g]$	Finite form of the gauge anomaly
$G_{(i)}(x)$	Gauss law constraints
$S_{ij}[A](x, y)$	Schwinger terms
$\Omega_n^m(A; g_1, ...g_n),\ \omega_n^m(A, \zeta_1, ..., \zeta_n)$	m-form n-cochain for the gauge group, algebra
$G^p,\ G^{p+1}$	Generators of a contractible free differential algebra
$\not{s}$	Sum of BRST and de Rham operators
$a \equiv dg\, g^{-1}$	Right-invariant form on G
$\omega^{D+1}(g)$	Non-trivial CE $(D+1)$-form on a compact G

References

Abraham, R. and Marsden, J. (1978): *Foundations of Mechanics* (2nd ed.), Addison Wesley

Abraham, R.; Marsden, J. and Ratiu, T. (1988): *Manifolds, tensor analysis, and applications*, Springer Verlag

Abud, M.; Ader, J.-P. and Gieres, F. (1990): *Algebraic determination of covariant anomalies and Schwinger terms*, Nucl. Phys. **B339**, 687-710

Achúcarro, A.; Evans, J.M.; Townsend, P.K. and Wiltshire, D.L. (1987): *Super p-branes*, Phys. Lett. **198 B**, 441-446

Adler, S. L. and Dashen, R. F. (1968): *Current algebras and applications to particle physics*, W. A. Benjamin

Aitchison, I.J.R. (1987): *Berry phases, magnetic monopoles, and Wess-Zumino terms or how the skyrmion got its spin* (reprinted in Shapere and Wilczek (1989)), Acta Phys. Pol. **B18**, 207-235

Aldaya, V. and de Azcárraga, J.A. (1978): *Variational principles on r-th order jets of fibre bundles in field theory*, J. Math. Phys. **19**, 1869-1880

Aldaya, V. and de Azcárraga, J.A. (1980): *Geometric formulation of classical mechanics and field theory*, Rivista Nuov. Cim. **3**, fasc. 10

Aldaya, V. and de Azcárraga, J.A. (1982): *Quantization as a consequence of the symmetry group: An approach to geometric quantization*, J. Math. Phys. **23**, 1297-1305

Aldaya, V. and de Azcárraga, J.A. (1985a): *Cohomology, central extensions and dynamical groups*, Int. J. Theor. Phys. **24**, 141-154

Aldaya, V. and de Azcárraga, J.A. (1985b): *A note on the covariant derivatives in super-symmetry*, J. Math. Phys. **26**, 1818-1821

Aldaya, V. and de Azcárraga, J.A. (1987): *Group foundations of quantum and classical dynamics*, Fortschr. der Phys. **35**, 437-473

Aldaya, V. and Navarro-Salas, J. (1990): *Quantization on the Virasoro group*, Commun. Math. Phys. **126**, 575-595

Aldaya, V.; Navarro-Salas, J. and Navarro, M. (1991): *Dynamics on the Virasoro group, 2D gravity and hidden symmetries*, Phys. Lett. **260B**, 311-316

Aldaya, V.; Navarro-Salas, J. and Navarro, M. (1992): *Hidden symmetries in field theory*, Contemp. Math. **132**, 1-25

Alekseev, A.Yu.; Madaichick, Ya.; Faddeev, L. and Shatashvili, S.L. (1988): *Derivation of anomalous commutators in the functional integral formalism*, Theor. and Math. Phys. **73**, 1149-1151 (Russian original **73**, 187-190 (1987))

Alekseev A. and Shatashvili, S. (1989): *Path integral quantization of the coadjoint orbits of the Virasoro group and 2-d gravity*, Nucl. Phys. **B323**, 719-733

Álvarez, O. (1985): *Cohomology and field theory* in *Symposium on anomalies, geometry and topology*, W.A. Bardeen and R.A. White eds., World Sci., pp. 3-21; *Topological quantization and cohomology*, Commun. Math. Phys. **100**, 279-309

Álvarez, O.; Singer, I.M. and Zumino, B. (1984): *Gravitational anomalies and the family's index theorem*, Commun. Math. Phys. **96**, 409-417

Álvarez-Gaumé, L. (1983): *A note on the Atiyah-Singer theorem*, J. Phys. **A16**, 4177-4182; *Supersymmetry and the Atiyah-Singer theorem*, Commun. Math. Phys. **90**, 161-173

Álvarez-Gaumé, L. (1986): *An introduction to anomalies*, in *fundamental problems of gauge field theory*, G. Velo and A. S. Wightman eds., Plenum Press, pp. 93-206

Álvarez-Gaumé, L.; Della Pietra, S. and Moore, G. (1985): *Anomalies in odd dimensions*, Ann. Phys. **163**, 288-317

Álvarez-Gaumé, L. and Ginsparg, P. (1984): *The topological meaning of non-abelian anomalies*, Nucl. Phys. **243**, 449-474

Álvarez-Gaumé, L. and Ginsparg, P. (1985): *The structure of gauge and gravitational anomalies*, Ann. Phys. **161**, 423-490 (Erratum: *ibid* **171**, 233 (1986))

Álvarez-Gaumé, L.; Sierra, G. and Gómez, C. (1990): *Topics in conformal field theory*, in Brink et al. (1990), pp. 16-184

Álvarez-Gaumé, L. and Witten, E. (1983): *Gravitational anomalies*, Nucl. Phys. **B234**, 269-330

Arnold, V.I. (1978): *Mathematical methods of Classical Mechanics*, Springer Verlag

Aschieri, P. and Castellani, L. (1993): *An introduction to noncommutative differential geometry on quantum groups*, Int. J. Mod. Phys. **48**, 1667-1706

Asorey, M. (1992): *Conformal invariance in quantum field theory and statistical mechanics*, Fortschr. Phys. **40**, 273-327

Asorey, M. and Boya, L.J. (1979): *Electromagnetism without monopoles is possible in nontrivial U(1)-fibre bundles*, J. Math. Phys. **20**, 2327-2329

Atiyah, M.F. (1979): *Geometry of Yang-Mills fields*, Lezioni Fermiane, Accad. Nazionale dei Lincei & Scuola Normale Superiore, Pisa (reprinted in Atiyah (1988), pp. 77-173)

Atiyah, M.F. (1984): *The Yang-Mills equations and the structure of 4-manifolds* in *Durham symposium on global Riemannian geometry*, Ellis Horwood Ltd. (reprinted in Atiyah (1988), pp. 11-17)

Atiyah, M.F. (1988): *Collected works*, vol. 5 (*Gauge theories*), Oxford Univ. Press

Atiyah, M.F. (1989): *Topological quantum field theories*, Pub. Math. Inst. Hautes Études Sci. **68**, 175-186

Atiyah, M.F. and Bott, R. (1984): *The moment map in equivariant cohomology*, Topology **23**, 1-28

Atiyah, M.F. and Hitchin, N. (1988): *The geometry and dynamics of magnetic monopoles*, Princeton Univ. Press

Atiyah, M.F.; Hitchin, N.J. and Singer. I.M. (1978): *Self-duality in four-dimensional Riemannian geometry*, Proc. Roy. Soc. Lond. **A362**, 425-461

Atiyah, M.F. and Jeffrey, L. (1990): *Topological lagrangians and cohomology*, J. Geom. Phys. **7**, 119-136

Atiyah, M.F. and Jones, J.D.S. (1978): *Topological aspects of Yang-Mills theory*, Commun. Math. Phys. **61**, 97-118

Atiyah, M.F. and Singer, I.M. (1968): *The index of elliptic operators I, II, III*, Ann. Math. **87**, 484-530, 531-545, 546-604

Atiyah, M.F. and Singer, I.M. (1984): *Dirac operators coupled to vector potentials*, Proc. Nat. Acad. Sci. USA **81**, 2597-2600

Bacry, H. (1967): *Leçons sur la théorie des groupes et les symétries des particules élémentaires*, Gordon and Breach

Bacry, H. and Lévy-Leblond, J.-M. (1968): *Possible kinematics*, J. Math. Phys. **9**, 1605-1614

Bailin, D.B. and Love, A. (1993): *Introduction to gauge field theory*, Inst. of Phys.

Bak, D. (1993): *Two cocycle arising from the Chern-Simons form*, Phys. Lett. **311**, 130-136

Balachandran, A.P. (1988): *Wess Zumino terms and quantum symmetries*, in *Conformal field theory, anomalies and superstrings*, C.K. Chew et al. eds., World Sci.

Balachandran, A. P.; Marmo, G.; Nair, V. P. and Trahern, C. G. (1982): *Non-perturbative proof of the non-Abelian anomalies*, Phys. Rev. **D25**, 2713-2716.

Balachandran, A.P.; Marmo, G.; Skagerstam, B.S. and Stern, A. (1980): *Magnetic monopoles with no strings*, Nucl. Phys. **B162**, 385-396

Ball, R.D. (1989): *Chiral gauge theory*, Phys. Rep. **182**, 1-186

Bardeen, W.A. and Zumino, B. (1984): *Consistent and covariant anomalies in gauge and gravitational theories*, Nucl. Phys. **B244**, 421-453

Bargmann, V. (1954): *On unitary ray representations of continuous groups*, Ann. Math. **59**, 1-46

Bartocci, C.; Bruzzo, U. and Landi, G. (1990): *Chern-Simons forms on superfiber bundles*, J. Math. Phys. **31**, 45-54

Barut, A.O. and Raczka, R. (1980): *Theory of group representations and applications*, PWN-Pol. Sci. Pub.

Baumann, K. (1992): *Quantum fields in 1+1-dimension carrying a true ray representation of the Poincaré group*, Lett. Mat. Phys. **25**, 61-73

Bayen, F.; Flato, M.; Fronsdal, C.; Lichnerowicz, A. and Sternheimer, D. (1978): *Deformation theory and quantization: I. Deformation of symplectic structures; II. Physical applications*, Ann. Phys. **110**, 61-110, 111-151

Becchi, C.; Rouet, A. and Stora, R. (1976): *Renormalization of gauge theories*, Ann. Phys. (N. Y.) **98**, 287-321

Belavin, A.A.; Polyakov, A.M.; Schwartz, A.S. and Tyupkin, Yu.S. (1975): *Pseudoparticle solutions to the Yang-Mills equations*, Phys. Lett. **59B**, 85-87

Belavin, A.A.; Polyakov, A.M. and Zamolodchikov, A.B. (1984): *Infinite conformal symmetry in two-dimensional quantum field theory*, Nucl. Phys. **B271**, 333-380 (reprinted in Itzykson et al. (1988))

Benn, I.M. and Tucker, R.W. (1987): *An introduction to spinors and geometry with applications in physics*, Adam Hilger

Berezin, F.A. (1966): *The method of second quantization*, Academic Press

Berezin, F.A. (1979): *Differential forms on supermanifolds*, Sov. J. Nucl. Phys. **30**, 605-608

Berezin, F.A. (1987): *Introduction to superanalysis* (A.A. Kirilov, ed.), Reidel

Berline, N. and Vergne, M. (1982): *Classes characteristiques équivariantes. Formule de localisation en cohomologie équivariante*, C. R. Acad. Sci. Paris **295**, 539-541

Berline, N. and Vergne, M. (1983): *Zeros d'un champ de vecteurs et classes characteristiques équivariantes*, Duke Math. J. **50**, 539-549

Bernard, C.W.; Christ, N.H.; Guth, A.H. and Weinberg, E.J. (1977): *Pseudoparticle parameters for arbitrary gauge groups*, Phys. Rev. **D16**, 2967-2977

Bernat, P.; Conze, N.; Duflo, M.; Lévy-Nahas, M.; Rais, M.; Renouard, P. and Vergne, M. (1972): *Représentations des groupes de Lie résolubles*, Dunod

Birmingham, D.; Blau, M.; Rakowski, M. and Thompson, G. (1991): *Topological field theory*, Phys. Reports **209**, 129-340

Bleecker, D. (1981): *Gauge theory and variational principles*, Addison-Wesley

Bonora, L. (1990): *Anomalies and cohomology* in *Anomalies, phases, defects...* Bregola, M.; Marmo, G. and Morandi, G. (eds.), Bibliopolis, 183-240

Bonora, L. and Cotta-Ramusino, P. (1983): *Some remarks on BRS transformations, anomalies and the cohomology of the Lie algebra of the group of gauge transformations*, Commun. Math. Phys. **87**, 589-603

Borel, A. (1955): *Topology of Lie groups and characteristic classes*, Bull. Am. Math. Soc. **61**, 397-432

Bott, R. (1959):*The stable homotopy of the classical groups*, Ann. Math. **70**, 313-337

Bott, R. and Seeley, R. (1978): *Some remarks on the paper by Callias*, Commun. Math. Phys. **62**, 238-245

Bott, R. and Tu, L.W. (1986): *Differential forms in algebraic topology*, Springer Verlag

Boulware, G. D.; Deser, S. and Zumino, B. (1985): *Absence of three-cocycles in the Dirac monopole problem*, Phys. Lett. **153B**, 307-310

Boya, L.J. (1991): *The geometry of compact Lie groups*, Rep. Math. Phys. **30**, 149-162

Boya, L.J.; Cariñena, J.F. and Mateos, J. (1978): *Homotopy and solitons*, Fortschr. der Phys. **26**, 175-214

Boya, L.J. and Mateos, J. (1980): *The kink: real Hopf bundle and Morse theory*, J. Phys. A **13**, L285-L288

Boya, L.J. and Sudarshan, E.C.G. (1991): *Rays and phases in quantum mechanics*, Found. Phys. Lett. **4**, 283-287

Braaten, E.; Curtright, T. L.; Zachos, C. K. (1985): *Torsion and geometrostasis in nonlinear sigma models*, Nucl. Phys. **B260**, 630-688

Brandt, F.; Dragon, N. and Kreuzer, M. (1990a): *Completeness and nontriviality of the solutions of the consistency equations*, Nucl. Phys. **B332**, 224-249

Brandt, F.; Dragon, N. and Kreuzer, M. (1990b): *Lie algebra cohomology*, Nucl. Phys. **B332**, 250-260

Brauer, R. and Weyl, H. (1935): *Spinors in n dimensions*, Am. J. Math. **57**, 425-449. Reprinted in H. Weyl's *Selecta*, Birkhäuser (1955), pp. 431-454

Brink, L.; Friedan, D. and Polyakov, A.M. (eds.) (1990): *Physics and mathematics of strings*, V. Knizhnik memorial volume, World Sci.

Brink, L. and Henneaux, M. (1988): *Principles of string theory*, Plenum

Brown, K.S. (1982): *Cohomology of groups*, Springer Verlag

Brown, L.S. (1992): *Quantum field theory*, Cambridge Univ. Press

Calabi, L. (1951): *Sur les extensions des groupes topologiques*, Annali di Mat. Pura ed Appl. **XXXII** 295-370

Callias, C. (1978): *Axial anomalies and index theorems on open spaces*, Commun. Math. Phys. **62**, 211-234

Cangemi, D. and Jackiw, R. (1992): *Gauge invariant formulation of lineal gravities*, Phys. Rev. Lett. **69**, 233-236

Carey, A.L.; Grundling, H.; Hurst, C.A. and Langmann, E. (1995): *Realizing 3-cocycles as obstructions*, J. Math. Phys. **36**, 2605-2620

Cariñena, J.F. and Ibort, L.A. (1988): *Noncanonical groups of transformations, anomalies and cohomology*, J. Math. Phys. **29**, 541-545

Cariñena, J.F.; del Olmo, M.A. and Santander, M. (1981): *Kinematic groups and dimensional analysis*, J. Phys. **A14**, 1-14

Cariñena, J.F. and Santander, M. (1981): *Semiunitary projective representations of the complete Galilei group*, J. Math. Phys. **22**, 1548-1558

Carmeli, M; Huleihil, Kh. and Leibowitz, E. (1989): *Gauge fields: classification and equations of motion*, World Sci.

Cartan, H. and Eilenberg, S. (1956): *Homological algebra*, Princeton Univ. Press

Chari, V. and Pressley, A. (1994): *A guide to quantum groups*, Cambridge Univ. Press

Chern, S.-S. (1942): *On integral geometry on Klein spaces*, Ann. Math. **43**, 178-189

Chern, S.-S. (1979): *Complex manifolds without potential theory (with an appendix on the theory of characteristic classes)* (2nd ed.), Springer-Verlag

Chern, S.-S. and Simons, J. (1974): *Characteristic forms and geometric invariants*, Ann. Math. **99**, 48-69

Chevalley, C. (1946): *Theory of Lie groups*, Princeton Univ. Press.

Chevalley, C. and Eilenberg, S. (1948): *Cohomology theory of Lie groups and Lie algebras*, Trans. Am. Math. Soc. **63**, 85-124

Choquet-Bruhat, Y. (1989): *Graded bundles and supermanifolds*, Bibliopolis

Choquet-Bruhat, Y.; De Witt-Morette, C. (1989): *Analysis, manifolds and Physics II (Applications)*, North Holland

Choquet-Bruhat, Y.; De Witt-Morette, C. and Dillard-Bleick, M. (1982): *Analysis, Manifolds, and Physics* (revised edition), North Holland. An extensive list of errata to this edition is included in Choquet-Bruhat and De Witt-Morette (1989)

Chu, B.-Y. (1974): *Symplectic homogeneous spaces*, Trans. Am. Math. Soc. **197**, 145-159

Coleman, S. (1982): *The magnetic monopole fifty years later*, Erice int. school of subnuclear physics

Coleman, S. (1985): *Aspects of symmetry (selected Erice lectures)*, Cambridge Univ. Press

Coleman, S. and Mandula, J. (1967): *All possible symmetries of the S matrix*, Phys. Rev. **159**, 1251-1256

Connes, A. (1990, 1994): *Géometrie non commutative*, Intereditions; *Non-commutative geometry*, Acad. Press.

Coquereaux, R. and Pilch, K. (1989): *String structures in loop bundles*, Commun. Math. Phys. **120**, 353-378

Cornwell, J. F. (1984): *Group theory in physics: Lie groups and their applications*, Vol. II, Acad. Press.

Cornwell, J. F. (1989): *Group theory in physics: Supersymmetries and infinite-dimensional algebras*, Vol. III, Acad. Press.

Corwin, L.; Ne'eman, Y. and Sternberg, S. (1975): *Graded Lie algebras in mathematics and physics (Bose-Fermi symmetry)*, Rev. Mod. Phys. **47**, 573-603

Dadashev, L. A. (1986): *Axiomatics of Galileo-invariant quantum field theory*, Theor. and Math. Phys. **64**, 903-914

Daniel, M. and Viallet, C. M. (1980): *The geometrical setting of gauge theories of Yang-Mills type*, Rev. Mod. Phys. **52**, 175-197

D'Auria, R.; Fré, P. and Regge, T. (1980): *Graded-Lie-algebra, cohomology and supergravity*, Riv. Nuovo Cim. **3**, fasc. 12

de Azcárraga, J.A.; Gauntlett, J.P.; Izquierdo, J.M. and Townsend, P.K. (1989): *Topological extensions of the supersymmetry algebra for extended objects*, Phys. Rev. Lett. **63**, 2443-2446

de Azcárraga, J.A.; Izquierdo, J.M. and Macfarlane, A.J. (1990): *Current algebra and Wess-Zumino terms: a unified geometric treatment*, Ann. Phys. **202**, 1-21

de Azcárraga, J.A. and Lukierski, J. (1982): *Supersymmetric particles with internal symmetries and central charges*, Phys. Lett. **113B**, 170-174

de Azcárraga, J.A. and Lukierski, J. (1983) *Supersymmetric particles in N = 2 superspace: phase-space variables and Hamiltonian dynamics*, Phys. Rev. **D28**, 1337-1345

de Azcárraga, J.A. and Townsend, P.K. (1989): *Superspace geometry and classification of supersymmetric extended objects*, Phys. Rev. Lett. **62**, 2579-2512

Delius, G.W.; van Nieuwenhuizen, P. and Rodgers, V.G.J. (1990): *The method of coadjoint orbits: an algorithm for the construction of invariant actions.* Int. J. Mod. Phys. **A5**, 3943-3983

De Witt, B. (1992): *Supermanifolds* (2nd ed.), Cambridge Univ. Press

Di Francesco, P.; Mathieu, P. and Sénéchal, D. (1997): *Conformal field theory*, Springer

Dijkgraaf, R. and Witten, E. (1990): *Topological gauge theories and group cohomology*, Commun. Math. Phys. **129**, 393-429

Dirac, P.A.M. (1931): *Quantized singularities in the electromagnetic field*, Proc. R. Soc. London **A133**, 60-72

Dixmier, J. (1957): *L'application exponentielle dans les groupes de Lie resolubles*, Bull. Soc. Math. France, **85**, 113-121

Dixon, J.A. (1990): *Supersymmetry is full of holes*, Class. Quantum Grav. **7** 1511-1521

Dolan, L. (1984): *Extended Kac-Moody algebras and exact solvability in hadronic physics*, Phys. Reports **109**, 1-94

Donaldson, S.K. and Kronheimer, P.B. (1990): *The geometry of four-manifolds*, Clarendon Press

Dubois-Violette, M.; Henneaux, M.; Talon, M. and Viallet, C.-M. (1992): *General solution of the consistency equation*, Phys. Lett. **B289**, 361-367.

Dubois-Violette, M.; Talon, M. and Viallet, C.-M. (1985): *BRS algebras. Analysis of the consistency equations in gauge theory*, Commun. Math. Phys. **102**, 105-122

Dubrovin, B.A.; Fomenko, A.T. and Novikov, S.P. (I 1992, II 1985, III 1990): *Modern geometry -methods and applications. Part I: the geometry of surfaces, transformation groups and fields* (1992); Part II: *the geometry and topology of manifolds* (1985); Part III: *Introduction to homology theory*, Springer Verlag (1990)

Duff, M.J. (1994): *Twenty years of the Weyl anomaly*, Class. and Q. Grav. **11**, 1387-1404

Duff, M.J. (1996): *M theory: the theory formerly known as strings*, Int. J. Mod. Phys. **A11**, 5623-5642; *Supermembranes*, hep-th/9611203

Duff, M.J. and Lu, J.X. (1992): *A duality between strings and membranes*, Class. and Q. Grav. **9**, 1-16

Duistermaat, J.J. and Heckman, G.J. (1982): *On the variation in the cohomology of the symplectic form of the reduced phase space*, Inv. Mat. **69**, 259-268 (Addendum (1983): ibid. **72**, 153-158)

Dunne, G.V. and Trugenberger, C.A. (1990): *Covariant Gauss law commutator anomaly*, Phys. Lett. **248B**, 305-310

Dyson, F.J. (ed.) (1966): *Symmetry groups*, W.A. Benjamin

Ebin, D.G. and Marsden, J.E. (1970): *Groups of diffeomorphisms and the motion of an incompressible fluid*, Ann. Math. **92**, 102-163

Eguchi, T.; Gilkey, P. and Hanson, A.J. (1980): *Gravitation, gauge theories and differential geometry*, Phys. Rep. **66**, 213-393

Eilenberg, S. (1948): *Extensions of general algebras*, Ann. Soc. Pol. de Mathém. **21**, 125-134

Eilenberg, S. (1949): *Topological methods in abstract algebra. Cohomology theory of groups*, Bull. Am. Math. Soc. **55**, 3-37

Eilenberg, S. and Mac Lane, S. (1947a): *Cohomology theory in abstract groups I, II*, Ann. Math. **48**, 51-78, 326-341

Eilenberg, S. and Mac Lane, S. (1947b): *Algebraic cohomology groups and loops* (Cohomology theory in abstract groups III), Duke Math. J. **14**, 435-463

Faddeev, L.D. (1984): *Operator anomaly for the Gauss law*, Phys. Lett. **145B**, 81-84

Faddeev, L.D. (1986): *Can theories with anomalies be quantized?* in G. Gibbons *et al.* (1986), pp. 41-53

Faddeev, L.D. and Shatashvili S.L. (1985): *Algebraic and Hamiltonian methods in the theory of non-abelian anomalies*, Theor. and Math. Phys. **60**, 770-778

Faddeev, L.D. and Shatashvili, S.L. (1986): *Realization of the Schwinger terms in the Gauss law and the possibility of correct quantization of a theory with anomalies*, Phys. Lett. 167B, 225-228

Faddeev, L.D. and Slavnov, A.A. (1991): *Gauge fields, an introduction to quantum theory* (2nd ed.), Addison-Wesley (Russian ed. (1986))

Felsager, B. (1981): *Geometry, particles and fields*, Odense Univ. Press

Ferrara, S.; Gatto, R. and Grillo, A.F. (1973): *Conformal algebra in spacetime and operator product expansion*, Springer Tracts in Mod. Phys. **67**

Flanders, H. (1963): *Differential forms with applications to the physical sciences*, Academic Press

Freed, D.S. and Uhlenbeck, K. (1984): *Instantons and four-manifolds*, Springer Verlag

Freedman, M.H. and Luo, Feng (1989): *Selected applications of geometry to low-dimensional topology*, University Lecture Series **1**, An. Math. Soc.

Friedan, D. and Windey, P. (1984): *Supersymmetric derivation of the Atiyah-Singer index and the chiral anomaly*, Nucl. Phys. **B235**, 395-416

Fujikawa, K. (1980a): *Path integral for gauge theories with fermions*, Phys. Rev. **D21**, 2848-2858

Fujikawa, K. (1980b): *Comment on chiral and conformal anomalies*, Phys. Rev. Lett. **44**, 1733-1736

Fujikawa, K. (1986): *Path integral quantization of gravitational interactions -local symmetry properties*, in *Quantum gravity and cosmology*, H. Sato and T. Inami eds., World Sci., 106-169

Fuks, D.B. (1986): *Cohomology of infinite dimensional algebras*, Consultants Bureau (Russian ed. (1984))

Fulton, T.; Rohrlich, F. and Witten, L. (1962): *Conformal invariance in physics*, Rev. Mod. Phys. **34**, 442-457

Galindo, A. (1967): *Lie algebra extensions of the Poincaré algebra*, J. Math. Phys. **8**, 768-774

García-Pérez, P.L. (1974): *The Poincaré-Cartan invariant in the calculus of variations*, Symposia Math. (Roma) **14**, 219-246

García Prada, O.; del Olmo, M. and Santander, M. (1988): *Locally operating realization of connected transformation Lie groups*, J. Math. Phys. **29**, 1083-1090

Gibbons, G.; Hawking, S.W. and Townsend, P.K. (1986): *Supersymmetry and its applications: Superstrings, anomalies and supergravity*, Cambridge Univ. Press

Gieres, F. (1988): *Geometry of supersymmetric gauge theories*, Lect. Notes in Phys. **302**, Springer-Verlag

Gilkey, P.B. (1984): *Invariance theory, the heath equation, and the Atiyah-Singer theorem*, Math. Lecture Series **11**, Publish or Perish

Gilmore, R. (1974): *Lie groups, Lie algebras, and some of their applications*, J. Wiley

Ginsparg, P. (1986): *Applications of topological and differential geometric methods to anomalies in quantum field theory*, in *New perspectives in quantum field theories*, J. Abad, M. Asorey and A. Cruz eds., World Scientific, pp. 29-83

Ginsparg, P. (1990): *Applied conformal field theory*, in *Fields, strings and critical phenomena* (Les Houches 1988), E. Brézin and J. Zinn-Justin eds., North-Holland, pp. 1-168

Göckeler, M. and Schücker, T. (1987): *Differential geometry, gauge theories, and gravity*, Cambridge Univ. Press

Godbillon, C. (1969): *Géometrie différentielle et mécanique analytique*, Hermann

Goddard, P. and Mansfield, P. (1986): *Topological structures in field theories*, Rep. Progr. Phys. **49**, 725-781

Goddard, P. and Olive, D. (1978): *Magnetic monopoles in gauge field theories*, Rep. Progr. Phys. **41**, 1357-1437

Goddard, P. and Olive, D. (1986): *Kac-Moody and Virasoro algebras in relation to quantum physics*, Int. J. Mod. Phys. **11**, 303-414

Goddard, P. and Olive, D. (eds.) (1988): *Kac-Moody and Virasoro algebras: A reprint volume for physicists*, World Sci.

Goldschmidt, H. and Sternberg, S. (1973): *The Hamilton-Cartan formalism in the calculus of variations*, Ann. Inst. Fourier **23**, 203-267

Gómez, C. (1983): *On the topological origin of non-abelian anomalies*, Salamanca Univ. preprint 06/83, unpublished

Gómez, C.; Ruiz-Altaba, M. and Sierra, G. (1995): *Quantum groups in two-dimensional physics*, Cambridge Univ. Press

Gray, A. (1992): *Lie groups* (mimeographed notes)

Green, M.B.; Schwarz, J.H. and Witten, E. (1987): *Superstring theory I, II*, Cambridge Univ. Press

Greub, W.; Halpering, S. and Vanstone, R. (I 1972, II 1973, III 1976): *Connections, curvature and cohomology* (vol. I: *de Rham cohomology of manifolds and vector bundles*; vol. II: *Lie groups, principal bundles and characteristic classes*; vol. III: *Cohomology of principal bundles and homogeneous spaces*), Academic Press

Greub, W. and Petry, H.-R. (1975): *Minimal coupling and complex line bundles*, J. Math. Phys. **6**, 1347-1351

Gribov, V. N. (1978): *Quantization of non-abelian gauge theories*, Nucl. Phys. **B139**, 1-19

Grossman, B. (1985,1986): *A 3-cocycle in quantum mechanics*, Phys. Lett. **152B**, 93-97; *Three-cocycle in quantum mechanics, II*, Phys. Rev. **D33**, 2922-2929

Grossman, B.; Kephart, T.W. and Stasheff, J. (1989): *Solutions to gauge field equations in eight dimensions: conformal invariance and the last Hopf map*, Phys. Lett. **220**, 431-434

Gruber, B. and O'Raifeartaigh, L. (1964): *S Theorem and construction of the invariants of the semisimple compact algebras*, J. Math. Phys. **5**, 1796-1804

Guichardet, A. (1980): *Cohomologie des groupes topologiques et des algèbres de Lie*, Cedic/F. Nathan

Guillemin, V. (1973): *Cohomology of vector fields on a manifold*, Adv. in Math. **10**, 192-220

Guillemin, V. and Sternberg, S. (1984): *Symplectic theories in physics*, Cambridge Univ. Press

Guillemin, V.; Lerman, E. and Sternberg, S. (1996): *Symplectic fibrations and multiplicity diagrams*, Cambridge Univ. Press

Guo, H.-y.; Wu, K. and Wang, S.-k. (1985): *Chern-Simons type characteristic classes*, Commun. in Theor. Phys. (Beijing) **4**, 113-122

Haag, R.; Lopuszański, J.T. and Sohnius, M. (1975): *All possible generators of supersymmetries of the S-matrix*, Nucl. Phys. **B88**, 257-274

Hall, M. (1959): *The theory of groups*, Macmillan

Hamermesh, M. (1962): *Group theory and its applications to physical problems*, Addison-Wesley

Heil, A.; Kersch, A.; Papadopoulos, N.; Reifenhäuser, B. and Scheck, F. (1990): *Structure of the space of reducible connections for Yang-Mills theories*, J. Geom. Phys. **7**, 489-505

Helgason, S. (1978): *Differential Geometry, Lie groups and symmetric spaces*, Acad. Press

Henneaux, M. (1985): *Hamiltonian form of the path integral for theories with a gauge freedom*, Phys. Rep. **126**, 1-66

Henneaux, M. and Mezincescu, L. (1985): *A σ-model interpretation of the Green-Schwarz covariant action*, Phys. Lett. **152B**, 340-342

Henneaux, M. and Teitelboim, C. (1992): *Quantization of gauge systems*, Princeton Univ. Press

Henniart, G. (1985): *Les inégalités de Morse [d'après Witten]*, Astérisque **121-122**, 43-61

Hermann, R. (1970): *Vector bundles in mathematical physics I, II*, W.A. Benjamin

Hilton, P.J. (1953): *An introduction to homotopy theory*, Cambridge Univ. Press

Hilton, P.J. and Stammbach, U. (1970): *A course in homological algebra*, Springer Verlag

Hirzebruch, F. (1966): *Topological methods in algebraic geometry* (3rd ed.), Springer Verlag

Hochschild, G. (1951): *Group extensions of Lie groups I, II*, Ann. Math. **54**, 96-109, 537-551

Hochschild, G. (1965): *The structure of Lie groups*, Holden-Day

Hochschild, G. and Mostow, G.D. (1962): *Cohomology of Lie groups*, Ill. J. Math. **6**, 367-401

Hochschild, G. and Serre, J.P. (1953a): *Cohomology of group extensions*, Trans. Am. Math. Soc. **74**, 110-134

Hochschild, G. and Serre, J.P. (1953b): *Cohomology of Lie algebras*, Ann. Math. **57**, 591-603

Hodge, W.V.D. (1941): *The theory and application of harmonic integrals*, Cambridge Univ. Press (reissued 1988)

Holt, D.F. (1979): *An interpretation of the cohomology groups $H^n(G, M)$*, J. Algebra **60**, 307-320

Horváthy, P.A. (1984): *Étude géometrique du monopole magnétique*, J. Geom. Phys. **1**, 39-78

Hou, B.-Yu; Hou, B.-Yuan and Wang, P. (1986): *How to eliminate the dilemma of the three-cocycle*, Ann. Phys. **175**, 172-185

Houard, J.C. (1977): *On invariance groups and Lagrangian theories*, J. Math. Phys. **18**, 503-516

Houard, J.C. (1980): *Une représentation intégrale des cocycles des groupes de Lie*, C.R. Acad. Sc. Paris, **290**, 61-64; *An integral formula for cocycles of Lie groups*, Ann. Inst. Henri Poincaré A **XXXII**, 221-247

Hull, C.M. and Spence, B. (1991): *The geometry of the gauged sigma-model with Wess-Zumino term*, Nucl. Phys. **B353**, 379-426

Humphreys, J. E. (1972): *Introduction to Lie algebras and representation theory*, Springer-Verlag

Husemoller, D. (1966): *Fibre bundles* (2nd ed.), Springer Verlag

Ibort, L.A. and Rodríguez, M.A. (1993): *Integrable systems, quantum groups, and quantum field theories*, (Salamanca 1992) NATO ASI series **C409**, Kluwer

Inamoto, T. (1993): *Quasi-invariant lagrangians on Lie groups and the method of coadjoint orbits*, J. Math. Phys. **34**, 649-673

İnönü, E. (1964): *Contractions of Lie groups and their representations*, in *Group theoretical concepts in elementary particle physics*, F. Gürsey ed., Gordon and Breach, pp. 391-402

İnönü, E. and Wigner, E. P. (1953): *On the contraction of groups and their representations*, Proc. Nat. Acad. Sci. **39**, 510-524

Jack, I.; Jones, D.R.T.; Mohammedi, N. and Osborn, H. (1990): *Gauging the general σ-model with a Wess-Zumino term*, Nucl. Phys. **B332**, 359-379

Jackiw, R. (1980a): *Dynamical symmetry of the magnetic monopole*, Ann. Phys. **129**, 183-200

Jackiw, R. (1980b): *Introduction to Yang-Mills theory*, Rev. of Mod. Phys. **52**, 661-673

Jackiw, R. (1984): *Quantization of physical parameters*, Comm. Nucl. Part. Phys., 141-156

Jackiw, R. (1985a): *Topological investigations in quantized gauge theories*, in Treiman et al. (1985), 211-359

Jackiw, R. (1985b): *Three-cocycle in mathematics and physics*, Phys. Rev. Lett. **54**, 159-162; *Magnetic sources and 3-cocycles (comment)*, Phys. Lett. **154B**, 303-304

Jackiw, R. (1987): *Chern-Simons terms and cocycles in physics and mathematics*, in *quantum field theory and quantum statistics*, vol. II, A. Batalin, C.J. Isham and G.A. Vilkovisky eds., Adam Hilger, pp. 349-378

Jackiw, R. (1990): *Two dimensional conformal transformations represented by quantum fields in Minkowski spacetime*, in Brink et al. (1990), pp.317-355

Jackiw, R. (1993): *Higher symmetries in lower dimensional models* in Ibort and Rodríguez (1993), 298-316

Jackiw, R. and Manton, N.S. (1980): *Symmetries and conservation laws in gauge theories*, Ann. Phys. **127**, 257-273

Jackiw, R. and Rajaraman, R. (1985): *Vector-meson mass generation by chiral anomalies*, Phys. Rev. Lett. **54**, 1219-1221; 2060(E); **55**, 2224(C)

Jackiw, R. and Rebbi, C. (1977): *Spinor analysis of Yang-Mills theory*, Phys. Rev. **D16**, 1052-1060

Jackson, J.D. (1975): *Classical electrodynamics*, J. Wiley

Jacobson, N. (1962): *Lie algebras*, Dover (reprinted 1979)

Jaffe, A. and Quinn, F. (1993): *Theoretical mathematics: toward a cultural synthesis of mathematics and theoretical physics*, Bull. Am. Math. Soc. **29**, 1-13 (1993). The correspondence that ensued from this essay is in vol. **30**, pp. 178-211

Jimbo, M. (1990): *Yang-Baxter equation in integrable systems*, Adv. Series in Math. Phys. vol. 10, World Sci.

Jo, S.-G. (1985): *Commutators in an anomalous non-abelian chiral gauge theory*, Phys. Lett. **163B**, 353-359; *Commutator of gauge generator in non-abelian chiral theory*, Nucl. Phys. **B259**, 616-636

Kac, V.G. (1977): *Lie superalgebras*, Adv. in Math. **26**, 8-96

Kac, V.G. (1985): *Infinite dimensional Lie algebras*, Cambridge Univ. Press

Kahan, T. (ed.) (I 1960, II 1971, III 1972): *Théorie des groupes en physique classique et quantique: (I) structures mathématiques et fondements quantiques; id.id.: (II) applications en physique classique; id.id.: (III) applications en physique quantique*, Dunod

Kanno, H. (1989): *Weil algebra and geometrical meaning of BRST transformation in topological quantum field theory*, Z. Phys. **C43**, 477-484

Kastler, D. and Stora, R. (1986): *A differential geometric setting for BRS transformations and anomalies I, II*, J. Geom. Phys. **3**, 437-482, 483-505

Kastrup, H.A. (1966): *Conformal group in spacetime*, Phys. Rev. **142**, 1060-1071

Kelnhofer, G. (1991): *Determination of covariant Schwinger terms in anomalous gauge theories* Z. Phys. **C52**, 89-96

Kelnhofer, G. (1993): *On the geometrical structure of covariant anomalies in Yang-Mills theory*, J. Math. Phys. **34**, 3901-3917

Kephart, T.W. (1985): *Safe groups and anomaly cancellation in even dimensions*, Phys. Lett. **151B**, 267-270

Kirillov, A.A. (1976): *Elements of the theory of representations*, Springer Verlag

Knapp, A.W. (1988): *Lie groups, Lie algebras and cohomology*, Princeton Univ. Press

Knizhnik, V.G. and Zamolodchikov, A. B. (1984): *Current algebra and Wess-Zumino model in two dimensions*, Nucl. Phys. **B247**, 83-103 (reprinted in Itzykson et al. (1988))

Kobayashi, S. and Nomizu, K. (1963, 1969): *Foundations of differential geometry I & II*, J. Wiley

Kobayashi, M.; Seo, K. and Sugamoto, A. (1986): *Commutator anomaly for the Gauss law operator*, Nucl. Phys. **B273**, 607-628

Kobayashi, O.; Yoshioka, A.; Maeda, Y. and Omori, H. (1985): *The theory of infinite-dimensional Lie groups and its application*, Acta Applicandae Math. **3**, 71-106

Kostant, B. (1970): *Quantization and unitary representations*, Lect. Notes in Math. **170**, 87-207

Koszul, S.L. (1950): *Homologie et cohomologie des algèbres de Lie*, Bull. Soc. Math. France LXXVIII, 65-127

Krichever, I.M.; Olshanetski, M.A. and Perelomov, A.M. (1986): *Wess-Zumino Lagrangians in chiral models and quantization of their constants*, Nucl. Phys. **B264**, 415-422

Kulkarni, R.S. (1975): *Index theorems of Atiyah-Bott-Patodi and curvature invariants*, Les Presses de l'Université de Montréal

Kurosh, A.G. (1960): *The theory of groups (vol. II)*, Chelsea Pub. Co.

Lang, S. (1972): *Differential manifolds*, Addison-Wesley

Laursen, M.L., Schierholz, G. and Wiese U.-J. (1986): *2- and 3-cocycles in 4-dimensional SU(2) gauge theory*, Commun. Math. Phys. **103**, 693-699

Leinaas, J.M. (1980): *Topological charges in gauge theories*, Fortschr. der Phys. **28**, 579-631

Leïtes, D.A. (1975): *Cohomology of Lie superalgebras*, Funk. Analiz. **9**, fasc. 4, 75-76

Leïtes, D.A. (1980): *Introduction to the theory of supermanifolds*, Usp. Mat. Nauk, **35**, 3-57

Lévy-Leblond, J.M. (1963): *Galilei group and non-relativistic quantum mechanics*, J. Math. Phys. **4**, 776-788

Lévy-Leblond, J.M. (1969): *Group-theoretical foundations of classical mechanics: the Lagrangian gauge problem*, Commun. Math. Phys. **12**, 64-79

Lévy-Leblond, J.M. (1971): *Galilei group and Galilean invariance*, in group theory and its applications, Vol. II, E.M. Loebl ed., Acad. Press, pp. 221-299

Lichnerowicz, A. (1988): *Applications of the deformations of the algebraic structures to geometry and mathematical physics*, in *Deformation theory of algebras and applications*, H. Hazewinkel and M. Gerstenhaber eds., Kluwer, pp. 855-896

Lichnerowicz, A. and Medina, A. (1988): *On Lie groups with left-invariant symplectic or Kählerian structures*, Lett. Math. Phys. **16**, 225-235

Mack, G. and Salam, A. (1969): *Finite-component field representations of the conformal group*, Ann. Phys. **53**, 174-202

Mac Lane, S. (1963): *Homology*, Springer Verlag

Mac Lane, S. (1979): *Historical note*, J. Algebra **60**, 319-320

Madore, J. (1981): *Geometric methods in classical field theory*, Phys. Rep. **75**, 125-204

Mañes, J.; Stora, R. and Zumino, B. (1985): *Algebraic study of chiral anomalies*, Commun. Math. Phys. **102**, 157-174

Mañes, J. and Zumino, B. (1986): *Non-triviality of gauge anomalies*, in Gibbons et al. (1986), 3-20

Marathe, K. B. and Martucci, G. (1989): *The geometry of gauge fields* J. Geom. and Phys. **6**, 1-106

Marathe, K. B. and Martucci, G. (1992): *The mathematical foundations of gauge theories*, North-Holland

Mariwalla, K. (1975): *Dynamical symmetries in mechanics*, Phys. Rep. **20C**, 287-362

Marmo, G.; Morandi, G.; Simoni, A. and Sudarshan, E.C.G. (1988): *Quasi-invariance and central extensions*, Phys. Rev. **D37**, 2196-2205

Marmo, G.; Saletan, E.J.; Simoni, A.; Vitale, B. (1985): *Dynamical systems*, J. Wiley

Marsden, J. (1992): *Lectures on mechanics*, London Math. Soc. Lecture notes **174**, Cambridge Univ. Press

Mathai, V. and Quillen, D. (1986): *Superconnections, Thom classes, and equivariant forms*, Topology **25**, 85-110

Michel, L. (1964): *Invariance in Quantum Mechanics and group extension*, in *Group theoretical concepts and methods in elementary particle physics*, F. Gürsey ed., Gordon & Breach, pp. 135-200

Mathai, V. and Quillen, D. (1986): *Superconnections, Thom classes, and equivariant forms*, Topology **25**, 85-110

Michel, L. (1964): *Invariance in Quantum Mechanics and group extension*, in *Group theoretical concepts and methods in elementary particle physics*, F. Gürsey ed., Gordon & Breach, pp. 135-200

Mickelsson, J. (1985a): *Comment on three-cocycle in mathematics and physics*, Phys. Rev. Lett. **54**, 239-242

Mickelsson, J. (1985b): *Chiral anomalies in even and odd dimensions*, Commun. Math. Phys. **97**, 361-370

Mickelsson, J. (1989): *Current algebras and groups*, Plenum Press, N. Y.

Milnor, J. (1969): *Morse theory*, Annals of Math. Studies **51**, Princeton Univ. Press

Milnor, J. (1976): *Curvatures of left invariant metrics on Lie groups*, Adv. in Math. **21**, 293-329

Milnor, J. W. (1984): *Remarks on infinite dimensional Lie groups*, in *Relativity, groups and topology II*, B. S. De Witt and R. Stora eds., North-Holland, pp. 1007-1057

Milnor, J. W. and Stasheff, J. (1974): *Characteristic classes*, Princeton Univ. Press

Minami, M. (1980): *Quaternionic gauge fields on S^7 and Yang's $SU(2)$ monopole*, Progr. Theor. Phys. **63**, 303-321

Mitter, P.K. and Viallet, C.-M. (1981): *On the bundle of connections and the gauge orbit manifold in Yang-Mills theory*, Commun. Math. Phys. **79**, 457-472

Morandi, G. (1992): *The role of topology in classical and quantum physics*, Lect. Notes in Phys. monographs 7, Springer-Verlag

Morozov, A. Yu. (1986): *Anomalies in gauge theories*, Sov. Phys. Usp. **29**, 993-1039

Morozov, A. Yu.; Niemi, A.J. and Palo, K. (1992): *Symplectic geometry of supersymmetric quantum field theories*, Nucl. Phys. **B377**, 295-338

Nakahara, M. (1990): *Geometry, topology and physics*, Inst. of Physics

Nash, C. (1991): *Differential topology and quantum field theory*, Academic Press

Nash, C.; O'Connor, D. and Sen, S. (1993): *Central extensions of sphere groups and their algebras*, J. Math. Phys. **34**, 3269

Nash, C. and Sen, S. (1983): *Topology and Geometry for Physicists*, Academic Press

Nelson, P. and Álvarez-Gaumé, L. (1985): *Hamiltonian interpretation of anomalies*, Commun. Math. Phys. **99**, 103-114

Nepomechie, R.I. (1985): *Magnetic monopoles from antisymmetric tensor gauge fields*, Phys. Rev. **D31**, 1921-1924

Niemi, A. and Semenoff, G.W. (1985): *Quantum holonomy and the chiral gauge anomaly*, Phys. Rev. Lett. **55**, 927-930

Novikov, S. P. (1982): *The Hamiltonian formalism and a many-valued analogue of Morse theory*, Uspekhi Math. Nauk. **37**, 3-49 (Russian Math. Surveys, 37, 1-56). See also Novikov's appendix in Dubrovin et al. (1990)

Okubo, S. and Patera, J. (1985): *Cancellation of higher-order anomalies*, Phys. Rev. **D31**, 2669-2671

Olive, D. (1990): *Introduction to conformal field theory and infinite dimensional algebras*, in *Physics, geometry and topology*, H.C. Lee ed., NATO-ASI **238**, Plenum, pp. 241-261

Olive, D. (1996): *Exact electromagnetic duality*, Nucl. Phys. **45A** (Proc. Suppl.) 88-102

Ol'shanetskiĭ, M. A. (1982): *A short guide to modern geometry for physicists*, Sov. Phys. Usp. **25**, 123-129

O'Raifeartaigh, L. (1965): *Lorentz invariance and internal symmetry*, Phys. Rev. **139**, B1052-B1062

O'Raifeartaigh, L. (1986): *Group structure of gauge theories*, Cambridge Univ. Press

Parthasarathy, K.R. (1969a): *Multipliers on locally compact groups*, Springer Verlag

Parthasarathy, K. R. (1969b): *Projective unitary antiunitary representations of locally compact groups*, Commun. Math. Phys. **15**, 305-328

Pauri, M. and Prosperi, G. M. (1966): *Canonical realizations of Lie symmetry groups*, J. Math. Phys. **7**, 366-375

Penrose, R. (1991): *On the cohomology of impossible figures*, Structural Topology **17**, 11-16

Percacci, R. and Rajaraman (1989): *Constrained Hamiltonian structure of the chirally gauged Wess-Zumino-Witten model*, Int. J. Mod. Phys. **A4**, 4177-4202

Polyakov, A.M. (1987): *Quantum gravity in two dimensions*, Mod. Phys. Lett. **A2**, 893-898

Polyakov, A.M. (1990): *Gauge transformations and diffeomorphisms*, in Brink *et al.* (1990), pp. 1-15

Polyakov, A.M. and Wiegmann, P.B. (1984): *Goldstone fields in two dimensions with multi-valued actions*, Phys. Lett. **141B**, 223-228

Pontryagin, L.S. (1966): *Topological groups*, Gordon and Breach

Preskill, J. (1984): *Magnetic monopoles*, Ann. Rev. Nucl. Sci. **34**, 461-530

Preskill, J. (1991): *Gauge anomalies in an effective field theory*, Ann. Phys. (N.Y.) **210**, 323-379

Pressley, A. and Segal, G. (1986): *Loop groups*, Clarendon Press

Quillen, D. (1989): *Algebra, cochains and cyclic cohomology*, Pub. Math. Inst. des Hautes Études Sci. **68**, 139-174

Rabin, J. (1987): *Supermanifold cohomology and the Wess-Zumino term of the covariant superstring action*, Commun. Math. Phys. **108**, 375-389

Racah, G. (1950): *Sulla caratterizzazione delle rappresentazioni irriducibili dei gruppi semi-semplici di Lie*, Lincei Rend. Sc. fis. mat. e nat. **VIII**, 108-112

Racah, G. (1951): *Group theory and spectroscopy*, Princeton lectures. Reprinted as CERN yellow report 61-8 and in Ergeb. Exact. Naturwiss. 37, pp. 28-84 (1965)

Ragoucy, E. and Sorba, P. (1992): *Extended Kac-Moody algebras and applications*, Int. J. Mod. Phys. **A7**, 2883-2971

Rai, B. and Rodgers, V.G.J. (1990): *From coadjoint orbits to scale invariant WZNW type actions and 2D quantum gravity action*, Nucl. Phys. **B341**, 119-133

Ramond, P. (1989): *Field theory, a modern primer*, Addison-Wesley

Rittenberg, V. and Wyler, D. (1978): *Generalized superalgebras*, Nucl. Phys. **B139**, 189-202

Ryder, L.H. (1980): *Dirac monopoles and the Hopf mapping*, J. Phys. **A13**, 437-447

Salam, A. and Strathdee, J. (1978): *Symmetry and superfields*, Fortschr. der Phys. **26**, 57-142

Sagle, A.A. (1961): *Mal'čev algebras*, Trans. Am. Math. Soc. **101**, 426-458

Saletan, E. J. (1961): *Contractions of Lie Groups*, J. Math. Phys., **2**, 1-21

Samelson, H. (1952): *Topology of Lie groups*, Bull. Am. Math. Soc. **58**, 2-37

Santaló, L. A. (1976): *Integral geometry and geometric probability*, Addison-Wesley

Saunders, D.J. (1989): *The geometry of jet bundles*, Cambridge Univ. Press

Scheunert, M. (1979): *Generalized Lie algebras*, J. Math. Phys. **20**, 712-720

Schücker, T. (1987): *The cohomological construction of Stora's solutions*, Commun. Math. Phys. **109**, 167-175

Schwarz, A.S. (1993): *Quantum field theory and topology*, Springer Verlag

Schwarz, J.H. (1997): *The M theory five-brane*, hep-th/9706197; *The status of string theory*, hep-th/9711029

Schwiebert, C. (1990): *A simple algebraic derivation of Schwinger terms from chiral anomalies*, Phys. Lett. **241B**, 223-228

Schwinger, J. (1959): *Field theory commutators*, Phys. Rev. **3**, 296-297

Segal, G. (1981): *Unitary representations of some infinite-dimensional groups*, Commun. Math. Phys. **80**, 301-342

Seiberg, S. and Witten, E. (1994): *Electromagnetic duality, monopole condensation and confinement in N = 2 supersymmetric Yang–Mills theory*, Nucl. Phys. **B426**, 19-52 (Erratum: ibid **B430**, 485-486)

Serre, J.-P. (1953): *Groupes d'homotopie et classes de groupes abéliens*, Ann. Math. **58**, 258-294

Shanahan, P. (1978): *The Atiyah-Singer index theorem*, Lect. Notes in Math. **638**, Springer-Verlag

Shapere, A. and Wilczek, F. (eds.) (1989): *Geometric phases in physics*, World Sci.

Shirokov, Iu. M. (1960): *Space and time reflections in relativistic theory*, Nucl. Phys. **15**, 1-12

Siegel, W. (1983): *Hidden local supersymmetry in the supersymmetric particle action*, Phys. Lett. **128B**, 379-399

Singer, I.M. (1978): *Some remarks on the Gribov ambiguity*, Commun. Math. Phys. **60**, 7-12

Singer, I.M. (1985): *Families of Dirac operators with applications to physics*, Astérisque (hors série), 323-340

Slanski, R. (1981): *Group theory for unified model building*, Phys. Rep. **79**, 1-128

Śniaticki, J. (1980): *Geometric quantization and quantum mechanics*, Springer-Verlag, N. Y.

Sohnius, M. (1985): *Introducing supersymmetry*, Phys. Rep. **128**, 39-204

Sonoda, H. (1986): *Berry's phase in chiral gauge theories*, Nucl. Phys. **B266**, 410-422

Souriau, J.M. (1969): *Structure des systèmes dynamiques*, Dunod

Spivak, M. (1979): *Differential geometry* vol. V (2nd ed.), Publish or Perish

Stasheff, J. (1992): *Homological (ghost) approach to constrained Hamiltonian systems*, Contemporary Math. **132**, 595-609

Steenrod, N. (1951): *The topology of fibre bundles*, Princeton Univ. Press

Sternberg, S. (1983): *Differential geometry*, Chelsea Pub. Co.

Stora, R. (1977): *Continuum gauge theories* in New developments in quantum field theory and statistical mechanics, M. Lévy and P. Mitter eds., Plenum Press, pp. 201-224

Stora, R. (1984): *Algebraic structure and topological origin of anomalies*, in Progress in gauge field theory, G. t'Hooft ed., Plenum Press

Stora, R. (1986): *Algebraic structure of chiral anomalies*, in New perspectives in quantum field theories, J. Abad, M. Asorey and A. Cruz eds., World Sci., pp. 309-342

Sullivan, D. (1977): *Infinitesimal computations in Topology*, Inst. des Haut Étud. Sci., Pub. Math. **47**, 269-331

Takahashi, Y. and Kobayashi, M. (1978): *Origin of the Gribov ambiguity*, Phys. Lett. **78B**, 241-242

Takhtajan, L.A. (1989): *Lectures in quantum groups*, in Introduction to quantum group and integrable massive models of quantum field theory, Mo-Lin Ge and Bao-Heng Zhao eds., World Sci., pp. 69-197

Teitelboim, C. (1983): *Gravitation and Hamiltonian structure in two spacetime dimensions*, Phys. Lett. **126B**, 41-45

Teitelboim, C. (1986): *Monopoles of higher rank*, Phys. Lett. **B167**, 69-72

Thierry-Mieg, J. (1980): *Geometrical reinterpretation of the Faddeev-Popov ghost particles and BRS transformations*, J. Math. Phys. **21**, 2834-2838

Thirring, W. (1978, 1979): *Classical dynamical systems, classical field theory* (Vols. 1 and 2 of *A course in mathematical physics*), Springer Verlag

Townsend, P.K. (1977): *Four lectures in M-theory*, Proc. of the 1996 Trieste Summer School on H.E.P. and Cosmology, hep-th/9612121

Townsend, P.K. (1989): *Three lectures on supermembranes* in Superstrings '88, M. Green et al. eds., World Science

Townsend, P.K. (1993): *Three lectures on supersymmetry and extended objects* in Ibort and Rodríguez (1993), pp. 317-345

Trautman, A. (1977): *Solutions of the Maxwell and Yang-Mills equations associated with Hopf fiberings*, Int. J. Theor. Phys. **16**, 561-565

Trautman, A. (1984): *Differential geometry for physicists*, Bibliopolis

Treiman, S.B.; Jackiw, R.; Zumino, B. and Witten, E. (eds.) (1985): *Current algebra and anomalies*, World Sci. (an excellent collection of review articles and papers)

Tsutsui, I. (1989a): *On the origin of anomalies in the path integral formalism*, Phys Rev. **D40**, 3543-3546

Tsutsui, I. (1989b): *Covariant anomalies in cohomology approach*, Phys. Lett. **B229**, 51-54

Turing, A.M. (1938): *The extensions of a group*, Compositio Math. **5**, 357-367

Turner-Laquer, H. (1992): *Invariant affine connections on Lie groups*, Trans. Am. Math. Soc. **331**, 541-551; *Invariant affine connections on symmetric spaces*, Proc. Am. Math. Soc. **115**, 447-454

Tuynman, G.M. and Wiegerinck, W.A.J.J. (1987): *Central extensions in physics*, J. Geom. Phys. **4**, 207-258

Van Nieuwenhuizen, P. (1983): *Free graded differential superalgebras*, in *Group theoretical methods in physics*, M. Serdaroğlu and E. İnönü eds., Lect. Notes in Phys. **180**, 228-247

Van Nieuwenhuizen, P. (1984): *An introduction to supergravity and the Kaluza-Klein program*, in *Relativity, groups and topology II*, B.S. De Witt and R. Stora eds., North-Holland, pp. 823-932

Varadarajan, V.S. (1984): *Lie groups, Lie algebras and their representations*, Springer-Verlag

Viallet, C.-M. (1990): *Symmetry and functional integration*, in *physics, geometry and topology*, H. C. Lee ed., Plenum, pp. 435-460

Von Westenholz, C. (1986): *Differential forms in mathematical physics*, North-Holland

Voronov, T. (1992): *Geometric integration theory on supermanifolds*, Sov. Sci. Rev. C. Math. Phys. **9**, 1-138, Harwood

Wang, H.C. (1958): *On invariant connections over a principal bundle*, Nagoya Math. J. **13**, 1-19

Wang, P. (1989): *Several subtle questions in topological field theories*, Ann. Phys. **196**, 71-88

Warner, F. (1971): *Foundations of differentiable manifolds and Lie groups*, Scott Foreman

Weiss, E. (1969): *Cohomology of groups*, Academic Press

Wess, J. (1986): *Anomalies, Schwinger terms and Chern forms*, in *Proc. of the 5th Adriatic Meeting on Particle Physics*, M. Martinis and I. Andric eds., World Science (1987)

Wess, J. and Bagger, J. (1991): *Supersymmetry and supergravity* (2nd ed.), Princeton Univ. Press

Wess, J. and Zumino, B. (1971): *Consequences of anomalous Ward identities*, Phys. Lett. **37B**, 95-97

West, P. (1990): *Introduction to supersymmetry and supergravity*, World Science

Weyl, H. (1931): *The theory of groups and quantum mechanics*, Dover (1950)

Weyl, H. (1946): *The classical groups*, Princeton Univ. Press

Whitehead, G.W. (1983): *Fifty years of homotopy theory*, Bull. Am. Math. Soc. **8**, 1, 1-29

Wightman, A. S. (1960): *L'invariance dans la mécanique relativiste*, in *Relations de dispersion et particules élémentaires*, C. D. Witt and R. Omnès eds., pp. 159-226, Hermann

Wigner, E. P. (1964): *Unitary representations of the inhomogeneous Lorentz group including reflections*, in *Group theoretical concepts and methods in elementary particle physics*, F. Gürsey ed., Gordon and Breach, pp. 37-80

Williams, D. and Cornwell, J.F. (1984): *The Haar measure for Lie supergroups*, J. Math. Phys. **25**, 2911-2932

Witten, E. (1982a): *Supersymmetry and Morse theory*, J. Diff. Geom. 661-692

Witten, E. (1982b): *An SU(2) anomaly*, Phys. Lett. **117B**, 324-328

Witten, E. (1983a,b): *Global aspects of current algebra*, Nucl. Phys. **B223**, 422-432 (reprinted in Treiman et al. (1985) and in Shapere and Wilczek (1989)); *Current algebra, baryons and quark confinement*, Nucl. Phys. **B223**, 433-444 (reprinted in Treiman et al. (1985))

Witten, E. (1985): *Global gravitational anomalies*, Commun. Math. Phys. **100**, 197-229

Witten, E. (1988): *Coadjoint orbits of the Virasoro group*, Commun. Math. Phys. **114**, 1-53

Witten, E. (1988b): *Topological quantum field theory*, Commun. Math. Phys. **117**, 353-386; *Topological sigma models, ibid.* **118**, 411-449

Witten, E. (1991): *Introduction to cohomological field theories*, Int. J. Mod. Phys. **A6**, 2775-2792

Witten, E. (1992a): *Two dimensional gauge theories revisited*, J. Geom. Phys. **9**, 303-368

Witten, E. (1992b): *On holomorphic factorization of WZW and coset models*, Commun. Math. Phys. **144**, 189-212

Woodhouse (1992): *Geometric quantization*, Oxford Univ. Press

Wu, S. (1993): *Cohomological obstructions to the equivariant extension of closed invariant forms*, J. Geom. Phys. **10**, 381-392

Wu, T. T. and Yang, C. N. (1975a): *Some remarks about unquantized non-abelian gauge fields*, Phys. Rev. **D12**, 3834-3844

Wu, T. T. and Yang, C. N. (1975b): *Concept of non-integrable phase factors and global formulation of gauge fields*, Phys. Rev. **D12**, 3845-3857

Wu, T. T. and Yang, C. N. (1976): *Dirac monopoles without strings: monopole harmonics*, Nucl. Phys. **B107**, 365-380

Wu, Y.-S. and Zee, A. (1985): *Cocycles and the magnetic monopole*, Phys. Lett. **152B**, 98-102

Wybourne, B.G. (1974): *Classical groups for physicists*, Wiley

Yamagishi, H. (1987): *A space-time approach to chiral anomalies*, Progr. Theor. Phys. **78**, 886-907

Yang, C. N. (1977): *Magnetic monopoles, fibre bundles, and gauge fields*, Ann. N. Y. Acad. Sci. **294**, 86-97

Zakrzewski, W.J. (1989): *Low dimensional sigma models*, Adam Hilger

Zamolodchikov, A.B. (1986): *Infinite additional symmetries in two-dimensional conformal quantum field theory*, Theor. Math. Phys. **63**, 1205-1213 (Russian original **65**, 374-359, (1985); reprinted in Itzykson et al. (1988))

Zinn-Justin, J. (1989): *Quantum field theory and critical phenomena*, Oxford Univ. Press

Zumino, B. (1985a): *Chiral anomalies and differential geometry*, in Treiman et al. (1985), 361-391

Zumino, B. (1985b): *Cohomology of gauge groups: cocycles and Schwinger terms*, Nucl. Phys. **B253**, 477-493

Zumino, B. (1988): *The geometry of the Virasoro group for physicists* (Cargèse, 1987), in M. Lévy et al. eds., NATO-ASI series B (physics) **173**, pp. 81-98, Plenum

Zumino, B.; Wu, Y-S. and Zee, A. (1984): *Chiral anomalies, higher dimensions, and differential geometry*, Nucl. Phys. **B239**, 477-507

Zwanziger, D. (1989): *Action from the Gribov horizon*, Nucl. Phys. **B321**, 591-604

Index

Note: numbers in italics refer to words that appear in footnotes.